Practical Aviation Security

Practical Aviation Security
Predicting and Preventing Future Threats
Second Edition

Jeffrey C. Price, MA
Jeffrey S. Forrest, PhD

AMSTERDAM • BOSTON • HEIDELBERG • LONDON
NEW YORK • OXFORD • PARIS • SAN DIEGO
SAN FRANCISCO • SINGAPORE • SYDNEY • TOKYO

Butterworth-Heinemann is an imprint of Elsevier

ELSEVIER

Acquiring Editor: Pamela Chester
Senior Editorial Project Manager: Amber Hodge
Project Manager: Priya Kumaraguruparan
Designer: Greg Harris

Butterworth-Heinemann is an imprint of Elsevier
225 Wyman Street, Waltham, MA 02451, USA
The Boulevard, Langford Lane, Kidlington, Oxford OX5 1 GB UK

First edition 2009
Second edition 2013

British Library Cataloguing-in-Publication Data
A catalogue record for this book is available from the British Library

Library of Congress Cataloging-in-Publication Data
Price, Jeffrey C.
   Practical aviation security : predicting and preventing future threats / Jeffrey C. Price, MA, Jeffrey S. Forrest, PhD. — Second edition.
      pages cm
   Includes bibliographical references and index.
1. Aeronautics—Security measures. I. Forrest, Jeffrey S. II. Title.
   TL725.3.S44P746 2013
   387.7068′4–dc23
                                                                            2012050432

ISBN: 978-0-12-391419-4

For information on all Butterworth-Heinemann publications
visit our web site at store.elsevier.com

Printed and bound in United States of America
13   14   10   9   8   7   6   5   4   3   2   1

Working together to grow
libraries in developing countries

www.elsevier.com | www.bookaid.org | www.sabre.org

ELSEVIER   BOOK AID
             International   Sabre Foundation

Dedication

In memory of Robert K. Mock, who was our colleague, professor, mentor, and friend. We also dedicate this book to all the victims of air terrorism, their families, and their friends.

Contents

Security Tools and Considerations   212

Conclusion   219

6. Introduction to Screening   221

    Objectives   221

    Introduction   221

    Evolution of Screening in the United States   222

    Screening under the Transportation Security Administration   227

    Screening Checkpoint Operations and Design   233

    Screening: Terminal Operations and Profiling   241

    Conclusion   254

7. Passenger and Baggage Screening   257

    Objectives   257

    Screening Passenger and Carry-On Baggage   257

    Screening Checked Baggage   266

    Special Issues in Screening   278

    Conclusion   285

8. Commercial Aviation Aircraft Operator Security   287

    Objectives   287

    Introduction   287

    Airline Security: Historical Context   288

    Aircraft Operator Standard Security Program   289

    Title 49 CFR Part 1544—Aircraft Operator Security:
        Air Carriers and Commercial Operators   290

    Law Enforcement Operations Related to Airline Security   301

    Aircraft Security Requirements   312

    Airline Employee Safety   313

Foreword from the First Edition

Today as I was about to board the commuter train to my job in Manhattan, an older gentleman approached me on the platform saying, "I noticed the pin on your lapel. I was there too." It is a small pin that at first glance appears to be two streaming American flags hung side by side in vertical fashion. The chairman of my agency presented this to employees and colleagues seven years ago in commemoration of an event that has since defined our lives and professions, wrenched our industry, and absorbed our national attention and resources. It was an event, as this gentleman's remark showed, that continues to bring strangers together in remembrance and renewed commitment to protect what is precious to a free society.

The work that the aviation industry and government have undertaken to ensure the security of air travel and commerce now informs other industry and business sectors of our economy. Indeed, security is now a pervasive element of business and social life around the world. The challenge of aviation professionals and government officials is how to achieve this security while preserving the vitality of the industry and the liberties of our citizens.

Jeff Price and Jeffrey Forrest's thorough text on aviation security instructs us in the history of assaults against civil aviation, which is nearly as old as civil aviation itself, while underscoring the changing nature of the assaults and the evolution of our government's response to these threats. The book documents the dramatic restructuring of transportation security oversight and regulation following the terrorist attacks of September 11, 2001, and provides us with a detailed understanding of the laws and regulations, the agencies and stakeholders, and the operating procedures and technical resources now brought to bear in the security of aviation.

Much has been accomplished to protect commercial flight from hostile intent, and time has proven the resiliency of the aviation system and our nation. There is, of course, still more to do, and we will look to those who follow us, perhaps some who are reading this book, to continue the vigilance and stewardship of this great industry.

May this little lapel pin that symbolizes the World Trade Center long remind us of those we lost in New York, NY, Arlington, VA, and Shanksville, PA, on September 11, 2001, and inspire us with confidence in the endeavor to pursue tolerance and understanding by drawing the world closer when we exercise our freedom and right to travel.

Jeanne M. Olivier, A.A.E.

General Manager, Aviation Security and Technology
The Port Authority of New York and New Jersey

Foreword

The morning of September 11, 2001 (9/11) will go into history as a time that changed the way we live for many years to come. The bold and cruel nature of the terrorist attacks that led to the death of thousands of innocent people demonstrated how vulnerable we are when the enemy ignores the "rules of war." We were caught unprepared as the terrorists attacked one of the most sensitive and fragile technological systems we rely on—aviation. For the sake of accuracy, we should underscore that the attacks were not against aviation per se, but instead used aviation as a means to damage targets symbolizing our way of life and our sources of societal power—free enterprise, military infrastructure, and a democratically elected government.

Terrorists have targeted aviation many times in the past. However, prior to 9/11 hijacked airplanes had not been used as weapons against ground targets in such dramatic and strategic ways. Terrorists have been more attracted to using aviation than other modes of transportation. Nevertheless, many voices doubt as to whether the enormous investments in protecting the aviation system are reasonable since in the best case we can only divert terrorism from aviation to other targets. These are valid concerns that responsible governments must address. In this regard, governmental decision making creates policies and laws that not only protect global aviation but also establish national priorities.

A fundamental premise for this book is that the protection of aviation should be a national priority. On the morning of 9/11, the U.S. government and its security forces had relatively little knowledge of how to protect aviation from terrorism. In the United States, the pre-9/11 era was characterized with aviation security as a low national priority. Aviation security was perceived as an unproductive budget item. Therefore, many decisions related to aviation security were based on trade-offs related to costs rather than effectiveness. Additionally, there was no attempt to create an effective security system that would challenge the increasingly sophisticated methods of global terrorism. As a result, the United States lacked a knowledge base to defend itself against the 9/11 attacks. Since 9/11, the United States has partially addressed this weakness by importing knowledge and expertise from other countries, such as Israel and the United Kingdom. This book will help to minimize this concern by greatly enhancing the reader's knowledge and expertise related to U.S. aviation security.

In the first few years after 9/11, aviation security focused mostly on improving the screening of passengers and baggage. This policy resulted in a few critical misconceptions, including the belief that passenger checkpoints and screening of baggage can inclusively protect us from all threats to aviation security. Recently, the Transportation Security Administration has recognized the need for other security measures at airports like perimeter protection, access control, and suspicious behavior detection (among many other methods and strategies). In the spirit of adopting new security measures, we must always be reminded of the cliché that any security system is as strong as its weakest point. Therefore, developing proactive and comprehensive methods and strategies for use in our aviation security system is a continuous challenge.

This is the reason why I compliment the authors of this text for rightly choosing a broad approach to addressing aviation security. Mr. Price and Dr. Forrest did a great service to worldwide aviation security by covering the subject in a very comprehensive manner, from the history of attacks against aviation to the detailed discussion of today's threats, from mitigation tactics to the different technological systems. The detailed discussion of different strategies from *profiling* to *everybody is equal* and various solutions that try to keep the system effective without jeopardizing the social values we try so hard to protect is a critical discussion. The reader will find all these issues well handled by the authors, and as a result be able to perform with greater understanding. This is the first time that interested readers can find a concise guide to aviation security. The industry requires this kind of book that will lead readers not only to a better understanding of this very complicated subject, but will also resolve many of the myths that surround aviation security.

- Rafi Ron

CEO, New Age Security Solutions (NASS)
Director of Security at Tel Aviv's Ben-Gurion International Airport (1997–2001)

Preface

Since the first edition of this text, we have seen a shift in the philosophy of the Transportation Security Administration (TSA) and how the industry is approaching aviation security. Shortly after 9/11, the traveling public was treated to a one-size-fits-all approach to security. All individuals were perceived to have the same level of risk, regardless of past history, age, behaviors, or other indicators. Within the past two years, the TSA has shifted the focus to risk-based security (RBS). RBS is based on the fundamental belief that the vast majority of individuals present no risk to the aviation system, and therefore, different levels of scrutiny should be incorporated into the process. The TSA has also significantly increased its intelligence collection and dissemination capabilities, and airport and airline operators continue to develop technologies and processes to protect aviation from the ever-evolving threat.

This book came about as a response to the changes brought about because of 9/11 and the lack of a comprehensive aviation security resource that practitioners, new and old, could look to. It is written by a lifetime aviation security practitioner and pilot (Price) and an academician and pilot (Forrest) to bring both practical strategies and comprehensive and well-thought-out explanations to airport, airline, and government aviation security personnel. We recognize that one cannot embark on such an endeavor, particularly within the fast-moving field of aviation security, without taking the risk that certain information will be out of date by the time the book reaches publication. To that end, we encourage you to visit the companion website at *http://www.leadingedgestrategies.com* for updates and additional information on changes that have taken place in the ever-evolving aviation security industry.

A challenge to writing about airport security is doing it without providing critical information to your adversaries. However, we believe the potential benefits of sharing information and knowledge related to airport security with industry members outweighs this risk. We also think it would be a disservice not to help inform and educate industry about basic strategies for protecting its infrastructure and, more important, the traveling public. For these reasons, we decided that the goal for this text is to serve as a resource for those interested in gaining a better understanding of airport security and applying practical strategies to improve aviation security processes.

The basics of airport security—the practical strategies—are not difficult to comprehend. Indeed, it is probable that your adversaries are already familiar with most of the common airport and airline security strategies. This dilemma is similar to challenges faced by police agencies. For example, police departments might hesitate to share with the public information that could deter theft. In this example, police agencies may fear that by releasing this information, burglars will better understand what defenses they need to overcome to commit a robbery. On the other hand, information shared by the police may help the public learn better ways to secure their property and potentially lower the risk of a robbery. We recognize that even with these strategies, criminals continue to seek ways to overcome various forms of security. That is why it is important for security experts to share new or improved methods for implementing security strategies. Planning and implementing airport and airline security procedures are constantly evolving, requiring continual long-term evaluation and implementation.

Even with the most current knowledge and security procedures, airports and airlines will remain threatened by criminal or terrorist activities. Perpetrators will consistently seek new paths to overcome security methods. We refer to this as pursuing a *strategy of least resistance* that the criminal or terrorist will exploit to accomplish the goals. This is an important premise for those concerned with airport and airline security. In this text, a history of air terrorism and related lessons learned will demonstrate that most attacks on aviation were and continue to be simple strategies developed by perpetrators to overcome established security methods with the least amount of effort.

In addition to the effect security procedures have on criminals or terrorists, the value of the perpetrator's goal or target will also affect where and how the perpetrator will strike. Aviation is a high-value target. It is the economic lifeblood of global commerce; global societies now depend on aviation to conduct business. Business travelers constitute a significant portion of the air-traveling public. Since the 9/11 downturn, vacationers have returned to the skies as a primary mode for traveling to resort destinations, and there is strong demand for same-day or next-day delivery of high-value cargo. The reliability, safety, and security of air transportation are critical to global economies.

Industry professionals and others, such as the traveling public and the media, would prefer a single blanket-security strategy or "silver bullet" addressing all aspects related to planning and implementing aviation-related security programs—especially airport security. The reality of aviation or airport security is that it is a highly dynamic and complex system of layers containing policies, strategies, tools, and processes. Each layer is designed to prevent, deter, or enable response to a particular attack or set of attacks. Providing layers of security is a well-established security strategy. In his book *America the Vulnerable*, Stephen Flynn (2004) described layered security as the constant application and evolution of multiple security measures designed to provide high levels of deterrence. Flynn also recommended that security measures should balance the probability of criminal activity in relation to the value of the target. In this regard, the more valuable the target, the more need for increased layers of security. The TSA commonly refers to these layers as "concentric rings of security."

Airport and aircraft security managers must understand that a single layer of security will not be 100% effective in blocking criminal or terrorist activity. However, through proper planning and implementation, multiple layers of security initiatives may provide a probability of nearly 100% effectiveness against these threats. Therefore, we believe that each layer of security, performed to its fullest level of effectiveness and combined with other layers of security, will result in the highest level of deterrence obtainable.

A major goal in aviation or airport security should be that security agents in each layer should strive for maximum effectiveness in their area of responsibility. Essential to embracing this goal is an understanding by all personnel that attacks will probably occur regardless of how many layers of deterrence are in place and how effective those layers are. If the target or goal is sufficiently valuable, the criminal or terrorist[1] will continue to evaluate or attempt new strategies of least resistance to achieve the goals.

Despite numerous security improvements implemented globally after the 9/11 attacks in the United States, criminals and terrorists continue to disrupt or destroy facets of the global aviation system. Terrorists view aviation as too valuable of a target to ignore. This is a grave realization when one considers that many in the industry argue that aviation security processes remain faulty, even when considering post-9/11 security-related initiatives.

Accepting the premise that attacks will continue to occur regardless of what we do is usually unacceptable. We do not argue that there should be acceptable losses and that we should continue to tolerate these losses. Rather, we believe losses will continue to occur regardless of the security system. As security advocates, we should design and implement practical and effective security measures to help mitigate or minimize these losses. These efforts are analogous to safety in commercial aviation. Although occasional aircraft losses occur (as a result of pilot, maintenance, or manufacturing errors, as well as other causes), the industry strives to reach a zero loss rate, while simultaneously acknowledging that the realization of this goal is unlikely.

Another precarious assumption made by some aviation security practitioners is to consider aviation security as being restricted to addressing acts of terrorism and therefore only focusing security efforts on preventing terrorist acts. The aviation security system must also deal with crimes against aviation and those using aviation to facilitate other forms of crime, such as drug trafficking, human smuggling (illegal alien and human slave trafficking), and the transportation of stolen property. Airports and airlines are businesses, sometimes large corporations, and experience crimes similar to those affecting nonaviation companies, such as employee theft, workplace violence, and cyber-threats. Airports and airlines are entities with a large general population passing through their facilities and onto aircraft every day. In this environment, there are numerous opportunities for criminals.

Many government agencies have offered various definitions for *terrorism*. However, law enforcement agencies and the U.S. District Attorney's Office determine whether a particular act is a form of terrorism. For example, the attempted bombing of American Airlines Flight 63 by Richard Reid (a.k.a. the shoe bomber) was considered an act of terrorism. In contrast, the shooting by a single perpetrator of three individuals waiting in line at the El Al ticket counter at Los Angeles International Airport in 2002 was considered under the

---

[1]We use the terms *crimes against aviation* and *criminal and terrorist acts* interchangeably with the commonly used phrase *acts of unlawful interference*.

criminal definition of murder (i.e., not terrorism). The aviation security practitioner should not be distracted with whether the event he or she is trying to prevent is rooted in terrorism or criminal activity. Rather, the security practitioner should focus on developing and implementing practical systems, measures, and procedures to prevent all forms of attack and criminal activity.

As individuals concerned with aviation security, our primary goal should be to do everything possible to ensure that the layers of security for which we are responsible are effective. There is an old saying in military law enforcement: "Not on my watch." This adage implies that, as individuals, we are not in control of everything but do acknowledge our responsibilities for those factors we do control. In this spirit, we must routinely create and implement practical strategies for managing these factors in ways that continuously improve aviation security.

- Jeffrey Price and Jeffrey Forrest

Acknowledgments

This book would not have been possible without the contributions and sacrifices of the following individuals.

First to our wives, Jennifer Price and Betsy Forrest, for their understanding and all their extra duties while we worked late nights and went on out-of-town research trips. Jeff Price would also like to thank his kids for their understanding that Dad couldn't always be there for playtime, and his parents, Zig and Dianne Price, who suffered through reading thousands of pages of first drafts. We would like to thank our colleague, Jennifer Caine, for her extensive editorial assistance on this text. Special thanks regarding manuscript preparation is also provided to Dawn Escarcega, Stephanie Horchreder, and Rene Victor Sabatini.

Very special thanks to Pam Chester from Butterworth-Heinemann for believing in this book, and to Kelly Harris, our developmental editor, who saved us both and made sure we got this project done.

And to the following people who contributed their time and expertise—in no particular order: Philip Baum, Jeanne Olivier, Lori Beckman, Mark Nagel, Brad Dalton, Teakoe Coleman, Sean Broderick, Barbara Cook, Colleen Chamberlin, Carter Morris, Rebecca Morrison, Spencer Dickerson, Jim Johnson, Jennifer Klass and the staff at the American Association of Airport Executives, John Duvall, Bernie Wilson, Michelle Freadman, Carrie Harmon, Robert Olislagers, Joram Bobasch, Mike Pilgrim, Norm Dawkin, Rafi Ron, Rob Hackman, Rob Rottman, Tim Barth, Tom Kinton, Jim Simmons, Jennifer Caine, Steve Davis, Lynne Gunter, Craig Williams, Tanya Sweeney, Huw Farmer, Mike Lanam, James Hiromasa, Dennis Treece, Al Graser, John Costigan, David Adams, Nahum Liss, Anthony Robbins, Chuck Burke, Charlotte Bryan, Yael Liberman, Eran Sala, Jim Slevin, Cathleen Berrick, Tinamarie Seyfer, Gerry Berry, Gary Smedlie, James Henderson, Rick Nelson, Michal Morgan, Wesley Fue, Pat Alhstrom, Dave Bassett, Al Meyers, Lynne Georold, Deven Smith, Brad Westoff, and all the attendees of the Airport Security Coordinator and Airport Certified Employee-Security courses that I have conducted and from whom I have learned so much.

For help in developing the second edition, we would like to thank the following individuals: Duane McGray, Executive Director of Airport Law Enforcement Agencies Network, Christopher L. McLaughlin, Doug Hofsass, Chuck Guffey, Martin Daniels, the Air Line Pilots Association, J. David Rigsby, and Amber Hodge, Senior Editorial Project Manager for Butterworth-Heinemann.

- Jeffrey Price and Jeffrey Forrest

Companion Website

Updates and other related materials are available online at: *http://www.leadingedgestrategies.com.*

1

Overview of the Aviation Industry and Security in the Post-9/11 World

Objectives

This chapter provides an overview of the national aviation system and an introduction to the development and environment of aviation security since the September 11, 2001, terrorist attacks (9/11). Readers will gain insight to the national aviation system and its importance to society. A fundamental framework describing the roles of airports, aircraft operators, and regulatory agencies involved in sustaining effective aviation security is presented. Fundamental to modern aviation security strategies and methods are lessons learned from the 9/11 attacks. Aviation security practitioners and students of aviation security should have at least an elementary understanding of the circumstances surrounding 9/11. Therefore, a case study describing the events of 9/11 and integrating concerns of aviation security is also provided.

Introduction

Throughout most of the history of aviation, terrorists and criminals have used aircraft and airports to conduct many forms of unlawful activity. Examples include special-interest groups or terrorists using aviation to gain geopolitical attention and criminals using commercial or general aviation to smuggle drugs. In these cases, aviation has provided a public stage for the former and an expedient distribution channel for the latter. Aviation is essential to sustaining the economic viability of world commerce, the movement of people and cargo, and the flow of information and knowledge throughout society. Therefore, it is essential that those responsible for protecting the aviation industry are proactive in developing and implementing strategic and tactical systems that are effective in helping to mitigate criminal and terrorist activity.

The aviation industry is composed of a series of overlapping operational areas or "systems of systems" that security personnel must protect. Examples include the management of passenger needs, such as parking, baggage check-in, screening (Figure 1.1), and other requirements (health concerns, guarding secured areas, etc.). Those responsible for each area must work in harmony to maintain aviation as an effective form of global transportation. Evaluating effectiveness in aviation security requires a variety of methods—from ratios used to develop metrics (baggage throughput, passenger flow, etc.), to extensive security evaluations conducted by various government agencies and private corporations.

FIGURE 1.1 Passengers go through security screening at Denver International Airport.

Aviation is an effective and efficient mode of transportation affecting worldwide social and economic stability. As such, aviation is a target that both terrorists and criminals highly desire.

The ability of aviation to move people and property faster than competing forms of transportation is essential to its economic viability. The Internet and related technologies such as videoconferencing and telecommuting provide additional options to transport information, knowledge, or products and services. The advantage aviation has over rail, trucking, and watercraft is speed, whereas its advantage over videoconferencing is that people still generally prefer face-to-face communication.

Sustained criminal or terrorist activity on aviation could cause a shift in passenger demand from airline travel, of which business travel is a significant percentage, to alternate forms of interaction or travel, such as videoconferencing or privately owned or chartered aircraft. These types of changes in demand for transporting people, cargo, or information could present airlines with serious economic challenges.

If business travelers switch to alternate modes of travel, commercial airlines will have to increase the cost of tickets to those passengers (usually leisure travelers) who cannot afford business rates. As costs increase, leisure travelers may not be able to afford air travel, resulting in more "staycations," which do not require air travel. Airlines would then have to raise prices to compensate as more leisure flyers switch to ground transportation.

Additionaly, many industries, such as the hotel, rental, and tourist industries, rely heavily on air transportation for their businesses to be successful. Repeated attacks on aviation could lead to a significant restructuring of commercial aviation.

Aviation Industry: An Overview

Despite its complex nature, the aviation industry's primary infrastructure consists of aircraft operations, airports, and supporting agencies. Many types of aircraft are used in various operations around the world. These are commonly categorized as commercial service, private operations (i.e., general aviation, or GA), and military operations. Airports are usually categorized as commercial service, general aviation, private, or military.

Aviation Security and Responding to Threats

The terrorist attacks in the United States on September 11, 2001 (9/11), were designed to damage global security and the U.S. economy—an economy reliant on aviation. A critical strategy for responding to terrorist threats is to moderate the response so it is appropriate and does not cause further deterioration to the economy or stability of a society. Terrorist organizations understand that they usually do not have the forces or resources to defeat an enemy in a traditional military conflict. Therefore, terrorists operate more asymmetrically, striking in ways that cause targeted countries or societies to incur loss of life, economic damage, changes in policies, or other effects. These attacks are usually designed with the hope that countries or societies will overreact in ways that further diminish the ability to protect or sustain safety and economic viability. Terrorists also know that with each subsequent attack, the targeted populace gives up more of its freedom through changes in laws and policies or by accepting that intrusion into private lives is unavoidable and required. In these ways, terrorists can cause economic and social degradation within nations and societies.

In addition to appropriate responses to terrorist or criminal attacks, those charged with protecting aviation must ensure that strategies and technologies remain current and viable for defending against new threats. Security practitioners employing outdated strategies and tactics create opportunities for terrorists to use these systems to their advantage. For instance, the 9/11 attacks showed ingenuity and were organized using modern technologies (e.g., the Internet) to defeat what was then a 1970's aviation security system.

Flawed management in designing and implementing modern security systems can also create opportunity for criminals or terrorists. For example, industry is responding to terrorism by investing billions of dollars in research and development for improved explosives detection equipment. Of concern is that much of this technology has not undergone extensive testing before deployment. Although technologies used in various explosives detection equipment[1] are mostly valid, the mean time before failure for these

---

[1]For example, a gas chromatograph (GC) or a programmed thermal desorber (PTD).

technologies is often low when used in a nonairport environment. The deployment of the PTDs proved to be ineffective when the units were not resilient enough to handle the airport environment. Many early-model explosive detection devices were not designed or tested to sustain the day-to-day usage at commercial service airport screening checkpoints. The industry is still catching up in developing other security strategies, such as employee security awareness and passenger profiling—strategies that other nations have used since the 1980s—and, perhaps most critically, the ability of our intelligence agencies to penetrate groups that represent credible threats. The increased focus on tracking and either arresting or eliminating known threats, such as the attack on and subsequent death of Osama bin Laden and Anwar al-Awlaki, have weakened one of aviation's primary enemies, al-Qaeda, but others will take their place.

In response to the challenges in aviation security, the U.S. National 9/11 Commission on Terrorist Attacks upon the United States (the 9/11 Commission)[2] and other security and policy experts developed and recommended strategies that should be effectively and efficiently implemented. However, implementing new aviation security methods has traditionally been reactive rather than proactive. Those charged with protecting the aviation industry must adopt a philosophy of sustained proactive improvement. Therefore, a premise to the knowledge presented in this text is as follows:

> *In aviation security, we must not stop moving forward in implementing proactive forms of security—our foes are committed to their cause and we must be exceedingly committed to ours.*

Although time is of the essence in commerce, it is critical in the aviation industry. Passengers must travel safely and efficiently. However, reasonable compromises need to be made for the system to continue functioning. A terrorist attack may be devastating and certainly tragic. Nevertheless, it would be even more tragic to allow such an attack to cause further economic and social damage by impeding global aviation. The ability of a security system to appropriately respond and quickly recover is a fundamental principle in planning and managing aviation security systems.

As discussed, aviation is vital to the survival and growth of the world's economy (Figure 1.2). Aviation is also a symbol of prosperity and stability among societies. An essential responsibility of those charged with protecting aviation is to balance responses to threats with the requirement to facilitate safe, secure, and efficient transportation of passengers.

---

[2]The 9/11 Commission was "an independent, bipartisan commission created by congressional legislation and the signature of President George W. Bush in late 2002. It was "chartered to prepare a full and complete account of the circumstances surrounding the September 11, 2001, terrorist attacks, including preparedness for and the immediate response to the attacks". The Commission is also mandated to provide recommendations designed to guard against future attacks" (see *www.9-11commission.gov*).

FIGURE 1.2 Aviation security embraces the protection of passengers, airports, airlines, and the national aviation system.

Aviation Economics

According to a study by DRI-WEFA, Inc. (2002), the economic impact of civil aviation within the United States exceeds $900 billion and provides 11 million jobs. This represents 9% of the U.S. gross domestic product (FY 2000). Commercial aviation is responsible for 88% of this contribution, and general aviation for 12% (p. 7). Commercial airlines carry between 600 and 800 million people per year and move 20–25 billion tons of cargo. Additionally, more than 160 million people travel on general aviation aircraft every year within the United States (DRI-WEFA, Inc., 2002).

The United States is geographically (and socially) oriented to rely on air travel. Unlike many countries with well-organized systems that enable efficient travel internally and internationally by rail (and thus have less of a need for air travel), the United States has no national high-speed railway system. Additionally, the United States shares borders with Canada and Mexico only, further limiting opportunities for international commerce by land. As a result, the United States relies on air travel to enable commerce with the rest of the world.

Although most of the world's freight still moves via sea, the demand of *just-in-time* goods delivery is increasing. The primary advantages of just-in-time shipping are that it allows for shorter production and development cycles and eliminates excessive inventory. DRI-WEFA, Inc. (2002) also provides this summary of the influence of aviation on the world economy:

> *Aviation is the primary means for economic growth with a significant influence on the quality of life of populations around the globe. Aviation facilitates the world economy and promotes the international exchange of people, products, investment, and ideas. Indeed, to a very large extent, civil aviation has enabled small community and rural populations to enter the mainstream of global commerce by linking such communities with worldwide population, manufacturing, and cultural centers. (p. 3)*

The impact of 9/11 on the aviation industry was devastating. Overall, the aviation industry experienced direct losses of $330 million per day (Kumar et al., 2003). Not included in this figure are losses to rental car, hotel, and tourist industries dependent on the aviation industry for customers. In addition to the direct loss of revenue, airlines and airports had to bear an increased cost of doing business to meet government-mandated security improvements. The costs to airports of purchasing new security technology and reconfiguring terminals to accept the larger security screening devices have caused airport operators to shift money from other airport capital improvement projects, such as runway and taxiway maintenance and expansion, to security measures. Additionally, airport operators lost revenue-producing space to accommodate new Transportation Security Administration (TSA) equipment and personnel, and with the deployment of advanced imaging technology (AIT), the so-called "body imagers," airports have had to sacrifice even more space to make way for larger equipment and longer passenger screening lines.

The national airspace system is both economically and operationally fragile. A delay of just 10 minutes can cost the airport and airlines significant money. Gate assignments at airports are often scheduled back to back, meaning that if one flight does not depart on time, the next flight will probably also depart late. This domino effect influences many operational considerations. In 2000, delays averaged 12 minutes per flight, totaling 142 million hours of passenger delays, and nearly $10 billion in associated costs to the U.S. aviation industry (DRI-WEFA, Inc., 2002, p. 7).

Commercial aviation is composed of a myriad of coupled systems (Figure 1.3). When one part is affected, it forces change in many of the other systems. When airport flight operations shut down because of weather or aircraft have to be rerouted around bad weather, the entire national airspace system is impacted. When compared to criminal activity or terrorism, weather is relatively predictable and airport operators know how to adjust operations for the kinds of weather in their region. More important, weather (with few exceptions) does not generally destroy essential components of the airport and scare the public from travel, as tends to happen with a terrorist attack.

The importance of protecting aviation as a driver of the global economy is clear. However, the various systems used to protect aviation can impede aviation's efficiencies.

FIGURE 1.3 Dozens of aircraft are stranded in Halifax, Nova Scotia, during the shutdown of the U.S. aviation system on 9/11.

Security managers must balance their efforts for protecting lives and infrastructure with policies and methods that enable aviation to provide commerce with efficient and reliable service.

Aviation Security Funding

The International Civil Aviation Organization (ICAO; see *www.icao.int*) recommends a reliable and consistent stream of monies to fund aviation security improvements and operations worldwide. In the United States, most commercial service airports receive revenue from multiple sources as outlined previously, and part of these monies goes toward funding aviation security at airports.

U.S. airports spend billions of dollars to implement new security regulations. Airport operators have also had Airport Improvement Program (AIP) funding, usually earmarked for runway and facility improvements and upkeep, redirected toward security projects. This financial burden creates a situation where airport operators must sometimes decide between safety- or security-related projects. Local Federal Aviation Administration (FAA) airport district offices (ADOs) play a critical role in deciding which airports in each state are allocated federal funding. Some airports receive money through entitlement programs that designate specific funding regardless of capital improvement needs, and an airport listed in the National Plan of Integrated Airport Systems (NPIAS) is eligible for federal *discretionary funding*, an appropriation of federal dollars based on the discretion of the ADO. A complex planning process precedes the apportionment of discretionary monies, often involving airport operators, the state, the federal government, and the FAA. Since the FAA is now effectively out of the security business, but still controls the AIP revenue stream, some airport operators have complained that money is less available for security-related projects than for safety and airport development. The TSA does not have a federal apportionment through a trust fund, the way the FAA does, to provide money to airports for security improvements and technologies.

Many airports also receive capital monies through their state aviation offices, which may collect a percentage of receipts on aviation fuel sold within the state. State aviation/aeronautical offices work with the airports to coordinate 5- or 6-year capital improvement programs and then decide which airports in the state should receive the funds (in addition to federal AIP funding). Some states participate in the Code of Federal Regulations (CFR) Part 156 State Block Grant Program (see *http://edocket.access.gpo.gov/cfr_2003/pdf/14cfr156.3.pdf*). The Block Grant Program states have the sole responsibility for deciding the distribution of federal funds for improvement projects at general aviation and nonprimary (i.e., very small) commercial service airports.

An important issue for airport operators has been the acquisition of explosive detection equipment and other federally mandated security upgrades. When a new explosive detection system (EDS) machine is installed for checked-baggage screening, it is not just the unit cost of the machine that is incurred. Facilities must often be redesigned to accommodate the additional weight and space of new detection equipment, and power requirements must be considered along with mean time between failure maintenance requirements. Since the initial deployment of these machines in the years following 9-11, airport operators are experiencing an entirely new dynamic as the machines are reaching their maximum effective operating life and must be replaced.

Aircraft operators have also been affected by aviation security mandates. For example, the required reinforcement of cockpit doors not only increases costs but also adds weight to the plane, thereby reducing the capacity for revenue-generating cargo and passengers. As a result, funding for aviation security requirements is now a significant part of airport and airline operating and capital budgets.

Aircraft Operations: An Overview

Aircraft operations encompass a wide range of flying categories—from personal travel, to military flight, to scheduled commercial airlines. An airline's purpose is to move people and goods from one place to another, safely and securely, while still making a profit. Airlines carry people, mail, and cargo, and are essential to the U.S. economy. Airlines also offer a variety of services, including private charter operations, carrying sports teams and rock stars (Figure 1.4).

Aircraft operations are generally classified as commercial service, general aviation, or military operations. Regardless of the aircraft type, all aircraft may be used for each of these types of operations. The Boeing 737, for example, though most frequently used commercially for airlines, is also used by the military to train navigators, and by the private sector as a business jet. If aircraft are used in a commercial capacity for the transportation of passengers and cargo for pay, these operations are commonly referred to as *commercial aviation.*

Individuals or companies can use an aircraft owned by the company or rent an aircraft to facilitate business. For example, a salesperson may use his or her privately owned airplane to service accounts. These types of operations are *general aviation*, or GA. However, if the salesperson takes a passenger and charges a fee, it is now commercial aviation even

FIGURE 1.4 A Ted B757 at the Las Vegas McCarron International Airport, one of the busiest airports in the world.

though both operations use the same aircraft. These brief examples should highlight that an aircraft operation is determined by how an aircraft is used, not by the appearance or design of the aircraft.

Figure 1.5 shows three aircraft—a Cessna 182 (foreground) used in flight training, a Cessna Citation used for business travel and charter operations, and a Boeing Business Jet (BBJ) used for private transportation by large corporations. The BBJ design is based on the Boeing 737 airframe and can easily be confused as a commercial airliner. It is not always possible through visual examination to determine whether an aircraft is being used for private transportation, as part of a public charter, or in commercial transportation. It is important to understand how an aircraft is being used, as various security measures are applied depending on the type of flight operation conducted.

Commercial service operations involve using aircraft to generate a profit. These are subclassified into scheduled service (the most common being airline operations) and public and private charter flights. Scheduled aircraft operations are similar to scheduled bus or train services; a person pays to occupy space or a seat for the purpose of transportation. The company (an aircraft operator) provides the scheduled service to make a profit. In comparison, public charter also seeks a profit by offering transportation to the client, but departure and arrival times vary and are often based on client needs and the availability and capability of the aircraft. Private charter is analogous to renting a limousine in that

FIGURE 1.5 Cessna 182, Cessna Citation, and Boeing Business Jet, each representing different types of aircraft used in general and commercial aviation, at Rocky Mountain Metropolitan Airport, CO, 2006.

the customer has purchased the exclusive use of the entire vehicle. In private charter, the purchaser pays for the right to his or her own transportation, to bring additional passengers or cargo, and to decide on the departure or arrival times.

Private operations are part of GA and may involve using an aircraft personally or as part of a business enterprise. In GA operations, aircraft are not *directly* used to generate profit but may enhance business opportunity as an efficient form of transportation. GA operations also include flight training, agricultural flying, soaring, experimental aircraft, and recreational flying—to name a few.

Military operations use aircraft to meet various needs specified by national governments. Military aircraft do not usually facilitate public or business transportation or generate profit, however, military aircraft frequently operate from commercial airports.

In the following sections, we will focus on the most well-known type of aircraft operation: the commercial airline.

Airline Management

Airline organizational structures vary depending on the size and history of the airline, its route structure (domestic or international), and the type of aircraft (regional or long haul). Air carriers are profit-making organizations typically owned by stockholders. The stockholders elect a board of directors, which designates a chairperson and outlines the goals and objectives for the airline. The board hires an administrative manager, a president, responsible for setting major policies and ultimately for the airline's success (Kane, 1998).

Within top-level airline management is the aircraft operator security coordinator (AOSC) who is the airline's key contact with the U.S. TSA, the U.S. Federal Bureau of Investigation (FBI), and other federal agencies. The AOSC oversees the implementation of security practices throughout the airline. Commercial service operators are regulated under U.S. Transportation Security Regulations (TSRs; Title 49 CFR Part 1544).[3] U.S. airlines must also adhere to the requirements in the Aircraft Operator Standard Security Program (AOSSP), a federally drafted document that outlines the practices airlines must follow to adhere to federal regulations.

An airline operations department is essential for managing effectiveness and efficiency. The operations department is responsible for delivering service to those demanding transportation or shipment of cargo. Within operations are the flight and ground operations departments. Flight and ground operations interact with the AOSC daily. The flight operations department is responsible for air carrier fleet safety, security, and efficiency. Flight operations consist of pilots, flight attendants, and dispatchers. Pilots directly operate the aircraft, whereas flight attendants provide essential safety and customer services on board the aircraft. Dispatchers generate the data needed before the flight (e.g., weight and balance, weather, and safety and security information) and communicate with the pilots on

---

[3]The TSA issues and administers TSRs, codified in Title 49 of the CFR, Chapter XII, parts 1500–1699.

company communication channels in flight. The ground operations department is responsible for maintenance and servicing of the aircraft. Through the ground operations department, the airline's ground security coordinators (GSCs) work with the flight department to disseminate pertinent security information.

Airline Economics

Airlines operate on very small profit margins, averaging between 1% and 2% compared to a U.S. industry average of 5% (Kane, 1998, p. 413).[4] Airline revenues are seasonal, with most air travel occurring in the summer. Although more than 60% of air travelers fly for personal reasons, the majority of passenger revenue for many airlines comes from the 26% of those who are frequent business travelers. The number of airline operations also fluctuates throughout the day, with most travelers flying before 10 a.m. or after 4 p.m. The nature of passenger flow through an airport, based on time of day and time of year, relates directly to staffing security personnel and deploying required security-related equipment (Kane, 1998).

The U.S. airline industry has very high capital and labor costs. Most airline equipment is financed through loans or stock rather than purchased. Airlines must generate a high cash flow to repay debt, buy fuel, service aircraft lease payments, and perform other operating expenses. More than a third of an airline's revenue is allocated to labor, with many airline workers unionized. As such, labor costs are among the highest of any industry. Airlines employ many highly specialized personnel including (Kane, 1998):

- Pilots
- Flight attendants
- Reservation, ticket, gate, and cargo agents
- Security personnel
- Cooks
- Administrative assistants
- Revenue planners
- Schedulers and dispatchers
- Accountants
- Lawyers
- Human resource managers
- Aircraft cleaners
- Baggage handlers

Airline bankruptcies are common. Airline profitability is often evaluated by the *load factor*, which is derived by dividing the number of seats generating revenue by the airline's available seats. On average, an airline needs to fill 70–80%[5] of its available seats on each

---

[4]For a complete description of the financial measurements used in airline operations, see Kane (1998).

[5]These figures are dynamic, especially during periods of rapidly escalating fuel prices.

flight (depending on the airline's particular expense structure) to generate a profit on that trip. The difference of just one or two seats filled by paying customers can be critical to an airline's ability to make a profit on a flight.

The cost of moving a passenger from point A to point B is often determined by calculating the cost per *revenue seat mile*. The cost per revenue seat mile represents the average cost to fly one full-fare-paying passenger occupying one seat a distance of one mile. For the "legacy carriers"[6] (e.g., United Airlines, Delta, American Airlines, U.S. Airways), this is $0.10–0.14 per seat mile. In comparison, the costs for low-cost or discount air carriers (e.g., Frontier Airlines, Southwest Airlines) range between $0.07 and $0.09 per seat mile. Factors contributing to higher costs include employee salaries, pensions, maintenance, in-flight food, airline ticket taxes, and the cost of aviation fuel.[7] An example of an airline cost factor related to security is a fee commercial-airline passengers pay as mandated in the Aviation and Transportation Security Act of 2001 (ATSA 2001).

Airline cost structures are complex. Each airline constantly watches the rates of its competitors. When an airline increases its rates, other airlines may not increase theirs. By not responding with a rate increase, these airlines wait to see if the competing airline will lose profitability by increasing its fares, as there is usually an inverse correlation between airline ticket prices and passenger load factors. For example, if the cost of jet fuel increases and an airline raises its fares to compensate, other airline revenue managers may keep fares low (even with the same increased costs) in the hope that their own airlines will see higher load factors to make up for the higher fuel costs (Kane, 1998).

In relation to aviation security, if airlines raise fares to cover costs associated with new security requirements, then passenger load factors may also decrease. Because of small profit margins, the airline industry is watchful of security practices that may increase costs. In an attempt to help mitigate the impact of security concerns on cost, the ATSA 2001 created the Registered Traveler Program, designed to accommodate frequent airline travelers (usually business travelers) with a personal identity system, enabling those passengers rapid passage through airport screening.

The airline industry is very sensitive to the fluctuations of the national economy. Both business and leisure travel drop when the national economy is declining. This was discussed at an aviation security conference hosted by the American Association of Airport Executives (AAAE) (Figure 1.6) in December 2001. At the conference, some participants warned that future attacks on aviation could trigger a "mode shift" to increased videoconferencing and online meetings. This shift in demand would affect not only airlines but hotel operators, rental car agencies, and airport financial models.

Functions within an airline have a tremendous interdependence; all departments are interrelated. When one component is impacted, the effects reverberate throughout the

---

[6]Legacy carriers are commonly described as large airlines operating out of *hub-and-spoke* airports. A hub-and-spoke airport is usually a large airport that serves as a connecting hub for indirect flights to smaller airports.

[7]With the cost of fuel spiking to record levels in 2008, airlines have had to heavily revise their cost structures to make up for the added fuel expense.

FIGURE 1.6 Logo of the American Association of Airport Executives, the nation's largest airport management trade association.

organization. For instance, when a flight is canceled or significantly delayed because of a bomb threat, many of the other flights for that airline are also affected. If flights were delayed or canceled each time a threat was issued (e.g., one by phone), one individual could call several airlines and shut down the U.S. aviation system. For this reason, bomb threats are vetted as thoroughly as possible before canceling or delaying a flight.

The national aviation airline route system is also highly sensitive—flight delays of just a few minutes can resound across the nation, multiplying the delay time for subsequent flights and even causing cancellations. Weather delays and temporary airport closures caused by security breaches can trigger the cancellation of dozens and sometimes hundreds of flights.

A security breach at an airport can also have an economic impact by disrupting aviation operations. Approximately 40 million passengers a year pass through Denver International Airport (DIA). If an individual accidentally (without criminal intent) breaches security by entering a concourse without proper screening, the airport must temporarily close while the concourse is evacuated. In such a case, it takes security and other airport and airline personnel around two hours to conduct a *resterilization* proce-dure, which encompasses a search for the individual breaching security and any prohibited items he or she may have brought into secure areas. A major impact of a security breach is rerouting inbound aircraft. This backs up aircraft movements across the country and may even affect inbound international flights. If an airport cannot accept an aircraft because of a security breach, the plane is held on the ground at its departure airport, slowed down en route, or placed into a costly holding pattern. During a breach, outbound aircraft are grounded and often evacuated and searched.

There are approximately 800 departures of commercial aircraft from DIA every day, and the average cost of a one-way flight was about $12,600 in 2005. This cost represents the total average cost for the airline to fly an airplane from departure to destination. Using these numbers, the economic impact of a breach in security can be shown in the following example. A security breach occurs at noon, and concourse evacuation and resterilization begins. At least 390 flights are already affected, placing more than $4.9 million of revenue at risk. Assuming successful resterilization and the perpetrator is either found or the federal security director (FSD) determines there is no risk to reopening the airport, all flights resume with at least a two-hour delay. If only half the original flights are canceled, the airport recognizes a net loss of $2.5 million. Across the United States, other airports are affected. By 5 p.m., an additional 200 flights out of Chicago are affected, and by 9 p.m., 800 total flights across the United States are affected.

Resterilization procedures result in long security screening lines, upset travelers, and lost revenue. There is also a loss of productivity as airport and airline personnel are called away from their primary job duties to assist in the resterilization process. Of course, the airlines lose significant revenue when security breaches occur.

Airports: An Overview

Excluding military airports, there are approximately 20,000 general aviation, private, and commercial service airports in the United States. About 420 of these airports are classified by the FAA as commercial service but also host general aviation operations. Nearly 6,000 airports are GA airports with approximately 300 of those hosting occasional small commercial service operations (Figure 1.7). The remaining 11,000-plus airports are private airstrips scattered throughout the United States.

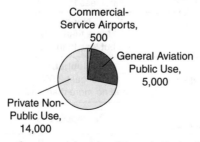

FIGURE 1.7 Approximate proportions of commercial service, GA, and private airports within the United States.

Airports support and provide a conduit to transportation. They also serve society in other ways. Sam Hoerter (2001), author of *The Airport Management Primer,* provides this description:

> *Think of an airport as a place where federal airspace and local roadways intersect. Think of it as a place where public infrastructure and private investment intersect. Think of an airport as a place where buyers and sellers meet in an open and ever-changing marketplace. In other words, think of it as a place forever caught between different worlds. Consequently, airport management requires a continuous effort to seek a balance among many competing forces. (p. 9)*

Adding to the challenges for airports as described by Hoerter (2001) is that airports must also operate safely and in a secure manner, while seeking to maintain balance between serving as a place to conduct commerce and public transportation and also complying with federal regulations.

The FAA classifies airports by categories of activities (Figure 1.8). Commercial service airports have at least 2,500 enplanements (passengers boarding) per year. Commercial service airports may be primary or nonprimary. Primary commercial service airports have more than 10,000 annual enplanements. Nonprimary airports have at least 2,500 but fewer than 10,000 enplanements. Cargo-service airports host aircraft with a yearly average landed weight of more than 100 million pounds. Airports can operate both as commercial service and as cargo airports.

Reliever airports are general aviation airports designated by the FAA to relieve congestion at commercial service airports. These often serve as an outlet for general aviation traffic to

Airport Classifications		Hub Type: Percentage of Annual Passenger Boardings	Common Name
Commercial-Service: Publicly owned airports that have <u>at least 2,500</u> passenger boardings each calendar year and receive scheduled passenger service §47102(7)	**Primary:** Have <u>more than 10,000</u> passenger boardings each year §47102(11)	**Large:** 1% or more	**Large Hub**
		Medium: At least 0.25% but less than 1%	**Medium Hub**
		Small: At least 0.05% but less than 0.25%	**Small Hub**
		Nonhub: More than 10,000 but less than 0.05%	**Nonhub Primary**
	Nonprimary	**Nonhub:** At least 2,500 and no more than 10,000	**Nonprimary Commercial-Service**
Nonprimary (Except Commercial-Service)		Not Applicable	**Reliever** §47102(18)

FIGURE 1.8 Airport categories. *(Source: FAA.)*

avoid the congestion of the commercial service facility. Approximately 13,000 airports are GA and they constitute the largest subcategory of airports in the United States. They can be public or privately owned and operated and can have up to 2,500 annual commercial service enplanements without being classified as a commercial service airport.

Commercial service airports are further classified by *hub* status.[8] Large-hub airports account for about 70% of the total annual passenger enplanements within the United States. Large-hub airports include Hartsfield-Jackson Atlanta International, Chicago O'Hare International, Los Angeles International, Dallas/Fort Worth International, and Denver International. There were 29 large-hub airports in the United States in 2010. Metropolitan Oakland International, Pittsburgh International, Portland International, Lambert-St. Louis International, and Cleveland-Hopkins International airports are examples of medium-hub airports (37 in 2010). Small-hub airports (72 in 2010) include Louisville International (Kentucky), Will Rogers Field (Oklahoma), El Paso International (Texas), and Albany International (New York). Nonhub (nonprimary) airports make up the largest number of commercial service airports, totaling 244 in 2010. These include airports such as Mahlon Sweet Field (Oregon), Fort Wayne International (Indiana), Joe Foss Field (South Dakota), and Juneau International (Alaska) airports. An additional 121 airports are classified as nonhub airports servicing small numbers of commercial traffic at the rate of fewer than 2,500 enplanements per year. Nonhub airports are predominantly GA operations (Table 1.1).

Table 1.1   Enplanements by Hub Type for 2000

Airport Type	Number of Airports	% of U.S. Enplanements	Examples
Large Hub (>1.0% of all enplanements)	29	68.0%	Sea-Tac, Denver, Salt Lake City, LAX, DFW, Atlanta, etc.
Medium Hub (0.25–1.0% of all enplanements)	37	20.0%	Portland, Anchorage, Reno, Cleveland, etc.
Small Hub (0.05–0.25% of all enplanements)	72	8.0%	Colorado Springs, Billings, Boise, Spokane
Non-Hub (>10,000 enplanements, but less than 0.05% of all enplanements)	244	3.0%	Aspen, St. George, Cheyenne, Kalispell, Hailey
Non-Primary Commercial Service (>2,500 enplanements and scheduled service)	121	0.1%	Pueblo, Moab, Pendleton, Eastsound, etc.
Relievers (>100 based aircraft or 25,000 annual itinerant operations—other criteria apply as well)	269	0.0%	Centennial, South Valley Regional, Paine Field, etc.
General Aviation	2,560	0.0%	Aurora State, Arlington Municipal, Nampa Municipal
Existing NPIAS airports	= 3,332	99.1%	

[8]A *hub* is an airport with a high frequency of arriving and departing flights along with a relatively high rate of connecting traffic.

Civilian commercial service airports may also be colocated with military airport facilities. Military airport facilities are not regulated by Federal Aviation Regulations (FAR; U.S. 14 CFR Part 1-199) or Transportation Security Regulations (TSR; U.S. Title 49 of the CFR, Chapter XII, parts 1500–1699). Civilian airport operators with joint military operations must geographically separate the civilian and military areas. For example, Colorado Springs Regional Airport is located on the south side of its primary runway and shares the runway with Peterson Air Force Base, which is located on the north side of the airfield (Figure 1.9). The areas are separated by the runway and by lines of demarcation painted on perimeter access roads.

Users of GA airports can include an eclectic mix of small training aircraft, experimental home-built planes, sailplane operations, helicopters, skydiving operations, and business-jet air traffic. As mentioned, some GA airports host limited commercial service operations. These small airline operations are usually characterized by seasonal or periodic flights rather than daily scheduled service. Many GA airports are small, single-runway facilities, whereas others equal the size of large commercial aviation airports with multiple runways, hundreds of thousands of annual operations, and large jet aircraft.

Of the approximate 13,000 GA airports, about 3,000 are part of the National Plan of Integrated Airport Systems (NPIAS; see *www.faa.gov/airports_airtraffic/airports/planning_capacity/npias*). NPIAS airports are designated as vital to the national airspace

FIGURE 1.9 Peterson Air Force Base in Colorado Springs shares runways with the Colorado Springs Municipal Airport.

system and thus eligible for federal funding for airport safety, capital improvements, and security projects.

Private entities or individuals own, construct, and maintain private airports. Use of a private airport is generally restricted to the owner or those granted permission by the owner. However, in the United States, any aircraft in an emergency may use any airport, including private airports.

Title 14 CFR Part 139 Certification of Airports assigns a different classification for commercial service airports for the purpose of compliance with the Airport Certification Manual (ACM; see *www.faa.gov/airports_airtraffic/airports/airport_safety/part139_cert/ media/sample_ACM.pdf*). The ACM outlines how the airport will comply with FARs to operate safely without specifically addressing security concerns. Under FAR Part 139, there are four classifications of commercial service airports: Class I airports encompass most large-, medium-, and small-hub airports; class II airports encompass small scheduled service and large charter operations; class III airports encompass small scheduled service operations only; and class IV airports encompass large charter operations.

Security classifications for airports are established by the TSA and fall into three types: complete, supporting, and partial. The type of security program required at a particular commercial service airport depends on factors such as the scope of the airport's operation, the geography (i.e., the airport is located near other critical infrastructure), the opinion of the TSA, and geopolitical circumstances. Complete security programs are generally required for large-, medium-, and small-hub airports. Supporting and partial security programs are generally required for nonhub airports.

Airport Ownership and Operation

Municipalities (e.g., cities and counties) own and operate the majority of publicly available service airports in the United States. Exceptions to this form of ownership are Alaska, Hawaii, and Rhode Island, which own all the civil airports within their boundaries.

A common misconception is that the FAA operates U.S. airports, as many airports have an FAA air traffic control tower (ATCT). ATCTs are staffed by FAA employees and coordinate the movement of aircraft in and around the runway of an airport. However, the movement of aircraft and ground vehicles within the "movement area" of an airport is generally the extent of the FAA's authority. The authority to close the airport, open and close runways, and other management functions is generally under the purview of the airport owner or operator, which is usually a municipality or an appointed authority. The relationship between the airport owner or operator and the FAA at a particular airport is defined in a letter of agreement (LOA).

City and county airports can be owned and managed through a variety of arrangements. A municipally operated airport is typically owned by a city or county and run as a department or division within the municipality's organizational structure by its employees (Wells and Young, 2004). In contrast, an airport may be operated by an airport authority, which is an entity created by state legislation, usually at the will of a city or county, and which acts as a separate governmental agency (Hoerter, 2001). The city or county retains airport ownership, but much of the management and planning decisions are turned over to the airport authority. According to Hoerter, airport authorities are the preferred method of ownership, because "the leadership is more focused on airport issues, the airport staff is less subject to political interference, and a metropolitan community can be better represented by the authority's governing body" (2001, p. 12).

Al Graser, during his tenure as general manager of John F. Kennedy International Airport and former chair of the AAAE Transportation Security Policy Committee, agrees with Hoerter's viewpoint and adds to it a perspective related to security. Graser believes that airport directors working for an airport authority have the benefit of consistency, which gives them the ability to develop industry relationships needed to work with the TSA and other government agencies, particularly in security issues. Graser also said that airport directors working directly for a city or county often campaign with the mayor or county commissioners for reelection, because if political situations change, the airport director is likely out of a job (Al Graser, personal communication, Dec. 2005).

Some airport authorities are responsible for entire aviation systems, such as the Los Angeles World Airports Authority (LAWAA). The LAWAA oversees the management and operation of Los Angeles International Airport, Ontario International Airport, Van Nuys Regional Airport, and Palmdale Regional Airport.

The *port authority* is another form of airport management structure. A port authority is a legally chartered institution, with management responsibility often extending beyond aviation and into several transportation modes. For example, the Port Authority of New York and New Jersey (PANY&NJ) represents John F. Kennedy International, LaGuardia International, Newark Liberty International, Stewart International Airport and Teterboro (a GA airport), Downtown Manhattan Heliport, and the New York seaport complex. PANY&NJ also manages various tunnels, bridges, bus terminals, the commuter ferry system, the AirTrain, and the PATH Rapid-Transit System commuter rail for New York and New Jersey.

Airport Security Operation

Airport directors or managers oversee airport operations. Airport staffing includes operations personnel responsible for managing airfield safety inspections and emergency response to incidents including compliance with the ACM; maintenance personnel who ensure the airfield lighting systems and other essential buildings, fleets, and field systems are operational; administrative personnel managing the budget, accounting, human resources, and information technology functions; public relations and marketing personnel; planning and engineering personnel; and security and law enforcement personnel who ensure compliance with the airport security program (ASP).[9] A large airport may have

---

[9]The ASP is a formalized plan required and approved by the TSA and used for implementing and managing security procedures at most U.S. airports.

hundreds or even thousands of employees. In contrast, only one individual may have management responsibility at a smaller GA airport.

Commercial service airports fall under Title 14 CFR Part 139, which requires an airport to meet a variety of standards related to airfield safety inspections, emergency response, wildlife management, public protection, and others. Flight operations at a commercial service airport are regulated under a variety of different regulations depending on the nature of the flight operation (scheduled air carrier, charter, private operation, etc.). Airports with 14 CFR Part 139 certificates are also required to adhere to Title 49 CFR Part 1542 Airport Security.

Most general aviation airports do not have specific federal regulations covering safety and security procedures; however, flight operations regardless of their location are regulated from a security perspective.[10] Three GA airports do have specific security regulations set forth by the federal government because of their proximity to Washington, DC (College Park, Washington Executive/Hyde Field, and Potomac Airfield). Some states have implemented their own regulations covering GA airports.

For many GA airports, the FAA provides funding in the form of federal grants for capital improvement projects. The FAA requires the recipient airport (known as a *sponsor*) to adhere to *grant assurances*. Grant assurances require the airport operator to adhere to certain safety standards and management policies (see *www.faa.gov/airports_airtraffic/airports/aip/grant_assurances*). However, the FAA has not normally provided funding to GA airports for security projects.

Commercial service airport operators prepare an ACM to demonstrate how their airport will comply with Title 14 CFR Part 139. ASPs are created by commercial service airports to demonstrate how they will comply with Title 49 CFR Part 1542. Aircraft operators have similar security programs to demonstrate compliance with federal transportation regulations.

At each commercial service airport, the airport security coordinator (ASC), who usually operates a security department or division, oversees compliance with federal regulations and maintains compliance with the ASP. ASCs are charged with conducting security exercises, providing fingerprint-based criminal-history record checks, issuing identification badges (commonly referred to as access media or credentials), ensuring there is an adequate law enforcement presence for response to incidents, and assisting the TSA, FBI, airlines, and airport tenants with security matters.

Airport Funding

Airports are funded by a variety of sources depending on the type of airport. Hoerter explained that "An airport can operate as a stand-alone enterprise because its revenues and expenses are related in a businesslike context; therefore, logical user fees can be

---

[10]Flight operations can be regulated under 14 CFR Parts 91, 121, 125, 127, or 135, and under Title 49 TSR CFR Parts 1544, 1546, 1548, or 1550.

charged for services rendered" (2001, p. 13). Commercial service airports are funded by leasing gates and administrative space to the airlines and other tenants, charging landing fees, and collecting passenger facility charges (PFCs), parking fees, and a percentage of concession sales. General aviation airports are primarily funded by leasing space to tenants and through the collection of a percentage of fuel sales. Understanding the revenue flows for an airport is important to planning and design decisions relating to security needs. Eliminating revenue-producing space to expand security checkpoints or include additional security equipment creates financial challenges for an airport, which must be offset with some other strategy.

For commercial and GA airports, part of the annual capital improvement funding comes from the FAA's Aviation Trust Fund (ATF).[11] The U.S. Congress reviews and approves airport funding through the AIP every few years (see *www.faa.gov/airports_ airtraffic/airports/aip*). Additional funding is also derived through legislation for security programs and is generated by passenger fees. Several U.S. airports receive funding from their city or county; however, the FAA generally encourages airports to be self-sufficient, generating enough of their own revenue to cover expenses.

Agencies and Organizations

Regulatory Agencies

In the United States, the primary regulating agency for aircraft and airport operations is the FAA, an administration of the U.S. Department of Transportation (DOT). It oversees the management and enforcement of CFR Title 14 Aeronautics and Space, including the rules governing airport safety certification and air traffic control.

Commercial service airports fall under 14 CFR Part 139 Certification of Airports. Part 139 requires an airport to meet a variety of standards such as airfield safety inspections, emergency response, wildlife management, and public protection. Flight operations at a commercial service airport are regulated under a variety of regulations depending on the nature of the flight operation. Airports operating under 14 CFR Part 139 Certificates[12] are also required to adhere to TSR Part 1542 Airport Security (see *www.tsa.gov/research/laws/ regs/editorial_1785.shtm*).

Before 9/11, the FAA presided over the security regulations pertaining to airport and aircraft operations. After 9/11, this function transferred to the newly created U.S. TSA. The TSA enforces the TSRs and is a division of the U.S. Department of Homeland Security (DHS).

---

[11]The ATF was formally known as the FAA Airport and Airway Trust Fund (AATF) and is designed to fund the nation's aviation system through various taxes (see *www.faa.gov/airports_airtraffic/trust_fund*).

[12]In general, 14 CFR Part 139 requires the FAA to issue airport operating certificates to airports that (1) serve scheduled and unscheduled air carrier aircraft with more than 30 seats, (2) serve scheduled air carrier operations in aircraft with more than 9 seats but fewer than 31 seats, or (3) the FAA administrator requires a certificate.

Industry Trade Organizations

Industry trade organizations play a significant role in influencing public aviation safety and security policies through lobbying efforts and various membership services. These organizations often provide training and research on issues related to aviation security. Aviation-related trade organizations include the Airlines for American (formerly known as the Air Transport Association; see *www.airlines.org*), which represents commercial airlines; the International Air Transport Association, which represents airlines throughout the world; the American Association of Airport Executives (AAAE; see *www.aaae.org*), which represents airport operators; Airports Council International, which also represents airport interests; the Aircraft Owners and Pilot's Association, which represents general aviation aircraft aviators; the National Business Aviation Association (NBAA; see *www.nbaa.org*), which represents business aircraft operators; and the National Agricultural Aviation Association (NAAA; see *www.agaviation.org*), which represents the operators of agricultural aircraft.

■ ■ ■ ━━

Case Study September 11, 2001 (9/11) Terrorist Attacks
September 11, 2001

On September 11, 2001, Mohammed Atta checked in for his flight to Boston. Atta was preselected under the Computer-Assisted Passenger Prescreening System (CAPPS) to undergo additional security measures. These measures were to ensure that his baggage did not board the aircraft without him. The positive passenger–baggage matching (PPBM) program is designed to deter placement of an explosive device in checked baggage, but not to deter hijacking.[13]

Four other hijackers—Khalid al Mihdhar, Majed Moqed, Nawaf al Hazmi, and Salem al Hazmi—also checked in at the American Airlines ticket counter at Washington Dulles International Airport. These terrorists were also selected for additional scrutiny under CAPPS, but again, only to ensure they boarded assigned aircraft. One of the hijackers had neither photo ID nor understood English, and the ticket agent labeled the passengers as suspicious.

When the 9/11 Commission interviewed screeners at access points used by the hijackers, none reported anything unusual about the hijackers. At Dulles Airport, some hijackers activated walk-through metal detector alarms and were subjected to additional screening by hand wand. During subsequent reviews, a screening expert seeing the closed-circuit video of the screening process testified that the quality of the hand wand process was marginal. Additionally, one of the hijackers was seen on video with an unidentified object clipped to his belt. At Newark, NJ, the departure point of United Flight 93, one of the hijackers was preselected for additional scrutiny under CAPPS. As with Boston, Newark's screening checkpoints did not have closed-circuit television (CCTV) surveillance, so it is difficult to determine what took place at the screening checkpoints.[14]

---

[13]PPBM is also ineffective against suicide bombers, but for decades it had effectively deterred many nonsuicide airline bombings.

[14]Although not mandated, CCTV monitoring of screening checkpoints is recommended by TSA and other aviation security experts.

Once hijackings began on 9/11, U.S. agencies had a difficult time communicating with each other. Information from hijacked aircraft to the FAA and from airline headquarters to the FAA was nonexistent or ineffective because of cockpit takeovers by the terrorists. One method pilots use to notify the FAA of hijacking is to broadcast an emergency code via a transponder.[15] This code is published in the publicly sold *Aeronautical Information Manual* (Federal Aviation Administration, 2008) and is taught in the most basic FAA pilot training curriculums. Because their flight training was conducted in the United States, the 9/11 hijackers would have known about the code and would have been familiar with the transponder. In three of the four hijacked aircraft, the transponders were turned off, making it difficult for the FAA to track the aircraft on radar.

The FAA and U.S. military agencies were not in communication throughout the entire 9/11 attack sequence. Defending the United States during the attacks was a mission of North American Aerospace Defense Command (NORAD). NORAD's mission was to protect the United States from external threats. NORAD explored the threat of terrorists using aircraft as weapons of mass destruction, but the threat was based on speculation and not on developed intelligence.

The U.S. government requires the FAA to work with the National Military Command Center (NMCC) in the Pentagon to manage hijackings in progress. Part of this process is to notify NORAD's Northeast Air Defense Sector (NEADS) in Rome, New York, to coordinate the intercept of hijacked aircraft by U.S. military aircraft. However, during 9/11, both the FAA and the NMCC assumed that a hijacked aircraft would not attempt to disappear off radar, that there would be time to address the hijacking, and that a hijacking would follow a traditional form (land and negotiate) rather than be enacted as a suicide mission. Communications between the FAA and NEADS were so inadequate that the military was using radar to search for American Airlines Flight 11 after the plane had struck the North Tower of the World Trade Center (WTC). The first indication NEADS had of the second hijacked flight, United Airlines 175, came about the same time as that plane hit the South Tower of the WTC.

NEADS did send two F-15 fighter jets from Otis Air Force Base in Falmouth, MA, but without an assigned target because of communication problems, they were ordered in a holding pattern off the coast of Long Island. A National Guard C-130 cargo plane was also requested to search for American Airlines Flight 77 heading toward its Pentagon target. Because the C-130 was unarmed, the aircraft could only attempt to collide with American Airlines Flight 77 (if it had been granted that action by proper authority).

After 9/11, there was considerable discussion by U.S. agencies and the federal government on who had authority to give an order to shoot down a commercial airliner. President Bush testified that he gave authorization to shoot down hijacked aircraft and that the order was passed to the vice president and through the NORAD command chain. This conversation between President George W. Bush and Vice President Dick Cheney took place between 10:10 a.m. and 10:15 a.m. Eastern Daylight Time on 9/11, just moments before United Airlines Flight 93 crashed in Shanksville, PA. The shoot-down order made its way through the NORAD command structure, and within 20 minutes flight controllers talking to interceptor aircraft had authorization.

The 9/11 Commission Report (Figure 1.10) outlines numerous other flaws in communication between the national military command authority and the FAA. The report stated:

The defense of U.S. airspace on 9-11 was not conducted in accord with preexisting training and protocols. It was improvised by civilians who had never handled a hijacked aircraft

---

[15]Referred to by pilots and air traffic controllers as "squawk 7500."

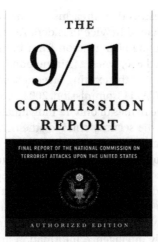

FIGURE 1.10 *The 9/11 Commission Report* outlined the structure and actions of the 9/11 attacks along with recommendations for improving aviation security.

that attempted to disappear, and by a military unprepared for the transformation of commercial aircraft into weapons of mass destruction. As it turned out, the NEADS air defenders had nine minutes' notice on the first hijacked plane, no advance notice on the second, no advance notice on the third, and no advance notice on the fourth. (9/11 Commission, 2004, p. 31)

Assumptions of what terrorists or criminals will do, based on a given situation and on the effectiveness of established communication protocols, are key concerns to aviation security practitioners. The assumptions made by stakeholders on 9/11 followed those of a "traditional hijacking," which proved fatal for thousands of U.S. citizens.

The 9/11 Commission provided insights on the changing nature of the terrorist threat to aviation. The 9/11 Commission suggested that the failure of terrorists to destroy the WTC in 1993 and the quick success of investigators and prosecutors in bringing the perpetrators to justice might have contributed to a widespread underestimation of future terrorist threats by U.S. officials and security personnel.

In addition to the notion of overconfidence by U.S. officials and security practitioners, other failures contributed to the 9/11 attacks. Historically, aviation security professionals have been a consumer of intelligence information rather than a source. Therefore, aviation security practitioners rely on intelligence information gathered, disseminated, and analyzed by other agencies such as the FBI and, before 9/11, the FAA. However, as the 9/11 Commission pointed out, the FAA is also not a traditional source of intelligence information related to threats and is just as much a consumer of that information as airport and aircraft operators. Therefore, breakdowns in the dissemination of intelligence information to entities that can take action are of vital importance to the future of aviation security.

The 9/11 Commission Report noted that over the previous three decades, dissemination of intelligence information from federal agencies was ineffective. During that time, there had not been a domestic hijacking in more than 10 years, and the commercial aviation industry was more concerned with explosives than firearms or other weapons. The Presidential Commission on Aviation Security, chaired by Vice President Al Gore in 1997, focused nearly all of its recommendations on protecting civil aviation from bombings even though the number of attempted hijackings exceeded 600, whereas airline bombings numbered fewer than 100.

The 9/11 Commission stated that the existing layered security system intended to prevent hijackings was "seriously flawed" (9/11 Commission, 2004, p. 83). At the time of the 9/11 attacks, the FAA's *no-fly list* contained the names of only 12 individuals who were not allowed on board U.S. civil aircraft. Furthermore, the security chief for the organization stated that he had not heard about the U.S. State Department's tipoff list of known and suspected terrorists. Apparently, the FAA did have access to the tipoff data but found it too difficult to use. As consumers of intelligence data, the FAA also lacked awareness of developing threats.

Various hijackers on 9/11 had taken flight lessons within the United States. *The 9/11 Commission Report* stated that in 2001, information about Middle Easterners taking flight lessons in the United States was not relayed to the FAA's 40-person intelligence unit. FAA Administrator Jane Garvey did not review daily intelligence information and was unaware of the hijacking threat information from her own intelligence unit.

Before 9/11, checkpoint screening was considered the most important layer of security. However, numerous Government Accountability Office (GAO) reports identified shortcomings in the screening process. Another security gap was FAA's policy of allowing knives with blades shorter than four inches onto commercial aircraft. A 1993 proposal to ban knives was rejected because of projected increased congestion at checkpoints. Although continuous and random secondary screening using EDS and explosive trace detection (ETD) technology was mandated after 1996, the policy was ignored to the point where it was operationally nonexistent. The report was further critical of airline opposition to procedures affecting the efficiency of flight operations.

Training and performance standards were part of the Gore Commission's recommendations, but not until the passage of the Airport Security Improvement Act of 2000 (ASIA 2000) were they placed into legislation. However, by 9/11 screener training and performance standards had still not been implemented. ASIA 2000 transferred the responsibility to the TSA, and screener training and performance standards are now required. Furthermore, the policy on prohibited items has changed. Knives shorter than four inches are no longer allowed on board commercial aircraft.

The final layer of airline security against hijackings used at the time of 9/11 was known as the Common Strategy, which instructed pilots and flight crews to cooperate with hijackers and land the plane as soon as possible, then let law enforcement take over to resolve the hijacking. The Common Strategy was not designed to counter suicide hijackings. This fundamental concept has changed to more active resistance against hijackers, as the assumption can no longer be made that future hijackers will follow traditional methods or purposes of hijacking.

We can learn important lessons by analyzing the planning of the 9/11 attacks. Khalid Sheikh Mohammed (KSM; Figure 1.11) began work on the future 9/11 attacks with Osama bin Laden in 1996. The two met in Afghanistan to discuss various attack methods against "potential economic and 'Jewish' targets in New York City" (9/11 Commission, 2004, p. 150). KSM had previously been involved in plots to assassinate President Clinton and to bomb U.S.-bound air cargo carriers by smuggling nitrocellulose on board. At the 1996 meeting, initial planning of the 9/11 attacks was formulated as KSM and bin Laden discussed plans to train pilots to fly into

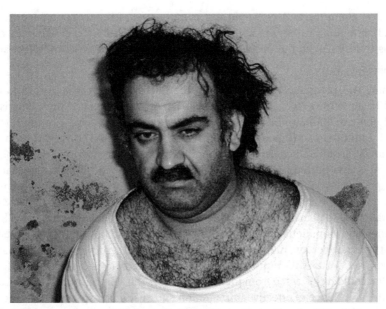

FIGURE 1.11 Khalid Sheikh Mohammed worked with Osama bin Laden to plan the 9/11 attacks.

U.S. targets. New York and Los Angeles were identified as economic centers of the United States, which factored into the decision to target those cities. Other forms of attack were considered, including hijacking planes and then blowing them up in flight (similar to the Pan Am 103 bombing). However, KSM felt that explosives were too problematic and that a novel form of attack was required.

By 1999, the plot by KSM and bin Laden was moving forward with a list of targets including the White House, Pentagon, World Trade Center, and U.S. Capitol. Suicide operatives were recruited. A second attack was also planned, which resembled Yousef's Operation Bojinka plot. In this secondary plot, suicide bombers would board U.S. aircraft on Pacific oceanic routes, possibly concealing bombs in their shoes. Both attacks (hijackings and airline bombings) were to be executed simultaneously, enhancing the psychological impact. This strategy is characteristic of the desire by terrorists for large physical and widespread psychological impact. However, by 2000, bin Laden canceled the second attack plan, declaring it too difficult to coordinate with the U.S. hijack operation. In late 2000, the U.S. National Security Council did warn the Justice Department about the second attack plan.

KSM spent much of 1999 training operatives on what he learned about the U.S. aviation system. He searched the Internet, collecting brochures and information on U.S. flight schools, reviewed airline timetables, bought movies depicting hijackings, and even purchased flight simulator software. Training also included lessons in English and how to read U.S. phone books and airline timetables. Although most of this material is publicly available, it is interesting to note the extent to which the attacks were planned to give the terrorists insight into the intelligence-gathering process. Part of the advance surveillance process included observing flight crews in action, noting particularly when the cockpit doors were opened and processes that would take place when flight crewmembers used the lavatory and when food was served to the flight crew. Some operatives would test components of the aviation security system by intentionally bringing box cutters onto commercial flights and by removing box cutters and other items while in flight, then observing the reactions of passengers and flight attendants.

Flight Training Begins

Once training was completed in Afghanistan, the hijackers began their move to the United States and began efforts to appear less radical. The hijackers shaved their beards and wore Western clothing in efforts to blend into U.S. society.

The 9/11 Commission outlined two abilities fundamental to the success of terrorist attacks: to travel and to transfer money. The plot required numerous international trips by planners and operators. Al-Qaeda hosted an office at the Kandahar Airport in Afghanistan that facilitated altering papers, visas, and passports. Operators were taught simple methods to alter their own passports and change documents while en route. It is estimated that $400,000 to $500,000 was spent to plan and carry out the 9/11 attacks. Al-Qaeda raised money through corrupt charities and donations from sympathizers in certain Gulf countries including Saudi Arabia, but it is not certain that all donors knew how their money was being used. The 19 hijackers were supported with funds wired from KSM, brought in as cash or held in foreign bank accounts that could be accessed from the United States. These findings of the 9/11 Commission highlight the importance of effective immigration and customs inspections as crucial components to national security that also affect aviation security. Although not part of the 9/11 attacks, the commission did note the importance of using fraudulent documents to facilitate an attack, when addressing the attempt by Ahmad Ressam (who used fraudulent documents) to attack Los Angeles International Airport in 2000.

Once in the United States, with their flight training started, the hijackers provided some interesting clues as to their intentions. A San Diego–based flight instructor of two of the operatives noted their unusual reaction when told they would have to start learning to fly in small aircraft. He believed they were joking when they requested to start their training in large planes, particularly Boeing-type aircraft. The instructor also noted that they were poor students and seemed more focused on learning how to operate the aircraft in flight rather than on how to land. A Florida flight instructor commented that two of the students, who were also hijackers, were rude and aggressive and at times fought with him over the controls. At one point, both those hijackers took the FAA exam to obtain the FAA instrument rating, and both failed. They were both very upset and stated that they had jobs waiting for them once flight training was complete.

Eventually, three of the hijackers did learn to fly well enough to obtain the FAA commercial pilot certificate and began flying large-aircraft simulators. One pilot, Hani Hanjour, obtained his flight certificates in Arizona between 1996 and 1997. He was identified as a pilot by the background information he provided while at an al-Qaeda training camp and referred to KSM in 2000. Hanjour returned to Arizona for refresher training and to learn how to fly multiengine aircraft. Difficulties with English made learning how to fly the multiengine planes a challenge, and his flight instructor recommended he discontinue. Hanjour then enrolled at the Pan Am International Flight Academy in Mesa, AZ, and began training on a Boeing 737 simulator.

A critical ingredient in the 9/11 attacks was the ability of the hijackers to fly the aircraft, which is why ATSA 2000 now places specific requirements on flight schools throughout the United States, now called the Alien Flight Training Program. Under the program, all flight students are now required to show proof of citizenship in the United States; legal aliens are required to have their fingerprints taken and their names checked against known terrorist watch lists.

In the early summer of 2001, the pilot hijackers began taking cross-country surveillance flights on U.S. air carriers. Some also took additional flight training including flights along the Hudson Corridor, "a low-altitude hallway along the Hudson River that passes New York landmarks like the World Trade Center," as well as through the Washington, DC, airspace (9/11 Commission, 2004, p. 242).

At a meeting with an al-Qaeda facilitator in Spain in the summer of 2001, Mohammed Atta discussed details of the attack, including orders that if the pilots were unable to reach their objectives, they were to crash the planes. Atta considered targeting a nuclear facility, but the other pilots disliked the idea because of airspace restrictions, which would make reconnaissance around such sites difficult and increase the possibility of the planes being shot down. Al-Qaeda leaders determined that a nuclear strike would not have equal psychological value to attacking the WTC, as the center symbolized the economic strength of the United States. Atta further advised the following (9/11 Commission, 2004, p. 245):

- On their surveillance flights, box cutters were not of concern to security screeners.
- The best time to access the cockpit was 10–15 minutes into the flight when the cockpit doors were normally open. Atta discussed taking a hostage to compel the flight crew to open the doors but he felt this would not be necessary, as they should be open or unlocked anyway.
- To use airplanes on long flights, as they would have more fuel on board.
- To hijack Boeing aircraft because he believed Airbus aircraft had an autopilot feature that prevented planes from being crashed intentionally into the ground.
- And the "muscle hijackers" had all arrived in the United States without incident.

Zacarias Moussaoui

KSM did testify that a "second wave" of attacks was planned, which included Zacarias Moussaoui (Figure 1.12). However, during the summer of 2001, he was too busy with the 9/11 operation to put much effort into its planning. Moussaoui had taken flight lessons in Norman, OK, and asked about flight simulation materials for a Boeing 747. He also was moving forward with an application to Pan Am International Flight Academy, bought knives, and inquired about the use of portable global positioning system (GPS) units for aviation (9/11 Commission, 2004, pp. 246–247).

Moussaoui's instructor at Pan Am notified authorities about his student's unusual behavior. Moussaoui paid the full balance of his Pan Am training in cash ($6,800), and it seemed odd to his instructor that Moussaoui was learning to fly large jets without any intention of obtaining a pilot's license. Immigration authorities arrested Moussaoui on August 16, 2001. However, bin Laden and KSM did not learn of his arrest until after 9/11. KSM indicated that had they heard about the arrest they likely would not have gone forward with the attacks.[16] This is an important point, as it demonstrates how unplanned events can deter an attack in the planning stages. Another important point is the flight instructor's suspicions of his student and his

---

[16]In 2006, Moussaoui plead guilty to six terrorism conspiracy charges and was sentenced to life in prison.

FIGURE 1.12 Zacarias Moussaoui took flight lessons in the United States in preparation for his role in the 9/11 attacks.

actions to report those suspicions to authorities. This was information that could have led to discovery or mitigation of the plan to attack on 9/11.

The System Is Flashing Red

During the summer of 2001, warning reports of planned terrorist attacks increased to their highest level since the millennium alert. In May 2001, the FAA was notified of a potential plot to hijack U.S. aircraft to free incarcerated terrorists in the United States. The FAA issued an information circular to U.S. air carriers warning of such attacks. During that same time, CIA Counterterrorist Chief Cofer Black warned National Security Adviser Condoleezza Rice that the current threat level was 7 on a scale of 1 to 10, as compared to an 8 during the millennium alert. In June, the CIA notified its station chiefs about intelligence, suggesting al-Qaeda was planning a possible suicide attack against a U.S. target. The Department of State notified all its embassies of the terrorist threat. By late June, a terrorist threat advisory warned of near-term "spectacular" attacks resulting in numerous casualties (9/11 Commission, 2004, p. 257). U.S. Central Command raised the force protection condition level for its troops to the highest level. By July, the FBI Counterterrorism Division issued a message to federal and state law enforcement agencies warning of increased threat reporting that indicated a potential attack against U.S. targets, but also stated that there was no credible threat of a pending attack within the United States. Also, many warnings circulating within the federal government suggested an attack was imminent at locations other than within the United States.

In July 2001, an FBI agent in the Phoenix field office sent a memo warning of a coordinated effort by bin Laden to send students to U.S. flight schools and recommended that specific actions be taken to further investigate this lead. The author of the memo stated that its subject was not related to suicide hijackings but more toward a potential bombing of U.S. airliners. No one at FBI headquarters saw the memo before 9/11 and the New York City field office took no action. The 9/11 Commission stated that if the memo had been read and followed up on, it may not have revealed the plot but may have made agents more sensitive to the threat when Moussaoui was arrested and therefore the matter taken more seriously.

By July 31, the FAA issued another information circular alerting aircraft operators of possible near-term terrorist operations in the Arabian Peninsula or Israel, but the circular also stated it had no credible evidence of specific plans to attack U.S. civil aviation.

Intelligence activity regarding potential attacks seemed to drop in August 2001. The intelligence community issued an advisory concluding that warnings of an al-Qaeda attack would continue indefinitely. From an intelligence perspective, although many were watching, few of the reported activities were being correlated. This sentiment is embodied in the 9/11 Commission's comment that "The September 11 attacks fell into the void between the foreign and domestic threats" (9/11 Commission, 2004, p. 263).

In an attempt to prevent future intelligence failures, on December 17, 2004, President Bush signed the Intelligence Reform and Terrorist Prevention Act of 2004. The act created the position of director of national intelligence (DNI). The DNI is now the president's primary intelligence adviser with authority over budget and personnel decisions throughout the 15-member U.S. intelligence community.

In the summer of 2001, the FAA did not issue any security directives that mandated increased security at checkpoints or on board aircraft. Mostly the FAA urged air carriers to "exercise prudence" and be alert. Before 9/11, the FAA presented a CD-Rom to air carriers and airport authorities describing the increased threat to civil aviation. The presentation mentioned the possibility of suicide hijackings but indicated that no known group was planning to implement this type of threat.

The FAA conducted 27 special security briefings for specific air carriers between May 1, 2001, and September 11, 2001. Two of these briefings discussed the hijacking threat overseas. None discussed the possibility of suicide hijackings or the use of aircraft as weapons. No new security measures were instituted.

Emergency Response

In describing the concept of emergency response, the 9/11 Commission (2004) stated:

> *Emergency response is a product of preparedness. On the morning of September 11, 2001, the last best hope for the community of people working in or visiting the World Trade Center rested not with national policymakers but with private firms and local public servants, especially the first responders: fire, police, emergency medical service, and building safety professionals. (p. 278)*

The 9/11 Commission Report touches on an important point of airport and aircraft operator security, which is the ability to both withstand an attack and recover from such an attack. The quicker an agency is able to recover from an attack, the less likely it is to be a target.

The WTC towers housed approximately 50,000 workers, and the entire complex exceeded the office space of downtown Atlanta, GA, and Miami, FL, combined (Figure 1.13). The towers provided a variety of features to mitigate fire and smoke damage; however, there was not a rooftop evacuation plan. Both roofs were sloped and cluttered with radiation towers. There

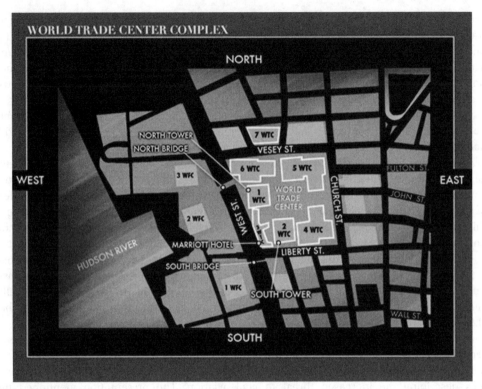

FIGURE 1.13 The WTC towers.

was a helipad on the South Tower, which did not meet FAA certification for use. During the building fires on 9/11, strong updrafts and a lack of visibility resulting from the rising smoke rendered helicopter rescues impossible. However, there were several rescues from the roof during the 1993 WTC bombing, which seemed to leave those who worked at the WTC with the impression that helicopter roof rescues were part of the evacuation plan—however, they were not.

In 1993, after a truck bomb exploded underneath the WTC towers in a parking garage, killing 6 and injuring more than 1,000, the towers lost power and communications, including emergency communications. The 9-1-1 emergency call systems were flooded, and the ensuing evacuation took four hours. These were important lessons learned, and the PANY&NJ (the offices of which were located within the WTC) invested more than $100 million to make physical, structural, and technological improvements. The improvements included enhancements to power sources, emergency exits, a computerized fire alarm system, and the creation of a fire safety director position. Fire safety teams composed of office workers with fire wardens and searchers were created, along with twice-per-year fire drills. A radio repeater system was also installed in the towers to assist fire department personnel with radio communications. On 9/11, after both towers had been struck and response efforts were under way, the repeater system for the radios that was installed after the 1993 bombing worked only partially (in both towers) depending on the channel used and the particular radio broadcasting or receiving.

The 9/11 Commission noted that the fire drills conducted in the WTC were not comprehensive enough to address such a massive fire as that caused by the aircraft on 9/11, and that tenants were never advised that a roof rescue was not part of the evacuation plan. During evacuation drills, tenants were advised to evacuate only a few floors below the fire. Apparently, WTC tenants were not the only individuals not informed of the lack of a roof rescue contingency. After the WTC North Tower was struck, a team from the New York City Police Department's Emergency Services Unit (ESU) requested to be picked up at the Wall Street heliport to assist in a roof rescue. The team was advised that they could not land on the roof because of flames and heavy smoke. Regardless, ESU units and three New York Police Department (NYPD) helicopters remained ready to try a roof rescue if the conditions changed.

The principal first responders on 9/11 were the Fire Department of New York (FDNY), the NYPD, the Port Authority Police Department (PAPD), and the Mayor's Office of Emergency Management (OEM). The PAPD law enforcement officers are cross-trained in fire suppression methods. Their radios only worked well within the vicinity of their command, and they lacked standard operating procedures on how to manage officers from multiple commands who would respond to an incident.

New York City's OEM was tasked with monitoring communications between the FDNY, NYPD, and other agencies, and to improve the city's response to major incidents. The OEM office was located in 7 WTC,[17] a building that became structurally unsound as a result of the 9/11 attacks and collapsed by 5 p.m. the afternoon of September 11. Some questioned the decision to locate the OEM offices so close to the site of a previous terrorist attack. This is an important lesson for aviation security practitioners regarding the location of their own incident command centers (Figure 1.14).

Mayor Rudolph Giuliani had previously issued a directive regarding control of New York's assets during an emergency. The directive stated that there would be an incident commander for each emergency and that OEM would act in an interagency support role. However, "FDNY and NYPD each considered itself operationally autonomous. As of September 11, they were not prepared to comprehensively coordinate their efforts in responding to a major incident. The OEM had not overcome this problem" (9/11 Commission, 2004, p. 285). On 9/11, Giuliani and his staff did respond to 7 WTC, and then, on seeing its condition and the hazard to their lives, quickly evacuated and reestablished OEM's function in a nearby church.

When American Airlines Flight 11 struck the North Tower, for the first 10 minutes, most of the city of New York, including the tenants in the towers and the deputy fire safety director, did not know what happened. The 9-1-1 emergency phone lines jammed, with some reporting that a commercial jet had struck the tower. Survivors in the North Tower gathered and prepared to evacuate, but many were trapped above the flames. Thick black smoke of burning jet fuel worked its way up through the upper floors and, with the prevailing winds, obscured visibility in the WTC South Tower. Although one fire safety director testified that a public evacuation announcement was made, many did not hear it because the intercom system was knocked out through much of the North Tower. Also, the NYPD aviation unit did not inform the 9-1-1 operators that rooftop rescues were not part of the plan, and therefore the 9-1-1 operators could not advise callers from inside the North Tower on which direction to go.

---

[17] 7 WTC also housed offices from the U.S. Secret Service, Department of Defense, the Securities and Exchange Commission, the Internal Revenue Service, and the Central Intelligence Agency.

FIGURE 1.14 The WTC on 9/11.

At 8:57 a.m. on 9/11, the FDNY ordered the South Tower evacuated. One company, Morgan Stanley, evacuated all of its employees, which occupied nearly 20 floors of the South Tower, on the decision of company security officials. This decision undoubtedly saved hundreds of lives. A public announcement from an unknown source in the South Tower was made, instructing tenants to remain on their floor or return to their offices. The 9/11 Commission concluded that several individuals were told it was safe to return to the South Tower. At 9:02 a.m., another public announcement was made ordering a general evacuation. This judgment was made because the North Tower crash was causing safety hazards for the South Tower, not on the supposition that there was another airplane targeting the South Tower. FDNY personnel also worried about their ability to fight the fire and whether water could be accessed in the upper floors.

FDNY personnel established a command post in the lobby of the North Tower and began working with PAPD and OEM personnel. Structural engineers advised them that, although a partial floor collapse was possible, no one anticipated the collapse of the entire building. The lobby was filled with debris and walking wounded as the first plane crash caused a fireball to shoot down the elevator shaft, blowing out windows on the ground floor. The explosion caused the PAPD onsite commanding officer to dive for cover. PAPD units began to respond after the first plane hit but lacked standard written procedures for responding to outside commands.

The OEM team had also responded and immediately activated its emergency operations center notifying numerous agencies from the FDNY, NYPD, Department of Health, and the Greater Hospital Association, in addition to the Federal Emergency Management Agency (FEMA), and requested five urban search and rescue teams.

It is significant to note that despite certain failures, within 17 minutes of the first plane striking the North Tower, the city of New York initiated a massive evacuation and the largest rescue effort ever, deploying more than 1,000 emergency responders, and made a critical decision—that they could not fight the fire.

At 9:03 a.m. EDT, United Airlines Flight 175 struck the South Tower. Many individuals killed in the South Tower died as they waited for overcrowded express elevators to take them out of the building after the order had come just one minute earlier to evacuate. A "lock release" order, which would unlock all the security doors in the South Tower and was part of the building's computerized security system, was transmitted to the Security Command Center, located in the North Tower. Damage to the software controlling the system prevented the order from being executed. The situation in the South Tower was exacerbated because of the 9-1-1 phone operators' lack of awareness of exactly what was occurring. Many evacuees calling the 9-1-1 phone emergency number were put on hold or given erroneous information.

In *The 9/11 Commission Report*, the FDNY made another important point. The FDNY was unable to receive information from NYPD helicopters, which had a bird's eye view of the situation. According to one FDNY chief, citizens watching TV had more information about what was going on than did the command team onsite.

At 9:37 a.m. EDT, American Airlines Flight 77 struck the west wall of the Pentagon. However, the ensuing incident response to the Pentagon was far less hectic than the response at the WTC. *The 9/11 Commission Report* highlights that in contrast to the WTC attacks, the Pentagon was not 1,000 feet up, was a single incident, and had relatively easy access. In addition, the incident command post had a complete view of the site, and there were no other buildings in the immediate area.

The 9/11 Commission outlined other items significant in regard to incident command on 9/11:

- Cell phones were of little value because of overloaded circuits.
- Radio channels were oversaturated.
- Problems with command, control, and communications occurred at both sites.
- Changes made by the PANY&NJ after the 1993 bombing improved the evacuation on 9/11, including dropping the general evacuation time from four hours to one hour.
- First responders played a significant role in successes of that day.
- Many of the first responders were actually private-sector citizens.
- NYPD and FDNY 9-1-1 operators were not a part of the plan (to the detriment of the entire incident).
- Civilians need to take personal responsibility for maximizing their probability of survival during a disaster.
- There were significant chain-of-command challenges between all responding agencies.

The 9/11 Commission

One of the most important post-9/11 aviation security–related publications was from The National Commission on Terrorist Attacks Upon the United States (9/11 Commission, 2004), known as *The 9/11 Commission Report* (see Figure 1.10). This report, commissioned by President Bush in 2002, outlined both the structure and actions of the 9/11 attacks and provided a broad look into the issues that contributed to the attacks. The 9/11 Commission provided a series of recommendations for fixing the aviation security system and the U.S. intelligence industries.

The mandate of the 9/11 Commission included the investigation of facts and circumstances relating to the terrorist attacks. This included the related roles of intelligence agencies, law enforcement, immigration and border control, commercial aviation, and congressional oversight. Nearly 1,200 individuals were interviewed and more than 2.5 million pages of documents reviewed. The 9/11 Commission learned that the enemy is "patient, disciplined and lethal" and that the "institutions protecting our borders, civil aviation and national security did not understand how grave [the] threat and did not adjust policies, plans and practices to deter or defeat it" (9/11 Commission, 2004, p. xvi).

With regard to the 9/11 attacks, the 9/11 Commission highlighted four fundamental failures in U.S. aviation security (2004, p. 339):

1. A failure of imagination.
2. A failure of policy.
3. A failure of capabilities.
4. A failure of management.

Failure of imagination ties directly to the performance of those in the position of protecting civil aviation. Without the imagination to conceive of different forms of attack, policies are not created, capabilities are not developed, and management is not trained, directed, and focused on new forms of attack.

The 9/11 Commission noted that they were making recommendations with the benefit of hindsight. However, analysis of the 9/11 attacks by the 9/11 Commission has created a norm for security policy makers and managers to interactively "imagine" or develop strategies to counter new forms of attack (Figure 1.15). When foresight and planning by security specialists stagnates, then the probability and seriousness of other forms of attack will increase.

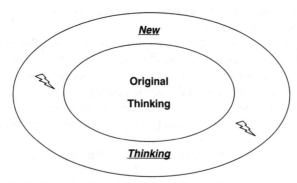

FIGURE 1.15 "Expanded thinking" as related to mitigating new forms of attack.

Before 9/11, there were several attacks along with intelligence data that, in retrospect, indicated that suicide attacks were highly probable. In 1994, a group of Algerian terrorists hijacked a plane and threatened to crash it into the Eiffel Tower in Paris, France. Also in 1994, a private plane crashed on the south lawn of the White House. In 1995, an accomplice of Ramzi Yousef told Philippine authorities of a plot to crash a plane into CIA headquarters (9/11 Commission, 2004, p. 345). However, when the Gore Commission[18] studied aviation security in 1997, the possibility of a suicide hijacking was not among its conclusions.

Al-Qaeda had used suicide truck-bomb attacks and had used a ship bomb to attack the *U.S.S. Cole*.[19] In 1998, a presidential daily briefing highlighted a potential plan for al-Qaeda to hijack a plane to free prisoners. Various U.S. intelligence reports also mentioned the possibility of an aircraft filled with explosives being flown into a U.S. city. In August 1998, another report mentioned the intent of Libyan operatives to crash a plane into the World Trade Center. In addition, Richard Clarke (the White House counterterrorism adviser) chaired a security analysis in which a stolen business aircraft loaded with explosives would be flown with the intention of crashing into and destroying the White House. Pentagon officials suggested the military would respond to that threat by scrambling fighter jets from Langley Air Force Base, but admitted they would need clarification on the Joint Chiefs of Staff Rules of Engagement[20] before shooting down the plane (9/11 Commission, 2004, p. 345).

In 1999, the FAA's security office summarized potential threats from bin Laden but focused mostly on the probable hijacking of an aircraft to secure the release of prisoners. The FAA team stated that a suicide hijacking does not allow for dialog and would be a last resort. NORAD envisioned a form of suicide attack from a hijacked aircraft but made the assumption it would be coming from outside the United States and that they would have time to react to the threat (9/11 Commission, 2004, p. 346).

The 9/11 Commission noted that since the attack on the United States at Pearl Harbor in 1941, the intelligence community developed rigorous analytic methods to forestall an attack. The four-step process was essentially to do the following (9/11 Commission, 2004, p. 346):

1. Imagine how surprise attacks may be launched.
2. Identify telltale indicators that such attacks were being planned.
3. Develop and collect intelligence on such indicators.
4. Develop defenses to deflect the most dangerous possibilities or at least provide an early warning.

The first component, imagination, is critical to the rest of the intelligence process. The 9/11 Commission noted that because the type of attack that took place on 9/11 had not yet been conceived, it did not perform an analysis from the enemy's perspective, commonly referred to as *red teaming*. Without attack indicators, such as possible terrorists learning to fly large-jet aircraft, there was no system in place to identify and report such indicators. Also, without such indicators, the FBI and other law enforcement agencies were not focused on

---

[18]Formally titled the White House Commission on Aviation Safety and Security.

[19]The guided missile cruiser *U.S.S. Cole* was attacked by suicide bombers on October 12, 2000, while harbored in the Port of Yemen; 17 sailors were killed.

[20]Joint Chiefs of Staff Rules of Engagement are policies established for U.S. military services to determine appropriate actions or responses to various threats or conditions.

this type of activity, and, hence, no credible threats were established (9/11 Commission, 2004, p. 347). Threat information going to airport and aircraft operators was the same information that they had been receiving for years, which was that hijackings remain possible. No information indicated that suicide hijackings were being planned.

Fixing the Weaknesses in Aviation Security

The 9/11 Commission agreed that it is vital to sustain a fundamental principle of aviation security, which is a *layered security system.* The layered security system is a proactive approach to predicting terrorist activity and reacting with overlapping strategies, tactics, and methods in ways that mitigate possible deficiencies in any specific security layer (Elias, 2005, p. 3). The 9/11 Commission outlined five specific weaknesses in the aviation security system (Elias, 2005, p. 1):

1. Prescreening processes that focus on detecting potential aircraft bombers rather than potential hijackers.
2. Relaxed checkpoint screening and permissive rules regarding small knives.
3. A lack of in-flight security measures such as air marshals and reinforced cockpit doors.
4. An industry-wide strategy of complying with hijackers in a nonconfrontational manner.
5. A lack of protocols and capabilities for executing a coordinated FAA and military response to multiple hijackings and suicidal hijackers.

Items 3–5 were previously addressed in the ATSA 2000 legislation and in the Homeland Security Act of 2002. The 9/11 Commission placed emphasis on weaknesses 1 and 2 (Elias, 2005). The 9/11 Commission's recommendations for addressing weaknesses in aviation security are as follows:

1. Enhance passenger prescreening.
2. Improve measures to detect explosives on passengers.
3. Examine human factors issues at screening checkpoints.
4. Expedite deployment of inline baggage screening systems.
5. Intensify efforts to identify, screen, and track cargo.
6. Deploy hardened cargo containers.
7. Institute risk-based prioritization as the basis for transportation security policy and the creation of a strategic plan for aviation security.

At the time of the 9/11 Commission's recommendations, the existing passenger prescreening system CAPPS was based on certain characteristics surrounding the purchase of an airline ticket and, at the time, a traveler's response to certain questions asked by ticket agents. If a passenger was preselected, then he or she would be subjected to additional screening. This additional screening was used only to the extent of trying to deter an airline bombing, not a hijacking. The CAPPS program was managed by the airlines. The follow-up program, CAPPS II, has been tied up in controversy, specifically over the privacy issues of airlines providing passenger identification data to the U.S. federal government. The 9/11 Commission was careful to avoid endorsing a specific screening program but did state that some sort of comparison of traveler names with the names on various terrorist watch lists needed to occur. Some limited testing of elements of the CAPPS II program continues, but now under the name Secure Flight.

With respect to checkpoint screening, the 9/11 Commission focused on Ramzi Yousef's Operation Bojinka plot to bomb 12 U.S. airliners over the Pacific Ocean. The 9/11 Commission recognized the limitations of passenger and baggage screening devices and encouraged the development and implementation of newer technology with better detection abilities. Several U.S. airports are now testing newer methods of passenger and baggage screening including the use of EDS systems. The 9/11 Commission also focused on human factors issues related to the security screening workforce and the accelerated implementation of inline EDS systems.

The 9/11 Commission made several recommendations regarding the security of air cargo, including methods to track and screen potentially dangerous air cargo shipments, the use of at least one hardened cargo container on commercial airliners, and the protection of all-cargo aircraft. Subsequent to the 9/11 Commission's findings, the TSA has issued revised air cargo regulations.

Finally, the 9/11 Commission reiterated a long-term approach to protecting aviation and the greater intermodal transportation system by utilizing a risk-based analysis. By conducting risk assessment and developing an aviation security strategic plan, policies can be formed and funding can be directed at those areas of highest risk and highest loss of life and property, should an attack occur.

(The following case study is a synopsis of the key contributing factors and issues in aviation security management that were salient to the 9/11 attacks. Most of the content for this case was sourced from *The 9/11 Commission Report* [9/11 Commission, 2004].)

■ ■ ■

Conclusion

The aviation industry is essential to the viability of the U.S. economy, which makes it a prime target for terrorists. The complexity and size of the industry also make it an attractive environment for crime. Understanding the tenuous financial structure of airlines, the aviation industry aids the security practitioner in applying appropriate and practical security measures. Aviation security practitioners must deploy systems, measures, and procedures to counteract both terrorist and criminal perpetrators. To meet these challenges, aviation security practitioners employ layered security systems that are symbiotic with the global aviation industry.

The U.S. Congress establishes policy for protecting U.S. aviation. Federal regulators convert these policies into regulations, which are therefore established as accepted industry practices. Federal regulators implement and supervise these policies and regulations across all aircraft operators and airports.

The 9/11 Commission was tasked to assess facts surrounding the September 11, 2001, terrorist attacks. The 9/11 Commission analyzed and recommended new strategies for adoption within the United States. Because of the 9/11 Commission's work, the largest overhaul of aviation security in U.S. history was implemented. However, to thwart or reduce the risk of crime or terrorist activity, the strategies recommended by the 9/11 Commission must remain ephemeral in evolution and application. The following chapters in this text will help aviation security practitioners or students build on the recommendations made by the 9/11 Commission. Readers of this text will be better prepared to understand, develop, and apply strategies, tactics, and methods that are appropriate and practical to the future needs of aviation security.

Aviation security practitioners or students of aviation security should have a solid understanding of the nature and contributing factors regarding the attacks of September 11, 2001. Therefore, it is strongly recommended that readers review the preceding case study on 9/11.

References

ASIA, 2000. Airport Security Improvement Act of 2000, P.L. 106–528 Sec. 3.

ATSA, 2001. Aviation and Transportation Security Act of 2001, P.L. 107–71 Sec. 118.

DRI-WEFA, 2002. The National Economic Impact of Civil Aviation, retrieved July 6, 2008, from www.aia-aerospace.org/stats/resources/DRI-WEFA_EconomicImpactStudy.pdf.

Elias, B., 2005. Aviation Security-related Findings and Recommendations of the 9/11 Commission, retrieved July 5, 2008, from http://fpc.state.gov/documents/organization/46482.pdf2005.

Federal Aviation Administration, 2008. Aeronautical Information Manual, retrieved July 1, 2008, from www.faa.gov/airports_airtraffic/air_traffic/publications/atpubs/aim2008.

FY, 2000. Fiscal Year, 2000 - the reference is the DRI study.

Hoerter, S., 2001. The Airport Management Primer, second ed. American Association of Airport Executives, Mount Pleasant, SC.

Kane, R., 1998. Air Transportation, thirteenth ed. Kendall-Hunt, Dubuque, IA.

Kumar, J.N., Tzou, D., Xu, L., Chen, Y., Xia, J., June 2003. The economic impact of September 11, 2001 on the aviation industry, Paper presented at the 9th Annual GTTL (Global Trade, Transportation, and Logistics) Conference, University of Washington, Seattle, WA.

9/11 Commission, 2004. The 9/11 Commission Report. U.S. Government Printing Office, Wasington, DC, retrieved July 5, 2008, from www.9-11commission.gov/Commission.

Wells, A., Young, S., 2004. Airport Planning and Management, fifth ed. McGraw-Hill, New York.

2

Crime and Terrorism in Aviation
A Retrospective

Objectives

This chapter introduces the significant incidents, crimes, and forms of terrorism committed in the history of aviation. These cases are shown to have had significant effects on the evolution of aviation security. The chapter also describes the motives for various forms of hijackings and terrorism from the days of early flight to modern times. Historically, aircraft have been a prime target for attack with airports enduring less risk. Studying the history of these attacks on aircraft and airports is an essential task for improving the effectiveness of aviation security. Lessons learned from attacks are presented in this chapter and should be examined thoroughly by aviation security practitioners or students.

> *Of the available forms of theater, few are so captivating figuratively, as well as literally, as skyjacking. The boldness of the action, the obviousness of the danger, the numbers of people involved as hostages and therefore potential victims, the ease with which national boundaries can be traversed and international incidents created, the instantaneous radio and television linkages, all these combine to make this crime one of the most immediately attractive forms of terrorist action. (Wilkinson and Jenkins, 1999, p. 27)*

Introduction

In the decade after the 9/11 attacks, there have been continued attempts to bomb commercial aircraft, shoot down airliners with surface-to-air missiles, bomb terminal buildings with both improvised explosive devices and vehicle-born improvised explosive devices, and attempts to bomb commercial aircraft by shipping bombs via air cargo. It is apparent that despite the increased focus on preventing attacks on aviation, terrorists and criminals continue to see aviation as a high-value target.

Three fundamental strategies for attacking aviation include hijackings, bombings, and airport assaults. These threats have remained relatively unchanged since their earliest use. However, the methods, motivations, and purposes for conducting hijackings, bombings, and airport assaults have changed over time. The 9/11 attacks were, fundamentally, aircraft hijackings, and the purposes of the 9/11 attacks were significantly different from those of previous hijackings. The traditional objective for past hijackings was to land and negotiate. In these cases, the terrorists used hostages taken during hijackings to leverage arrangements or demands on a culture or society. In contrast, the 9/11 hijackers used airliners as guided missiles against civilian and military targets. Regardless of these differences, the fundamental mission of aviation security remains the same—to prevent or deter hijackings or attacks from occurring.

It is not the purpose of this text to review or document every known aviation security–related incident. Instead, this text emphasizes significant attacks on the aviation industry that resulted in new policies or practices or reinforced certain principles of aviation security. Examples of such principles include searching aircraft between departures or the practice of matching checked baggage with passengers on board the flight.[1] The website companion to this text highlights many of the details of the incidents discussed in this chapter (www.leadingedgestrategies.com).

By examining past attacks on aviation, it becomes apparent that one of the difficulties of aviation security is that acts to mitigate threats often create the next challenge. Criminals and terrorists become more creative, more daring, and more deadly as new technologies and strategies, such as baggage screening and air marshals, are introduced to prevent certain avenues of attack. Although closing vulnerabilities in the security system may reduce the number of criminal or terrorist attacks, it also increases the potential severity of future attacks. This was certainly demonstrated on 9/11. Malcolm Gladwell characterized this problem by stating, "Airport-security measures have simply chased out the amateurs and left the clever and the audacious" (2001, section 3).

Very few forms of crime or terrorism perpetrated in the aviation industry are new. Strategies and tactics used by criminals and terrorists to carry out hijackings, bombings, and airport attacks may and do change with time. However, we can learn much from past events to envision and prepare for future threats. In the spirit of trying to learn from the past and not repeat mistakes in the future, the ensuing sections are intended to provide aviation security practitioners with:

1. A basic overview of aviation security incidents.
2. Explanations of the impacts of aviation security's watershed events.
3. The relevance of lessons learned.

Historical Context on Acts of Air Terror

Criminal and terrorist acts against aviation started with the first hijacking in 1931 and the first airline bombing in 1933. Airport attacks including attacks on airline offices started in the late 1960s. The eras of attacks on aviation can be subdivided into four basic periods:

- 1930–1979
- 1980–1990
- 1990–2001
- Post September 11, 2001 (9/11)

---

[1]For an excellent review of aviation security incidents from the 1930s to the mid-1990s, examine David Gero (1997), *Flights of Terror,* published by Patrick Stephens Ltd., Somerset, UK.

Before the 1960s, a desire to escape from some form of persecution or prosecution or a hostage taking to extort money characterized most hijacker motives. Hijacking a flight to Cuba or Mexico from the United States was common and usually intended to provide an escape from prosecution for a crime. Extradition treaties between Cuba and the United States effectively ended this practice. Bombings were rare, and when they occurred they were usually motivated by insurance fraud.

By the 1960s, hijackings were turning deadly and soon became standard operating procedure for Middle Eastern terrorist groups who often used the tactic to leverage hostages for the release of political prisoners and to call attention to their cause. In the United States and overseas, hijackings became a dangerous endeavor—more so for the hijackers than the hostages. For a period of time, at the rate of about 10 hijackings per year, the majority of hijackings ended with the shooting deaths of the hijackers by airline security officers, local police, FBI agents, or military personnel. Several hijackings ended in negotiated surrenders, while very few ended in the safe passage of the hijackers to a country with a nonextradition policy.

By the 1980s, U.S. airlines routinely became targets of hijackings and terrorism from various factions in the Middle East. Of these attacks, two airline bombings made significant headlines because of their high death toll: the bombing of Air India Flight 182 and Pan Am Flight 103. Also, in 1987, an airline employee brought a gun onto a company aircraft and downed the plane, killing all 44 on board. Combined, these incidents would result in the passage of the Aviation Security Improvement Act of 1990, which set the foundation for security practices in the United States throughout the 1990s.

Throughout the 1990s, hijackings and bombings continued primarily overseas. The crash of TWA Flight 800 off Long Island, New York, in 1996, which was not attributed to an attack, resulted in the White House Commission on Safety and Security and the passage of the Aviation Security and Anti-terrorism Act of 1996. The terrorist attacks on 9/11 would signify the first major aviation security incident in more than 10 years.

Subsequent to 9/11, there have been other forms of attack on aviation, such as man-portable air-defense systems (MANPADs) and suicide bombings. An abundance of legislation continues to reshape the aviation security system,[2] resulting from the massive loss of life and negative economic impact of 9/11.

Overall, hijackings have been the preferred method of air terror, constituting nearly 90% of all attacks. However, with the exception of 9/11, airline bombings have taken more lives. As such, the majority of aviation security regulations and policies focus on protecting aircraft. Only about 5% of all attacks on aviation have been attacks on or at an airport. Since 9/11, the number of attempted hijackings has fallen significantly while airline bombings continue to be attempted.

Regardless, in a study conducted by Ariel Merari, professor of psychology at Tel Aviv University and one of Israel's leading academic experts on terrorism, he concluded that

---

[2]Since 9/11, several significant legislative acts have passed that have affected aviation security policy and practices, including the Homeland Security Act of 2002 and the National Intelligence Reform Act of 2004.

the effectiveness of aviation security measures had little effect. Merari concluded that the average hijacker had an 81% chance of actually seizing control of an airliner, whereas the success rate of bombing an airliner was 76% (Wilkinson and Jenkins, 1999). Furthermore, Merari stated that it was not widespread negligence or indifference by security personnel that were the problems, but a lack of foresight and the limitations of getting actionable intelligence to airport and aircraft operators. This is important, because the 9/11 Commission's findings stated there was a "failure of imagination" and noted significant problems in obtaining intelligence information from the Federal Aviation Administration (FAA) and then transferring that information to airport and aircraft operators who could implement protective measures.

When carefully studied, previous attacks on aviation and nonaviation targets can sometimes provide insights into future methods of attack. As an example, the suicide bombings committed on two Russian airliners in August 2004 were not the first instances of suicide bombings. In June 1975, 29 years before the Russian airliner bombings, 63 passengers and 5 flight crewmembers were killed when a saboteur carried a bomb on board a Philippine Airlines aircraft and detonated it in the lavatory. In this case, the pilot survived and managed to land safely. Three years later, the same airline experienced another suicide-bombing attempt. However, this time the bomb only killed the saboteur when it blew the attacker out of the aircraft at 24,000 feet (Gero, 1997). Even the 9/11 terrorist attacks were not without precedent. The concept of hijacking an aircraft and crashing it into a target on the ground, or at least into the ground, was not new, with at least two attempts in the early 1970s and another attempt in 1994.

Even nonterrorist small-scale events should cause security practitioners to take notice. In two separate incidents, criminals shot passengers waiting in ticket lines in the Los Angeles International Airport and the New Orleans International Airport. Practitioners should focus on the tactics and strategies that will mitigate similar future attacks. Security professionals should also expand the implications and outcomes of such an attack if perpetrated by an organized, trained, and heavily armed group. Contingencies can then be developed to handle both small- and large-scale forms of attacks by organized terrorists.

A traditional mistake the aviation security industry seems to make is that it often does not learn from close calls or near tragedies. In 1999, Ahmed Ressam attempted to detonate an improvised explosive device at Los Angeles International Airport. It was only through the curiosity and diligence of a Port Angeles customs inspector that Ressam was caught. The attempt did nothing to fundamentally change the protections already in place at U.S. airports. One may ask if, had the attempt been successful, airport operators would have been given new directives concerning the protection of their facilities.

Finally, threats from individuals working inside the aviation system represent one of the most dangerous forms of attacks on aviation. The possibility that an airport or airline employee will be the primary or supporting actor in an attack on aviation is frightening from several perspectives. First, aviation employees are entrusted with protecting passengers from both safety and security risks. Second, aviation employees have greater knowledge about how the aviation security system operates, and in some cases may have detailed knowledge of how law enforcement handles hijackings, bombings, and other attacks. Third, aviation employees have approved access to airports and aircraft and thus may bypass many of the airport's security layers.

Some of the most significant aviation security incidents, including the hijacking of TWA Flight 847, the downing of PSA Flight 1771, and the bombing of Pan Am Flight 103, involved individuals working within the airport system. Although these events occurred in the 1980s, some of the very first hijackings also involved airport and airline employees, either as a facilitator or as the assailant.

1930–1960: The First 30 Years

The first recorded hijacking occurred on February 21, 1931, in Arequipa, Peru. Armed revolutionaries approached Byron Rickards and demanded use of his aircraft (a Ford tri-Motor). However, Rickards refused the use of his plane for several days. Finally, the revolutionaries informed Rickards that their uprising had been successful and he was free to go, provided he flew one of their members to Lima, Peru. This, arguably, was a hijacking because the revolutionaries forced the unauthorized use of an aircraft.[3] The hijacking demonstrated a fundamental precept in aviation security—criminals and terrorists will assess new technologies and determine if those technologies can improve their chances for success. In the early 1990s, Osama bin Laden assessed the possibility of using agricultural aircraft to deliver chemical or biological weapons. More recently, terrorists have been assessing the use of helicopters to commit terrorist acts or assist in such by using them for target surveillance or to provide access to lower levels of airspace such as in and around cities, restricted from larger aircraft (Lichtblau, 2005). Additionally, the Defense Nuclear Detection Organization (DNDO) has been interested in terrorists attempting to bring a nuclear device into the United States using a general aviation aircraft.

The first airline bombing occurred in 1933, when a United Airlines Boeing 247 was destroyed en route from Cleveland, OH, to Chicago, IL, killing all seven individuals on board.[4] Investigators determined that a nitroglycerin-based explosive detonated by a timing device destroyed the aircraft. In this case, no one was prosecuted for the attack (Gero, 1997). At the time of the 1931 hijacking in Arequipa, Peru, and the aforementioned Boeing 247 bombing, there were no passenger or baggage screening requirements. Passengers arrived at the airport and boarded the aircraft without any security screening.

[3]Air piracy is defined under 49 U.S.C. 1472i (reference (p)) as any seizure or exercise of control, by force or violence, or threat of force or violence, or by any other form of intimidation and with wrongful intent, of an aircraft within the special aircraft jurisdiction of the United States. *Hijacking* is a modern term for air piracy.

[4]There may have been an earlier bombing in March 1933 over Dixmude, Belgium, when an Armstrong Whitworth Argosy trimotor aircraft caught fire and crashed, killing 12 passengers and 3 crewmembers. Investigators determined that a fire started in the rear of the aircraft in a passenger's suitcase or in the lavatory. However, sabotage was never proven conclusively (Gero, 1997).

Regulations addressing these types of aviation security concerns would not begin until 1971, nearly 40 years later.

In 1949, a pilot and hijacker worked together to hijack a Hungarian Airline flight. In 1958, airline flight crewmembers hijacked two aircraft and flew the planes to the United States. In 1966, a hijacker on a flight en route from Santiago de Cuba to Havana turned out to be the aircraft's second officer.

In the 1940s and 1950s, commercial aviation was still in its infancy and crashes, while not frequent, still occurred. Airports installed insurance kiosks inside their terminal buildings so that passengers could purchase insurance on themselves before departure. However, criminals used insurance kiosks to commit insurance fraud. At the time, the forensic abilities of investigators and law enforcement agencies were not always capable of determining if an airplane crash was an accident or an intentional event. The airline bombings in 1949 marked the beginning of a series of insurance frauds involving commercial aircraft.

Aircraft-related bombings-for-insurance scams reached great significance in 1955 with the bombing of United Airlines Mainliner Flight 629. A bomb destroyed Flight 629 while departing Denver's Stapleton International Airport to Portland, OR. All 44 on board perished, including Daisie Eldora King, the mother of John Gilbert Graham. An investigation determined that Graham had placed dynamite inside his mother's luggage in an attempt to claim more than $37,000 in life insurance money.[5] Grant's trial determined him guilty of murder. Shortly after the Graham bombing, another individual, Julian Frank, insured himself in an attempt to will money to his relatives. Frank blew up his flight by carrying dynamite within his carry-on bag.

Insurance-related attacks can also be construed as "passenger-dupe" scenarios. This form of attack uses a passenger that unknowingly brings explosives onto an airplane in his or her luggage. The passenger-dupe scenarios eventually led to the development of "first-level" passenger profiling. This profiling policy introduced two of the most common questions asked by security personnel in air travel en route before 9/11:

1. Has anyone unknown to you asked you to carry an item on this flight?
2. Have any of the items you are traveling with been out of your immediate control since the time you packed them?

Between 1947 and 1953, there were 23 hijackings worldwide. Europeans seeking various forms of political asylum committed most of these hijackings. In the United States, between 1930 and 1967, 12 hijackings were attempted; 7 of these attempts resulted in successful hijackings (Moore, 1976). Providing firearms and related training to airline pilots in the United States became an additional layer of security designed to prevent

---

[5]Graham was executed in the Colorado gas chamber in 1957. The story may have been the inspiration for the 1970 Arthur Haley novel and 1973 movie *Airport*. The actual investigation and details of the entire incident are covered excellently in Andrew Field's 2005 book, *Mainliner Denver: The Bombing of Flight 629* (Boulder, CO: Big Earth Publishing).

hijackings. While the arming of pilots after 9/11 is a heavily contested industry issue, the first recorded case of an armed pilot shooting a hijacker occurred on July 6, 1954. A 15-year-old stormed the cockpit of an American Airlines DC-6 at the Hopkins Municipal Airport in Cleveland, OH (Gero, 1997). The captain of the flight was carrying a .380-caliber pistol in his flight bag, as required by U.S. postal regulations, and shot and killed the hijacker (Moore, 2001). In the early days of mail carriage, postal carriers were required to be armed, which included pilots of aircraft transporting mail.

1960–1980: The Early "Jet Age"

Fidel Castro's rise to power coupled with the lack of passenger or baggage screening requirements in the 1960s and early 1970s established a period of prolific hijackings. More than 240 hijackings or attempted hijackings were related to flying to or from Cuba between 1960 and 1974. This increase in hijackings resulted in the U.S. Congress passing the Anti-hijacking Act of 1974, mandating passenger and carry-on baggage screening. Even after the act went into effect, there were still 60+ hijacks or hijacking attempts involving Cuba between 1974 and 1989.

In 1961, Antuilo Ramierez Ortiz used a gun to force the flight crew of a National Airlines' jet to divert to Cuba, and thus became the first U.S. hijacker (Department of State, 2006). Soon after, President John F. Kennedy ordered armed law enforcement officers into the skies, thus creating the first air marshal program. U.S. Congress also approved legislation mandating the death penalty or 20 years in prison for hijacking an aircraft.[6] In 1971, the U.S. regulations covering aviation were amended to include Federal Aviation Regulation Part 107 Airport Security, which made airports responsible for protecting the airfield from unlawful intrusion.

The years between 1968 and 1973 marked the peak of hijackings and use of anti-hijacking measures. During that time, the U.S. Department of Transportation (DOT) estimated that 364 total hijackings occurred worldwide. The majority of hijackings throughout this period were for one of three reasons: political asylum, release of prisoners, or financial gain. In 1968, 19 of 22 hijackings concluded in Cuba with the majority hijacked while on intrastate routes in the United States. The high rate of hijackings caused the FAA to create a task force to study methods to deter future hijackings. Their findings resulted in the first hijacker profile, along with limited use of metal detectors to screen passengers. During this time, several airlines voluntarily began using walk-through metal detectors (WTMDs), more commonly referred to as magnetometers. The hijacker profile was a set of behaviors that hijackers would generally follow after hijacking an aircraft, specifically to allow the aircraft to land and let negotiations begin.

---

[6]Originally organized under the U.S. Marshal Service as sky marshals. In 1985, the air marshal program was reestablished under the FAA.

By 1969, the number of hijackings reached an epic proportion. In 1970, hijackings to Cuba reached 32 for the year. By the late 1960s and early 1970s, the world was moving into the "jet age," bringing about faster aircraft that could carry more passengers and more fuel, meaning hijackers could now fly farther and faster and had more hostages on board. Then, the following three attempted hijackings occurred in the early 1970s that are eerily reminiscent of the motivation and tactics of the 9/11 hijackers.

- On March 17, 1970, on board Eastern Airlines Flight 1320 en route from Newark, NJ, to Boston, MA, passenger John DiVivo entered the cockpit with a gun and ordered the crew to continue flying until the plane ran out of fuel and crashed. The flight crew fought back, and First Officer James Hartley, who was mortally wounded during the altercation, managed to disarm DiVivo and then shot him. Captain Robert Wilbur was injured during the fight but still managed to safely land the plane.[7]
- In 1972, three alleged rapists (including one escaped convict) took over a Southern Airways DC-9 as it departed Birmingham, AL. The hijackers demanded $10 million dollars and directed the plane back and forth over the country. At one point, the hijackers threatened to crash into a nuclear facility in Oak Ridge, TN. The 31 passengers were held for 29 hours. The first officer was shot and wounded before the ordeal ended (Moore, 1976, pp. 13–14).
- In 1974, another individual would attempt to hijack an aircraft—this time with the intent to crash it into the White House. Samuel Byck, using a stolen pistol, shot and killed Maryland airport police officer Neal Ramsburg at Baltimore–Washington International Airport. Byck then boarded a Delta Airlines DC-9 and ordered the pilots to take off and fly low toward Washington, DC. Byck intended to crash the plane into the White House in an attempt to assassinate President Richard Nixon. When the pilots refused to take off, Byck shot both pilots, killing the first officer, and then ordered a passenger to help the pilot fly the plane. An FBI agent, using the fallen police officer's .357-caliber pistol, fired through a window in the aircraft's door and killed Byck.

Another characteristic of hijackings during the 1960s and 1970s was that there were often only one or two hijackers who used guns, grenades, bombs, and in some occasions simply the threat of having a bomb to take over a flight. Response to hijackings was similarly straightforward—in El Paso, TX, officers attempted to shoot out the tires of a hijacked Continental Airlines jet.

A key lesson that U.S. law enforcement agencies learned from numerous hijackings was to keep hijacked aircraft on the ground. Once a hijacked aircraft is airborne, the crisis is more ephemeral and risky. Hijacked aircraft can land outside a country's jurisdictional authority, create hazards to air navigation, and serve as a weapon of mass destruction. An airborne hijacked aircraft is essentially a roving crime scene in progress, with huge dynamics and random elements at play.

---

[7]DiVivo was convicted and sent to prison, where he committed suicide in October 1970.

Middle East and Asia

In the Middle East, there were few bombings in the 1960s and 1970s, but when an aircraft was bombed it usually resulted in the complete loss of the aircraft and all on board. Hijackings in the Middle East occurred less frequently than what was experienced in the United States, yet they resulted in greater loss of life and overall destruction. Middle Eastern hijackings focused on extorting the release of prisoners or on delivering a political message rather than flight to another country.

The first hijacking of a commercial flight in the Middle East was on an unlikely target—the Israeli airline El Al (Gero, 1997). On July 23, 1968, three gunmen hijacked El Al Flight 426 en route from Rome, Italy, to Tel Aviv. These gunmen claimed to be members of the Popular Front for the Liberation of Palestine (PFLP). The aircraft was forced to land in Algiers where negotiations began with the hijackers demanding the release of certain Arab prisoners. A news crew was already on hand when the plane landed in Algiers, demonstrating the power of a hijacking as a tool to attract worldwide attention. No one was killed during the hijacking, and within a few days the hijackers released all the prisoners even though Israel did not release any Arab prisoners. As a result of this hijacking, Israel implemented the strictest security measures on El Al of any air carrier and also adopted a retaliation policy toward those groups who seek to harm Israeli citizens (Gero, 1997).

Six months later, terrorists again attacked an El Al flight while on the ground at Athens, Greece. Two terrorists armed with automatic weapons and hand grenades boarded the airliner. One person was killed, another injured, and substantial damage was done to the aircraft. Because of these attacks and hijackings, El Al began implementing the practice of escorting taxiing flights with armed personnel in vehicles. Additionally, in retaliation for the attack, Israeli commandoes raided the airport in Beirut, Lebanon, and destroyed a dozen Lebanese-registered aircraft (Gero, 1997). As of this writing, there has never been another successful hijacking of an El Al airliner (Walt, 2001).

Attacks in the Middle East on commercial aviation were not restricted to hijackings. In 1969, gunmen sprayed an El Al jet taxiing at the Zurich-Kloten airport with machine gun fire. An armed passenger on board the flight opened one of the aircraft doors and returned fire (Gero, 1997).

From 1969 to 1970, there were a few other Middle East–related hijackings, mostly for extortion or transportation, along with a couple of airline bombings, but nowhere near the rate of hijackings that were occurring in the United States during the same period. There were several other hijackings throughout Europe and Asia, mostly in the Soviet Union or its associated territories.

Dawson's Field

On September 6, 1970, teams of Palestinian hijackers departed from three separate airports with the intent to hijack three aircraft en route from Europe to the United States. Members of the PFLP intended to land the aircraft at a remote airfield and hold the passengers hostage in an attempt to negotiate the release of other PFLP members held in jails throughout Europe and in Israel. The first hijack attempt was conducted on board El Al Flight 219 from Amsterdam to New York on a B-707 when hijackers Patrick Arguello and Leila Khaled attempted to take over the aircraft. Khaled had prior experience hijacking aircraft and had previously undergone plastic surgery to change her appearance. In an interview for the documentary *Hijacked* (Ziv, 2006), Khaled stated that she knew the airline would have armed guards, but that her team believed the guards would be afraid because the hijackers were armed with hand grenades and guns (Ziv, 2006).

Before takeoff, the El Al security chief and airline station manager identified four suspicious passengers. The date of ticket purchase, last-minute arrival times at the airport, sequential numbering of two of the passports, and other anomalies (not expanded on in the reference) caused suspicion. The captain of the flight was consulted about the suspicious passengers and decided to allow Khaled and Arguello on board but denied boarding for the other two. Twenty minutes into the flight, Khaled and Arguello initiated a hijacking using grenades and pistols. They threatened the flight attendants in an attempt to get the flight crew to open the cockpit door. A security officer was already jump-seating in the cockpit. However, Captain Lev decided that rather than opening the door, he would execute a "negative G pushover" of the 707 by suddenly pushing the aircraft control stick forward, causing anyone and anything not strapped down to quickly float to the top of the aircraft cabin. In the subsequent confusion, passengers and security personnel overwhelmed the hijackers. An air marshal shot and killed Arguello (a Nicaraguan-American) and passengers subdued Khaled. The aircraft made an emergency landing at the London Heathrow International Airport where Khaled was arrested.

Earlier the same morning (September 6, 1970), hijackers took over TWA Flight 74 (a Boeing 707) from Frankfurt, Germany, to New York, and Swissair DC-8 from Zurich, Switzerland, to New York. The two aircraft totaled more than 300 hostages, mostly American citizens (Ziv, 2006). After they had been removed from the El Al flight, the two hijackers bought tickets and boarded Pan Am Flight 93, a B-747, from Amsterdam, the Netherlands, to New York. The Pan Am captain conducted a pat-down search of the passengers but did not find their grenades and pistols, which were hidden in the groin area of each hijacker. Further details on the search are not available.

Shortly after takeoff, the two Palestinians successfully hijacked the flight. The B-707 and DC-8 were flown to an abandoned airstrip at Dawson's Field, Jordan (renamed Revolution Airport by the PFLP). The B-747 was flown to Cairo, Egypt, where it was wired with explosives while still in flight. All of the passengers and crew evacuated immediately upon landing, and within minutes the aircraft exploded. Egyptian police arrested the hijackers.

Then, on September 9, a BOAC VC-10 was hijacked as leverage to negotiate the release of Leila Khaled, the hijacker who had been caught earlier in the week attempting to hijack the El Al flight. The VC-10 was also brought to Dawson's Field, and all 500 passengers were confined inside the aircraft for several days. Six days later, the passengers were deplaned and the terrorists used explosives to destroy the three aforementioned aircraft at Dawson's Field while TV cameras recorded the event. The various airlines involved lost millions of dollars in assets. Authorities released Khaled about a month after the BOAC VC-10 hijacking, in a hostage trade deal in exchange for the passengers on the VC-10. This example was characteristic of the type of hijack response strategy used in the United States. In an interview years later, Khaled said that she was under orders not to injure or kill any passengers.

A premise in hijackings before 9/11 was that hijackers used hostages and aircraft as bargaining tools. The assumption was that hijackers were more interested in an outside cause (escape, extortion, political message) than in using the aircraft as a guided missile. Thus, before 9/11 the goal of flight crews during a hijacking was to land the aircraft so authorities on the ground could take over negotiations. This assumption does not work in a post-9/11 world where aircraft are used as weapons of mass destruction (WMD) and hostages are merely victims (or obstacles) to the end result.

Airport Attacks

Airport attacks are the third major form of attack against the global aviation system. The focus of the aviation security system has historically been to protect aircraft. Airports have not received as much attention, partially because airports have not been the frequent targets of terrorists, the way aircraft have. However, airports provide shelter, services, and various levels of security to aircraft while they are on the ground. Airports are where passengers and airline employees make the transition from land transportation to flight (and back again), holding hundreds and in some cases thousands of individuals at a time, and are the transit point for more than 30,000 flights a day in the United States. A disruption or shutdown of a U.S. commercial-service airport can result in disruptions throughout the national airspace system. Airports are public facilities where screening is not required for entry into a U.S. airport, which thereby increases the possibility for anyone to enter the facility with guns, grenades, or other explosive devices and attack a large citizenry. In addition, there are multiple avenues of access to an aircraft that bypass traditional screening processes, such as catering and maintenance. Airports are national assets, essential to the proper function of the aviation security system, and thus deserve considerable protection.

The most common form of airport attack has largely been an individual or group armed with automatic weapons and explosives (usually hand grenades) storming the public area of an airport terminal or in some cases the airfield. One of the most significant airport attacks occurred in 1972 at Lod International Airport in Tel Aviv, Israel. In this case, three terrorists of the Japanese Red Army recruited by the PFLP opened fire in the baggage claim area killing 26 people and injuring dozens more (BBC News, 1972). In 1973, Palestinian terrorists shot their way through the Rome, Italy, airport, destroying an aircraft. More than 30 people died in that attack.

In the United States, an airport shooting led to a hijacking in 1972. In the attack, bank robbers shot their way onto an Eastern Airlines jet at the Houston Intercontinental Airport (Gero, 1997). The bank robbers, a father and his two sons, killed a ticket agent along with a bank manager and a police officer (at the bank) and wounded another airline employee in the process.[8]

---

[8]The hijackers forced the flight crew to take them to Cuba. The hijackers returned to the United States 3 years later and were sentenced to 50 years in prison.

In the Middle East, there were several occurrences of terrorist teams that would storm an airfield, start shooting, and toss explosives at aircraft parked on the ramp. In 1973, Palestinian terrorists destroyed a Pan Am B-707 in Rome, Italy, by tossing incendiary bombs into the parked aircraft. The fire caused the deaths of 30 passengers and a flight attendant.

Another form of airport attack is the use of bombs placed in airport public-use lockers. In 1974, a bomb blast killed two skycaps and several others at Los Angeles International Airport. The bomb was placed in a locker between the Pan Am and Korean Airline ticket counters (UPI, 1974). In 1975, 11 people died and 75 were injured when a bomb exploded in a public locker at LaGuardia Airport in New York (Springer, 2002). Forensics teams determined the explosive force was the equivalent of 25 sticks of TNT or plastic explosive, triggered by an alarm clock. The use of public locker facilities was discouraged for several years before authorities permanently discontinued their use in the mid-1990s. Lockers in the sterile area are routinely used but are usually restricted to individuals with higher Department of Homeland Security (DHS) security levels.

Other Notable Lessons from the 1970s

Several noteworthy incidents occurred at the end of the 1970s. On November 15, 1979, a mail parcel containing a bomb exploded in the cargo hold of American Airlines Flight 444 en route from Chicago, IL, to Washington, DC. Authorities later traced the bomb, which had a barometric trigger switch, to Ted Kaczynski (a.k.a. the Unabomber) (Khan, 1999). This was the first time in the United States that a bomb was placed on board an aircraft as air cargo. Today, air cargo security has been the focus of much attention and rulemaking; however, there have been very few aircraft bombings where the bomb was put on board using air cargo. Terrorists have placed most bombs on aircraft using carry-on strategies (e.g., duping a passenger or using suicide bombers) or in checked baggage. The 2010 Yemen Air Cargo plot demonstrated one of the most recent attempts at placing bombs on commercial aircraft via cargo, and has caused additional scrutiny and a rush to implement preventative measures to keep bombs from being introduced to passenger aircraft via air cargo.

Criminals and terrorists are highly adaptive and imitative. In 1971, when D. B. Cooper (Figure 2.1) parachuted out of an airplane he had just hijacked and held for ransom, 19 other parachute-from-an-airplane hijacking attempts followed. Boeing, the manufacturer of the aircraft experiencing the parachute hijackings, installed the Cooper Vane, which prevented the rear air-stairs from being lowered in flight, thus eliminating the ability of a hijacker to parachute to safety. The Cooper Vane did not stop hijackings but rather motivated hijackers to develop new tactics (Gladwell, 2001).

When improvements in passenger and baggage screening made boarding concealed guns and grenades more difficult, hijackers discovered alternative methods, such as having cleaning and catering personnel bring weapons and explosives on board aircraft. Strategies such as positive passenger–baggage matching have prevented bombers from checking a bag with a bomb hidden inside without personally boarding the aircraft.

FIGURE 2.1 Artist's rendition of D. B. Cooper.

In response, terrorists adjusted their strategies and recruited suicide bombers who were willing to die for their cause, or they placed bombs inside the suitcases of unsuspecting passengers. This example demonstrates the challenge of aviation security to implement policies and strategies that prevent the existing threat, while also anticipating how future threats will be developed and implemented.

Aviation was not safe from the world of "narco-terrorism" as demonstrated in 1989 when an Avianca B-727 was destroyed en route from Bogata to Cali, Colombia, killing all 107 people on board, including 5 police informants. Medellin cartel drug lord Pablo Escobar orchestrated the bombing as part of a campaign to eliminate informants, police, and politicians (Marshall, 1997). The bombing brought attention to the already heated drug war between the Colombian government and the drug cartels. Throughout the 1980s and into the 1990s, drug traffickers used aviation to transport narcotics. Although the use of general aviation aircraft for the transportation of drugs is well known, there was also a substantial amount of narcotics smuggled on board commercial airliners (commercial airliners continue to be used extensively for smuggling and transporting illegal narcotics throughout the world). Local drug lords have frequently paid airline employees with access to the cargo and baggage holds of commercial aircraft to smuggle narcotics onboard. Air carrier security managers soon realized that the ease with which drugs could be smuggled on board demonstrated the ease with which bombs could also be placed on board.

In March 1979, a hijacker in Tucson, AZ, allowed the passengers to deplane during a fuel stopover. Unbeknownst to the hijacker, the flight crew escaped through the cockpit windows by using emergency escape ropes. Seven years later, a similar incident occurred where the flight crew escaped during the hijacking but with slightly different results. On September 5, 1986, 17 people died on Pan Am Flight 73, a B-747, during an attempted hijacking by terrorists dressed to resemble airport security guards. With 379 passengers onboard, the aircraft sat on the tarmac of a Karachi, Pakistan, airport. When the terrorists fired the first shots, the captain and his flight crew used the emergency escape ropes to leave the aircraft.[9] Their escape was concurrent with FAA recommendations for flight crew actions during a hijacking, which was essentially to leave the plane if able and thereby "disable" the aircraft (Wallis, 1993).

1980–1990: Aviation Security Policies Are Set

Several devastating attacks on aviation characterized the decade of 1980–1990. These attacks caused significant changes to security policies and practices both in the United States and around the world. Airline bombings occurred at unprecedented levels of frequency and fatalities. Hijackings turned even deadlier, and there were several more attacks on airports. By 1990, new policies had been implemented, and overall attacks on aviation decreased significantly throughout the 1990s. This extended lull in activity may have contributed to a false sense of trust in the aviation security system leading up to the 9/11 attacks.

The 1980s started with a new round of hijackings from the United States to Cuba setting a new record of 18 in 1980. In the first hijacking of that year, flight attendants exercised their ability to egress from a flight. When the Delta Airlines L-1011 was on the ground in Havana (after being hijacked during its flight from New York to Atlanta, GA) and the hijacker was in the cockpit, the flight attendants and most of the passengers escaped through a floor hatch. The hijacker then demanded that the aircraft take off, but authorities had parked a truck in front of the plane to prevent it from moving (Gero, 1997).

In what was becoming a popular and rather effective technique, many of the hijackings in 1980 were implemented by individuals who threatened to have an explosive but did not actually have one. Among the more "entertaining" threats was one hijacker whose bomb turned out to be a piece of soap. Another hopeful hijacker had two red sticks with the word "TNT" scribbled on the side. During this time, hijackers developed a new technique—they would use a flammable liquid, such as gasoline or rubbing alcohol, and then hijack the flight by spreading it around and threatening to light it. This technique was used numerous times in the early 1980s. Today, flammables are prohibited on board commercial aircraft and newer technologies are being developed to determine if the contents of bottles that passengers bring on board are indeed dangerous. When new passenger and carry-on baggage screening requirements made it more difficult to smuggle a weapon or explosive onboard, hijackers began bluffing to passengers and crews that they either had access to a bomb or gun, in an attempt to hijack the flight. Hijackings in the United States hit another

---

[9]Eventually, 22 people, including 2 Americans, would die during the standoff.

peak in 1983, with an estimated 15 that year. The FAA restarted the air marshal program around 1985, about the same time hijackings in the United States ended, but internationally, they continued to rise.

Two other security incidents are noteworthy as they involved personnel with authorized access to the aircraft or airport property. On May 3, 1986, a bomb destroyed a Sri Lanka Lockheed L-1011 while the plane was on the ground at Colombo Airport, killing 20. Authorities arrested an airport customs officer, sympathetic to a separatist movement, and charged the officer with sabotaging the aircraft. The positive identification and monitoring of airport and airline personnel are important concerns to aviation security practitioners, especially because those with authorized access have caused destruction and deaths in the past, and their access privileges allow them to routinely bypass many of the existing security systems designed to protect aircraft.

The 1980s also brought more airport attacks. In 1981, terrorists planted three bombs at JFK airport in New York. One was booby-trapped inside a briefcase and left inside a men's restroom, killing a young man when it exploded. Two other bombs, one in a public terminal area and one in a women's restroom, were discovered before they exploded (NYPD, 2006). These bombs were planted by the Armed Forces of National Liberation (FALN), a Puerto Rican nationalist group.

Then, on December 27, 1985, armed gunmen attacked two airports simultaneously in an active shooter scenario. The public terminal areas of the Rome and Vienna airports were stormed by revolutionaries most likely associated with the Abu Nidal[10] terrorist organization (BBC, 1985). These attacks targeted passengers waiting at the El Al ticket counter with the use of gunfire and grenades. In these attacks, 18 people died and more than 120 were injured. To this day, the terminal building at the Leonardo daVinci International Airport is heavily guarded with tactical police officers carrying submachine guns and wearing body armor. Officers patrol both from a catwalk above the public area of the terminal and on the passenger level.

In 1986, Anne Marie Murphy was attempting to take a flight from London Heathrow International Airport to Tel Aviv on El Al. She was traveling to meet the parents of her fiancé who, she did not know, was secretly a Syrian intelligence agent. Her fiancé told her not to allow her bag to be searched but did not tell her what was in it. Through the profiler interview process that is conducted on all passengers before boarding an El Al flight, security officials determined that something was amiss and searched her luggage. They found plastic explosives hidden in the lining of the suitcase, and validated, to an extent, the security process of behavioral profiling.

The 1980s saw four significant events in the history of aviation security, which involved TWA Flight 847, Air India Flight 182, PSA Flight 1771, and Pan Am Flight 103.

---

[10]Abu Nidal fronted the Fatah Revolutionary Council, generally considered the world's most feared terrorist organization before the rise of al-Qaeda, responsible for over 900 deaths in more than 20 countries. Nidal was found dead under mysterious circumstances in Iraq in August 2002 (BBC, 1985).

TWA Flight 847

On June 14, 1985, Shiite Muslim terrorists hijacked TWA Flight 847, a Boeing 727 en route from Athens, Greece, to Rome, Italy. The hijacking of Flight 847 would provide several lessons in the handling of hijacking incidents for years to come. Making the flight even more noteworthy was the capability of mass media to bring the episode in real time to the viewing audience. The hijacking became an event on the evening news for several weeks, with stirring images of terrorists leaning out the cockpit window threatening the pilot with a gun.[11]

"We've got a hijack," announced the second officer, aboard TWA Flight 847 from Athens to Rome. Those chilling words began one of the most famous hijackings in the history of aviation. At its conclusion, this hostage crisis spanned thousands of miles of airspace, involved several governments, directly affected the operations at three major airports, turned a flight attendant into a folk hero, resulted in the death of a U.S. serviceman, and gave rise to Hezbollah.

While in flight, hijackers pounded on the cockpit door and started kicking out the lower panel of the door, while the second officer Christian Zimmerman, tried to hide the emergency fire ax stored in the cockpit. He did not know what weapons the hijackers had but did know that he did not want to provide them with another one (Testrake and Wimbish, 1988). In 1985, the concept of suicide hijackers was not considered, and the tactic of using deadly force to protect the flight deck was not part of the common strategy to handle hijackings. The flight crew finally opened the door after being informed by flight attendant Uli Derickson that the hijackers were now beating several passengers.

Two hijackers, armed with guns and grenades, stormed inside, demanding to travel to Algiers. Captain Testrake commented that his immediate concern was that one of the hijackers was holding hand grenades and that both hijackers were very nervous. The hijackers also wanted access to the radio, which was still out of range of Beirut, but that did not stop them from broadcasting their demand of the release of 800 prisoners being held in Israel.

Throughout the first several minutes of the hijacking, the assailants kept control by running up and down the aisles, yelling and punching passengers at random. When the hijackers left Captain Testrake and his crew alone in the cockpit for a few moments, the crew notified air traffic control of their situation and then tried to find a map and compute fuel requirements for the detour to Algiers. Because none of the flight crew spoke Arabic and neither of the hijackers spoke English, there was a substantial communication gap. A third hijacker who did speak English had missed the flight and was arrested in Athens earlier that day (the hijackers also demanded his release). However, Captain Testrake soon learned that one of the hijackers spoke German and, fortunately, so did his chief purser, Derickson. When Captain Testrake told the hijacker there was not enough fuel, the hijacker thought the captain was trying to trick them but was soon convinced that

---

[11]The ordeal is retold by Testrake and Wimbas (1988) in their book *Triumph over Terror on Flight 847*, which also provides good lessons for anyone involved in a hostage taking.

it was true. The hijackers then ordered the flight crew to head to Beirut instead. Captain Testrake was grateful because he knew the plane was safer on the ground, particularly if one of the grenades detonated (Testrake and Wimbish, 1988).

There are some interesting points to make regarding this hijacking:

1. The hijacking occurred very quickly and violently.
2. The hijackers inflicted physical pain and mental fear on passengers as a successful strategy for gaining access to the flight deck.
3. The pilots hid the fire ax in the cockpit.
4. The flight attendant was immediately involved in the hijacking, both as a reluctant advocate for the hijackers' requests to enter the flight deck and as the chief communicator.
5. The pilots knew the safest place for the aircraft was on the ground.

En route to Beirut, the hijackers noticed the empty holder that once held the crash ax. They demanded to know where it was, and Zimmerman told them the plane was not equipped with one. The hijackers then kicked out the doorknob on the cockpit door to prevent the flight crew from locking them out. Also en route the hijackers kept themselves busy by beating passengers, mostly with a chair arm they had ripped from Zimmerman's seat. Their favorite targets for beatings were a passenger who was a member of the U.S. military, Zimmerman, and on occasion, First Officer Mascera. However, Captain Testrake was never hit, which he later speculated may have been a sign of respect that he was the captain or the fact that he was flying the plane. The hijackers also made the passengers sit in an uncomfortable position with their heads down on their laps. Whenever a passenger looked up to stretch, the hijackers beat the passenger back into position.

Approaching Beirut, the tower controllers closed the airport and refused Testrake permission to land. At the time of this hijacking, the International Civil Aviation Organization (ICAO) did not classify a hijacked aircraft as an "aircraft in distress." An aircraft classified as in distress would have received protection and services provided by international treaties, such as allowing the hijacked aircraft to land and rendering all available assistance.

Only after repeated pleas by Testrake, who at one point did declare that his aircraft was in distress, did controllers allow the plane to land. Testrake taxied to the refueling area, but when the fuel trucks did not arrive quickly, the hijackers resumed beating Robert Stethem (Figure 2.2), a passenger who was a member of the U.S. Navy. Finally, the fuel trucks arrived, and the beating ceased, for the time being. Testrake attempted to negotiate for the release of the women and children. Although the hijackers would not agree with his request, Derickson's natural negotiating skills did succeed, and soon 19 women and children were released to Beirut authorities (Testrake and Wimbish, 1988).

The plane took off, heading to Algiers, but the flight crew ran into the same problem in trying to land, as Algerian flight controllers temporarily refused landing permission. They eventually relented, and Algerian military forces quickly surrounded the plane. The hijackers negotiated for hours with Algerian officials over the radio about the need for

FIGURE 2.2 Petty Officer Robert Stethem.

more fuel. When negotiations were failing, the hijackers beat more passengers until finally a fuel truck arrived.

One "humorous" moment did occur when the fuel truck driver refused to fuel the plane unless paid with a Shell Oil credit card. Derickson surrendered her Shell Oil gas credit card and $6,000 of Jet-A fuel was charged to her account (which TWA later covered).

In Algiers, the hijackers continued beating passenger Stethem, but they allowed 21 more passengers to go free for "humanitarian reasons." Loaded with fuel, the plane headed back to Beirut. While en route, Captain Testrake got up to stretch but immediately had a hijacker waving a gun in his face. Testrake yelled back at the man that he was an old grandfather and that his back hurt. The hijacker backed down. "It was the same old pattern," stated Testrake. "If you spoke sharply to them, they could act almost like school kids who had been caught in some form of mischief and were sorry. But at other times they could be unfeeling and brutal ... as we were soon to find out" (Testrake and Wimbish, 1988, p. 88).

Approaching Beirut, the flight crew was again not given permission to land, and the tower controllers turned off the runway lights. This time the hijackers threatened to crash the plane into the presidential palace unless granted permission to land, which it eventually was. The hijackers ordered the flight crew to stay off the radio out of fear that the U.S. Delta Force rescue team was en route.[12] Captain Testrake stated that

[12]While the aircraft was making its roundtrips between Beirut and Algiers, U.S. Delta Force operatives were mobilized in Cyprus, ready for an assault on the plane; however, the assault was not approved.

even if Delta Force had attempted a rescue, he still feared that people would be killed because one of the hijacker's carelessness with their hand grenades, often pulling the pins out and popping them back in. Delta Force was never green-lighted to execute the operation.

Once on the ground in Beirut, more fruitless arguing between the hijackers and Beirut officials took place until one of the hijackers shot Stethem and threw his body to the tarmac. It was also reported by author Don Mann that a Navy SEAL sniper team had one of the hijackers in their sights, but was never given the order to shoot (Mann, 2011). The murder got another Palestinian faction involved, the more moderate Amal, and once back in Algiers 14 more terrorists associated with Amal boarded the aircraft in the dark of night.

At one point the hijackers identified passengers with Jewish-sounding names and separated them from other passengers. This separation was a frequent occurrence on previous hijackings as there is a long history of animosity by Palestinian organizations against citizens from the United States and Israel. When the hijackers asked flight attendant Derickson's assistance in identifying U.S. and Jewish citizens on the aircraft, she attempted to hide several of the incriminating passports (Jewish and U.S.). A doctor was also allowed on board to check the passengers' health. The terrorists released the remaining women, children, and male hostages from countries other than Israel and the United States, and flight attendant Derickson was also released in Algiers along with the other flight attendants. Derickson's calmness and actions became a model for flight crew performance during a hijacking.

Also boarding the flight in Algiers was the third hijacker who had been arrested in Athens, Ali Atweh (a.k.a. Ali Atwa). The crew then headed the plane back to Beirut. En route, Testrake, who had overheard the hijackers talking about going to Tehran, Iran, tried talking with air traffic controllers to obtain information from the U.S. Department of State on the ramifications of going to Iran. Eventually, Captain Testrake was asked by air traffic control to fly to the Larnaca airport in Cyprus, where the U.S. Delta Force was staged and ready to conduct an assault. Testrake also had the chance to broadcast to controllers in Tel Aviv and let them know how many hijackers there were onboard, how they were armed, and what was going on. However, the hijackers soon entered the cockpit and the plane ended up on approach to Beirut, again.

Upon landing in Beirut, the flight crew secretly shut down two of the aircraft's engines and convinced the hijackers that the aircraft could no longer fly. More terrorists came on board as the plane sat on the tarmac in Beirut. Eventually they even worked out a shift schedule for guarding the plane and its occupants. The standoff continued for 18 more days, upon which all hostages were released.

Captain Testrake's actions during the crisis serve as a model of professional conduct in the way he took care of his passengers and flight crew and how he calmly handled the lengthy hijacking. Ali Atwa, the terrorist responsible for the hijacking, was still at large in 2006 and on the Federal Bureau of Investigation's most wanted list.

Authorities learned several lessons from the hijacking of TWA Flight 847:

1. All attempts should be made to keep the hijacked aircraft on the ground.
2. Flight crew should be trained in emergency safety and security operations.
3. Flight crew should be trained in crisis management during a hostage or security incident.
4. Despite Captain Testrake's desire that a rescue not be attempted, flight crews must understand that during a hijacking situation they are effectively hostages, and as such with the duress involved and the lack of other information, they cannot be counted on to decide what actions regarding a rescue should be taken. However, flight crews or passengers should attempt to provide information to law enforcement agencies that may assist in the decision-making process.
5. Flight crews should take all available information into account when making decisions throughout the entire event. Although standard operating procedures (SOPs) should be trained, it is the flight crew who must make numerous decisions about whether to follow the SOPs or decide if it is safer to deviate from policy.

The hijacker responsible for passenger Stethem's death was caught and sentenced to life imprisonment in West Germany, and a navy destroyer, the *U.S.S. Stethem,* was named in honor of Petty Officer Robert Stethem (Figure 2.2).

■ ■ ■ ▬▬▬▬▬▬▬▬▬▬▬▬▬▬▬▬▬▬▬▬▬▬▬▬▬▬▬

Perspectives on Hostage Situations

Hostages must be cared for because they provide bargaining power for terrorists and criminals. They have food and water needs, toilet needs, and, many times, medical needs. Children and the elderly usually require additional attention. Often these needs can work to the favor of negotiators seeking to win the release of as many hostages as possible. Even hostages in good physical condition pose their own dilemma for hijackers. Although these individuals may be able to survive the longest in confinement, they are also the most capable of overpowering the hostage takers.

A dead hostage reduces bargaining power, and killing hostages does not promote empathy for the hijackers. When all of the hostages are dead, the situation changes from a hostage scenario to a siege scenario, thereby providing more options to law enforcement and antiterrorism agencies. Therefore, terrorists and criminals can sometimes be convinced to release hostages in trade for needed items, such as fuel, food, and water (for themselves and the hostages), and other supplies. The end goal is to secure the release of as many hostages as possible and keep the aircraft on the ground.

▬▬▬▬▬▬▬▬▬▬▬▬▬▬▬▬▬▬▬▬▬▬▬▬▬▬▬ ■ ■ ■

Air India Flight 182

The flight of Air India 182 was a bellwether to the international community to pay attention to aircraft bombings—Air India Flight 182 was actually two bombings committed simultaneously on two international flights departing Canada. On June 23, 1985, a Boeing

747 en route from Vancouver, Canada, to London exploded at 31,000 feet off the southern tip of Ireland, killing all 329 on board including 60 children (Jiwa and Hauka, 2006). Around the same time, two baggage handlers at the Tokyo Narita Airport were killed and four others wounded as they transferred baggage from Canadian Pacific Flight 003 to Air India Flight 301 bound for Thailand. Both flights had originally departed from Vancouver.

The B-747 was last observed on radar near the Irish coast when it disappeared at 7:15 a.m. local time (Ireland). The debris spread across several miles of the Atlantic Ocean where water depths reach 7,000 feet. Searchers recovered the bodies of 132 victims. They also recovered enough wreckage including the cockpit voice recorder and flight data recorder for investigators to conclude that the aircraft's destruction was due to a bomb in the forward cargo hold (Gero, 1997).

Authorities suspect that the bomb was loaded on Flight 182 while on the ground in Vancouver. Connecting the Narita Airport bombing with the Air India bombing was the fact that the same man had made bookings on the two flights—CP Flight 003, heading to Tokyo, and CP Flight 60, the flight that would eventually connect with Air India and be designated as Flight 182. According to Salim Jiwa, an investigative reporter for the Vancouver *Province,* both explosions were planned to occur on the ground, with the intent to only destroy the aircraft and not cause fatalities (Jiwa and Hauka, 2006). However, the bomber misjudged the amount of time Flight 182 would remain on the ground in Toronto.[13]

In Vancouver, four days before the explosions, a man purchased tickets on each flight. On June 23, his bags were checked and loaded. The individual checked in for CP Flight 60, bound for Toronto, but requested that his bag be checked through to Air India Flight 182. The gate agent explained that because he was only listed as a stand-by on the Air India flight, it was not possible to honor his request because of the baggage/passenger reconciliation policy, also known as positive passenger–baggage matching. The man persisted, and finally the gate agent gave in to his demands; the bag was checked through contrary to company policy (Gero, 1997). Because of personnel and equipment shortages at Air India, the reconciliation process may not have occurred anyway.

Ultimately, these bombings caused significant changes in the way checked baggage would eventually be handled in Canada. The events also changed Canada's view of its role in aviation security. "The Air India tragedy had an enormous effect on Canada and the Canadian government," wrote author Rodney Wallis (1993) in *Combating Air Terrorism.* Wallis added:

The country had been living in an age of innocence until the incident occurred. Canada had followed a noncontroversial political role vis-à-vis the world scene. No known terrorist organization had a dispute with the Canadians and everyone

---

[13]Jiwa would go on to author *Margin of Terror: A Reporter's Twenty-Year Odyssey Covering the Tragedies of the Air India Bombing* (Key Porter Books, 2007, Toronto, Canada).

believed the country was remote from any possible terrorist activity. The Air India disaster brought home to Canada and the world that these criminal acts need not be confined to the geographic location of the real disputes ... but that any land can be used if it suits the purpose of the criminals. (p. 7)

After the Air India bombings, screening checked bags became a standard practice in Canada. The country implemented a sustainable and affordable five-level screening process. All checked bags are first screened by conventional X-ray equipment, which can automatically clear about 60% of the bags, and are then loaded on the aircraft. The remaining bags continue to be examined through the use of additional X-ray machines, visual inspection, and explosive trace and computerized tomography technologies, until security personnel are satisfied that the bags contain no hazardous items or explosives.

PSA Flight 1771

In 1987, U.S. Air placed employee David Burke on unpaid leave pending an investigation into his alleged theft of $68 from the flight attendant's on-board liquor fund, but the airline did not confiscate Burke's airline identification badge. These provided Burke with the means to access airport facilities and, specifically, to bypass the passenger and baggage screening process, a standard procedure for airline and airport personnel at the time.

Burke's supervisor commuted to work by air every day to San Diego from the Los Angeles area, and Burke knew which flight he would be on. Three weeks after his suspension, on December 7, 1987, Burke bypassed screening using his employee ID and boarded the flight carrying a .44-magnum pistol. At approximately 4:15 p.m. local time, the air traffic control center in Oakland, CA, noted that PSA Flight 1771 was squawking an in-flight emergency code from its transponder and the pilot reported gunshots had been fired in the cabin—a postaccident investigation by the FBI determined that the shots were from Burke killing his supervisor in the cabin. Burke then entered the cockpit and killed the flight crew, then himself. The aircraft nosed over and exceeded the speed of sound as it dived uncontrollably toward the ground. The high-speed dive overstressed the BaE-146's airframe, and it broke apart. Forty-four people, including Burke, died in the ensuing crash. Witnesses on the ground watched the aircraft nose over as it plunged to the ground, crashing into a hillside near San Luis Obispo, CA.

The PSA incident led to rule changes requiring airport and airline employees who have been suspended or terminated from employment to have their airport access immediately confiscated. The rule changes also mandated that airport and airline personnel who access the sterile area via the screening checkpoints must undergo the screening process. It did not mandate the screening of those same personnel if they accessed the sterile area from the air operations area or security identification display area (SIDA).

Pan Am Flight 103

Before 9/11, the tragedy of Pan Am Flight 103 was the deadliest act of aviation terrorism ever committed. Although the bombing occurred over the skies of Lockerbie, Scotland,

FIGURE 2.3 Memorial to the victims of Pan Am Flight 103 located at Syracuse University, Syracuse, NY.

many of the passengers were U.S. citizens (Figure 2.3), which resulted in further scrutiny to the U.S. aviation security system. On December 21, 1988, a Pan Am Boeing 747 was bombed over Lockerbie, Scotland. The explosion killed 259 passengers and crew, plus 11 citizens of Lockerbie, Scotland, when debris rained down onto their homes.

The case of Flight 103 began in Malta, where a Boeing 727 labeled as Pan Am Flight 103 flew to Frankfurt, Germany. From there, the plane flew to London, England, landing at the London Heathrow International Airport, and parked next to a Pan Am B-747. The bags from the B-727, along with the flight number "103," were transferred to the B-747. The Department for Transport (DfT), the U.S. equivalent of the Department of Transportation, required that all baggage leaving the United Kingdom be reconciled with a passenger on board the flight. The same baggage reconciliation procedure was also in effect at Pan Am and should have taken place in Frankfurt, Germany (Wallis, 1993). The practice was in line with ICAO guidance after the Air India bombing in 1985. However, in the case of Pan Am Flight 103, the procedure was not followed at either location.

At approximately 6:25 p.m., 25 minutes later than its 6:00 p.m. departure, Pan Am's Clipper "Maid of the Seas" Flight 103 took off from London Heathrow International Airport. The aircraft entered Scottish airspace and climbed to its assigned cruising altitude of 31,000 feet en route to New York's JFK International Airport. A bomb containing 450 grams (16 ounces) of plastic explosive was detonated in the forward cargo hold underneath the first-class cabin. The aircraft broke up quickly, with the nose separating from the fuselage in as little as three seconds. The explosion and resulting crash killed everyone on board—259 passengers and crewmembers and an additional 11 people were killed by falling debris in the town of Lockerbie. The dead included 35 students from Syracuse University returning from studying abroad in London. Considering that the flight was en route to the United States and the large number of American victims, the bombing was a tragedy for the United States, as well as for the United Kingdom. It would result in sweeping changes to the aviation checked-baggage screening practices in the United Kingdom. British aviation security officials have referred to Pan Am Flight 103 as "England's 9/11."

Leading up to the disaster, Pan Am airline officials and U.S. authorities knew that bombs intended for use against aircraft had been manufactured in West Germany. There was also a specific bomb threat against Pan Am, and in October 1988, West German police uncovered a bomb-making factory in an apartment building with one bomb in the assembly phase. Police determined that a large Toshiba cassette radio was to be used to conceal the explosive charge, the timing device, and a barometric triggering device, which clearly suggested that terrorists would use the device on an aviation target. On December 5, 1988, the U.S. Embassy in Helsinki received a detailed bomb threat against a Pan Am flight operating between Frankfurt and the United States that would take place in the next few weeks (Wallis, 1993). The threat gave details as to the perpetrators and the method. Although Finnish authorities decided that the threat was a hoax, the FAA passed along the warning to Pan Am and other U.S. airlines. Authorities did not warn passengers of the threat. This is a considerable point of contention with the families of the victims of Pan Am Flight 103.

A long-standing policy in aviation security is that authorities should not broadcast bomb threats against airlines to the public. The rationale is that if bomb threats were made available to the public, passengers would cancel their reservations on that flight. Therefore, the act of making a threat, particularly on a widespread basis, could theoretically shut down the national airspace system. Making bomb threats public provides perpetrators with the publicity they are seeking and causes financial harm to the airline (Gero, 1997). A study of bomb threats conducted between the late 1970s and early 1980s showed that in more than 10,000 cases of bomb threats, an actual bomb was never found (Wallis, 1993).

Additionally, the positive passenger–baggage match procedure may have caught the bag containing the bomb, as the bag did not have a passenger on board to match. However, during the postincident investigation, Pan Am declared that the FAA had given the airline an exemption from the required baggage reconciliation procedure, a point the FAA disputed. Also, had the bag been X-rayed, in accordance with international security policies at the time, the Semtex explosive may not have been detected, as conventional X-ray machines could not detect that type of explosive at the time.

The use of the barometric pressure trigger is also significant in the Pan Am Flight 103 attack. In this case, the barometric trigger was a tool known to exist in the aviation terrorist bomber arsenal. Some airports used pressure chambers to predetonate any explosives using barometric pressure triggers. Terrorists quickly learned of this technology, and in the case of Pan Am Flight 103, the barometric trigger was intended to work in conjunction with the timed triggering device. In the bomb that was discovered in West Germany, the barometric device acted as a safeguard against the predetonation of the bomb. The barometric chamber would trigger the timing device, thus preventing the bomb from detonating if the aircraft sat on the ground because of an unplanned delay. Plus, if the device were subjected to a pressure chamber, it would simply arm the timing device, rather than detonate the bomb, thus making detection very difficult. In the accident investigation of Pan Am Flight 103, authorities determined the triggering device was a simple timer and that a barometric device was not used.

The bombers of Pan Am Flight 103 may have intended to have the bomb detonate over the water, thus reducing the available evidence of a terrorist attack. However, the bomb detonated while the aircraft was over the ground. The delay at London Heathrow International Airport may have contributed to the bomb detonating over land, or the bombers may not have known the actual flight path the aircraft would take and made the assumption that it would travel directly over water after departing.

The complete accident investigation was conducted with more than 1,000 Scottish police officers collecting more than 10,000 pieces of evidence. The debris spread across an 88-mile corridor.[14] Eventually Abdel Basset Ali al-Megrahi, who had been handed over by Mohammar Gadaffi, a Libyan intelligence officer and the head of security for Libyan Arab Airlines (LAA), was charged and convicted with 270 counts of murder for the crime. He was sentenced to life in prison. Megrahi served a portion of a 27-year sentence but was released by the Scottish government on compassionate grounds due to being diagnosed with terminal prostate cancer; he returned to Tripoli in 2009.

Three countries were involved in the fallout from Pan Am Flight 103. In the United Kingdom, Minister of Transport Paul Channon had already experienced numerous rail, sea, and air disasters during his tenure (Wallis, 1993). His department was criticized for failing to pass along warnings to airports and airlines in a timely manner. However, Omar Malik, former British Airways captain and chairman of the British Airline Pilot's Association, writing on aviation security before Lockerbie, noted that the DfT was underresourced and that advisers often lacked expertise in airline security. Malik went on to say that the DfT focused primarily on issuing enactments (rules and advice) to airlines and airports. Malik noted poorly phrased and composed enactments; for example, "[airlines] should reasonably satisfy themselves," which left such enactments open to interpretation

[14]In the ensuing accident investigation, the coroner uncovered some chilling facts. The accident analysis showed that within three seconds of the blast, the aircraft had been torn apart. Further, forensic pathologist Dr. William G. Eckert, who conducted autopsies on the victims, believed that more than 140 passengers and the pilot himself were still alive as the aircraft hit the ground. There were 243 passengers on board and 16 crewmembers, led by the pilot, Captain James MacQuarrie, first officer Raymond Wagner, and second officer Jerry Avritt. Thirty-five students from Syracuse University were also on board, flying home from an overseas study program in London. Five members of the Dixit family, including 3-year-old Suruchi Rattan, were flying to Detroit from New Delhi. Suruchi became forever associated with an anonymous note left with flowers at a Lockerbie memorial site: "To the little girl in the red dress who made my flight from Frankfurt such fun. You didn't deserve this."

(Wallis, 1993). Most often, airlines selected the interpretation that was the most convenient, not necessarily the interpretation that provided the highest level of security. Additionally, most enactments were advisory in nature and, as Malik pointed out, airlines do not feel bound to follow advice, because advice is not enforceable. Malik's observations should be given careful consideration, particularly by those in an aviation security regulatory capacity (Wilkinson and Jenkins, 1999).

In the United States, newly elected President George H. Bush established a Commission on Aviation Security and Terrorism. As a practical matter, the commission focused on the destruction of Pan Am Flight 103. In Scotland, the country held a fatal accident inquiry (FAI). Both the presidential commission's findings and the findings of the FAI were particularly damning to the aviation security system in the 1980s.

In the United States, the commission concluded the following:

1. The U.S. civil aviation security system is seriously flawed.
2. The Federal Aviation Administration is a reactive agency.
3. Pan Am's apparent security lapses and the failure of the FAA to enforce its own regulations followed a pattern that existed for months before and after the tragedy.
4. The destruction of Pan Am Flight 103 was preventable had stricter baggage reconciliation measures been in place to stop any unaccompanied checked bags from boarding the flight in Frankfurt.

The 182-page report recommended a top-to-bottom revamping of the government's airline security apparatus. The report stated that positive passenger–baggage matching is the bedrock of any heightened civil air security system. However, airline lobbyists worked to keep the process out of the domestic air carrier security program for another 11 years because of the perception that it would slow down flight operations (Wallis, 2003). The commission called for the creation of an assistant secretary position for transportation security within the FAA, which would later become known as the associate administrator for aviation security and would focus on security and intelligence issues. The commission also called for research and development of explosives detection systems to be a top priority and for special federal security managers to be appointed and employed at the nation's busiest commercial-service airports. This presidential commission resulted in the Aviation Security Improvement Act of 1990.

Just weeks after the Pan Am Flight 103 bombing, the FAA implemented new security measures for airlines that flew out of Western Europe and the Middle East, including positive passenger–baggage matching. However, even FAA administrator T. Allan McArtor admitted that the new procedures would not have detected the plastic explosive that brought down Pan Am Flight 103 (McAllister and Parker, 1988). This type of response is what security expert Gavin deBecker, in his book *Fear Less*, termed a "Category Two" response to a security threat (deBecker, 2002). DeBecker declared that all security precautions fall into two categories: a Category One response is implemented to reduce risk, and a Category Two response is implemented to reduce anxiety. DeBecker stated that both types of responses have meaning and purpose, but they are not the same. For example, after the 1999 Columbine High School shootings in Colorado, many other school officials installed closed-circuit TV (CCTV) cameras around their high schools. However, at the time of the shootings, Columbine High School had numerous security cameras, which did not prevent the shootings. DeBecker speculated that Category Two responses are used in the United States because the American public seems to say, "they've taken steps, looks good," and then goes "back to sleep." Aviation security practitioners should be diligent in the types of security measures being implemented and mindful as to whether the procedures are intended to actually reduce risk or simply to reduce anxiety. In the aviation industry, Category Two responses get the public flying again, whereas Category One responses help protect passengers from acts of aggression.

Author Rodney Wallis noted in his book *Combating Air Terrorism* that "Many government civil aviation officials around the world have been apt to issue directives with little or no effort being made to ensure their terms are understood" (1993, p. 37). Furthermore, Wallis pointed out in his post-9/11 book *How Safe Are Our Skies?* that the baggage bomb has been the chosen weapon of the saboteur for more than two decades (Wallis, 2003). Although the policy of positive passenger–baggage matching does not deter suicide bombers, it does at least require bombers to commit to their own death or that of an accomplice.

Pan Am Flight 103 taught authorities the following lessons:

1. Known threat information must be distributed to those who can act upon the information, and those who have the ability to act have the responsibility to act, regardless of government regulations.
2. When a new threat against aviation occurs, regardless of its location, policies and procedures must be reviewed throughout the aviation community, with the understanding that certain locations and airlines will have higher risk levels because of various geopolitical factors. Higher-risk flights should incorporate additional security measures such as the use of additional screening.
3. Security policies and procedures should be adhered to, such as the practice of positive passenger–baggage matching.
4. Government regulators, in a position of both providing guidance on adhering to regulations and advising on security policies and procedures, should have an understanding of airport and airline operations and, further, should focus on providing specific information that is clearly worded and enforceable.
5. There was a pattern of the FAA, which was charged with protecting aviation from sabotage, of not enforcing regulations consistently.
6. Good security is part of fiscal responsibility and results in cost savings for the company. Although the Pan Am bombing was a huge tragedy in the loss of life, the failure of the airline to provide a secure aircraft for its passengers, just as it has the responsibility to provide a safe aircraft, ended up costing the airline billions of dollars. Considering insurance companies paid the lawsuits, insurance for aircraft operations also increased substantially, thereby increasing the operating cost for all other airlines.

7. Policies and procedures can become relaxed over time when there is not an apparent threat.
8. Security measures must be evaluated to determine if they are intended to actually reduce risk or to reduce the public's anxiety about flying.
9. Technologies in place to detect weapons and explosives should actually detect the threats that are prevalent at the time (i.e., conventional X-ray machines used to detect explosives could not detect the types of plastic explosives that were becoming more common in aviation sabotage).
10. There were numerous warnings of a potential bomb being placed aboard a commercial aircraft, even specifying the airline and the type of device. Warnings should be heeded and given careful consideration.

The last lesson, and an appropriate closure to our discussion of the tragedy of Pan Am Flight 103, was highlighted by Omar Malik in the following narrative:

> *Finally, aviation's long tradition of safety consciousness had not, in 1988, been extended to security consciousness and commitment ... given the remoteness of the probability of a terrorist attack on any specific flight, many of [the pilots] adjusted their priorities accordingly. Responsibility for these attitudes rested not with such managers but with the senior managers of the airline who determine corporate culture ... staff attitudes to security are as important as security regulations and procedures. Regulations and procedures are the building blocks of security; personnel attitudes are the cement ... which holds the blocks together."*
>
> (Wilkinson and Jenkins, 1999, p. 121)

1990–2001: A False Sense of Security

By 1990, a decade of deadly terrorist attacks on aviation came to a close. This period was followed by more than 10 years of relative calm in global aviation security. With the lack of any significant or successful attacks in the United States and on the aviation security system, travelers and some within the aviation security community slipped into a false sense of security.

The mid-1990s did bring a few notable attacks and incidents, some on aviation and some that were not directly related to aviation but still affected aviation policy. Examples include the crash of TWA Flight 800, the bombing of the World Trade Center (WTC) in 1993, and the bombing of the Murrah Federal Building in Oklahoma City, OK, in 1995.

The attack on the WTC, Murrah Federal Building, and a similar vehicle bomb attack on the U.S. Air Force Khobar Tower barracks in Saudi Arabia acted as the catalysts for the *300-foot rule*, which was put into place at U.S. airports in 1995. The 300-foot rule creates a clear zone around airport terminals and air traffic control facilities where vehicles are not allowed to park or be unattended. The distance was based on a Bureau of Alcohol, Tobacco, Firearms, and Explosives blast analysis. This rule caused major problems for airport operators around the country, because most airports were not designed naturally with a 300-foot buffer. As part of this rule's enforcement, authorities rapidly positioned barricades around the airport and officials towed vehicles within 300 feet of terminal buildings and checked the vehicles using canine explosive detection teams. Airport revenues also suffered as travelers moved to offsite parking locations or had friends and associates drop them off at the airport.

The WTC, Murrah Federal Building, and U.S. Air Force Khobar Tower attacks also prompted many airports to conduct blast analysis studies and incorporate protective measures into their building construction and design. One change is credited with helping to save lives when American Airlines Flight 77 struck the Pentagon on 9/11. After the bombing of the Murrah Federal Building in 1995, many government structures, including the Pentagon, replaced their glass windows with a higher-strength safety glass. Military leaders and Pentagon renovation specialists credited the blast-resistant windows, incorporating laminated glass and the steel structure that supported them, with saving many lives. Pentagon renovation program communications specialist Brett Eaton stated, "The new blast-resistant window system installed in Wedge 1 supported the floors directly above the impact for approximately 30 minutes after the attack, allowing hundreds of people to flee to safety" (DuPont, 2003, p. 2). Some airports, such as Ben Gurion International Airport in Israel, also use various types of safety glass in their terminal structure.

FedEx Flight 705

On April 7, 1994, Auburn Calloway, a pilot for FedEx, attempted to hijack a flight from his own airline. Practitioners can learn several lessons from this attempt, in terms of the reactions of the flight crew and the importance of employee security measures including background checks, workplace violence education, and early-intervention programs.

Calloway was catching a ride on FedEx Flight 705, a common practice known as "jump-seating," which is a known perk of working for an airline or air cargo operator. Believing that he would soon be fired, Calloway brought a guitar case containing several hammers, a knife, and a spear gun. Calloway was also a martial arts expert. He divorced in 1990, had two children, and had been having difficulty making his financial commitments. FedEx officials had recently informed him that there was a discrepancy in his pilot records, which could have soon resulted in his termination of employment. He was asked to stop by his supervisor's office and discuss the problem at his next available opportunity.

Faced with significant financial obligations and now the potential loss of his job, Calloway took out a life insurance policy on himself; next he plotted to attack a flight crew of a FedEx flight and then crash the plane into the ground. The large life insurance policy would take care of his family, but part of the success of the plan meant making the crash look like an accident. There is also some speculation that he considered crashing the aircraft into the FedEx headquarters building in Memphis, TN, but that would have not assisted in his attempt to make the crash look accidental. Calloway's attempt on FedEx Flight 705 was thwarted by the sheer willpower of the three-person flight crew, who, although severely wounded, fought Calloway for nearly 30 minutes while continuing to fly the plane to a successful emergency landing in Memphis.

Calloway was scheduled to be part of the Flight 705 crew. However, he and his fellow crewmembers went past their allowable duty time, so another crew was assigned. Calloway's initial plan was to fly on Flight 705 as the second officer, thereby reducing the number of people he would have to fight—the male pilot and the female first officer. As the second officer, Calloway would also have the control of the cockpit voice recorder (CVR) breaker switch, which could have enabled him to shut off the CVR, thus hindering crash investigators in determining the cause of an accident. It is suspected that Calloway attempted to turn off the CVR anyway, as second officer Andy Anderson noted twice that he observed the CVR breaker switch in the off position.

When Calloway boarded the flight, he was not searched. Air cargo pilots do not undergo screening the way many airline pilots do when they access the sterile area. The cargo area of most airports is located away from the passenger terminal facility, therefore, there is not a sterile area. Most cargo pilots are only required to go through a security-controlled door to access the tarmac and aircraft.

Shortly after takeoff as the aircraft completed its climb to cruise altitude, Calloway entered through the open cockpit door and hit the second officer several times in the head with a hammer. He then quickly turned and struck first officer Jim Tucker in the head with a hammer. Calloway next turned his attention to the captain, who began wrestling with Calloway for control of the hammer. Calloway then left the cockpit momentarily but returned with the spear gun. The second officer grabbed the protruding spear, and he and the captain pushed Calloway out of the cockpit and into the galley. The first officer, despite a severe head injury, took over the controls of the aircraft.

While at the controls, the first officer began a series of violent flight maneuvers meant to keep Calloway off balance and out of the cockpit. Eventually, the captain and second officer were able to temporarily subdue Calloway, long enough for the captain and first officer to switch places and get the plane turned back toward Memphis.

Calloway recovered slightly and continued fighting the two flight crewmembers until the plane landed at Memphis. All three flight crewmembers survived, but their injuries ensured they would never fly again.

There are several important points to this incident:

1. Calloway was serving as a flight crewmember after he had been notified of a discrepancy that could result in his termination from employment.
2. Calloway was a martial arts expert and had the advantage of surprise and the ability to strike his victims several times before they could fight back; however, the three men were still able to fight and subdue him. No matter how bad their injuries, they never gave up the fight.
3. Calloway used weapons and tools that he hoped would not have caused investigators to suspect foul play. However, in numerous crashes and bombings, investigators have still been able to find evidence of foul play, even when the aircraft was at the bottom of the ocean, as in Air India Flight 182. If Calloway meant for a crash to appear as an

accident, it is likely that there would have been enough evidence from the plane crash, or certainly the will he left on his bed, to suspect foul play.

4. The first officer used the aircraft and extreme maneuvering to keep control of the situation. Although many airline flight procedure manuals state that pilots are not to use violent maneuvers during an air piracy attempt, it appears to either be a natural reaction, a preferred form of self-defense by pilots, or the most available and efficient means to attempt to control what is happening. In interviews conducted with several pilots for this textbook, they all stated that regardless of what "the book" says, they would not hesitate to use extreme maneuvering to keep hijackers off balance. In 1970, an El Al Airline pilot used this tactic to avoid a hijacking, and even the hijacker on United Airlines Flight 93 used this tactic in an attempt to keep the passengers from storming the cockpit on 9/11.

5. A violent blow to the head with a hammer does not necessarily end the fight. All the flight crewmembers and Calloway himself were hit repeatedly in the head with hammers, but all continued to fight.

6. Commercial aircraft are constructed so that during an emergency it is easy to get out, but not as easy to get in without a boarding bridge or air stairs. Since 9/11, some airport law enforcement agencies have developed "raider" trucks, specially designed pickup trucks that have air stairs, which can rapidly unfold as the truck approaches an aircraft door. This allows law enforcement personnel the ability to rapidly access the aircraft, as compared to using the slower-moving standard air stair vehicles.

Calloway was charged with air piracy. He pleaded temporary insanity, but a jury disagreed and sentenced him to life in prison without the possibility of parole.

Operation Bojinka (a.k.a. the Manila Air Plot)

Operation Bojinka was actually a series of terrorist attacks planned to occur in 1995. The attacks included the bombing of 12 U.S. airliners over the Pacific Ocean, a plot to assassinate both Pope John Paul II and President Bill Clinton, and a plot to crash a Cessna general aviation aircraft filled with explosives into CIA headquarters. The key player in the Bojinka plots was the mastermind behind the 1993 bombing of the World Trade Center, Ramzi Yousef (Figure 2.4).

After the WTC bombing, Yousef fled to Pakistan and continued to evade capture for the next two years. Also known as the Manila Air Plot, Yousef had planned for bombers to board 12 Asia-based flights bound for the United States, assemble and arm their devices, then debark at stopovers. The bombs would be timed to detonate when the aircraft were over water. Yousef had already tested airport security measures by smuggling nitroglycerine in contact lens bottles secreted in his shoes. In fact, the 9/11 Commission used Yousef's test as an example of the inadequacy of airport metal detectors and X-ray machines to detect the bombs that Yousef intended to use.

Had Yousef been successful, 12 airplanes, each carrying an average of 250 passengers, would have been destroyed, killing more than 3,000 people, which would have nearly matched the total number of fatalities in the 9/11 attacks. Yousef's next step was to test

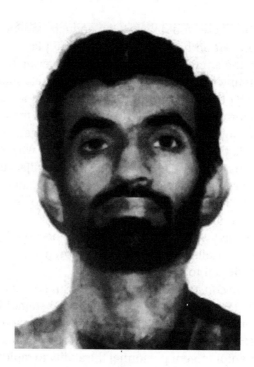

FIGURE 2.4 Ramzi Yousef.

his device on an actual airplane. On December 11, 1994, while aboard a Tokyo-bound Philippines Airlines flight, Yousef planted a bomb under the seat with a digital watch timer, mainly to see if the timing device worked. Yousef again made it through airport security concealing the bomb and bomb-making materials. The aircraft flew from Manila to Cebu, where Yousef, who assembled the device in the airplane lavatory, departed the plane upon landing, leaving the bomb under seat 26 K. Later, after the plane had again taken off, the bomb detonated, killing the Japanese businessman sitting in 26 K and injuring 10 others. The plane managed to make a safe emergency landing in Okinawa. Yousef also bombed a theater in the Philippines as another test.

Although his experiment worked, the airline bombing plot was thwarted when an associate of Yousef's accidentally set fire to the kitchen in his apartment while his associate was mixing chemicals. Police conducting the investigation went through laptop computers that had been seized from the apartment, along with other evidence that revealed the details of the plot. More details were soon revealed, such as Yousef's tie-in with fellow 9/11 conspirator and al-Qaeda operative Khalid Sheikh Mohammed, who was also involved in the plot, but neither he nor Yousef were at the lab when it burned. He and Yousef had also planned to bomb U.S.-bound cargo flights by sending jackets containing nitrocellulose[15] as cargo.

---

[15]Nitrocellulose is a highly flammable compound, which is also a form of smokeless gunpowder often used for magician's flash paper and in early photographic film processing. It also has medicinal uses but can be fashioned into a propellant or low-order explosive. For nitrocellulose to operate as an explosive, it must first be detonated by a high-order explosive such as a detonator or blasting cap.

Yousef was later arrested in Pakistan, brought to the United States, and stood trial in the 1993 bombing of the WTC.[16] An interesting sidebar to this story is that in the original WTC bombing the explosive device included cyanide gas, which Yousef had included in the hopes of causing massive casualties. The cyanide gas evaporated in the truck bomb explosion. This is evidence of an attempt by a terrorist to use a chemical weapon of mass destruction in the United States.

The case of the Manila Air Plot highlights the importance of flight crews searching an aircraft before departure and ensuring that individuals who depart an aircraft take their belongings with them. If they are returning to the same flight, flight crews must ensure that all deboarded passengers do return. Although this does not prevent suicide bombers, it does help to prevent "leave-behind" bombs from being placed on an aircraft. The issue of inadequate airport screening equipment in detecting the threats of the time would be addressed a year later after the crash of TWA Flight 800.

TWA Flight 800

On July 17, 1996, TWA Flight 800 departed JFK International Airport in New York bound for Paris. Shortly after takeoff, the Boeing 747 crashed off the coast of Long Island, NY (Figure 2.5). Although the crash was determined not to be caused by a bomb or missile strike, it did result in significant changes to the U.S. aviation security system.

FIGURE 2.5 Investigators reconstruct TWA Flight 800.

[16]Yousef was convicted and is currently serving a life sentence at "Supermax," in Florence, CO.

Within moments after the crash, dozens of security experts were on TV resolute in their conviction that the crash was the result of a terrorist attack, either a bombing or a surface-to-air missile. This theory was further supported when investigators found residue from explosives of the same type that had been used in a previous bombing. After a 16-month investigation, the FBI concluded that the explosives residue was left from a canine security training exercise and that the surface-to-air missile some witnesses said they spotted was most likely a trail of fuel igniting as the aircraft exploded. Eventually, the National Transportation Safety Board (NTSB) investigators would suspect a faulty centerline fuel tank issue as the cause of the explosion.

In the length of time it took to ascertain that the crash was not related to a security incident, the U.S. government took action as if it were a security issue. President Bill Clinton soon established the White House Commission on Aviation Safety and Security, led by Vice President Al Gore and commonly referred to as the Gore Commission.

Flight Crew Suicides

Airline flight crews undergo some of the most in-depth background checks, which often include psychological evaluations, as part of the hiring process. However, like most individuals, pilots experience life-related problems—and these problems include the risk of suicide. Sometimes a suicidal individual will want to murder others before, or while, taking his or her life. Unfortunately, when these problems manifest within a pilot, he or she has the ability to affect the fate of many others. These types of incidents, regardless of whether they are classified as terrorism, are the most difficult to thwart because the perpetrator already has access to the airport, the aircraft, the cockpit, and the knowledge of how to fly and control the plane.

In 1997, a Boeing 737 operated by Silkair with 104 passengers and crew on board departed Jakarta-Soekarno-Hatta International Airport in India. While cruising at 35,000 feet, the plane promptly began a rapid descent, breaking up over the Musi River Delta. NTSB investigators believed that the captain may have switched off both flight recorders (voice and data) and nosed the aircraft over when the first officer left the flight deck. The captain may have been struggling with several personal and financial issues at the time (Flight Safety Foundation, 1997).

In 1999, a Boeing 767 operated by EgyptAir departed Los Angeles International Airport with 217 passengers and crew on board. The flight was bound for Cairo, Egypt, with a stopover at Kennedy Airport in New York. Approximately 30 minutes into the flight, the relief first officer took the controls and the primary first officer and captain both left the flight deck. According to the NTSB's investigation, the relief first officer made several control inputs resulting in the aircraft going into a severe dive and soon exceeding the plane's maximum airspeed. The relief first officer repeated the phrase "I rely on God" several times. When the captain returned, he asked what was happening but the relief first officer kept repeating the phrase. Although there was apparently an attempt by the captain to recover the aircraft from the dive, it crashed into the Atlantic Ocean south of Nantucket Island, MA (Flight Safety Foundation, 1999).

There have been other suicides by flight crew, including another incident in 1999 when an Air Botswana captain boarded an ATR-42 turboprop aircraft, took off with no passengers or other flight crew on board, and flew around the airport demanding to speak to the country's president. The pilot threatened to crash into the other parked aircraft and eventually did so, eliminating the Air Botswana airline fleet.

Unfortunately, there is little that passengers on board a commercial airliner can do to prevent a pilot from committing suicide. By the time most passengers determine something is wrong with their flight, the aircraft would likely be heading toward the ground. This type of aviation threat, which includes the FedEx Flight 705 attack, is best handled with thorough background checks and educating managers and personnel on workplace violence and early-intervention programs.

The Millennium Bomber

In December 1999, Ahmed Ressam was arrested at the Port Angeles, WA, dock when he attempted to smuggle a vehicle-borne improvised explosive device (VBIED) through the Customs area. He intended to set off the device at the international arrivals Bradley Terminal Building at Los Angeles International Airport.

He was caught when a suspicious Customs agent, Diana Dean, noticed that Ressam was acting "hinky." As inspectors took a closer look at his vehicle, they noticed several green bags filled with a white-powder substance. Suspecting narcotics, they initiated a pat-down search of Ressam, at which point he wrestled out of their custody and took off running. He was caught after a short foot chase and later sentenced to 22 years in prison after being convicted for trying to plant a bomb at Los Angeles International Airport.

During the search of his trunk, before inspectors realized they were dealing with a bomb, one inspector vigorously shook a jar containing a brown liquid. Upon shaking the jar he noticed Ressam, now in custody in a police car, suddenly ducked down. Later, it was discovered that the jar contained nitroglycerine, intended to be used as a triggering device for the VBIED. Although there was virtually no way of knowing what was in Ressam's trunk, this is an important lesson for anyone responsible for conducting a vehicle search. The 300-foot rule was instituted in the 1990s in response to the Oklahoma City and WTC bombings, and Ressam's attempt demonstrates that the use of a VBIED against an airport has been considered and attempted.

September 11, 2001

On the morning of September 11, 2001, 18 operatives from the al-Qaeda terrorist network hijacked four commercial airliners over the continental United States. Two of the airliners struck the north and south towers of the World Trade Center in New York City. Another airliner struck the side of the Pentagon in Washington, DC. The fourth airliner was retaken by the passengers and crashed in a field near Shanksville, PA. The U.S. aviation system was immediately shut down. Nearly 6,000 aircraft were ordered to land immediately at the closest airfields. Only military and essential emergency or government flights (like Air

Force One) were allowed into the air for the next several days. The attack triggered U.S. offensive operations in Afghanistan and Iraq and sweeping changes to the aviation security system, both in the United States and overseas.

The following timelines are taken from *The 9/11 Commission Report*:

- *7:59 a.m. (EDT)*. American Airlines Flight 11, a Boeing 767, departs from Boston Logan Airport heading to Los Angeles with 5 hijackers on board. The leader of the 9/11 attacks, Mohammad Atta, and another hijacker had flown from the Portland Jetport in Maine earlier that morning to Boston.
- *8:38 a.m.* The Boston air traffic control center notifies the North American Aerospace Defense Command's (NORAD) Northeast Air Defense Sector (NEADS) that American Airlines Flight 11 had been hijacked. With 81 passengers, 9 flight attendants, and 2 pilots on board, the Boeing 767 diverted from its assigned course and headed eastbound toward New York City.
- *8:46 a.m.* NEADS scrambles fighter jets to intercept American Airlines Flight 11. However, just 40 seconds later, the airliner crashes into the North Tower of the WTC. By 10:29 a.m. the North Tower (1 WTC) completely collapses.
- *8:14 a.m.* United Airlines Flight 175, a Boeing 767, departs from Boston Logan Airport bound for Los Angeles with 5 hijackers on board. The 9/11 report estimates the aircraft was hijacked between 8:42 a.m. and 8:46 a.m. At 8:52 a.m., a flight attendant on board notifies United Airlines national headquarters that the flight has been hijacked, and it turns toward New York City. By 8:55 a.m., the FAA in New York suspects a hijacking.
- *9:03 a.m.* United Airlines Flight 175, with 56 passengers, 7 flight attendants, and 2 pilots, crashes into the South Tower of the WTC (2 WTC), killing all on board and more in the tower. The South Tower collapses at 9:50 a.m.
- *8:20 a.m.* American Airlines Flight 77, a Boeing 757, departs from the Washington Dulles Airport bound for Los Angeles, with 5 hijackers on board. The 9/11 report estimates the hijacking occurred between 8:51 a.m. and 8:54 a.m. The aircraft turns south, then back toward Washington, DC. At 9:05 a.m., American Airlines is aware that Flight 77 has been hijacked. At 9:25 a.m., the FAA orders all nonmilitary aircraft to land immediately and cancels all flights.
- *9:32 a.m.* Dulles Tower controllers spot a fast-moving flight on radar (American Airlines Flight 77). At 9:34 a.m., NEADS is advised that American Airlines Flight 77 is missing. Three minutes later, American Airlines Flight 77 crashes into the Pentagon, killing everyone on board, including 58 passengers, 4 flight attendants, and 2 pilots, plus an additional 125 people in the Pentagon.
- *8:42 a.m.* United Airlines Flight 93, a Boeing 757, departs from the Newark Liberty International Airport bound for San Francisco, with 4 hijackers on board. At 9:24 a.m., United Airlines sends a text message via ACARS[17] to all of its flights, including Flight 93, to be alert for possible cockpit intrusions. Flight 93 responds to the message asking for

---

[17]ACARS stands for Aircraft Communication Addressing and Reporting System. It is a digital data link system for transmitting text messages between aircraft and ground stations.

clarification. The 9/11 report estimates the Flight 93 hijacking occurred at approximately 9:28 a.m., just minutes after the flight received the text message warning. There are 40 passengers on the flight, including a pilot, a judo expert, and an off-duty federal agent. Several passengers make cell phone and in-flight telephone calls to family members. Passengers are advised by family members of the crashes at the WTC and the Pentagon and elect to take action against the hijackers.

- *9:57 a.m.* The revolt by the passengers begins, and at 10:03 a.m. United Airlines Flight 93 crashes, inverted and at high speed, into a field near Shanksville, PA. All aboard are killed, but there are no ground casualties. It is speculated that the target of Flight 93 was either the White House or the U.S. Capitol. There is some speculation that NEADS may have shot down Flight 93. However, the closest military aircraft, an F-16, was nearly 100 miles away when Flight 93 crashed. At the time of the crash, NEADS was not aware that Flight 93 had been hijacked.

The hijackers selected transcontinental flights, which assured full or nearly full fuel loads. Rick Rescorla was the head of security at the Morgan Stanley Bank in the WTC. Before 9/11, he had researched scenarios that involved loading an airplane with explosives and crashing it into the WTC. According to Rescorla, aircraft such as the Boeing 767 have enough fuel on board (23,000 gallons at capacity) to create a large enough explosive force that additional explosives were not necessary to be able to bring down the buildings. Rescorla was killed in the attacks; however, many of Morgan Stanley's employees were saved by his emergency evacuation exercises and plans.

All of the flights carried light passenger loads, making it easier to control the relatively smaller number of people. Additionally, the hijackers had previously flown on the flights they would eventually hijack and took flights in general aviation aircraft around New York to familiarize themselves with the airspace and their targets. The hijackers piloting the aircraft were trained at flight schools in the United States.

Evidence indicates that the hijackers advised passengers that they were to remain calm and that the airplanes were returning to the airport to have their demands met, thus creating the belief by passengers that a "traditional hijacking" was taking place. This mindset may have contributed to the lack of action by crewmembers and passengers on the first three flights. The passengers on Flight 93 had the benefit of knowing that the intentions of the hijackers were not just to land the plane and start issuing demands, but more likely to turn Flight 93 into a guided missile.

Other evidence indicates the hijackers were cleared through screening checkpoints with knives permitted by security or that were concealed. Before 9/11, the United States allowed knives less than four inches in length into the passenger cabin. On Flight 93, some passengers indicated that one of the hijackers carried a gun. It is not known if this gun was real or a toy plastic gun. The hijackers may have also displayed a fake bomb, using clay and other materials that would normally be allowed on a commercial flight.

The U.S. government shut down the national aviation system for several days after the attacks. The only aircraft flying in U.S. airspace were military, certain essential government aircraft, aircraft fighting wildfires in California at the time, and some medical and humanitarian flights. When the airports did reopen and the planes took to the skies once again, Army National Guard troops were stationed in airport terminals, metal detectors were turned to their highest sensitivities, and security queuing lines were backed up for hours as screeners methodically checked each traveler. The United States, Canada, and the ICAO all focused on similar and appropriate responses to the 9/11 attacks.

Within months of 9/11, the U.S. Congress passed the Aviation and Transportation Security Act of 2001. Canada revised its security programs, and the ICAO revised its aviation security guidance manual. The United States also passed several subsequent legislative measures including the Air Transportation Safety and Stabilization Act, which provided money to U.S. air carriers for financial losses suffered during the attacks.

The terrorist attacks of 9/11 forever changed the way the world's public and, especially, flight crews will perceive and react to future hijackings. There have been at least 35 hijackings (through 2012) since 9/11, and in nearly all cases the hijackers were stopped by onboard security personnel, passengers, crewmembers, or law enforcement when the aircraft landed. Although many pre-9/11 hijackings also ended similarly, and with the safe release of most of the hostages, the pre-9/11 strategy of complying with hijacker requests has now changed to active resistance. This new approach to stopping attacks against aviation was demonstrated shortly after 9/11.

Post-9/11 Attacks

Richard Reid (a.k.a. the Shoe Bomber)

On December 22, 2001, just three months after the terrorist attacks on 9/11, Richard Reid (Figure 2.6) attempted to detonate an explosive concealed in his shoe while on board American Airlines Flight 63, en route from Paris to Miami, FL. He was caught when a flight attendant observed him attempting to light a match on the tongue of his shoe. She intervened and was pushed aside by the much larger 6-foot-4-inch Reid. Another flight attendant yelled for help and was assisted by several passengers who responded and managed to wrestle Reid to the floor. Two doctors on the flight used sedatives to subdue him as the aircraft diverted to Boston Logan International Airport.

A day earlier, Reid had attempted to board the same flight but was turned away by a suspicious gate agent. He was referred to security representatives for additional questioning. The security personnel were privately employed by the airport and not allowed to conduct a search. When they also became suspicious of Reid, he was turned over to local police officials. He was again questioned but not searched. Reid's profile fit several suspicious passenger elements. He was on an international flight without any checked baggage; no available means of supporting himself such as traveler's checks, cash, or credit cards; and very little carry-on baggage. Perhaps the most salient cause for suspicion was that he was a British citizen, traveling on a passport issued in Brussels, Belgium, and flying from France to the United States. Regardless of these indications for suspicion, authorities

FIGURE 2.6 Richard Reid.

released Reid, who quickly booked passage on the next day's flight. On December 22, Reid returned to Charles deGaulle International Airport. He managed to pass undetected through an airport metal detector with plastic explosives concealed in his shoes and boarded American Airlines Flight 63.

Authorities found the plastic explosive triacetone triperoxide (TATP) and a detonator hidden in the lining of Reid's shoes. This resulted in U.S. passengers having to remove their shoes for separate screening at security checkpoints.

Reid was convicted on terrorism charges and sentenced to life in prison. The incident pointed to two important issues in aviation security:

1. The use of passenger questioning for the purposes of assessing risk.
2. The inability of conventional X-ray equipment and metal detectors to detect the types of explosives currently in use.

The incident also points to the importance of training flight crewmembers in both security awareness and self-defense techniques, as well as the strategy of passenger intervention.

MANPAD Attacks (Kenya and Iraq)

The use of a surface-to-air missile against a civilian aircraft has occurred several times in the course of aviation's history. Surface-to-air missiles are a legitimate threat to today's commercial airliners. Since 1938, there have been approximately 80 incidents related to shooting down a commercial airliner. There are commonalities among commercial aircraft shootdowns:

1. Most occurred during a time of war over a country experiencing armed conflict.
2. The aircraft was shot at while it was still on the ground.
3. The aircraft was shot down by a military aircraft.

The first shooting after 9/11 of MANPADs on a commercial airliner that was not associated with a war occurred on November 28, 2002. Al-Qaeda operatives shot two SA-7 shoulder-launched surface-to-air missiles at an Israeli-flagged Arkia commercial airliner as it departed Mombasa, Kenya, with 271 passengers on board. Both missiles missed, and the Israeli government responded by equipping its El Al Airline fleet with antimissile flare-defense systems.[18]

The November 22, 2003, shooting of a surface-to-air missile into a DHL cargo aircraft as it departed Baghdad, Iraq, again focused U.S. attention to the threat of MANPADs (Figure 2.7). The Airbus A-300 cargo plane was hit with an SA-7 missile (another missile

FIGURE 2.7 DHL flight after being hit by a surface-to-air missile.

[18]The attack was conducted simultaneously with a suicide car bombing attack against the Paradise Hotel in Mombasa that killed 14 people.

was fired but missed), striking the trailing edge of the aircraft's left wing. Although severely damaged, including the loss of much of the hydraulic flight control system, the flight crew was able to land the aircraft.

On May 22, 2002, Patrick Gott, of Pensacola, FL, allegedly angry because people had ridiculed his turban, opened fire with a shotgun and invoked the name of Allah at the Southwest Airlines ticket counter inside the Louis B. Armstrong New Orleans International Airport. One individual was killed and another wounded in the incident. Bystanders tackled the man, preventing him from continuing the attack.

On July 4, 2002, Hesham Mohamed Hadayet opened fire at the Los Angeles International Airport (LAX) international arrivals terminal building, killing two and wounding four others before El Al security agents killed him. Armed with a .45-caliber pistol, a 9-mm pistol, a knife, and extra ammunition for the firearms, Hadayet entered the international terminal and began shooting passengers waiting in line at the El Al ticket counter. Two armed El Al security agents responded, along with LAX police. Within moments, Hadayet stabbed one of the El Al agents, who then opened fire, killing the assailant. This incident highlights the importance of having properly armed and trained law enforcement personnel in the public-use areas of an airport.

Russian Airliner Bombings

In August 2004, explosives carried onboard by terrorists brought down two Russian airliners. The bombers were two Chechen women who boarded separate flights at Russia's Domodedovo International Airport with explosives concealed in their braziers. This incident is now referred to as the "Black Widows."

Volga-Avia Express Flight 1303, a Tupelov-134, crashed as the result of a bomb 26 minutes after departing the airport, killing all 43 passengers and crew. Minutes later, Siberia Airlines Flight 1047, a Tupelov-154, also crashed as the result of a bomb, killing 46 passengers and crew. During the check-in process at the airport, both women were referred to a police captain for further scrutiny but no devices were found. The police captain was sentenced to seven years of incarceration for negligence. A ticket agent was also charged with accepting a bribe to allow one of the bombers, who did not have proper identification, to board the plane. She also received a jail term. In the subsequent crash investigations, the explosive material RDX (also known as Hexogen) was found in the wreckage of both aircraft. Chechen militants had used RDX previously.

The X-ray machines and metal detectors used in Russia are similar to those used in the United States for many years and cannot detect many of the current explosives in use by terrorist organizations today. After this incident, Russia required travelers to remove bulky clothing, shoes, and belts for X-rays, and subjected passengers to a profiling interview to identify higher-risk travelers requiring additional scrutiny.

London Bomb Plot

On August 9, 2006, authorities in Great Britain arrested 21 individuals suspected of plotting to detonate liquid explosives onboard several commercial aircraft departing from the

United Kingdom and bound for the United States. The number of operatives involved in the plot is significant, as is the fact that many of them were U.K. citizens. As a result of the threat, air traffic at airports throughout the United Kingdom came to a temporary standstill. This bomb plot was similar to the strategies used in the Manila Air Plot attempted by Ramzi Yousef in 1995.

The U.S. government's response to this incident provides three important considerations for aviation security policy makers and practitioners. The first is the *concept of thinking then acting* (versus acting then thinking). The second is the *concept of preventative maintenance*. The third is the *concept of the systematic approach* to aviation security.

The first reaction of the DHS to the London bomb plot was to increase security threat levels to red for aircraft traveling from the United States to the United Kingdom, and to orange for all other domestic flight operations. Despite the fact that the DHS stated there is not a known threat to U.S. aviation from potential bombers, the next reaction was to prohibit the carriage of all forms of liquids or gels from the cabin. The rationale was to prevent potential bombers from carrying explosive components onto an aircraft, then assembling the devices in aircraft lavatories. However, this strategy was designed to only protect the aircraft, leaving airports wide open to another form of devastating terrorist attack, as the no-liquids policy immediately caused thousands of passengers to once again congregate in airport ticket and screening queues. The potential impact of a suicide bomber, particularly a VBIED-style attack, would be devastating in these areas of large passenger populations.

The no-liquids policy is an example of acting, then thinking. The failure of government to do the preventative maintenance necessary in aviation security systems has trained security agencies to react to, and then think and roll back if later deemed appropriate, certain security measures. Preventative maintenance in this case means continually upgrading aviation security programs to match the new threats as they develop. For example, metal detectors and X-ray machines have been a part of aviation security since the 1970s. However, the threats prevented by these technologies have changed over time. After Pan Am Flight 103, officials stated that the X-ray machines in use could not detect Semtex, the plastic explosive used in the bombing. However, the machines were not replaced with ones that could detect explosives until 13 years later. In 1995, Ramzi Yousef smuggled nitroglycerine through security checkpoints, and the 9/11 Commission noted that, again, the X-ray systems at the time could not detect these chemicals. The United States responded in 1996 with the limited use of explosive trace detection equipment. Even with this equipment, the majority of passengers continued to go through the metal detectors and their luggage through X-ray machines that still could not detect the explosives terrorists were using in attacks throughout the world.

One controversial strategy to determining what items or materials should be restricted is *not to take away* items from travelers that could potentially be made into explosives. Laptop computers and cell phones could be used as timing and power devices, but they are also necessary tools for today's business traveler. DVD players, MP3 players, and portable video games help keep passengers entertained on long flights, but they too could be used as a timing or power device for an explosive. The solution to this security risk is to improve the screening process to be able to handle the profile of today's business and recreational flier, not take away every object that could possibly be used to trigger a bomb or as a weapon.

The London bomb plot presented a slightly new threat to airport security, liquid-based explosives. Therefore, new security measures that do not inhibit the safe flow of passengers must be developed. Passenger and carry-on checkpoint screening practices must be upgraded with effective procedures and better equipment. The establishment of profiling methods should reduce the number of people receiving additional scrutiny. X-ray machines need to be upgraded to stand-alone Computed Tomography (CT) explosive detection systems, and metal detectors must be upgraded to explosive trace detectors or full-body scanners. A continuing investment to improve security systems to be able to defeat existing and developing threats is a better long-term strategy, which will help avoid temporary solutions, such as carry-on baggage restrictions, that disrupt the transportation system.

The third lesson in response to the London bomb plot is for government regulators to understand that aviation security is part of a larger and more complex security system. It is imperative that those responsible for ensuring aviation security recognize that security is made up of a "system of systems." Airport security is linked to a series of interconnected security systems. When one aspect is changed, one or more related outcomes will occur that will affect other components of the system.

Another challenge present in the liquid-bomb plot is that authorities suspected that the plot was moving forward several days, and possibly even weeks, prior to putting a stop to it. This is a controversial but necessary course of action. To stop every potential terrorist attack when it is first detected may only capture low-level operatives, while allowing high-level operatives to escape, learn from their mistakes, and make another, possibly successful attempt. Identifying and tracking attackers and planners as the plot progresses allows law enforcement to identify higher-level operatives, such as cell leaders and potentially bomb makers, resulting in the arrest and possible termination of the cell entirely.

Unfortunately, this strategy does not always work as the Bureau of Alcohol, Tobacco, Firearms, and Explosives found out during Operation Fast and Furious, when the agency lost track of numerous firearms agents they were tracking in an attempt to capture higher-level weapons traffickers. Possibly hundreds of the weapons were lost and later linked to crimes, including the shooting death of a U.S. Border Patrol agent, Brian Terry, in 2010. Had authorities lost track of the liquid-bomb plotters, several aircraft and thousands of U.S. citizens could have died in the subsequent attacks.

In their usually adaptive methods, there is evidence that at least one of the terrorists involved in the liquid-bomb plot was planning on taking his wife and children on the aircraft with him, to reduce suspicion by not being a single passenger. This represents a high level of dedication on the part of the operative, who was not only willing to sacrifice his own life as an airline suicide bomber, but to also sacrifice the lives of his family.

Passengers affect security systems when forced to check more bags because of security concerns. Baggage management and checked-baggage security systems are designed to handle a certain number of bags per year. Dumping millions more bags into those systems results in flight delays as aircraft loaders wait for the bags to be processed. Furthermore, more lost bags, increased financial loss to the airlines, and more people standing in airport arrival halls result from increased restrictions to carry-on baggage. The government must understand that sudden or highly restrictive policies related to carry-on baggage could threaten the economic viability of airlines and the national economy. Additionally, with laptop computers and other expensive electronic equipment being placed in checked baggage, more of it gets stolen. Laptop theft is particularly damaging as the information on the laptop can often be exploited for a variety of purposes, such as identity theft and corporate espionage.

The frequent business traveler generates much of an airline's revenue. The standard profile of a business traveler is to take only carry-on bags. Waiting for checked baggage to arrive can add hours of time to an airline trip. Hours spent waiting for bags to arrive have a direct financial impact on the ability of a businessperson to make money. Business travelers select public or private charter aircraft when the time to check in or depart an airport becomes excessive. This change in transportation vendors has a significant economic impact on airlines.

As another alternative strategy to traveling with long delays at airports, business travelers may select not to travel but rather conduct business through another mode of communication. In this case, the entire travel industry, including airlines, hotels, and rental car agencies, will lose money when business travelers decide to Web conference or teleconference rather than fly. Security processes must continue to enable the advantages of air travel while providing a reasonable level of security and an acceptable level of risk.

Northwest Flight 253 (a.k.a. Underwear Bomber)

On December 25, 2009, a Nigerian national, Umar Farouk Abdulmutallab, attempted to detonate plastic explosives hidden in his underwear while on board Northwest Airlines Flight 253, en route from Amsterdam to Detroit, MI. The bomb failed to detonate, catching fire instead and burning the bomber, and passengers stopped Abdulmutallab from continued attempts.

After the attack, the Transportation Security Administration (TSA) took an initially silly approach to stopping future attacks by requiring passengers to keep everything off of their laps during the final hour of flight. This process made no sense, because what is the difference in detonating a device during the last hour of flight, versus at any other time during the flight? This "rule" was quickly dismissed and the United States then stepped up its use of the pat-down screening technique at the security checkpoints, and accelerated the deployment of the body imaging technology.

General Aviation Aircraft—IRS Building

In February 2010, a Texas man, upset with the Internal Revenue Service (IRS), took his own small aircraft and crashed it into the IRS building in Austin, TX, killing himself and an IRS service manager. While not terrorist related the incident did bring brief attention to the lack of security at general aviation airports. The attack was a suicide-by-airplane-style attack that most security measures would not have normally prevented, as the individual owned the aircraft and had lawful access to both it and the airport where it was based.

Yemen Air Cargo Plot

Two desktop printers, with plastic explosives concealed in their printer cartridges, were loaded onto cargo flights out of Yemen on October 29, 2010. Al-Qaeda in the Arabian Peninsula (AQAP) took responsibility for the plot, which was intended to take down commercial aircraft carrying the printers as cargo. Saudi Arabia intelligence agencies discovered the plot, notified U.S. authorities, and the bombs were soon found during routine stopovers at East Midlands Airport in the United Kingdom, and Dubai, on separate cargo planes bound from Yemen to the United States. Although Congress mandated that cargo shipped on commercial airliners be screened by August 2010, that requirement did not yet extend to international flights. In 2012, Congress mandated that by the end of the year, all cargo on U.S. flights and inbound from international destinations is required to be screened by the aircraft operator.

Moscow Domodedovo Airport Terminal Bombing

In January 2011, 35 people were killed and more than 100 injured in a suicide bomb attack at Moscow's Domodedovo airport, in the international arrivals terminal. Domodedovo was also the launch point for the female Chechen suicide bombers who took down two Russian airliners in 2004. After the 2004 attack, Russia installed body imaging devices into their security checkpoints, but the bomb detonated in the public area of the terminal, prior to the checkpoints. Numerous other suicide bomb attacks had previously taken place throughout Moscow prior to the airport bombing.

Underwear Bomber II

In May 2012, another underwear bomb attack was thwarted by U.S. intelligence agencies and the plot never posed a serious threat to aviation security. The bomb was allegedly made by Yemen-based Ibrahim Hassan al-Asiri, the same bomb-maker who made Abdulmutallab's device and the Yemen air cargo bombs, but was apparently an upgraded version from the first underwear bomb in that it did not contain metal and would not likely have triggered an airport metal detector. However, an Advanced Imaging Technology (AIT) machine may have identified the bomb had an individual attempted to bring it through a U.S. screening checkpoint, but those remain in limited use internationally.

Since the first edition of this book was published, attacks on aviation continue throughout the world with several targeted at the United States. A review of virtually any edition of

Aviation Security International magazine will reveal incidents of sabotage, unruly passengers, attacks, and even the occasional attempted hijacking. A few notable incidents include the following:

- In 2003, a gun was found artfully concealed in a teddy bear at the Orlando International Airport in Florida. The bear was a gift from a stranger to a child two days earlier.
- In 2006, security screeners discovered a gun concealed in a laptop computer at the Austin-Bergstrom International Airport in Texas. Also in 2006, a knife was found concealed in a child's car seat.
- In June 2007, a sport utility vehicle loaded with propane tanks was driven into the terminal building and set ablaze at Glasgow International Airport. The attack was largely unsuccessful at causing massive amounts of damage to the building, but it did call attention to the possibility of an airport vehicle-borne improvised explosive–style attack.
- In January 2008, a 16-year-old boy was arrested on board a Southwest Airlines B-737. The teenager had handcuffs, rope, and duct tape, and the intention of hijacking the plane to crash it into the Hannah Montana concert in Louisiana.

Whether it is the well-coordinated attacks of 9/11 and Pan Am Flight 103, attacks by disgruntled employees, or the unrealistic visions of a 16-year-old boy, aviation security practitioners must continue to focus on strategies that prevent and mitigate any attack or threat while continuing to move people through the aviation system.

Conclusion

Policy makers and practitioners of aviation security face the challenge of deciding where the most dangerous threat lies and taking measures to prevent those threats from becoming real attacks. Funding for security measures must be guided by risk analysis, historical data, and a careful and continuing assessment of the developing threats throughout the world. While the United States has made significant strides in dismantling terrorist networks and eliminating key operatives—through the liberal use of drone attacks, American military forces, and particularly U.S. Special Forces Command personnel—new threats continue to surface. Terrorist cells operate like the drug smugglers of the 1980s and 1990s, as loosely organized groups sharing a common goal, relying on subversion and quickly replacing members when necessary.

Taking costly and rigorous security measures when there has only been a minimum of casualties may not receive the necessary public support. However, waiting too long until hundreds or thousands have died will result in accusations of being reactionary and negligent by failing to take the necessary precautions ahead of time. In fact, this pattern is clear when looking at the history of aviation security, which has tended to take action only following a significant loss of life. Canada, England, the United States, and Russia have all had their 9/11 s—their watershed events that finally caused a change in the way security practices are managed.

Current events and the dissemination of various types of information continue to fore-shadow potential future attacks on aviation. Security policy makers and practitioners have not always demonstrated the implementation of strong policies or procedures for detecting or acting on these warnings. As a global society, we tend to pay little attention to the unsuccessful attempts, even when those failures were caused by chance or poor planning by the attackers. As an example, the history of commercial aviation foreshadowed the events of 9/11 in that numerous incidents of hijackers attempting to take over control of an aircraft for use as a WMD had occurred. However, until the attacks of 9/11, using a commercial aircraft as a WMD had not been successful.

Other information, even in the form of entertainment, can also help to predict potential attacks by terrorists or criminals. Before 9/11, author Tom Clancy wrote in *Debt of Honor,* about a commercial airline pilot flying his plane into the U.S. Capitol as a suicide attack (Clancy, 1995). In 2011, Clancy and coauthor Peter Telep, in their book *Against All Enemies*, presented a plot to attack several U.S. commercial aircraft using surface-to-air missiles smuggled through the porous U.S.– Mexico border. Author Nelson DeMille has also presented plausible attacks on aviation in *The Lion's Game*, which came out one year before 9/11, an excellent thriller that depicts a chemical weapons attack against passengers on a jet coming into JFK, and again in his 2004 book, *Nightfall*, which presents a possible alternate scenario to the downing of TWA Flight 800 by surface-to-air missile.

As individuals interested in aviation and airport security, the preceding examples should force us to ask: *What attacks are foreshadowed that should be identified and mitigated today?* In recent years, the world experienced its first MANPAD threat against a civilian airliner, and airports have again come under attack by lightly armed gunmen. Perhaps Tom Clancy indirectly foreshadowed these attacks on aviation when in *The Teeth of the Tiger* (Clancy, 2003) he described a story in which armed gunmen attack various U.S. shopping malls with weapons and explosives. It does not take immense effort to realize the extensive damage and loss of life that would occur if such an attack were to take place during the busy holiday season in a terminal full of travelers. The commercial success of these works of fiction represent real attack possibilities and that the public and terrorists are still fascinated with threats to aviation.

Consider the tactics employed in Iraq, Afghanistan, and Israel by suicide bombers using vehicles containing explosive devices as WMDs. The unsuccessful plan to detonate a large-scale vehicle containing explosives next to an airport terminal building by Ahmed Ressam, the Millennium Bomber, was a significant event that may foreshadow future attempts by terrorists to cause destruction and death. Based on the history of aviation security incidents, could these unsuccessful attacks be simple precursors to much more devastating attacks to come?

References

Andrew, F., 2005. Mainliner Denver: The Bombing of Flight 629. Big Earth Publishing, Boulder, CO.

BBC News, May 16, 1972. 1972: Japanese Kill 26 at Tel Aviv Airport, retrieved July 26, 2006, from http://news.bbc.co.uk/onthisday/hi/dates/stories/may/29/newsid_2542000/2542263.stm1972.

BBC News, 1985. 1985: Gunmen Kill 16 at Two European Airports, retrieved July 26, 2006, from http://news.bbc.co.uk/onthisday/hi/dates/stories/december/27/newsid_2545000/2545949.stm1985.

Clancy, T., 1995. Debt of Honor. The Berkley Publishing Group, New York.

Clancy, T., 2003. The Teeth of the Tiger. G. P. Putnam's Sons, New York.

deBecker, G., 2002. Fear Less. Little Brown and Company, New York.

Department of State (DOS), 2006. Significant Terrorist Incidents, 1961–2003: A Brief Chronology. U.S. Department of State, retrieved July 14, 2008, from www.state.gov/r/pa/ho/pubs/fs/5902.htm.

DuPont, 2003. Blast-resistant Windows at Pentagon Credited with Saving Lives, retrieved Aug. 9, 2006, from www.dupont.com/safetyglass/lgn/stories/2111.html2003.

Flight Safety Foundation, 1997. Aviation Safety Network Accident Description: Silkair Flight 185. Flight Safety Foundation, Alexandria, VA.

Flight Safety Foundation, 1999. Aviation Safety Network Accident Description: EgyptAir Flight 990. Flight Safety Foundation, Alexandria, VA.

Gero, D., 1997. Flights of Terror. Patrick Stephens, Somerset, UK.

Gladwell, M., Oct. 1, 2001. Safety in the Skies. The New Yorker, retrieved July 31, 2006, from www.newyorker.com/fact/content/articles/011001fa_FACT.

Jiwa, S., Hauka, D.J., 2006. Margin of Terror. Key Porter Books, Toronto, Canada, retrieved July 31, 2006, from www.flight182.com2006.

Khan, K., Sept. 15, 1999. U.S. v. Kaczynski, Court TV, retrieved July 14, 2008, from www.courttv.com/trials/unabomber1999.

Lichtblau, E., Mar. 14, 2005. Security Report on U.S. Aviation Warns of Holes. The New York Times, retrieved July 14, 2008.

Mann, D., 2011. Inside Seal Team Six. Little, Brown and Company, Boston.

Marshall, D., July 9, 1997. DEA Congressional Testimony: Cooperative Efforts of the Colombian National Police and Military in Anti-narcotic Efforts, and Current Initiatives DEA Has in Colombia, retrieved July 14, 2008, from www.usdoj.gov/dea/pubs/cngrtest/ct970709.htm1997.

McAllister, B., Parker, L., 1988. Security Rules Tightened for U.S. Airlines Abroad. Washingtonpost.com, retrieved Aug. 7, 2006, from www.washingtonpost.com/wp-srv/inatl/longterm/panam103/stories/faa123088.htm1988.

Moore, E., 2001. Hero in the Cockpit. Houston Chronicle, retrieved July 26, 2006, from www.chron.com/cs/CDA/printstory.mpl/metropolitan/10874672001.

Moore, K.C., 1976. Airport, Aircraft and Airline Security. Security World, Los Angeles.

NYPD, 2006. Counter Terrorism: History, New York Police Department, retrieved August 2, 2006, from www.nypd2.org/nyclink/nypd/html/ctb/history.html2006.

Springer, J., 2002. LaGuardia Christmas Bombing Remains Unsolved 27 Years Later. CNN.com, retrieved August 2, 2006, from http://archives.cnn.com/2002/LAW/12/24/ctv.laguardia2002.

Testrake, D., Wimbish, J., 1988. Triumph over Terror on Flight 847. Kingsway Publications, Eden, UK.

UPI, Aug. 6, 1974. 4 killed in LAX Bombing. Post Advocate, retrieved via Los Angeles Fire Department Historic Archive, Aug. 2, 2006, from www.lafire.com/famous_fires/1974-0806_Explosion-LAX/1974-0807_Explosion-LAX.htm.

Wallis, R., 1993. Combating Air Terrorism. Brassey's, New York.

Wallis, R., 2003. How Safe Are Our Skies? Praeger, London.

Walt, V., 2001. Unfriendly Skies Are No Match for El Al. USA Today, retrieved July 15, 2008, from http://usatoday30.usatoday.com/news/sept11/2001/10/01/elalusat.htm.

Wilkinson, P., Jenkins, B., 1999. Aviation Terrorism and Security. Frank Cass, London.

Ziv, I., Producer 2006. Hijacked. [motion picture]. Paramount Entertainment, Los Angeles.

3

Policies and Procedures
The Development of Aviation Security Practices

Objectives

This chapter provides an overview of the development of aviation security policies in the United States and within other nations. We introduce the roles of the United Nations and the International Civil Aviation Organization (ICAO) in affecting aviation security policy and legislation. The ICAO's Annex 17 and other related programs are illustrated in relation to airport and aircraft operator security. The chapter also highlights key changes in aviation security created by legislation passed after the Aviation and Transportation Security Act of 2001.

Introduction

Security policies in the United States and abroad have largely been created through public reaction to specific security incidents. Airlines started using air marshals after numerous hijackings in the 1960s, and the U.S. government created laws governing air piracy. By the late 1960s and early 1970s, the frequency of hijackings exceeded the ability of the air marshals to handle them. As a result, passenger and carry-on baggage screening were adopted using technology designed to counter those specific threats. Hijackers were known to use guns and hand grenades in their attacks, so metal detectors and X-ray machines were used to screen for such items. When aircraft were bombed in the 1960s and 1970s, dynamite was often the preferred explosive, which can be identified by an X-ray operator. Plastic explosives were not commonly used at that time, so security policy makers did not focus on the detection of plastics. Even as plastics became the explosive of choice in the 1980s and 1990s, airlines did not adjust and continued using conventional X-ray machines and metal detectors, neither of which can reliably detect plastic explosives.

In the 1980s, new security strategies were developed in response to the Air India Flight 182 and Pan Am Flight 103 bombings. Two notable hijackings in the 1980s, TWA Flight 847 and PSA Flight 1771, spurred additional policy and procedural development, specifically in the areas of airport access control and credentialing (i.e., badging and background checks). The 1990s brought more security legislation, and more policies and procedures as legislators responded to the bombing of the World Trade Center in 1993, and in 1995 to the Murrah Federal Building bombing in Oklahoma City and Operation Bojinka.[1] In 2001,

[1]Operation Bojinka was a plot to destroy 12 U.S. airliners over the Pacific Ocean using bombs placed under passenger seats. See Chapter 1 for more details.

the United States implemented the largest aviation security system overhaul ever in response to the September 11, 2001, terrorist attacks. The 9/11 terrorist attacks were a watershed event for global aviation security.

Unfortunately, the industry as a whole did not keep up with the evolving terrorist threat until after 9/11, and debate continues over whether the United States should, or could, have done more to prevent the 9/11 hijackings. Policy makers face a dilemma in determining the development of national security policy: If there has not been a major incident to incite massive change (and massive government spending), then the populace criticizes the government for wasting money. If, however, there is a significant incident, the populace criticizes the government for not acting soon enough. Without the events of 9/11 and regardless of the intelligence available beforehand, it is likely that there would have been little public support for the massive changes and spending the U.S. government subsequently implemented.

International Civil Aviation Organization

When security practitioners need guidance and help, they can often look to the ICAO. The ICAO was established at the Convention on International Civil Aviation in Chicago on December 7, 1944. Its purpose is to "secure international cooperation and highest possible degree of uniformity in regulations and standards, procedures, and organization regarding civil aviation matters" (ICAO, 2006). Most of the security regulations in place at the world's airports can be traced either to ICAO Annex 17 or to the ICAO *Security Manual*. Although ICAO's guidelines are not mandated, they are commonly practiced throughout the U.S. aviation security system. Over time, the U.S. Federal Aviation Administration (FAA) contributed significantly to the development of ICAO policies.

The ICAO is a specialized agency of the United Nations charged with the administration of the principles laid out at the convention, which includes ensuring the safe and orderly growth of international civil aviation throughout the world. The ICAO's mission also includes developing standards related to airport, airway, and air navigation facility development, and to promote the safety of flight in international air navigation. The assembly meets at least once every three years. Each contracting state is entitled to one vote, and decisions of the assembly are decided by a majority. There are now 188 member states.

Related to aviation security is Annex 17 to the Chicago Convention and the Standards and Recommended Practices (SARPs) for safeguarding international civil aviation.[2] The ICAO publishes a security manual that contains guidance on the interpretation and implementation of SARPs. In the wake of the terrorist attacks on September 11, 2001, the ICAO developed an Aviation Security Plan of Action for strengthening aviation security worldwide.

A significant concept developed at the Chicago Convention in 1944 was that of an "aircraft in distress." It is a fundamental humanitarian principle to assist another

[2]SARPs was developed by the ICAO Council.

contracting state when an entity of that state is in danger. This principle has long required vessels from other countries on the high seas to respond to vessels in distress, no matter the reason for that distress, and render all available assistance. The Chicago Convention felt that this "law" should include aircraft from contracting states and therefore included "aircraft in distress" as part of the convention. The committee drafting the convention did not provide descriptors of what constituted "distress," thus allowing countries to come up with their own definitions. The freedom of a contracting state to decide what constitutes distress has been criticized. This policy was questioned in the 1985 hijacking of TWA Flight 847, when flight controllers in Beirut and Algiers repeatedly denied permission for the hijacked flight to land.

ICAO Annex 17: Security

Annex 17 to the ICAO Convention—Safeguarding International Civil Aviation Against Acts of Unlawful Interference—addresses findings and guidance developed over the years related to aviation security practices. The best practices are provided in the ICAO's *The Security Manual for Safeguarding Civil Aviation Against Acts of Unlawful Interference.*

Annex 17 set the first definitions in the industry for terms such as *screening, regulated agent,* and *security-restricted area.* Annex 17 also established guidance on key security issues including the following:

- Measures and procedures to prevent unauthorized access to the airfield
- Development of training programs
- Isolation of security-processed passengers
- Inspection of aircraft for concealed weapons or other dangerous devices
- Prisoner transport
- Law enforcement officer checked-baggage transport
- Cargo and mail screening
- Incorporation of security considerations into airport design
- Background checks for aviation employees
- Passenger/baggage reconciliation
- Security measures for catering supplies and operators

And, after 2001:

- Access control standards
- New standards for passenger, carry-on, and checked-baggage screening
- In-flight security personnel
- Protection of the cockpit

An interesting contrast can be made in comparing ICAO standards and international security practices and aviation security practices in the United States. The following demonstrates that contrast: On August 25, 2006, dynamite was found in the checked-baggage hold of a Continental Airlines flight as it arrived into the Houston/Bush Intercontinental

Airport from Buenos Aires, Argentina. After this discovery, newspaper services referenced a U.S. Department of Homeland Security study. The study found that many countries, while exceeding ICAO standards for checked-baggage screening, do not meet the U.S. standards. Internationally, checked-baggage screening systems consist of five levels of scrutiny: conventional X-ray, operator review of X-ray imagery, explosive detection system (EDS) review, operator review of EDS imagery or explosive trace detection (ETD) testing, and physical search or denial of loading.

Although the accuracy of the findings is not in question, it should be noted that the United States did not screen checked baggage on domestic flights *at all* before 9/11. When the United States started screening checked baggage following the Aviation and Transportation Security Act of 2001 (ATSA 2001) legislation, the baseline for screening was established at a higher standard.

There are four major ICAO conventions that have significantly impacted aviation security policy making. The Tokyo Convention in 1963 set the stage for many subsequent aviation security policies. The convention addressed unlawful acts committed onboard an aircraft that affect the safety of flight and, most important, allow the pilot in command to take reasonable measures to protect those onboard, including the ability to have a passenger restrained or removed from the flight. The convention also addressed how countries should handle a hijacked aircraft. Specifically, contracting states are obligated to take appropriate measures to restore or maintain control of the aircraft to the pilot in command. These appropriate measures extend only to the limit of what the country can feasibly and legally do (Abeyratne, 1998, p. 151). The convention required contracting states that receive a hijacked aircraft to permit the aircraft and all passengers and crew to proceed to their destination, and further specified that the country in which the aircraft is registered has jurisdiction over a hijacked aircraft. The issue of jurisdiction is a continuing problem.

By 1970, hijackings worldwide were reaching epidemic proportions. Thus, the Hague Convention that year made hijacking a distinct offense, calling for severe punishment of hijackers (Abeyratne, 1998, p. 157). The convention defined the unlawful seizure of an aircraft (hijacking) as committed by any person(s) who "unlawfully, by force, or threat thereof, or by any other form of intimidation, seizes, or exercises control of, that aircraft, or attempt to perform any such act; or is an accomplice of a person who performs or attempts to perform any such act" (Abeyratne, 1998, p. 157). The convention also obligated the states to return a seized aircraft to the country in which the aircraft is registered, including its passengers and cargo, without delay. The Hague Convention also addressed jurisdiction over hijacked aircraft. Jurisdiction goes first to the contracting state in which the aircraft is registered. If that state refuses or is unable to respond, jurisdiction passes to the contracting state where the aircraft first lands.

Issues related to prosecution were more difficult to address because many contracting states refuse to extradite to countries that have the death penalty. Although the Hague Convention required contracting states to make hijackings punishable by severe penalties and to either extradite or prosecute hijackers, some nations chose to ignore this requirement. The Hague Convention also did not list the penalties that should be imposed for hijacking or specify policies against extradition or a refusal to prosecute hijackers (or prosecute with severe penalties). This made some contracting states safe havens for hijackers and bombers.

The Hague Convention only addressed hijackings committed on an aircraft in flight, which did not cover acts of sabotage committed on the ground including at an airport. In 1971, the Montreal Convention expanded the focus to cover any attack on an aircraft, regardless of whether it occurred in flight. It also expanded coverage to air navigation facilities, including airports.

An aircraft is defined as being "in flight" from the moment all external doors are closed until such time when any door is opened for disembarkation. This definition remains in effect. The Montreal Convention added to the list of perpetrators of an attack on an aircraft those who are not actually onboard the aircraft, such as saboteurs, bombers, and those who help facilitate the attack.

The Montreal Convention defined five types of aviation offenses:

1. Committing an act of violence against a person onboard an aircraft in flight that endangers the safety of the flight.
2. Destroying an aircraft or causing damage to an aircraft, rendering it incapable of flight, or endangering its safety while in flight.
3. Placing or causing to be placed a device or substance likely to destroy or damage an aircraft.
4. Destroying or interfering with air navigation facilities.
5. Communicating false information that interferes with the safety of a flight (including bomb threats).

The convention also required all contracting states to implement passenger and carry-on baggage screening, and a national security agency to be in place at all major airports. This provision requiring deployment of a national security force is the first international guidance to come forth regarding airport perimeter security at nonmilitary airports. In the United States, this requirement is normally addressed by local police agencies.

ICAO Post-9/11

The 33rd session of the ICAO assembly met in September–October 2001 and adopted Resolution A33-1, Declaration on Misuse of Civil Aircraft as Weapons of Destruction and Other Terrorist Acts Involving Civil Aviation. This resolution directed the ICAO council and the United Nations secretary general to consider the establishment of an audit program relating to airport security arrangements and civil aviation security programs. It also directed the council to convene an international high-level, ministerial conference on aviation security with the objective of strengthening ICAO's role in the adoption of standards and recommended practices in the field of aviation security and in the auditing of their implementation.

A central element of ICAO's global aviation security strategy was the Aviation Security Plan of Action, which includes regular, mandatory, systematic, and harmonized audits to enable the evaluation of aviation security in all member states (ICAO, 2006).

The plan addresses the need to identify and assess global responses to new threats and take action to protect airports, aircraft, and air traffic control centers. The ICAO provides significant guidance on conducting aviation security audits. This includes the execution of an ICAO audit program, in which the ICAO will assist a contracting state, airport, or aircraft operator in developing programs to address deficiencies in aviation security capabilities. The program, the ICAO Universal Security Audit Program (USAP), is part of a larger effort to establish a global aviation security system.

The USAP audit is conducted in a way that states have the opportunity to monitor, comment, and respond to the audit. The results of USAP audits understandably remain confidential as sensitive security information. See Figure 3.1 for an illustration of the USAP audit process.[3]

To assist states in implementing the standards contained in Annex 17, the ICAO also developed a training program for aviation security. The program includes instruction on the following:

- The protection of aircraft, including preflight precautions, aircraft searches, and control of access to aircraft
- Access control, including physical security measures, background checks, personnel identification system design, and vehicle permits
- Quality control, including security inspections and audits, security tests, and training of security staff
- Airport design as it relates to security, including minimizing the effects of an explosion on people and facilities
- Managing responses to unlawful acts
- Security equipment (e.g., WTMDs, X-ray machines) and explosives
- Search and evacuation guidelines
- Surface-to-air missiles
- Incident command and emergency operations
- Model airport and aircraft operator security programs
- Dangerous goods

The ICAO also recommends the following:

- The commercial explosives material should be tagged with chemical taggants, which make it easier to detect with ETD machines.
- Information on the actions and movements of known terrorists should be shared with other member states.

---

[3]ICAO offers more information on the USAP audit at *www.icao.int/icao/en/atb/asa/usap_principles.htm*.

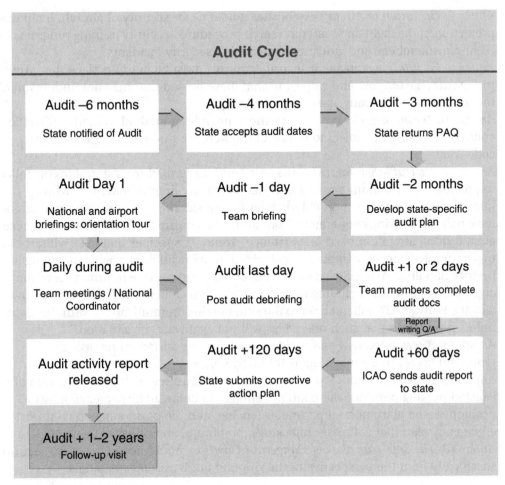

FIGURE 3.1 USAP audit process.

- Information regarding the use of forged travel documents and information relating to arms and explosives smuggling should be shared.
- Countries must prosecute suspected bombers or extradite them.

Additional ICAO Annexes Addressing Security

Although Annex 17 focuses specifically on aviation security, it also contains several attachments that cross-reference other ICAO annexes related to security. These annexes address different security standards and also provide the following security guidance:

- *Annex 2: Rules of the Air* allows air traffic control to direct aircraft under unlawful interference to deviate from their approved flight path and to be granted priority handling.

- *Annex 6: Operation of Aircraft* establishes guidance for security of aircraft, including protection of the flight deck, aircraft search procedures, security training programs for flight crewmembers, and guidance in reporting security incidents.
- *Annex 9: Facilitation* states that security controls must be put into place that protect the aircraft, passengers, and cargo, but that these procedures must not unduly inhibit the movement of aircraft, passengers, and cargo.
- *Annex 10: Aeronautical Telecommunications* provides a method to notify air traffic control of an aircraft subject to unlawful interference through the transponder code 7500.
- *Annex 11: Air Traffic Services (ATS; U.S., Air Traffic Control)* directs air traffic controllers to provide the maximum amount of assistance to an aircraft under unlawful interference. Annex 11 is particularly relevant considering the difficulty some aircraft have had in receiving cooperation from air traffic control facilities to land a hijacked aircraft or an aircraft in need of assistance. Annex 11 specifies that "ATS will provide maximum assistance to an aircraft subjected to unlawful interference including priority handling. ATS will also notify and continue to provide information to the appropriate agencies, including rescue coordination centers, and other aircraft in the vicinity. Further, ATS will not refer to the situation in communications with the aircraft unless it is certain that such references will not aggravate the situation."
- *Annex 13: Aircraft Accident and Incident Investigation* addresses reporting requirements for incidents of unlawful interference.
- *Annex 14: Aerodromes* addresses airport security measures such as the creation of an isolated parking position where aircraft subject to unlawful interference may be positioned; the placement of perimeter fencing; and the positioning of an airport emergency plan that addresses hijackings, bombings, and other security incidents.
- *Annex 18: The Safe Transport of Dangerous Goods by Air* addresses air cargo security, specifically from the perspective of the shipment of hazardous materials.

The ICAO spelled out the duties and role of air traffic controllers during a hijack situation. These roles were played out on 9/11, not just with the hijacked aircraft but also in handling the nearly 4,000 additional aircraft that were still airborne.

Air traffic controllers are responsible for ensuring that an aircraft under unlawful influence receives greater separation from other aircraft. Controllers have greater leeway in handling the flight, deciding the best way to provide assistance, which may include enlisting the aid of other air traffic services and providing flight crew with information on suitable airports for landing, minimum safe altitudes, and weather.

In a hijacking situation, controllers will attempt to determine the number of individuals onboard or remaining and the presence of hazardous materials. Whenever there is a suspected or known bomb threat, air traffic controllers must promptly respond to requests by pilots, which may include air navigation procedures and services along the route of flight and at airports of intended landing. Controllers must try to expedite all phases of a flight, transmit information pertinent to the safe conduct of the flight to other agencies, monitor and plot the progress of the flight, coordinate transfer of the flight to other air traffic control facilities without requiring transmissions or other responses from the aircraft, advise adjacent ATS units of the progress of the flight, and notify the aircraft operator, rescue coordination centers, and designated security authorities.

If the flight crew does not know that a bomb threat has been received, air traffic controllers must advise the flight crew immediately of the threat, determine the intention of the flight crew, and provide clearances for new destinations without delay. Affected aircraft on the ground must be advised to remain as far away as possible from other aircraft and facilities. If possible, the threatened aircraft should vacate the runway and proceed to an isolated parking area. It is the controller's responsibility to direct the pilot in command off the runway and to the appropriate parking area. However, air traffic controllers will not provide advice or guidance concerning an explosive device (Price, 2004, pp. 30–31).

U.S. Aviation Policies

The Federal Aviation Act of 1958 was the foundation of the policies, procedures, and regulations that the aviation industry currently practices. However, it was not until the late 1960s that aviation security became an issue.

One of the first American presidents to take action on aviation security was John F. Kennedy when he began assigning air marshals to flights in 1961. In 1971, airports were brought into the aviation security program with the creation of the Federal Aviation Regulation Part 107 Airport Security, which placed the responsibility of protecting the air operations area of an airport from unauthorized access. Whereas many security regulations are created through specific acts, other policies and regulations are often created through presidential executive orders or as amendments attached to funding legislation. Since 1970, Congress has consistently addressed aviation security issues, largely in response to specific terrorist incidents. What follows is a short description of some of the significant U.S. congressional acts passed in response to air terrorism. Because this text is a practical explanation of security procedures, the following descriptions should provide a framework and extract some lessons for current and future policy makers.

Traditionally, the implementations of solutions related to aviation safety occur after problems have been identified. When the aviation industry recognizes a problem with an airplane's design, manufacturing, maintenance, operation, training, or other nonsecurity-related factor that is causing accidents and incidents, the industry moves rapidly to solve the problem and reestablish the trust of the traveling public. Several government agencies oversee aviation safety issues, including the FAA, the National Transportation Safety Board (NTSB), the Occupational Safety and Health Administration (OSHA), and the Environmental Protection Administration (EPA). The aviation safety industry also has several incident and accident reporting procedures and publicly available data. Safety incidents are analyzed and compared with similar incidents to look for trends and patterns, and solutions are recommended and implemented. When there is an aircraft accident, the FAA and NTSB conduct an investigation to determine the causal factors. When necessary, the NTSB makes recommendations to the industry to make changes, leading to the prevention of further accidents.

The environment for aviation security varies from that of the aviation safety area in many ways. For example, public reporting for aviation security incidents is not as formalized as in the aviation safety situation. Often, when there is an aviation security incident, industry looks to the government to provide solutions and funding. When there is a major aviation security incident, a presidential commission is often established. A commission's findings often result in action and changes to the aviation security system, but are sometimes dependent on various public-interest organizations and the law-making process.

Anti-Hijacking Act of 1971

One of the first acts to address aviation security in the United States, and bring the United States up to ICAO standards on the prevention of an unlawful seizure of an aircraft, was the Anti-Hijacking Act of 1971. The act included punishment of hijackers by either the death penalty or life in prison. The bill provided a maximum penalty of only five years should the hijackers surrender during the hijacking, thus providing hostage negotiators some leverage in persuading hijackers to surrender. The act mandated passenger screening but not the screening of carry-on baggage.

Anti-Hijacking Act of 1974

In response to numerous hijackings and bombings, the U.S. Congress enacted the Anti-Hijacking Act of 1974. This act authorized the president to suspend air service to foreign nations that encouraged hijacking and authorized the secretary of transportation to restrict the operation of foreign air carriers in the United States. The act also required the screening of all passengers and all property by "weapons detecting" technology. The legislation further enabled airlines to refuse to transport anyone not consenting to a search, and mandated a law enforcement presence at the nation's commercial-service airports. This policy remains in effect.

One of the most important requirements of the Anti-Hijacking Act of 1974 was that it placed the responsibility for passenger screening onto the air carrier (or its designated contractor). This decision was criticized throughout the following years as airlines sought security screening companies that were the lowest bidder rather than the most effective. This decision also went against common international practices of placing all security responsibilities, including passenger screening, onto the airport operator.

Act to Combat Terrorism (1978)

The 1978 Act to Combat Terrorism added the requirement for airports to notify travelers to foreign airports that were deemed dangerous for use by Americans. It further authorized the U.S. president to suspend air service to those countries. This requirement can now be found in Title 49 CFR Part 1542 Subpart D: Public Advisories.

Aviation Security Improvement Act of 1990

After nearly 20 years of bombings and hijackings, the United States embarked on its first major legislation to change the aviation security system. In response to the downing of PSA Flight 1771 and the bombing of Pan Am Flight 103, Congress passed the Aviation Security Improvement Act of 1990. This act revised Title 14 CFR Part 107 Airport Security to include more comprehensive regulations on personnel identification systems and airport access control systems. It also required airport and airline personnel (including pilots) to undergo passenger screening in certain circumstances.

As with many responses to security threats, some actions may seem counterproductive or have no significant effect in deterring attacks. A portable radio was found to be part of the bomb used to take down Pan Am Flight 103. From this discovery, the U.S. government started requiring travelers to turn on their laptop computers and other portable electronic devices at screening checkpoints. The question remained, why not simply X-ray the devices as is done today? At the time, laptop computers were new and there was a fear of damaging the computer if X-rayed. The practice of turning on a laptop was intended to show that the computer was a working computer and not a shell hiding a weapon or explosive; however, it is still possible to construct a working improvised explosive device within a working radio or computer. Soon after the downing of Pan Am Flight 103, screeners at London Heathrow International Airport discovered a detonation device inside a working calculator, which also lessened the logic that a bomb could only be concealed inside a nonworking electronic device. The batteries and other electronics within many approved electronic devices (at the time, mostly laptop computers and radio/cassette players) can be used as a triggering or powering mechanism for an explosive device. Restricting laptops and cell phones from being carried on an aircraft impacts the advantages inherent in aviation, particularly for business travelers who are responsible for a significant portion of an airline's revenue.

The act addressed the capabilities of the access control systems[4] at the nation's airports and personnel identification systems. Federal Aviation Regulation Part 107.14 Access Control was expanded to require computerized access control systems that keep a record of who has authorized access to doors, gates, or other access points. These systems must be able to immediately deny access to people who do not have authorized access, distinguish airfield access authority for aviation personnel allowed access to only certain areas,

---

[4]Before 1990, many airport access control systems were not able to restrict airfield access to only those employees authorized to have access. Door locks were often simple cipher locks, and there was no way to track who accessed a particular door. Some emergency exit doors, connected to a centralized computer system, could distinguish between employee accesses and would alarm if forced or left open. Cipher-locked doors were often easy to access as the codes were common knowledge throughout the airport employee population. In some cases, employees (or others) could figure out the code either by watching another employee go through the door, or even in some cases, the door code was etched or written on the wall adjacent to the door.

and immediately prevent individuals without approved access media from accessing the airfield. These access control standards are still in place and embodied in Title 49 CFR Part 1542.207(a). Because most security doors leading out to the airfield are also fire alarm doors, fire regulations require a crash bar that allows egress from the building in the case of an emergency.

Airport access media systems underwent an overhaul in 1990. Before the 1990 act, access media (personnel identification badges or airport ID badges) were easy to duplicate, often requiring just a simple typewriter, an instant camera, and a laminator. In some cases, coding was placed on the magnetic strip on the back to enable an employee to open certain airfield access doors. Title 49 CFR Part 107.14 raised the standard for personnel identification systems, making them more difficult to duplicate.

Although the act also required flight crews to undergo screening, the concept of "employee screening" including flight crewmembers is a difficult issue. After 1990, employee screening was, and still is, required for those using the screening checkpoints to access the *sterile area* concourses or piers. However, many employees, including flight crews at their home-based airport and who have approved access media, bypass the screening checkpoints and use other doors leading to the airfield. Employees can then access the sterile area beyond the checkpoint by using airside doors that lead into the concourses. Employees with such access are considered to be "screened" as they have passed a fingerprint-based criminal-history record check (CHRC) process. If that same person accesses the sterile area by way of the screening checkpoint, he or she is still required to be screened. This dichotomy has existed since 1990 in the U.S. aviation system.

Employee screening has also been a contentious issue for pilots since the passage of the rule. Pilots argue that if they desire to crash the plane, they can do so by simply flying it into the ground. Also, they question the effectiveness of taking away small knives and sharp objects from them, when there is a fire ax in the cockpit. In 2008, the Transportation Security Administration (TSA) initiated a test program to allow certain pilots to bypass screening.

Aviation Security and Antiterrorism Act of 1996

In July 1996, TWA Flight 800 departed John F. Kennedy International Airport en route to Charles DeGaulle International Airport in Paris, France. The aircraft exploded shortly after takeoff, just off the coast of Long Island, killing all 229 aboard. The investigation took longer than 18 months, during which time federal investigators looked at a variety of theories, from mechanical mishap to terrorist attack. During the investigation, the exact cause of the crash was uncertain. Numerous security experts claimed that the crash was the result of a terrorist attack and that some bomb residue was discovered in the wreckage, which further drove the bomb theory.[5] As a result of these claims, President Bill Clinton

---

[5] A subsequent investigation showed that the bomb residue was likely the result of earlier, routine, K-9 bomb dog detection training.

established the White House Commission on Aviation Safety and Security. Chaired by Vice President Al Gore, the Commission is commonly referred to as the Gore Commission.

One of the most significant findings of the Gore Commission was that the federal government should consider aviation security to be a national security issue and provide substantial funding for capital improvements. The Gore Commission stated that "The Commission believes that terrorist attacks on civil aviation are directed at the United States, and that there should be an ongoing federal commitment to reducing the threats that they pose" (Gore, 1997). With this statement, the commission formally established a proactive policy toward aviation security.

The Gore Commission report established that aviation security is essentially a government responsibility. This was a welcome relief to the airline industry, which had been responsible for providing aircraft screening since the early 1970s. A General Accounting Office (GAO) report on aviation security echoed this sentiment (GAO, 2001, p. 4), and the recommendation resulted in the development of ATSA 2001 after 9/11.

Many of the commission's findings would help to formulate the Aviation Security and Anti-Terrorism Act of 1996. Of the more than 50 recommendations, the following became policy after 1996.

Airports established a security consortium consisting of airport security and law enforcement personnel as well as representatives from the FAA, Federal Bureau of Investigations (FBI), and other airport stakeholders. The consortium's goals are to stay abreast of evolving threats to aviation and to develop strategies and recommendations to counteract such threats. The consortium continues to exist.

The act required airports to conduct vulnerability assessments, which was supported by the FAA Authorization Act of 1996 that required the FAA and FBI to conduct joint threat and vulnerability assessments on security at least every three years at each airport determined to be at high risk. The Department of Transportation (DOT), FAA, and FBI also established formulas to determine if an airport was at high risk. Such formulas are classified as security-sensitive information and vulnerability assessments are still a routine part of any airport security program (ASP).

The act required fingerprint-based CHRCs for all screeners and all airport and airline employees with access to secure areas; however, they were required for only those screeners and airport and airline employees who could not meet the requirements of a 10-year employment history check known as an *access investigation*.[6] This limitation resulted in only a fraction of employees undergoing the CHRC process. This requirement was modified in the Aviation Security Improvement Act of 2000 to include all screeners,

---

[6]Before 1996, U.S. aviation employees did not undergo mandatory background checks—not even for outstanding criminal warrants. The 1996 legislation called for individuals with gaps in their employment history, which may indicate time out of the country or time in prison, to submit their fingerprints to the airport or airline. The prints were sent to the FBI to check against a list of 26 disqualifying felonies, a process that often took up to three months. Only a small percentage of personnel had to undergo the process. After 9/11, turnaround times were reduced to a few days through the expedited use of its Electronic Fingerprinting Image Print Server.

airport employees, and airline employees at Category X and Category I airports (most large- and medium-hub commercial-service airports).

Another flaw in the background check program was that anyone who had committed a felony but could still verify the previous 10 years of employment could easily lie on the application by stating he or she had never been convicted of a crime. This weak security procedure allowed people with questionable backgrounds to obtain approved airport and airline access media. A check of outstanding warrants was still not required for any aviation employee who could pass the 10-year access investigation. Individuals convicted of crimes not on the disqualifying offenses list, including all misdemeanors such as theft, could and did work within the security areas of an airport.

The Computer-Assisted Passenger Prescreening System (CAPPS) was developed as a result of the legislation. Based on certain indicators surrounding the purchase of an airline ticket, passengers were separated into low-risk and high-risk categories. High-risk passengers received additional scrutiny during screening.

The commission called for the deployment of EDSs and ETDs. More than 50 EDS systems and numerous ETD systems were subsequently deployed to commercial-service airports across the United States. The units were used randomly to screen passengers and their baggage, and for certain passengers who were identified through CAPPS. A GAO report subsequent to 9/11 revealed that the advanced explosives detection equipment was used to such a limited extent before 9/11 as to be minimally useful to aviation security.

The legislation significantly expanded the use of bomb-sniffing dogs with the FAA receiving funding for 114 new dog teams. It also set up the Aviation Security Advisory Committee (ASAC), a team of industry professionals brought together to advise the federal government on security issues. Several ASAC teams were involved in the development of aviation security regulations after 9/11. Also included in the act were the requirements to search commercial aircraft before each flight and increase inspections of air cargo through the Known Shipper program.

The Gore Commission spent a lot of time discussing the development of passenger profiling systems. This included encouraging the FBI, the Central Intelligence Agency (CIA), and the Bureau of Alcohol, Tobacco, and Firearms (BATF) to research known terrorists, hijackers, and bombers, and then develop profile indicators that could be tied to an automated passenger information system. The commission pointed to the success of profiling used by U.S. Customs, crediting the program with reducing passenger aggravation while increasing the detection of illegal substances being transported.

The Gore Commission was careful to point out that any profiling system should not be based on race, religion, or national origin. The commission also recommended that the airlines should not retain passenger name data information. Passengers should also be made aware of screening processes and be allowed to not participate and consequently not fly. It was further recommended that the Department of Justice periodically review profiling procedures and that an advisory board be established to discuss issues related to civil liberties arising from profiling methods. The Gore Commission believed that profiling becomes less effective with the development of more efficient screening technology.

Therefore, the commission recommended that profiling cease after EDS units are fully deployed. This policy ran contrary to the policies of many aviation security experts who viewed profiling as an integral part of the entire screening process. Many experts, then and now, believe that profiling should be combined with existing technologies, not used in lieu of them.

The 1996 legislation also brought about aggressive tests of existing security systems by more than 300 FAA security agents acting as *red teams*. Red team (adversary) testing includes frequent, sophisticated attempts by red team personnel to find ways to circumvent security measures, find weaknesses in the system, and anticipate what adversaries may attempt. However, although red teaming may have increased, its effectiveness was called into question in the following discussion by Andrew Thomas in his book *Aviation Insecurity*:

> *FAA testing at security checkpoints would involve placing a simulated explosive device in an uncluttered bag, oriented in such a way that it would be easily identifiable to even the most poorly trained X-ray machine operator. Conversely, the Red Team would disguise the testing device within a cluttered bag. (2003, p. 58)*

However, the red team test results were not admissible for civil enforcement proceedings against airport and airline operators. Thomas continued:

> *[T]he FAA would only make the airlines accountable for failures to detect easy-to-find explosives contained within standard testing protocols . . . the inability to detect a more concealed explosive from the Red Team would be disregarded. The only motivation for the airlines was to catch the standard items so as to avoid a letter of reprimand or a fine. Red Team testing in essence meant nothing. (2003, pp. 58–59)*

As a result of a Gore Commission recommendation, on October 9, 1996, Congress passed the Aviation Family Disaster Act of 1996 giving the NTSB the responsibility for aiding families of aircraft accident victims and coordinating the federal response to major domestic aviation accidents.

Airport Security Improvement Act of 2000

By 2000, several legislative actions designed to enhance the aviation security system had passed. However, the implementation of many of these was falling behind schedule. According to a National Academy of Sciences study, there were about 15,000 screeners working in the United States. The average pay for screening employees ranged from $5.25 per hour to $6.75 per hour with most screeners turning over every two months. There was one report that the turnover rate at one airport was about 400% a year (globalsecurity.org, 2001). Section 302 of the 1996 FAA Reauthorization Act directed the FAA to begin certifying screening companies, but the FAA did not issue a proposed rule to do this until January 2000. The Airport Security Improvement Act of 2000 (ASIA) directed the FAA to issue the final rule to certify screening companies by May 31, 2001, a deadline the FAA missed. The rule was to certify companies that provided security screening and improve the training and testing of security screeners. It would have required companies to develop uniform performance standards for providing security screening services (globalsecurity.org, 2001).

The Air Transportation Association (ATA; now known as Airlines for America, A4A) favored the proposed certification rule believing that the FAA should directly regulate screening companies to ensure full regulatory compliance and improve screener performance. The ATA/A4A also believed the rule would help to assess performance deficiency penalties against the screening company rather than against the airline (globalsecurity.org, 2001).

ASIA expanded the list of disqualifying offenses for security identification display area (SIDA) access and required all aviation employees to undergo a fingerprint-based CHRC. The legislation also allowed the aviation industry several years to implement the CHRC-for-all-personnel requirement. Access to the FBI's database was still not at a point where the CHRC process would be expedient. By 9/11, implementation of this requirement was still in its infancy.

ASIA also required the FAA to complete a rule that would hold individuals directly accountable for noncompliance with access control requirements, issue regulations requiring airport operators to have a security compliance program that rewards compliance, and ensure that airports and air carriers provide comprehensive and recurrent training while also teaching employees their role in airport security. These requirements would not be implemented until after 9/11. The act required the FAA to assess security improvements at air traffic control facilities and to make cyber-network security improvements (Price, 2003, p. 36). ASIA also concentrated on maximizing the use of EDS machines. It was noted in ASIA that the deployed machines were screening fewer bags in a day than they were designed to screen in an hour.

By the summer of 2001, the FAA was finalizing the FAR Part 107 Airport Security rewrite, incorporating many of the requirements of the ASIA 2000 legislation. On September 11, 2001, despite the 1996 and 2000 legislation, screening companies were still not being certified, performance standards or mandatory training for screeners still did not exist, nor were all airport and airline employees being subjected to a comprehensive CHRC.

Aviation and Transportation Security Act of 2001

The Aviation and Transportation Security Act of 2001 was the nation's first legislative response to the terrorist attacks on 9/11. Unlike previous legislation that had addressed aviation security, the 2001 act implemented policies that went beyond the prevention of another 9/11-style attack. In the past, whenever there had been a significant attack on aviation, the subsequent legislation focused on measures to prevent the kind of attack that had just occurred. ATSA 2001 addressed preventative hijacking strategies and virtually every other component of the aviation security system, including passenger and baggage screening, air cargo security, and general aviation security.

Much of the ATSA 2001 legislation was based on a GAO report issued on September 20, 2001, titled "Aviation Security: Terrorist Acts Demonstrate Urgent Need to Improve Security at the Nation's Airports," (GAO, 2001). The report summarized many of GAO aviation security recommendations for the past decade. These recommendations included improving airport access control systems, implementing screener training and performance standards, and tightening processes for law enforcement officer credentials. This last recommendation permitted state and local law enforcement officers to fly armed on civil aircraft. The ATSA 2001 legislation addressed numerous issues including:

1. Creating the TSA.
2. Requiring federal government takeover of the screening process.
3. Creating the position of federal security director (FSD) for each commercial-service airport.
4. Creating an oversight board for the TSA.
5. Reinvigorating the air marshal program.
6. Requiring fingerprint-based CHRCs for all aviation employees with SIDA access.
7. Requiring airports to develop screening programs for aviation employees before their entry into the SIDAs.
8. Requiring the screening of all checked baggage using EDS technology, including the use of the passenger/baggage reconciliation process.
9. Requiring that air cargo security be addressed.
10. Making it a federal crime to assault aviation employees with security duties. This includes screeners and any aviation employee with a SIDA access ID, because regulations require aviation employees to report violations of security regulations, thus making them part of the security process.
11. Creating a $2.50-per-passenger fee to help airports cover security expenses.
12. Developing technology to detect or neutralize chemical and biological weapons being carried on an aircraft.
13. Requiring cockpit doors on commercial passenger airliners to be reinforced.
14. Implementing security rules on charter flights for aircraft exceeding 12,500 pounds.
15. Requiring passenger manifests on international flights to be given to U.S. Customs before the aircraft lands in the United States.
16. Allowing five airports in the United States to conduct a pilot project using privatized screening workforces (that still must meet federal government training and performance standards).
17. Calling for the development of a 9-1-1 phone system for aircraft.

Since the passage of ATSA 2001, there have been numerous developments. The following discussion provides a summation of the changes in relation to several of the points listed.

Transportation Security Administration

ATSA 2001 created the Transportation Security Administration to regulate not only aviation security but other modes of transportation. The agency was initially placed in the Department of Transportation, but because of criticism that the TSA would become too closely affiliated with the FAA and questionable connections to the airline industry, on March 1, 2003, the TSA became an agency of the new U.S. Department of Homeland Security (DHS). The move was designed to mitigate the influence the aviation industry had over the DOT.

Federal Government Takeover of the Screening Process

The quandary over how to manage airline screeners after 9/11 evolved into three possible solutions:

1. Turn the existing private contract screeners into FAA employees, an idea that was quickly dismissed because the momentum after 9/11 was to take aviation security out of the hands of the FAA.
2. Place the screeners into an existing law enforcement agency such as U.S. Customs or the U.S. Border Patrol.
3. Place the screeners under a new federal agency.

Ultimately, the screeners were hired, trained, and employed by the TSA. This decision effectively established the U.S. government's responsibility for screening passengers and baggage at the nation's hundreds of commercial-service airports. ATSA 2001 mandated that the transition from private screening companies to TSA personnel must be completed by November 19, 2002, a deadline that it met. The government continues to operate and regulate its own screening services.

The TSA's screening function is an odd mix, using a government agency as both the operator and regulator of a function. The United States took the option to hire security personnel to work as federal government employees, rather than establishing rigid standards, then contracting out the security, which is international practice. Overseas, even before 9/11, the Argenbright Corporation, which held numerous private screening contracts in the United States, also provided similar services at several airports in Europe. Because of the higher government standards, Argenbright personnel overseas were highly trained and held to rigid standards, which they met. In the United States, the standard of performance as mandated by U.S. law and FAA policy was so low that the screening companies did not have to train their personnel or hold them to a higher standard. Because the airlines and not the government paid for the screening companies in the United States, there was little incentive to hire the "best and brightest," and more expensive, screening workforce. Despite the international standard, there appeared to be strong public sentiment that any form of screening that resembled the pre-9/11 model would be a return to the inadequate screening systems of the past.

The GAO examined security screening at five international airports known for having good screening practices and included some interesting points in their recommendations. These airports were in Belgium, Canada, France, the Netherlands, and the United Kingdom (GAO, 2001, p. 2). The GAO noted that responsibility for screening in these countries is placed with the airport authority or respective federal government, which then contracts out screening. Both Ben Gurion International Airport in Israel and London Heathrow International Airport are operated by airport authorities who contract out the security. In the case of Heathrow, the entire operation of the airport is managed by an external contracting service.

Creation of the Federal Security Director

Before 9/11, the FAA administered the security enforcement function through its civil aviation security field offices (CASFOs). After 1990, the largest commercial-service airports were assigned additional federal personnel. These individuals were known as federal security managers (FSMs) and were assigned to be a direct link between the airport operator and the FAA security branch in Washington, DC. An FSM was not part of the CASFO reporting structure. However, an FSM did have the authority to approve airport security processes. This created a conflict-based relationship between the "Cat X Agent," which was the CASFO agent responsible for overseeing the airport's compliance with its ASP, and the FSM, who had the authority to override the CASFO's guidance and directives, without being part of the reporting structure. ATSA 2001 attempted to resolve this conflict by assigning a new position titled the federal security director (FSD), who is at the operational head of an airport's federal security organizational chart.

Whereas the initial plan called for an FSD at each commercial-service airport, this plan was quickly downsized as the U.S. Congress realized that many commercial-service airports have very limited or seasonal service. FSDs are now at the large- and medium-hub airports in the United States. Other small-hub and nonhub airports are consolidated under regional FSDs.

Revitalization of the Air Marshal Program

On 9/11 there were approximately 33 active air marshals, mostly covering international flights. Although the actual number is classified, public estimates now range at 5,000+ for full-time employed air marshals.

Fingerprint-Based Criminal History Record Checks

This long-awaited policy was finally implemented with the proper technology and processes quickly being developed to enable all screeners and airport and airline employees to undergo a complete fingerprint-based CHRC. Through the use of computerized kiosks and the Transportation Security Clearinghouse (TSC) operated by the American Association of Airport Executives, results of the CHRC are now sent back from the FBI in a few days as opposed to weeks or months.

Screening Aviation Employees

ATSA 2001 did not set a deadline for implementing screening for aviation employees, and the proposed programs are not without a few problems. Many airports already screen a percentage of their workers, including those accessing the sterile area through screening checkpoints. However, to screen the entire workforce at each airport (including air crews) requires billions of dollars of infrastructure development and equipment acquisition, and a greatly increased annual personnel and operating cost. Presently, the fingerprint-based CHRC is considered "screening" for aviation employees. Although some airports around the world do screen their employees, many of those airports are designed to minimize the number of people in the SIDA, whereas airfields in the United States are not similarly designed. In 2008, the TSA embarked on an employee screening test program at several airports throughout the United States. That program is explained in more detail in Chapter 7.

Screening Checked Baggage

Because of the production and deployment limitations of expensive explosive detection equipment, most airports could not meet the requirement of 100% EDS checked-baggage screening by the congressionally imposed deadline. However, airports were allowed to "meet" the criteria by using passenger/baggage reconciliation procedures, ETD technologies, or K-9 bomb detection teams.

Air Cargo Security Procedures

Additional regulations covering air cargo security procedures were issued in August 2004 and went into effect in May 2006.

Security Fees

Passenger security fees continue to increase. As of this writing, passenger security fees are $5 per person per flight route, for a maximum of two routes on any given airline trip.

Chemical and Biological Weapons

Numerous companies are aggressively pursuing government research contracts to develop chemical, biological, nuclear, and radiological detection technologies. Much of this research is in conjunction with technologies being tested to detect chemical/biological/radiological/nuclear materials at other transportation modes such as rail and subway stations.

Cockpit Doors

Cockpit doors have been redesigned and reinforced to prevent forced entry. In many cases, cockpit doors allow viewing, either by peephole or closed-circuit television (CCTV), from the cockpit into the cabin areas. Some aircraft, by design, do not have cockpit doors. Air carriers using these types of aircraft are still required to restrict access to the flight deck.

General Aviation Charter Flights

The TSA's 12-5 rule addresses security procedures for private and public charters and cargo operations for aircraft above 12,500 pounds. The TSA's private charter rule addresses security procedures for privately chartered aircraft exceeding 45,500 kilograms (100,309.9 pounds).

Passenger Manifests on International Flights

Customs has used passenger manifest information such as name, date of birth, citizenship, gender, passport country of issuance, visa number, or resident alien card to cross-check names of passengers traveling to the United States against terrorist watch lists. In some cases, aircraft have been turned back or forced to land in a foreign country and debark certain passengers before continuing into the United States. Depending on the circumstances, authorities have allowed some aircraft with suspicious or wanted persons onboard to land within the United States. In these cases, authorities placed these individuals in custody. The program has evolved into the Advance Passenger Information System (APIS), which also includes the Crew Vetting Program (CVP), the master crew list (MCL), and the flight crew manifest (FCM). In Title 19 CFR Parts 122 and 178, airlines must transmit passenger manifest data to U.S. Customs and Border Protection no later than 15 minutes after the departure of an aircraft bound for the United States.[7]

Screening Partnership Program (a.k.a. Opt-Out)

The ATSA allowed five airports in the United States to conduct a pilot project using a privatized screening workforce, which was required to meet federal government training and performance standards. The private screening program, commonly called Opt-Out and referenced by the TSA as the Screening Partnership Program (SPP), continues to be successful according to several audit reports. The program experienced a slow start because of the lack of liability protection afforded to airports that elect to opt-out of their TSA-employed screeners. In 2005, Congress passed legislation giving airports the long-sought-after liability protection.

---

[7]As of February 2008, aircraft operators can either submit passenger manifests to APIS no later than 30 minutes prior to securing aircraft doors for departure or using the APIS Interactive Quick Query, which allows transmission of manifest information as each passenger checks in, up to but no later than the time aircraft doors are secured.

9-1-1 System for Aircraft

The technology to allow the use of cell phones on aircraft has already been developed and is in limited use in the business aviation community. Aircraft must be equipped with technology that reduces the strong cell phone signal to a lower level, which will better communicate with ground cell phone towers. This program continues to move forward. It has also created spin-off research, such as the ability to provide CCTV footage of the aircraft interior to personnel on the ground during an incident.

Air Transportation Safety and Stabilization Act (2001)

Just 11 days after 9/11, the government passed the Air Transportation and Stabilization Act (ATSSA), which earmarked $15 billion to the aviation industry, including $5 billion in grants and $10 billion in secured loans. The act also reimbursed the airlines for increased insurance premiums and limited compensation to the victims of 9/11 to $1.6 million each if the families waived their right to sue.

The importance of this act to aviation security practitioners, particularly airline security managers, is the liability limitations and the limits on victim compensation. Although the ATSSA was specific to the 9/11 attacks, it does offer some insight into how policy makers may address similar future issues. Of interest is that despite U.S. federal funding to the airlines, the major air carriers fired or furloughed nearly 20% of their staffs. This is perhaps the result of airlines exhibiting poor financial planning, taking advantage of federal subsidies, or demonstrating the extreme economic impact the airlines felt from the post-9/11 environment. Overall, at least 100,000 aviation employees lost their jobs as a result of 9/11.

Homeland Security Act of 2002

Shortly after 9/11, the United States government established the Aviation and Transportation Security Act of 2001. The signing was primarily as a stopgap in an attempt to fix the many existing problems in the aviation security system. However, policy makers were challenged and continue to be challenged with many questions such as the following: Who is the enemy? How do we make difficult choices regarding the distribution of resources for homeland security? How do we pursue foreign policy that goes after the terrorists while also preventing the emergence of new terrorists (Kean and Hamilton, 2006, p. 11)?

Much of the future policy regarding aviation security and homeland security stemmed from *The 9/11 Commission Report*. In other cases, policy makers would revisit previously commissioned reports and independent studies of security and terrorism in a search for answers on how to prevent attacks. One of those earlier commissions, the United States Commission on National Security in the 21st Century, better known as the Hart–Rudman Commission, became a focal point for developing homeland security policies. The Hart–Rudman report became a blueprint for the Homeland Security Act, which was passed in November 2002. The act brought in more than 22 government agencies from other departments including the TSA, the U.S. Coast Guard, the U.S. Customs Service, the Secret Service, the Federal Emergency Management Agency (FEMA), the FBI's National Infrastructure Protection Center (cyber-protection), the Federal Protective Service,[8] the Immigration and Naturalization Service (INS),[9] and a new Bureau of Citizenship and Immigration Services.

Some important considerations were made in the conclusions of the Hart–Rudman Commission. The first issue addresses homeland security as a local, state, and federal responsibility. It is a fundamental precept of emergency management that emergencies must be handled first at the local level and that municipalities must be prepared to respond to a certain level of disaster or incident. Once the local resources are overwhelmed, the state then has the responsibility to assist, and so on up to the federal government. The second point of the commission addresses disorganization in previous homeland security efforts. Each government agency at all levels had specific areas of responsibility for homeland security. These agencies were not centrally coordinated, nor functioning under a common vision.

The mission of the DHS is to prevent terrorist attacks within the United States, reduce the vulnerability of the United States to terrorism, and minimize the damage and assist in the recovery from terrorist attacks that occur within the United States. Included in DHS's primary responsibilities are information analysis and infrastructure protection; chemical, biological, radiological, nuclear, and related countermeasures; border and transportation security; emergency preparedness and response; and coordination (including the provision of training and equipment) with other executive agencies, state and local government personnel, agencies, and authorities, the private sector, and other entities (Kirkeby, 2006).

Besides the creation of the DHS, other significant actions resulted from the Homeland Security Act, including arming commercial pilots, creating a liability protection program for companies involved in antiterror technologies, and a lessening of the deadline for the implementation of EDS systems into airports.

Title XIV of the Homeland Security Act of 2002 (P.L. 107-296), the Arming Pilots Against Terrorism Act, established a program to deputize qualified volunteer commercial passenger airline pilots to serve as federal flight deck officers (FFDOs). FFDOs were assigned to defend the cockpit of an aircraft against acts of criminal violence or air piracy.

The Support Anti-terrorism by Fostering Effective Technologies Act of 2002 (commonly known as the SAFETY Act) recognized the need for new antiterrorism technologies and also recognized that the manufacturers of antiterrorism products and services would need certain liability protections to meet the nation's growing antiterrorism needs. The SAFETY Act provides liability protections for companies that produce antiterrorism technologies or provide antiterrorism services in two ways. SAFETY Act designation limits the liability of a company to 90% of its insurance coverage. SAFETY Act certification also eliminates the manufacturer or provider of antiterrorism technologies from federal lawsuits where the

---

[8]The Federal Protective Service is a police force that guards federal buildings.
[9]The INS split into immigration enforcement functions under Border Security.

product or service is involved. Insurance limits for aircraft operators were also addressed in the act. Title XII, Airline War Risk Insurance Legislation, extended the period during which the U.S. secretary of transportation may certify an air carrier as a victim of terrorism (and thus subject to the $100 million limit on aggregate third-party claims) for acts of terrorism from September 22, 2001, through December 31, 2003.

Title XVI of the Homeland Security Act corrected some areas of ATSA 2001 by requiring the FAA and the TSA to conduct research and development activities of antiterror technologies; it directed the TSA to issue regulations regarding the protection of sensitive security information, opened up screener qualifications to include naturalized citizens, and increased the fine to $25,000 (from $11,000) for air carrier operators who violate aviation security regulations.

Although not part of the Homeland Security Act itself, the government also addressed the issue of national crisis coordination through the creation of the U.S. Northern Command. This command is a unit of the Department of Defense, overseeing the federal response to natural and human-made disasters.

Vision 100: Century of Aviation Reauthorization Act

The Vision 100: Century of Aviation Reauthorization Act, passed on January 7, 2003, established or refined several programs related to aviation security—specifically in the areas of the government, aircraft operators, airport operators, and general aviation. Vision 100 directed the TSA to publish incidents and passenger complaints involving passenger and baggage screening. It also required the DHS to study the effectiveness of the transportation security system and required the government to establish the Air Defense Identification Zone (ADIZ) around Washington, DC. The act required the TSA to conduct security audits of foreign aircraft repair stations to ensure compliance with proper security measures by those companies engaged in the maintenance of U.S. aircraft.

Vision 100 created guidance, which is now included in TSR Part 1503, pertaining to enforcement actions against airport and aircraft operators who are in violation of security regulations. Vision 100 also focused on the computerized passenger screening system known as CAPPS 2, which is part of the ATSA 2001 legislation. Vision 100 placed several limitations on CAPPS 2, restricting it from being implemented until a thorough test phase was completed and both privacy concerns and the protection of passenger name records (PNRs) data could be ensured.

Aircraft Operators

Vision 100 required air carriers to provide basic flight crew security training to prepare crewmembers for potential threat conditions. The subject areas in training must include the following:

1. Recognizing suspicious activities and determining the seriousness of any occurrence.
2. Crew communication and coordination, and the proper commands to give passengers and attackers.

3. Appropriate responses to defend oneself and use of protective devices assigned to crewmembers.
4. Psychology of terrorists to cope with hijacker behavior and passenger responses.
5. Situational training exercises regarding various threat conditions.
6. Flight deck procedures or aircraft maneuvers to defend the aircraft and cabin crew responses to such procedures and maneuvers.
7. The proper conduct of a cabin search, including explosive device recognition. This requirement did not specify the nature of the training.

At some airlines this training was met with the use of a book, handouts, or a PowerPoint presentation. However, under Vision 100, aircraft operators must provide flight crew and crewmembers with the ability to take a voluntary course of advanced flight crew security training. The advanced training must include classroom and hands-on instruction. The subject areas must include the following elements of self-defense: deterring a passenger who might present a threat; advanced control, striking, and restraint techniques; training to defend oneself against edged or contact weapons; methods to subdue and restrain an attacker; the use of available items aboard the aircraft for self-defense; and appropriate and effective responses to defend oneself, including the use of force against an attacker. Vision 100 also extended the FFDO program to all-cargo aircraft pilots.

Airport Operators

Vision 100 expanded the list of airport capital improvement projects eligible for federal funding to include projects to replace baggage conveyer systems related to aviation security; projects to reconfigure terminal baggage areas to facilitate the installation of explosive detection systems; projects to enable the TSA to deploy explosive detection systems behind the ticket counter, in the baggage sorting area, or in line with the baggage-handling system; and other airport security capital improvement projects. Previous to Vision 100, airport security projects were not listed as being eligible for federal funding.

Vision 100 also created a funding source for airports needing security upgrades. The Aviation Security Capital Fund was created and funded to the level of $250 million, half of which is allocated to airport operators with the other half available as discretionary grants. The current allocation is 40% for large-hub airports, 20% for medium-hub airports, 15% for small-hub airports and nonhub airports, and 25% distributed by the TSA to any airport on the basis of aviation security risks.

General Aviation

Vision 100 set forth the policy directing the TSA to generate a program by which general aviation aircraft would be allowed back into Ronald Reagan Washington National Airport. General aviation flights had been restricted since the 9/11 attacks. Vision 100 also required the development of the air charter security programs for aircraft with a maximum certificated takeoff weight of more than 12,500 pounds (eventually becoming known as the 12-5 rule) and clarified the alien flight training program, setting forth regulations on foreign nationals desiring to fly an aircraft in excess of 12,500 pounds.

The act further required flight schools to conduct security awareness training for their personnel. This included aviation colleges, individual flight instructors providing flight training, and entities providing flight simulation training on aircraft exceeding 12,500 pounds.

The Intelligence Reform and Terrorism Prevention Act of 2004 (IRTPA) is best known for its creation of the director of national intelligence (a.k.a. the intelligence czar). The act played a key role in expanding or clarifying several aviation security programs, including those that involve air marshals, checked-baggage screening, man-portable air-defense systems (MANPADs), and air cargo. IRTPA pushed forward previous ATSA mandates including the establishment of a biometric uniform travel credential system for federal, state, and local law enforcement officers carrying weapons onboard an aircraft; the creation of the Transportation Worker Identification Credential (TWIC); and the development of Secure Flight (the program replacing CAPPS 2).

For in-flight protection, IRTPA directed the TSA's technology division to study the viability of devices or methods enabling crewmembers to discreetly notify pilots in the case of security breaches or safety issues in the cabin, and to report on the costs and benefits of using secondary flight deck barriers and whether such barriers should be mandated for all air carriers. A secondary flight deck barrier provides an obstruction to cockpit access from the cabin and is particularly useful when pilots must exit the cockpit to use the lavatory. Some airlines have voluntarily installed their own secondary barriers (those airlines using the barriers are classified as sensitive security information (SSI) but any frequent traveler will see them in use on occasion). Many airlines solve the problem procedurally by blocking access to the cockpit using a beverage cart held in place by flight attendants.

The act required the director of federal air marshals to promote operational practices that protect air marshal anonymity. However, a policy directive from the head of the federal air marshals allegedly required air marshals to dress in business professional attire rather than in a manner that would allow them to blend in with their environment. Little changed until the retirement of the air marshal director in charge at the time the policy was implemented.

In relation to the air marshal program, the act also required that air marshals be trained to address in-flight counterterrorism. Training related to weapons-handling procedures and tactics were also to be provided to U.S. federal law enforcement officers who fly while in possession of a firearm. Additionally, the air marshal service was to ensure that TSA screeners, federal air marshals, and federal and local law enforcement agencies in U.S. states that border Canada or Mexico receive training in identifying fraudulent identification documents, including fraudulent or expired visas or passports. The act also encouraged the U.S. president to pursue international agreements that would allow air marshals on international flights and enable air marshals to train foreign law enforcement personnel in in-flight security practices.

The act mandated the installation of security monitoring cameras in areas where checked-baggage screening is not publicly viewable.

General aviation was affected by the act, particularly the charter industry, which was now required to implement the airline passenger prescreening system to general aviation charter flights. Private charter operators are required to perform their own screening, after being trained by the TSA. The prescreening program compares the PNR data of individuals who rent, charter, or are passengers on an aircraft in excess of 12,500 pounds against the selectee and no-fly lists.

Air cargo security was an issue first addressed in ATSA 2001. However, the IRTPA gave cargo security a direction and a specific outcome by requiring the TSA to issue revised air cargo regulation and develop technology to better identify, track, and screen air cargo.

Implementing the Recommendations of the 9/11 Commission Act of 2007

The passage of the 9/11 Commission Act of 2007 amended the 2002 Homeland Security Act and required the nation to implement certain recommendations of the National Commission on Terrorist Attacks Upon the United States (i.e., the 9/11 Commission), including the mandate for 100% screening of air cargo and the creation of intelligence fusion centers and other requirements to improve intelligence sharing. It also established several avenues for the provision of grant money for homeland security–related programs. Related to aviation the Act required the government to strengthen the visa-waiver program and that the executive branch nominate a U.S. government official to serve as the director of the Human Smuggling and Trafficking Center.

This act would also greatly expand the TSA's explosives detection canine teams. Today, hundreds of canine teams are deployed to airports and transportation centers throughout the United States.

Other requirements include:

1. Clean up and provide a path of redress for misuse or errors on the no-fly and selectee lists, while pushing Secure Flight forward (this would also indirectly create the security threat assessment process).
2. Establish in DHS the Checkpoint Screening Security Fund and establish a passenger fee to fund the purchase, deployment, installation, research, and development of equipment to improve the ability of security personnel at screening checkpoints to detect explosives.
3. Extend funding for aviation security improvements including inline checked-baggage systems and other airport security technology improvements.
4. Fund several research and development programs on bomb detecting and disrupting technologies.
5. Direct the TSA administrator to develop a standardized threat and vulnerability assessment program for general aviation airports.
6. Conduct pilot programs to identify technologies to improve security at airport exit lanes.
7. Direct studies to improve and implement flight crewmember security procedures at screening checkpoints (to make it easier for flight crews to go through the screening

process) and push the administration to develop biometric-based identification for law enforcement officers flying on commercial aircraft (a reiteration of an ATSA 2001 requirement).

International Aviation Policies

This section should help security practitioners and aircraft operators understand certain international aviation security practices that may be useful when conducting operations in certain foreign countries.

European Union

After 9/11, the European Union (EU) passed Regulation (EC) No. 2320/2002 of the European Parliament and the European Civil Aviation Conference (ECAC) on December 16, 2002. This regulation established standards for aviation security at all EU airports. The regulation is based on standards contained in ICAO Annex 17, the recommendations of the ECAC, and proposals by the European Commission (EC). Regulation (EC) No. 2320/2002 calls for 100% screening of all aviation employees who have contact or interaction with screened passengers or baggage; mandatory aircraft inspections and other aircraft protection measures; checked (hold) baggage screening; cargo; courier and express parcels screening; air carrier catering and cleaning staff vetting; general aviation security regulations; staff recruitment and training; and security equipment standards.

Regulation (EC) No. 2320/2002 recognizes that one of the fundamental precepts of aviation security is the development of consistent international standards, recognizing that not all airports are of the same shape, size, and operational nature, and that each airport's unique qualities must be taken into account, particularly smaller airports, when establishing aviation security practices for each airport. The regulation standardized performance criteria for acceptance tests of aviation security equipment, detailed procedures for the protection of sensitive security information, and detailed criteria for the exemption of certain security measures where applicable.

Regulation (EC) No. 2320/2002 establishes a uniform civil aviation security program, approved by the appropriate federal government agency, which oversees compliance with regulations and has the power to increase security measures or exempt certain airports and air carrier operators from specific measures. It establishes a uniform civil aviation security program for air carrier operators and restricts access to sensitive security information, including performance criteria for security measures, and security equipment.

Within the field of airport security, Regulation (EC) No. 2320/2002 calls for airport operators to control access to the airside portion of the airport and that security controls be applied to passengers, baggage, mail, cargo, courier, catering stores, and supplies. Carry-on and checked-baggage screening must also be conducted.

Employees with airside access are subjected to a five-year background check, which is repeated approximately every five years. Employees with airside access are also required to wear approved access media (identification cards bearing the wearer's photo and other key information).

Regulation (EC) No. 2320/2002 requires screening of all airport and airline personnel, including the flight crew, before being allowed access to a security-restricted area. Of interest is that the regulation allows each country to develop its own definition of *security-restricted area* and *screening.* This flexibility in policy is important in determining how an employee is screened.

As mentioned, the screening of aviation employees, particularly flight crews, is a frequently debated topic. The mandatory screening of all aviation employees in a similar or the same manner as passengers are screened is common throughout the European Union. However, many European airports have different definitions for a security-restricted area, as compared to U.S. airports. Additionally, many European airports are constructed to minimize the number of employees who must access security-restricted areas. One specific requirement called for in the regulation is that any employee who conducts the screening of passengers or baggage must first be screened in the same manner that passengers are screened.

With respect to passenger and employee screening, Regulation (EC) No. 2320/2002 does specify additional random screening of personnel operating within security-restricted areas. For example, it is common practice at London Heathrow International Airport for passengers to be selected for additional security screening at the boarding gate. The random screening may include a physical search of carry-on belongings and the use of metal detecting hand-wands. This policy was met with resistance in the United States and was tossed out a few years after 9/11. Nevertheless, it is a commonly accepted best practice designed to prevent the introduction of a prohibited item that may have either slipped through the screening checkpoint or introduced by an aviation employee.

Although aviation employees are also screened, certain prohibited items (box cutters and knives) are still allowable within some of the shops and restaurants in the concourse sterile areas. It is conceivable that an employee acting in concert with a criminal or terrorist could pass along a prohibited item after the individual has cleared the screening checkpoint. This process is even doubly important in the United States where many aviation employees are not required to pass through a screening checkpoint before entering the sterile area (via an airside access door).

Physical security and patrols of the airfield, public terminal areas, and aircraft maintenance areas are standard practices under Regulation (EC) No. 2320/2002, as is perimeter fencing and CCTV monitoring for the protection of the airfield. Employees must also challenge other individuals not wearing an approved airport identification badge.

Regulation (EC) No. 2320/2002 covering aircraft operator security specifies that aircraft must be searched before being placed into service. A security check conducted for aircraft executing a turnaround or transit stop is also required. Each member country may decide what consists of a *search* versus a *check.* Further, access to the aircraft must be controlled at all times while the aircraft is in service. When not in service, the aircraft doors must be secured, the aircraft pulled away from any passenger boarding bridge, or tamper evidence applied to the doors, and a physical security patrol must be conducted of the vicinity around the aircraft every 30 minutes.

Passenger screening is conducted in much the same manner as in the United States, with passengers required to go through walk-through metal detectors and their carry-on bags undergoing scrutiny through an X-ray machine. Passengers causing alarm may be separated for additional scrutiny, such as a pat-down or hand-wand search. Carry-on bags that present suspect X-ray images may also be selected for testing by an ETD device. Some airports are upgrading their X-ray equipment and installing whole-body imaging technology.

Checked bags are subjected to the passenger/baggage reconciliation requirement. However, there is some latitude in the physical screening of checked bags. Bags can be checked by hand search, conventional X-ray, EDS, ETD, or the Primary Explosive Detection System (PEDS). The United States, however, calls for 100% screening of checked bags by EDS technologies.

Cargo security is addressed in Regulation (EC) No. 2320/2002 and focuses on the regulated agent as the key player in the cargo security process. The EU defines *regulated agent* as an agent, freight forwarder, or other entity who conducts business with an operator and provides security controls that are accepted or required by the appropriate authority in respect to cargo, courier, and express parcels or mail.

Cargo must be searched by hand or physical check, screened by X-ray equipment, subjected to a simulation (depressurization) chamber, or subjected to other means, both technical and biosensory (sniffers, trace detectors, explosive detection dogs, etc.). However, these security controls do not apply to a known consignor, referred to as a known shipper in the United States.

The carriage of mail is regulated in a manner similar to that used in the United States. Mail from an unknown shipper must be subjected to security controls such as screening, simulation chamber, or other means. Mail from a known shipper weighing less than a specific weight, containing life-saving materials such as human organs or high-value goods, carried on all-mail flights, or transshipment mail do not have to be screened. Air carrier mail such as internal dispatches or correspondence must be screened and controls applied to prevent the unauthorized introduction of other substances or items to company mail before it is shipped.

Airline catering companies require direct access to the aircraft cabin. Catering personnel often carry prohibited items such as box cutters in the performance of their duties and are responsible for delivering to the cabin food and beverages in sealed containers. For these reasons, catering services must be strictly supervised, both at the food preparation location to ensure that prohibited items are not introduced during food packaging, and at the aircraft delivery point to ensure that catering employees do not leave behind a prohibited item. Regulation (EC) No. 2320/2002 requires airline catering companies to have a security manager and a security program to ensure that high-quality employees are hired. The regulation requires that background checks are conducted on all employees and that the catering facility is protected from unauthorized access. Further, if the catering site is off airport property, the security manager must ensure that the catering vehicles are not tampered with during transit to the airport. Random screenings should also be conducted on catered goods, both during the delivery process and while in storage. Similar procedures apply to airline cleaning companies and their supplies, storage facilities, and vehicles.

As in the United States, general aviation aircraft operations are addressed in Regulation (EC) No. 2320/2002, but only with respect to preventing general aviation aircraft operations in commercial-service airports. Passengers disembarking from general aviation aircraft must be screened before they are allowed onto a commercial-service aircraft. General aviation aircraft must be parked as far away as possible from the commercial-service portion of an airfield, and security controls put into place to prevent the unauthorized access from the general aviation aircraft area to the commercial-service area. The regulation does not address security at general aviation airports.

ICAO guidance addresses the employment of security personnel including selection, knowledge, and training requirements. The EU recognizes the value of this guidance and similarly addresses specific standards in Regulation (EC) No. 2320/2002. In the United States, only law enforcement officers and security screening personnel must adhere to government standards for selection, knowledge, and training. Many other aviation job duties within the U.S. aviation security system do not fall under federal governmental guidance or controls. The EU does provide beneficial guidelines for the selection, knowledge, and training of security officials by requiring (1) extensive experience in aviation operations, (2) national certification, (3) knowledge of security systems and access control systems, (4) ground and in-flight aircraft operator security, (5) preboard screening, (6) baggage and cargo security, (7) aircraft security and searches, (8) weapons and prohibited articles, (9) overview of terrorism, and (10) other areas that may enhance security awareness.

The first requirement listed—extensive experience in aviation operations—is seen as critical in protecting aviation. The U.S. TSA was strongly criticized for placing individuals with little or no knowledge or experience in aviation operations into positions of creating and interpreting aviation security regulations, enforcing aviation security regulations, and developing operational standards and policies in the application of security. This resulted in numerous operational delays and the loss of millions of dollars in some places as aircraft stood idle while federal employees tried to make decisions and educate themselves on the nature of aviation operations. This initially created a great deal of friction between the TSA and airport and aircraft operators.

Over time, the TSA is slowly being infused with individuals who do have aviation experience, and those who come from nonaviation industries are being educated on the nature of airport and airline operations. Although the airport or aircraft operator security practitioner does not always exert great influence on the federal government's hiring decisions, the practitioner can make a special effort to help educate those within the federal structure about the economic impacts to the aviation industry when critical security decisions are being considered. Aviation security practitioners similarly should continue to educate themselves in the areas in which they regulate or operate.

Finally, the EU recognizes that the training of flight crew in security awareness, and the certification of security equipment such as metal detectors and X-ray machines, should be standardized throughout the European Union.

European Civil Aviation Commission

The European Civil Aviation Commission (ECAC) is a regional grouping of ICAO contracting states. Founded in 1955, the ECAC's objective is to promote the continued development of a safe, efficient, and sustainable European air transport system by standardizing civil aviation policies and practices among its member states. The ECAC promotes an understanding on policy matters between its member states and other parts of the world. The ECAC works in close conjunction with the ICAO and through the ECAC and the EU's rulemaking capability; the ECAC has more leverage than the ICAO over its member states, using resolutions, recommendations, and policy statements.

The ECAC establishes procedures generally well ahead of the ICAO and often draws on the experience of its members, many of whom are large industrial powers, when introducing or supporting new standards and recommendations (Wallis, 1993, p. 193). The ECAC is divided into two working groups: one group focusing on technology and the other on administration and operations. The United States and Canada both have permanent observer status in the ECAC. European and international aviation associations, such as the International Federation of Airline Pilots Association, will send their experts to ECAC meetings (Wallis, 1993, pp. 123–124).

One very important component in aviation security, domestic or international, is Annex 9—Facilitation (of Annex 17) of the ICAO Chicago Convention. Annex 9 states that security controls should be applied in a manner to continue the expeditious movement of personnel, baggage, and aircraft through the system. Security controls that are too burdensome may provide a higher level of protection or significantly reduce risk. However, onerous security controls may hinder the expeditious movement of personnel and cargo through the aviation system. There is a breaking point at which there is so much security that the advantages inherent in aviation are eliminated. Decisions about aviation policy should be carefully considered so that, ultimately, practices are not adopted that eradicate the industry.

The importance of the concept of facilitation cannot be overstated. In the early days of security measures, passenger and baggage screening meant passengers were hand-frisked and the contents of their bags were dumped onto a table for all to see. This violation of privacy was both unnecessary and at odds with the goal of making security practices effective and efficient. The introduction of walk-through metal detectors and X-ray machines were important facilitation steps that preserved the respect and privacy of the traveling public.

After 9/11, passengers waited for hours to undergo screenings and were subjected to a higher level of scrutiny. Initially, U.S. travelers were willing to accept the inconvenience of the longer wait times; however, patience wore thin after about six months. The TSA was quickly expected to fix the security problems and continue to keep screening wait times to a minimum. Wait times gradually became shorter until August 2006 when the liquid bomb plot in the United Kingdom was discovered. Again, screening wait times increased, though for a shorter period as TSA and airport operators adjusted quicker to the new restrictions placed on the carriage of liquids and gels. At Denver International Airport, the average wait times rose to record levels in the morning after the London bomb plot was discovered, but by that same afternoon they had resumed normal levels.

The lack of facilitation is also a contributing factor to in-flight passenger disturbances. The frequency of these types of passenger disturbances increased shortly after the ban of liquids and gels took effect. With the removal of bottled water, lip balm, and similar conveniences that help passengers cope with the physical, emotional, and mental stresses of air travel, passenger disturbances increased, and facilitation was reduced. Long-range security planning, such as the identification of new methods and tools used by terrorists and the implementation of protective strategies and technologies ahead of time, will assist with the goal of facilitation.

Aviation security practitioners can learn important lessons in facilitation from the ECAC. Being a smaller organization than the ICAO, the ECAC was able to better assist its members in the implementation of passenger/baggage reconciliation. Through its *Manual of Recommendations and Resolutions*, the ECAC provides guidance in the development of a national aviation security program, airport security, cargo security, and specifications of security screening technologies.

Canada

After 9/11, Canada also experienced a change of practices and organization in its aviation security system. Many of the U.S. and Canadian systems are similar in structure and operation with a few exceptions. In the United States, the TSA is both the regulator and the operator of the screening function. Canada separates these two functions, specifying Transport Canada as the regulator and the Canadian Air Transportation Security Authority (CATSA) as the operator.

Transport Canada is similar to the U.S. DOT in that the agency is the regulator and the policy maker for aviation security practices. However, in the United States the responsibility for aviation security has been shifted to the Department of Homeland Security to provide some separation between the TSA and the influences of transportation lobbying groups.

Operating since April 1, 2002, CATSA is a crown corporation (i.e., a state-controlled company or enterprise) reporting to Parliament through the minister of transport. CATSA is responsible for passenger (including airport and airline employees) and baggage screening; deployment, operation, and maintenance of explosives detection equipment; airport policing; financing the Royal Canadian Mounted Police (RCMP) to provide air marshal capabilities and the issuing of access media; and financing the provision of law enforcement officers for airport security. CATSA conducts security services at

89 airports across Canada with more than 4,000 contract employees. Similar to other common international practices, CATSA is charged with the provision of screening and other security services and contracts with private firms to provide these services.

Unlike the United States, where aviation security is largely funded from aviation revenues, CATSA is funded through parliamentary appropriations from the tax-funded Consolidated Revenue Fund. CATSA and its service providers rent their office and administrative spaces from airport operators, thus helping to support the local airport revenue base.

In the United States after 9/11, airport operators were directed to provide office and administrative space to the TSA at no charge. Considering that "space equals revenue," this was a considerable loss of revenue to many airports.

The Canadian aviation security model is similar to the U.S. model in that various agencies are given responsibility for protecting different elements of the system. In the Canadian model, no single agency is responsible for all facets of aviation security. However, Canadian airport authorities retain responsibility for perimeter security and apron, taxiway, and runway security, whereas airlines retain security for the check-in and departure areas. In-flight security is a shared responsibility of the airline, the RCMP, and CATSA.

CATSA also operates a program whereby airport and airline employees are randomly screened throughout the restricted areas by teams of two CATSA employees. The screening consists of a check of the employees' biometrically encoded restricted-area identification badge (i.e., approved access media) and the use of a portable metal detection device.

CATSA maintains a security communications center similar to the Transportation Security Operations Center (TSOC), located in Alexandria, Virginia to provide quick reaction to aviation security incidents and monitor new and existing threats. CATSA further recognizes the value of facilitation in the provision of its security services and focuses on four performance areas in the provision of security services, consistency of screening practices, effective business practices including measurement of metrics and incident reporting, cost effectiveness in resource allocation, and passenger service.

As in the United States, Canada is assessing several threats and new programs for their potential impacts to aviation security. Among the threats is the transportation of chemical/biological/radiological/nuclear (CBRN) materials, an attack using such substances, and vulnerabilities associated with air cargo. Among the new measures Canada is assessing or testing to reduce higher levels of passenger scrutiny are the Registered Traveler Program, the use of no-fly and selectee lists, behavioral observation (profiling), Secure Flight, alternative airport layouts, and self-service checkpoints.

CATSA does have a strategic security plan and works with Transport Canada on long-range funding for the financial stability of the agency. CATSA is also working on various screening staffing models in an attempt to achieve greater efficiencies in its screening operations.

Canada must address some different security challenges than those experienced in the United States—especially as related to airport design and the processing of passengers, and in customs and immigration practices. At the Vancouver International Airport, the international terminal is completely segregated by a large-window wall structure.

Passengers arriving from or departing to the United States must stay on one side of the glass while all other international passengers and their flights stay on the other side. The segregation takes place immediately after check-in. Customs and immigration are handled separately with different processing for Canadian citizens, U.S. passport holders, and all other nationalities.

The cruise ship industry has even affected the design of the Vancouver International Airport. To eliminate processing travelers arriving from the United States who are transitioning directly to U.S.-registered cruise ships, a special level was constructed within the Vancouver International Airport terminal building whereby U.S. travelers may go directly from their flights to busses that take them directly to their cruise ships. Their bags are also delivered to their ship directly, thereby reducing the number of bags that must be checked by customs. This results in passengers never "officially" entering or exiting the United States, thus reducing lines and congestion at immigration and customs.

Similar to the United States Visit Program, Canada is testing procedures to expedite the travel of frequent visitors to and from the country through CANPASS. Operated by the Canada Border Services Agency, CANPASS programs streamline customs and immigration clearance for low-risk, prescreened travelers. Commercial airlines allow CANPASS participants to pass quickly through Canadian customs and immigration at major Canadian airports, by meeting border clearance obligations through retina recognition technology. There are similar programs for corporate aircraft, private aircraft, personal watercraft, and remote border crossings.

Australia and the Asia-Pacific Rim

As with many other countries, the 9/11 attacks brought new policies and procedures to aviation security in Australia and the Asia-Pacific Rim countries. Aviation is particularly important to Australia as 99% of the country's international passenger movements are conducted by air. Australian airports enplane about 16.5 million international passengers and 30 million domestic passengers per year. The Australian Parliament took several measures in response to the U.S. 9/11 attacks, most recently passing the Aviation Transport Security Act 2004, in an attempt to align their security standards with those of ICAO. Before 9/11, Australia had not implemented many of the baseline ICAO standards; however, the country also has not experienced significant attacks on its aviation structure.

Australia's aviation system is the responsibility of the Transport Security Department of the Australian government and reports to the minister for Transport and Regional Services (under the Department of Transport and Regional Services, or DOTARS). The Aviation Security Branch in DOTARS is responsible for the administration of the aviation security provisions of the Aviation Transport Security Act 2004 and Australia's Aviation Transport Security Regulations. This is somewhat consistent with the ICAO recommendation that there be a national government entity responsible for aviation security. As in Canada and in the pre-9/11 United States, the Australian security department reports to an overall transportation agency versus a security or justice agency.

The Aviation Transport Security Act 2004 created the role of an inspector of Transport Security. The inspector is responsible for investigating security incidents, both in the aviation and maritime community. Many of the changes the act brought to Australian aviation security are similar to U.S. measures, including an air marshal force to protect aircraft in flight, adding the requirement of 100% checked-baggage screening, improving passenger and carry-on baggage screening technology, mandatory background checks for aviation personnel, increased funding for law enforcement officers at Australian airports, security measures for general aviation charter operations, and expansion of its canine explosives detection programs.

Australia also conducts a periodic audit titled the Aviation Security Risk Context Statement (AVSRCS). AVSRCS assesses current security risks and the environment of aviation security in the country.

Many of Australia's aviation security regulations mirror those of the United States, Canada, and the European Union as they follow the general ICAO principles for establishing an aviation security manager or director; implementing an ASP; establishing security-restricted areas and personnel identification systems; and passenger, baggage, and cargo screening practices. Significant attention is paid to conducting security risk assessments at Australian airports and aircraft operators, more so than in the other countries we have assessed.

Specifically, after 9/11 Australia's government focused on tighter controls on air cargo, screened all laptop computers, and screened all property and personnel entering sterile areas. In 2002, Australia's deputy prime minister for transport announced additional security measures including the screening of passengers and carry-on baggage at more airports, 100% checked-baggage screening for all international flights and domestic flights by the end of 2004, the placement of ETD technologies at screening checkpoints, the use of threat image projection (TIP) screening equipment, and an upgrade to aviation security at Christmas Island Airport and Cocos (Keeling) Islands Airport.

In 2003, Australia expanded its aviation security identification card (ASIC) to all airports where passenger screening is required and for access to other airport-related security-sensitive areas, such as fuel facilities and critical air control facilities. Australia reissued all ASICs using tamper-evident technology and required ASIC holders to undergo security screening by the Australian Security Intelligence Organization (ASIO) to supplement existing criminal history checks. Aircraft operators were also required to harden their cockpit doors and protect access to the flight deck.

The most recent activity related to aviation security in Australia is the Securing Our Regional Skies funding package. This initiative provides for Regional Australian Federal Police Protective Service Rapid Deployment Teams; new screening capability for 146 regional airports; a joint training and exercise program involving state, territory, and federal police; CCTV trials; improved security training for regional airline and airport staff; an aviation security public awareness campaign; and additional funding for hardened cockpit doors.

Aviation security in other developed nations of the Asia-Pacific Rim follows closely with ICAO guidance and in some areas exceeds international standards. In Japan, some airports conduct checked-baggage screening before passengers arrive at the ticket counter, which may help to prevent an explosive device from being detonated in the terminal building. Also, in 2004, the Tokyo Narita Airport was already testing liquid explosives detection technology to determine if liquids being taken onboard aircraft were flammable, without unduly inconveniencing the traveling public.

Conclusion

Aviation security policies are largely developed through the guidance of the International Civil Aviation Organization's Annex 17, the security standards and recommended practices of the ICAO, and the passage of key legislative measures. Historically, many of the guidelines, policies, and laws created by these organizations and processes have been reactive to a recent attack on aviation.

The ICAO is an outgrowth of the United Nations that regulates civil aviation operations around the world. However, the ICAO is very limited in its enforcement of those operations and its own standards. The ICAO is based in Montreal, Canada, and has initiated four key conventions covering aviation security policy measures (Tokyo, The Hague, Montreal pre-9/11, and Montreal post- 9/11).

The development of aviation security policy in the United States began in the early 1970s in response to many hijackings of civil aircraft. The major policy developments occurred in 1990 following the downing of Pan Am Flight 103 and PSA Flight 1771, the crashing of TWA Flight 800 in 1996, and again in 2001 after the 9/11 attacks. Policy making on aviation security has continued at a record pace since the passage of ATSA 2001, with new policies being created in various funding acts and legislation, Homeland Security legislation, and intelligence reform bills.

Internationally, aviation security largely follows ICAO guidance and is similar to U.S. aviation security policy making. Newly created legislation related to aviation security generally follows a recent terrorist incident or breach of security.

There has been a notable change in aviation security policy making since the 1970s. Previously, nations would only create new policies in light of a recent terrorist attack. After 9/11, there has been a trend toward proactive policy making in the United States and around the world.

New forms of attack must be constantly assessed, such as suicide bombings in a terminal building, MANPAD attacks, and radiological dispersion devices. However, there remain significant lessons that should be learned from the overall history of hijackings, bombings, and airport attacks on aviation. It is important to look forward in defending against new forms of attacks, but the fundamentals of protecting aircraft and airports from attack must always remain in place.

References

Abeyratne, R.I.R., 1998. Aviation Security: Legal and Regulatory Aspects. Ashgate, Brookfield, VT.

ATSA, 2001. Aviation and Transportation Security Act of 2001, P.L. 107–71 Sec. 110.

GAO, 2001. Aviation Security: Terrorist Acts Demonstrate Urgent Need to Improve Security at the Nation's Airports. U.S. General Accounting Office, no. GAO-01-1162T, retrieved July 9, 2008, from www.gao. gov/new.items/d011162t.pdf2001.

globalsecurity.org, 2001. Subcommittee on Aviation Hearing on Aviation Security and the Future of the Aviation Industry, retrieved Aug. 21, 2006, from www.globalsecurity.org/security/library/congress/ 2001_h/010921-memo.htm2001.

Gore, A., 1997. White House Commission on Aviation Safety and Security, Final Report to President Clinton, retrieved July 8, 2008, from www.fas.org/irp/threat/212fin~1.html.

ICAO, 2006. Background, Aviation Security Audit Section (ASA), International Civil Aviation Organization (ICAO), retrieved Aug. 15, 2006, from www.icao.int/cgi/goto_m_atb.pl?icao/en/atb/asa/Background. htm2006.

Kean, T.H., Hamilton, L.H., 2006. Without Precedent: The Inside Story of the 9/11 Commission. Knopf, New York.

Kirkeby, C., 2006. Homeland Security Act of 2002, Class Brain, retrieved July 18, 2008, from www.classbrain. com/artteenst/publish/homeland_security_act.shtml2006.

Price, J., 2003. Airport Certified Employee (ACE)—Security Module 1, vol. 1. American Association of Airport Executives, Alexandria, VA.

Price, J., 2004. Airport Certified Employee (ACE)—International Aviation Security, Module 1, vol. 2. American Association of Airport Executives, Alexandria, VA.

Thomas, A., 2003. Aviation Insecurity. Prometheus Books, New York.

Wallis, R., 1993. Combating Air Terrorism. Brassey's, New York.

The Role of Government in Aviation Security

Objectives

An essential responsibility for aviation security professionals is to understand and have a current working knowledge related to the functions of government in aviation security. This chapter introduces various roles and interactions that federal, state, and local government agencies have with airport security practitioners and other aviation-related law enforcement agencies. Readers will gain essential knowledge regarding the security functions of the Department of Homeland Security and the Transportation Security Administration. Special emphasis is placed on understanding the development and implementation of transportation security regulations (TSRs) and other relevant forms of legislation. Students will gain an understanding of how government policy and regulations in aviation security affect the job responsibilities of various airport security personnel. Some security regulations are addressed in this chapter, while others relating directly to airport, aircraft operator, and air cargo security are addressed in their own chapters.

Introduction

The International Civil Aviation Organization (ICAO) recommends that each nation should have a national government organization charged with providing internal national security. For example, this type of national security agency in the United Kingdom is the Home Office, which has long handled a wide range of security responsibilities, including customs and immigration enforcement. In the United States, the Department of Homeland Security (DHS) is charged with providing internal national security for aviation, customs and immigration, the U.S. president, specific federal facilities, and the U.S. coastline. Responding to major natural disasters or terrorist attacks is also a security-related responsibility of the DHS.

As the regulating agency for U.S. homeland security, the DHS oversees the U.S. Transportation Security Administration (TSA). In turn, the TSA regulates transportation security in the United States, which includes rail, trucking, shipping, and aviation. Since its establishment shortly after 9/11, the TSA's primary focus has been aviation security; while attacks on rail and maritime systems are possible, aviation has remained the primary target. The TSA provides direction to airports and aircraft operators on compliance with federal regulations related to aviation security, embodied in Title 49, Part 1500, of the U.S. Code of Federal Regulations. The regulations relate to airport and aircraft operator security, air cargo and general aviation security, and security for foreign air carrier operations operating in the United States. The TSA also oversees airline and airport regulatory compliance and conducts the screening at most U.S. airports.

Before 9/11, the U.S. Department of Transportation (DOT) through the Federal Aviation Administration (FAA) managed internal aviation security. The FAA employed a few hundred aviation security inspectors to oversee the security programs of airports and aircraft operators. Many FAA aviation security employees transferred to the newly formed TSA after 9/11. The TSA hired more than 50,000 employees in its first two years of operation; most were screeners hired for passenger and baggage screening. Originally structured under the DOT, the TSA moved to the DHS in 2002 to help insulate it from the lobbying pressures of the aviation industry and bring together those agencies charged with homeland security.

Within the United States, each state has a department or division of transportation. Most also have a division of aeronautics focusing on aviation system planning and coordination of airport and airspace issues with the FAA. States can provide guidance and direction to airport operators and often provide funding for airport capital improvement projects, independent of federal funding.

Some states, such as Massachusetts and Alaska, directly manage and operate their airports as part of the state transportation system, but this is the exception. Most airports are locally operated by a municipality or airport authority and managed by its employees.

Transportation Security Regulations

After 9/11, the U.S. federal aviation regulations governing aviation security were transferred to the TSA. These are now under Title 49 CFR, Part 1500 series. They are listed in Table 4.1.

Part 1500: Aviation Security

TSR Part 1500 contains general terms and abbreviations associated with transportation security regulations. Some key terms include:

Administrator means the under secretary of transportation for security identified in 49 U.S.C. 114(b) who serves as the administrator of the TSA.

Table 4.1   TSA Regulations Pertaining to Aviation Security

TSRs Covered in This Chapter	TSRs Covered in Chapter 5
Part 1500 Aviation Security	Part 1542 Airport Security
Part 1503 Enforcement Actions	Part 1544 Air Carrier Security
Part 1520 Sensitive Security Information	Part 1546 Foreign Air Carrier Security
Part 1540 General Rules of Security	Part 1548 Indirect Air Carrier Security (Air Cargo)
Part 1542 Airport Security	
Part 1550 General Aviation Operations	

Person means an individual, corporation, company, association, firm, partnership, society, joint-stock company, or governmental authority. It includes a trustee, receiver, assignee, successor, or similar representative of any of them.

Transportation Security Regulations (TSRs) mean the regulations issued by the TSA, in Title 49 of the Code of Federal Regulations, Chapter XII, which includes parts 1500–1699.

Part 1503: Enforcement Actions

Part 1503 covers enforcement and the process for opening and prosecuting a case against a regulated party. Outcomes can be fines to a guilty individual or entity or, rarely, criminal prosecution. Fines are based on the nature of the violation and variables such as the experience of the violator with the aviation system.

Violations can be of minimum, moderate, or maximum severity. Each level has its own structure of fines depending on whether the offender is an individual, an airport operator, an aircraft operator, or a cargo agent. Fines per incident for aircraft operators range from $2,500 for a minimum-level violation to up to $27,500 for a maximum-level violation. Those for airports and cargo agents range from $1,000 to $11,000 per incident, and fines for individuals range from $250 to $11,000 and higher, per incident.

Aggravating and mitigating factors are considered in determining the level of fine, including the potential risk to security associated with the violation, the nature of the violation (inadvertent, deliberate, negligence), the individual's or entity's history of violations, the violator's level of experience, and the attitude or disposition of the violator. The violator's character or demeanor may also influence the way violations are handled. Table 4.2 lists some common violations.

Aggravating factors in determining the severity of a violation for individuals include concealing a prohibited or illegal object; potential severity, type, or number of weapons or explosives held illegally; and evidence of intent to interfere with or test the aviation security system. Mitigating factors may include voluntary disclosure, whether the violator is a juvenile, and whether other penalties have been assessed by civil or criminal courts.

The TSA publishes an "Enforcement Guidance Sanction Policy" that fully articulates the various offenses, aggravating and mitigating factors, and the fine structure. It is available on their public website.

Aside from people attempting to bring prohibited items onto an aircraft, usually inadvertently, the most common violations stem from airport and aircraft operators who must uphold the regulatory standards of their security programs. These violations constitute a significant portion of those issued by the TSA. The TSA usually issues a violation after the TSA has brought the problem to the attention of the party for rectification and without an immediate enforcement action. Enforcement actions usually result if the problem is repeated or is of such an egregious nature that immediate enforcement is deemed necessary.

When an enforcement action is initiated, a TSA compliance inspector dispatches a letter identifying the specific violation and requesting an explanation and correction.

Table 4.2 Common Violations

Individuals	Airport Operators	Aircraft Operators	Air Cargo
Entering sterile areas without being screened	Failure to ensure airport security coordinator (ASC) fulfills duties	Refusal to carry federal air marshals	Failure to produce a copy of the security program
Failure to undergo secondary screening when directed	Failure to train ASCs	Failure to pay security fees	Failure to supply certification to the aircraft operator
Improperly entering security identification display areas (SIDAs) or airport operation areas (AOAs)	Failure to allow the TSA to inspect an airport	Failure to prevent unauthorized access to secured area or to aircraft	Failure to meet requirements for accepting cargo from an all-cargo carrier with an approved security program at a station(s) where cargo is accepted or processed
Improper use of access media	Failure to carry out a security program requirement	Failure to comply with requirements for carriage of an accessible weapon by an armed law enforcement officer	Failure to transport cargo in locked or closely monitored vehicles
	Failure to notify the TSA of changes in the security program		

The TSA then adjudicates the process. Violators have the right to appeal through an administrative law judge or the judicial process.

Part 1520: Sensitive Security Information

The Air Transportation Security Act of 1974 created sensitive security information (SSI) by allowing the FAA to create a method for sharing intelligence information with airport operators and air carriers. After 9/11, SSI was expanded into all forms of transportation and moved into Title 49 CFR Part 1520.

Part 1520 addresses the control and handling of SSI contained in documents such as the TSA-approved security programs (e.g., airport security program, Aircraft Operator Standard Security Program), TSA security directives, TSA information circulars, technical specifications of security equipment, vulnerability assessments, and certain records and maps. The rules governing the use and dissemination of SSI material apply to all regulated parties and anyone receiving SSI information. Common types of SSI documents at an airport are the airport security program (ASP), security directives (SDs), and information circulars (ICs). The ASP includes within its appendices security contingency plans and incident response plans.

SSI documents must be strictly controlled with distribution and dissemination restricted to those with an operational "need to know" (defined as needing the information to do the job, train others to do the job, or supervise those doing the job). SSI fits into a category considered "sensitive but unclassified." While SSI information does not meet the criteria for "classified" material, such as those materials used by the U.S. military and federal government (notably, "confidential," "secret," and "top secret"), it is not intended nor should be publicly disseminated.

SSI is defined as information obtained or developed that, if released publicly, would be detrimental to transportation security. SSI examples include the no-fly and the selectee lists, screening checkpoint standard operating procedures, ASPs, and the Aircraft Operator Standard Security Program.

SSI is information obtained or developed in the conduct of security activities, including research and development, the disclosure of which TSA has determined would be detrimental to the security of transportation.

The 16 categories of SSI, as related to aviation security, are:

1. Security programs and contingency plans (the ASP and the Aircraft Operator Standard Security Program).
2. Security directives (SDs).
3. Performance specifications (checkpoint or checked-baggage screening equipment).
4. Information circulars (ICs), which include information advisory in nature about a potential threat.
5. Vulnerability assessments.
6. Security inspection or investigative information (including incidents, violations, or inspections that might reveal a security vulnerability).
7. Threat information.
8. Security measures (access control measures recommended by the federal government, federal air marshal deployment and the operation of federal flight deck officers, or FFDOs).
9. Security screening information (screening procedures, no-fly and selectee lists, security screener tests and results, performance data from screening equipment, electronic images on TSA-owned screening equipment).
10. Security training materials (e.g., SIDA training records).
11. Identifying information of certain security personnel (individuals issued a SIDA badge, transportation security officers, federal air marshals, and FFDOs).
12. Critical aviation infrastructure asset information (any list identifying systems or assets, physical or virtual, that is vital to the aviation system, which the incapacity or destruction of such would have a debilitating impact on transportation security).
13. Systems security information, including any information involving the security of operational or administrative data systems operated by the federal government and critical to aviation security (including automated security procedures and systems).
14. Confidential business information (bid information submitted to DHS or DOT, trade secret or commercial information, or financial information requested by DHS or DOT).

15. Research and development.
16. Other information not otherwise described that TSA determines is SSI.

Anyone creating SSI must include an SSI header and footer. Even if only one sentence in a document or presentation includes SSI, every page must be marked as SSI. The following paragraph is the approved language for marking each page as SSI:

> *WARNING: This record contains Sensitive Security Information that is controlled under 49 CFR Parts 15 and 1520. No part of this record may be disclosed to persons without a "need to know," as defined in 49 CFR Parts 15 and 1520, except with the written permission of the Administrator of the Transportation Security Administration or the Secretary of Transportation. Unauthorized release may result in civil penalty or other action. For U.S. government agencies, public disclosure is governed by 5 U.S.C. 552 and 49 CFR Parts 15 and 1520.*

The ASP presents a good example of the critical nature of SSI documents. The ASP contains systems, methods, and procedures for how an airport conducts its security. The number of full-version copies should be kept to a minimum to reduce the potential that the information within the document will be compromised.

Before accepting SSI information, recipients should have signed a current version of the DHS Nondisclosure Form 11000-6. If there is an update to an SSI document such as an ASP, the original copies should be returned to the security manager for destruction, usually via shredding or controlled burning.

Airport and aircraft operator security personnel often must disseminate information contained in TSA-issued security directives, which are considered to be SSI, so it is essential that personnel understand the dissemination, marking, and control standards for SSI. When a SD is issued, often it contains data meant for multiple audiences, such as airport police, the badging and identification office, airport operations personnel, and others. To properly disseminate the SSI data within an SD, the airport or aircraft operator security manager often will create their own SSI memorandums, and include the information applicable from a SD to each individual work group (i.e., police, badging office, etc.).

Determining the distribution of the ASP can be politically sensitive for the security manager. Often people within an airport governing structure feel they should have a copy of the document, although they do not have a valid need for it. Typically, the airport director, airport operations manager, and security and law enforcement personnel receive copies of the ASP. Air carrier station managers may also receive a copy of the complete ASP.

Other airport tenants, vendors, and contractors should receive a desensitized version of the ASP. These edited ASP reports are sometimes referred to as participant manuals. Participant manuals include non-SSI information about the ASP important to airport tenants, vendors, and contractors. Information such as processes for filling out airport ID badge applications; point-of-contact information for airport security, operations, and law enforcement personnel; hours of operation for the security office; and procedures for deleting an employee's access from the security system when one of their employees no longer works for the company are typical topics included in the participant manual.

The federal security director has final authority regarding SSI and should be consulted regarding any questions related to its distribution. Most SSI data come from the federal government or is approved by the federal government, such as TSA approval of an airport or aircraft operator ASP.

Part 1540: General Rules of Security

Part 1540 covers the terminology of aviation security and individual accountability. This section addresses the security responsibilities of employees, passengers, and others; interference with screeners or any aviation employee with security duties; the carriage of weapons and explosives onboard and in checked baggage; and the inspection of FAA airman and medical certificates.

Anyone applying for access to a SIDA is prohibited under Part 1540.103 from falsifying an application or any other record related to the issuance of security access media. Anyone who has passed a criminal history record check (CHRC) and is subsequently arrested for a disqualifying crime must report this to the issuing agency.

Part 1540.105 requires travelers and aviation employees to conform to security regulations. It prohibits individuals from tampering with airport access control systems or illegally entering airport security areas. Before 9/11, anyone who accessed an airport's security area illegally could, at most, be charged with trespassing, with a possible subsequent investigation by the TSA. In most cases, trespassing is a misdemeanor. After 9/11, Part 1540 made trespassing within airport security areas a federal crime. Part 1540 prohibits access to airport security areas through an unapproved access point. Although aviation employees may have approved access, they are still required to enter and exit security areas through approved checkpoints.

It is also a violation to tamper, interfere, compromise, modify, or circumvent any security system, measure, or procedure. A violation of this sort might be a maintenance person working on or near a security access control door who, without permission, props the door open to facilitate work.

It is a violation under this section to falsify an airport ID badge or other access media. This extends to affixing stickers or personal mementos to airport and airline identification badges.

Intentionally violating a security regulation or policy to test the system is addressed in Part 1540. Authority to test security systems extends *only* to TSA security inspectors and those designated in the ASP, not to any individual with an approved airport or airline identification badge, including transportation security officers (i.e., TSA screeners) and flight crewmembers. The public, including the media, may not violate security policies and practices and claim immunity by declaring they were conducting a test of the security system.

Sometimes, the federal government may use other personnel such as red team[1] members, employees from the Government Accountability Office, and personnel under contract to the airport or aircraft operator to conduct vulnerability assessments, in which case the airport security coordinator may amend the ASP to include these individuals.

There have been occurrences of airport, airline, or government personnel who have violated security measures by believing that security protocols should not apply to them. These people, when caught, often use a "testing the system" statement as a defense. This is akin to an off-duty police officer declaring to an on-duty officer who has pulled him or her over for speeding that he or she was just "testing" the other officer. These situations have occurred with enough frequency to cause valid security concerns. Security systems can become weakened or fail if employees are exempted from following security measures.

Part 1540.107 requires individuals to submit to airport screening before entering sterile areas. This applies to passengers and aviation employees who access sterile areas through screening checkpoints.

Part 1540.109 protects against interference with screening personnel and extends to any aviation employee with security responsibilities, including anyone with security area access. Acceptance of such access ability brings with it the responsibility to report any suspicious activity and violations of security rules. This regulation also prohibits assaulting, threatening, or intimidating individuals with security duties.

Under Part 1540.111, it is a violation to have a prohibited item, once screening has begun, while in sterile areas or when attempting to board or be onboard an aircraft. Airport screening normally "begins" when an individual's carry-on bags enter the X-ray machine or when the person steps through the metal detector, advanced imaging technology (AIT), or other approved screening device. However, with the deployment of TSA behavior detection officers (BDOs), there is some debate as to when screening officially begins, since the BDOs normally operate in the public areas of the terminal building. Additionally, there is also some debate whether screening officially begins at the travel document stations where TSA personnel check the identification of passengers and their boarding documents. A final definition has not been clarified, but TSA has encouraged airports to post signs throughout the public areas warning individuals that they may be subject to screening. There was also debate at some airports where local concealed or open-carry weapons laws lawfully allowed the carriage of weapons in the public areas of a government facility, including airports, which has created more questions about where screening lawfully begins and when does an individual voluntarily submit to a screening process.

It is necessarily redundant to have restrictions regarding prohibited items when going through screening, in sterile areas, during boarding, or while on an aircraft. The reason for this is that there are small commercial-service operations and airports in the United States where flights are not screened and where sterile areas do not exist. Therefore, the

---

[1]*Red teaming* is an industry term for inspections and tests conducted outside of the normal testing parameters using tactics more in line with those of a criminal or terrorist.

regulation extends to cover these types of airports by prohibiting the possession of a prohibited item while boarding or onboard any aircraft. The rule extends to aviation employees. Airport and airline personnel who must carry prohibited items as part of the job, such as airline mechanics, must bring those items into the security areas and transport them to the worksite.

Title 49 CFR Part 1540 is the basis for the prohibited items list. Part 1540.111 prohibits the carriage of loaded firearms in checked baggage. Explosives and incendiaries are also prohibited; firearms must be declared to the air carrier during baggage check-in and stored in a hard-sided container with TSA-approved locks.[2] Before 9/11, some airlines would place large red stickers on the outside of firearm cases as a safety precaution, which also identified the firearm to thieves.

Because general aviation pilots do not carry airport-issued identification badges, the regulations provide a way for transportation security inspectors to determine whether a general aviation pilot has a legitimate need to be in the general aviation area of an airport. Part 1540.113 grants TSA personnel the right to request examination of the pilot's airman certificate (pilot's license) and required FAA medical certificate. However, it does not give the TSA the right to confiscate these documents. Instead, if the documents are to be confiscated, an official from the FAA's Flight Safety District Office must be called to do this.

How Regulations are Changed

Regulations related to aviation security are generally initiated through laws passed by Congress, such as the Aviation and Transportation Security Act of 2001. A government agency then drafts the regulations with or without public comment. In the case of the transportation security regulations passed after 9/11, there was little chance for public comment, but the majority of the security regulations were already in place under the jurisdiction of the FAA. Once in place, there are several ways that the federal government can change the regulations and policies related to aviation security. These include security directives, notices of proposed rulemaking, and amendments to airport and air carrier security programs.

Notice of Proposed Rulemaking

Besides the passage of laws, regulations are changed through the notice of proposed rulemaking (NPRM). The TSA drafts then issues a proposed rule to the U.S. Office of Management and Budget (OMB) for approval. If the OMB approves, then the rule is published in the *Federal Register*, a public document that describes what the U.S. federal government intends to do or is doing. The listing in the *Federal Register* provides a specified period for

---

[2]After 9/11, several companies began selling locks that include a universal access key lock in addition to the user's key or combination lock. TSA screening personnel have the universal access keys for the locks and thus may access an individual's bag without compromising the user's lock.

public comment.[3] After the TSA has received the comments and the comment period has ended, the agency will make modifications (if deemed necessary), then publish the final rule in the *Federal Register*. When a proposed regulation is published a second time in the *Federal Register*, it becomes law.

In emergencies, regulations can be drafted and immediately implemented. The temporary flight restrictions used by the federal government throughout the United States after 9/11 is an example of an emergency rulemaking. The benefit of the NPRM process is that it allows for input from an industry before making a regulation statutory, providing opportunity for feedback to proposed rules. This enhances the usefulness and viability of the proposed regulation.

Changing an Airport or Aircraft Operator Security Program

Because of the changing nature of the aviation industry, airport operators and federal officials may be required to change an airport or air carrier security program. These programs are generally changed through a permanent or temporary amendment to an ASP, which can be initiated by the regulated party (airport or airline) or the TSA, a notice of a changed condition that affects security, or a security directive issued by the TSA.

U.S. Regulation Part 1542.105 Approval and Amendments addresses the methods for amending an ASP. Amendments are generally permanent changes to the program versus a "changed condition affecting security," which is usually a temporary condition.

Many factors can cause an airport or air carrier to change its federally approved security programs. Airfield construction, new security threats, intelligence pointing to a potential attack on aviation, staffing problems at an airport, maintenance issues on airport and air carrier security systems, and even natural disasters can result in changes to security programs.

An airport or air carrier operator uses an amendment when the operator desires to include formally a new process in its approved security program. Amendments are often in response to changing systems, measures, or procedures, such as the format of the airport's access control system or badging system; the addition of new air carriers or access gates; changes to the geographical boundaries of the security areas; an appointment of a new airport security coordinator; or changes to the law enforcement requirements, contingency plans, or incident management procedures.

If an airport or air carrier desires to amend its ASP, it will draft the proposed change and then provide it to the TSA at least 45 days before the effective date. The TSA needs this time to review the proposed change(s) and any potential effects the change will have to the overall security posture. The TSA must respond within 30 days of receiving the proposed amendment, either by accepting or denying it. If denied, the operator has 30 days to appeal.[4]

---

[3]Often, industry organizations play a significant role in this phase of the rulemaking process as the organizations represent large memberships.

[4]As a best practice, when the airport security coordinator desires to amend the ASP, he or she commonly will hold discussions with the TSA to address any questions or concerns before formally submitting the amendment.

On occasion, airport and air carrier operators draft temporary amendments not intended to become part of the permanent ASP—for example, for airfield construction or emergency exercises.

During airfield construction, airport operators may be required to remove parts of the perimeter fence, install temporary access gates, employ temporary personnel to control airfield access, and allow contractors to hire personnel who might not undergo required airport badging. These changes would clearly not be allowed under an ASP, so the airport operator will endeavor to create an amendment consisting of alternative procedures and demonstrate that the security of the airport can be maintained. Alternatives may include allowing individuals to be escorted by badged personnel, who must stay onsite at all times construction is occurring. Another alternative is to cordon off the construction site using temporary fencing to deter nonbadged individuals from accessing the security areas.

Emergency exercises are another reason for a temporary amendment. During many aircraft emergency exercises, there are dozens and sometimes hundreds of volunteers and participants who do not have badged access authority. For this situation, the security coordinator may exclude part of the airfield from the other airport security areas for the duration of the exercise. The security coordinator must demonstrate in the amendment how the security of the remaining security areas will be maintained and how the exercise area will be restored to a secure condition after the exercise is complete.

The amendment process does not prohibit airport or air carrier operators from increasing their own security measures in response to imminent threats. An operator may add security measures without including them into the ASP. In fact, many airport and air carrier operators draft security programs to meet the minimum standards that the TSA will allow, but then exceed the standard in actual operating practices. Airports and airlines often want to meet higher security standards but understand that any system, measure, or procedure approved as part of their TSA-approved security program must be adhered to or the entity is in potential violation and may be fined for noncompliance.

If the TSA wants to amend an operator's security program, the operator has 30 days to comment on the proposed amendment. After this time, the TSA implements or rescinds the amendment. The amendment is good until it expires. The operator may appeal the amendment but must comply with the requirements and conditions of the amendment during the appeal. This type of amendment is generally not done to address an imminent threat but more a potential weakness in the system that could be eventually exploited.

Although many amendments pertain only to one airport, some TSA amendments apply to all security programs at all regulated entities. In 2007, the TSA attempted to issue an amendment to incorporate all of the procedures required in all of its previously issued security directives. The amendment also included additional procedures and was met with resistance from airport and air carrier operators. The TSA rescinded the proposed amendment later that year.

In the 25 years preceding 9/11, there were a handful of significant amendments to aviation security regulations. When the federal aviation regulations pertaining to security were transferred to the TSA, many of FAA amendments, which had never been formalized into regulations, were incorporated into the new TSRs. A few older FAA amendments were not embodied in the TSRs but remain in effect, including preventing the use of public storage lockers in nonsterile areas, requiring airport operators to continuously notify travelers not to leave baggage unattended, and requiring airport security coordinators to coordinate security efforts with air carriers.

The TSA can also issue emergency amendments to security programs. An emergency amendment is issued when there "is an emergency requiring immediate action with respect to safety and security in air transportation or in air commerce that makes procedures in this section contrary to the public interest," which allows an airport no review time and must be implemented when the TSA specifies (Part 1542.105(d)). Again, the airport may appeal the emergency amendment but must comply with its requirements and conditions in the meantime.

Changed Condition Affecting Security

Airports are dynamic environments and security challenges and changes can be frequent, particularly at the larger airports. Access control doors go out of service, vehicle gates malfunction, and people inadvertently enter the security areas. For some of these incidents, such as a breach of security, the airport operator initiates the appropriate law enforcement response (if necessary) and notifies the TSA immediately. A changed condition occurs when a gate goes out of service, or some other condition on the airport changes, causing a different condition to exist than what is described in the ASP. A changed condition affecting security (Part 1542.107) occurs when a security problem causes an airport or air carrier to go out of compliance with its security program and temporary measures cannot be implemented to maintain it at the level called for in the ASP. For instance, if the number of law enforcement officers available at an airport is reduced, such as might happen when police officers are supplied by a municipality and the municipality decides to reduce the number to below that required in the ASP. Under the regulations, the airport operator must notify the TSA verbally within 6 hours of the change and by written notice within 72 hours. The airport security coordinator must submit an amendment to the TSA within 30 days if the changed condition is projected to continue for more than 60 days.[5] The airport must also advise the TSA of actions taken to ensure that the same level of security is maintained during the change, along with a date or time the airport expects the problem to be resolved.

Changed condition notifications are not generally needed when a situation occurs to affect the security posture, but immediate action is taken to ensure that security is not compromised. An example would be a vehicle access gate going out of service. The airport

---

[5]Airport operators should remember that the TSA has discretion determining when the 30-day timeline begins. Some TSA personnel believe the 30 days begins when the change first occurred, whereas others believe the 30 days begins when the airport operator becomes aware that the change will last longer than 60 days. The best practice according to several TSA security inspectors is to keep the federal security director and TSA-compliance personnel in the information loop on any changed condition.

must first ensure that no unauthorized entry occurred. If it did, then it is an incident with mandatory TSA notification and action by the airport to find who may have entered the security areas. If there was no unauthorized entry and appropriate procedures were followed to restrict access (such as posting a security guard in contact with security personnel in the security operations center to verify those with authorized access), security is not compromised—although a different procedure is in place other than the procedure in the security program.[6] However, some TSA inspectors and federal security directors may require a changed condition notification anytime there is any unplanned alteration to the way the airport is carrying out its security measures that differ from what is articulated within the ASP, regardless of whether the same level of security is still maintained.

Intelligence and Intervention

In the June 22, 2007 issue of *Joint Pub 0-2*, intelligence is described as follows:

> *Intelligence is the product resulting from the collection, processing, integration, evaluation, analysis, and interpretation of available information concerning foreign nations, hostile or potentially hostile forces or elements, or areas of actual or potential operations. The term is also applied to the activity which results in the produce and to the organizations engaging in such activity.*

The intelligence community (IC) is a collection of executive branch agencies that work separately and together to conduct intelligence activities necessary for the conduct of foreign relations and to protect the national security of the United States (Interagency Threat Assessment & and Coordination Group, 2012).

Intelligence and intervention related to aviation security are most often the responsibility of national governments. In *How Safe Are Our Skies?* Rodney Wallis (2003) faulted the 9/11 attacks on the failure of U.S. government intelligence:

> *The government controls the intelligence services, and intelligence is vital if democracies are to be guarded against terrorism, whether land-based or airborne. There is every reason to believe the intelligence services failed the American people over the September 2001 attacks. It became clear immediately after the events that information had been circulating but had not been acted upon. Names of the hijackers were known to the CIA and FBI, yet these criminals were allowed to move freely within the United States and board aircraft with little or no special attention paid to them. (p. 19)*

---

[6]Some TSA personnel will still want verbal and written notification of any change in airport security. These issues are at the discretion of the federal security director and, in some cases, the transportation security inspectors.

The DHS manages security intelligence and intervention strategies. These efforts serve as the outermost layer of the U.S. aviation security system. DHS security functions include gathering intelligence, monitoring terrorist and criminal activities, and, if necessary, intervention before a potential attack on the United States. The DHS works with the TSA to notify airport and airline security managers of potential targets and terrorist strategies.

In the United States, the TSA and Federal Bureau of Invetigation (FBI) disseminate aviation security information to airports and airlines that enables them to implement precautionary security measures, often referred to as contingency plans. Each commercial-service airport has an FBI agent assigned to coordinate aviation security activities and share intelligence with the local airport and airline operators. The TSA's federal security director also receives intelligence information and works with the FBI and local airport and airline personnel to develop and implement contingency plans.

U.S. intelligence agencies and military services conduct most of the nation's intelligence gathering and early intervention related to aviation security. However, intelligence analysts are always careful to note that there is a distinction between *information* and *intelligence*, and the even more evasive *actionable intelligence*, which is information that enables U.S. military or law enforcement assets to take action (arrest or capture/kill). Further, the U.S. intelligence community consists of numerous organizations and agencies, often with overlapping missions. More often than not, intelligence trickles in looking more like tidbits of information, until eventually a picture begins to form of potential criminal or terrorist activity. Many agencies can have different pieces of a very large puzzle, and those pieces are not always accurate or shared across domains.

The intelligence cycle includes five steps:

1. Planning and direction
2. Collection
3. Processing and collation
4. Analysis and production
5. Dissemination

Planning and direction focuses on intelligence and law enforcement agencies on a particular direction or threat(s). Collection includes the collection of relevant information about the threat, from a variety of sources, which could include the Department of Defense (DOD), the Department of Justice (DOJ), DHS, other federal agencies such as the National Geospatial Intelligence Agency, the CIA, FBI, DEA, and state, local, and tribal organizations. Processing and collation attempts to take multiple pieces of information from these sources and put them together to form a better picture of terrorist or criminal activity. Analysis and production includes preparing reports to disseminate back to the intelligence and law enforcement agencies, and dissemination is the process of distributing the information to those entities. It is the aviation security practitioner who needs to be both part of the contribution of information to this system, and a consumer of its products.

Failure to review and analyze international or any off-airport security issue is negligent aviation security management. Airport operators frequently rationalize that there is little they can do at the local level to mitigate threats outside airport boundaries or their internal operations. Just as an airport operator would work to solve a safety issue or environmental problem on their airport, they can and should take preventive security measures when international security threats are detected or anticipated. These proactive measures include using increased levels of law enforcement personnel, increasing surveillance of the airport perimeter, implementing random increases in security measures either temporarily or permanently, or a variety of other strategies. It is essential that any aviation security manager practice due diligence and keep appraised of current and developing threats.

The analysis of intelligence and application of its meaning and results can contribute significantly to the protection of assets and personnel. There are six types of intelligence (Interagency Threat Assessment & Coordination Group, 2012):

1. Signals intelligence (SIGINT) is the exploitation of electronic emissions information, which is derived from four sources: electronic, communications, foreign and weapons-related command, and control signals. Email, radio intercepts, and cell phone monitoring typically fall into this category.
2. Imagery intelligence (IMINT) is the product of processing raw images, in the form of pixels, digits, or other forms, and the attempt to determine the time, date, and place that the imagery was obtained.
3. Measurement and signature intelligence (MASINT) is scientific and technical intelligence (metrics, angles, spatial, wavelength, etc.) derived from sensors to detect identifying distinctive features associated with the source.
4. Human-source intelligence (HUMINT) is derived from human beings who may be both sources and collectors of information, either by direct observation and the use of recruited agents and, in some cases, interrogation.
5. Open-source intelligence (OSINT) is unclassified information of potential intelligence value and is open to the general public.
6. Geospatial intelligence (GEOINT) is intelligence derived from imagery and geospatial information of physical features and geographically referenced activities on Earth (typically associated with satellite gather information).

There are five categories of intelligence available to the consumer of intelligence (Interagency Threat Assessment & Coordination Group, 2012):

1. Current intelligence: day-to-day events, new developments, related background, and an assessment of their significance to warn of near-term consequences. Presented in daily, weekly, and monthly publications.
2. Estimative intelligence: seeks to assess potential developments that could affect U.S. national security, beginning with facts, then exploring the unknown and the unknowable. Helps policymakers to think strategically about long-term threats.

National intelligence estimates produced by the National Intelligence Council and the director of national intelligence are the most authoritative written assessments of national security.

3. Warning intelligence: sounds an alarm or gives policy makers notice—urgent in nature and implies the need for action or response. Security directives are intended for warning intelligence.

4. Research intelligence: presented as in-depth studies as an underpinning to current and estimative intelligence.

5. Scientific and technical intelligence: information on technical developments and characteristics, and performance and technical capabilities of weapons and security systems. Bomb appraisal officers (BAOs) will often share this type of intelligence with TSOs so they can watch for new types of explosive devices.

Intelligence products that are typically available to front-line response personnel include situational awareness and threat reports and information reports such as an information intelligence report (IIR) or homeland intelligence report (HIR) that disseminate unevaluated intelligence within the IC. Intelligence assessments are a finished intelligence product resulting from analysis; threat assessments provide in-depth analysis related to a specific threat; intelligence bulletins are finished intelligence products used to detect trends in an article format; and a joint intelligence assessment is written by two or more agencies that address the same types of information that can be found in an intelligence bulletin issued by a single-agency (Interagency Threat Assessment & Coordination Group, 2012).

Two notable types of ongoing intelligence material that may be of interest to airport police and security personnel are *Roll Call Release* and *Terrorism Summary* (Interagency Threat Assessment & Coordination Group, 2012). *Roll Call Release* focuses on terrorist tactics, techniques, procedures, terrorism trends, and indicators of suspicious activities (Interagency Threat Assessment & Coordination Group, 2012). *Terrorism Summary* is a secret classified document that includes terrorism-related intelligence available to the National Counterterrorism Center (NCTC) (Interagency Threat Assessment & Coordination Group, 2012).

Handling Intelligence Information

"Controlled unclassified information" refers to information that does not meet the standards for National Security Classification under EO 12958, but is pertinent to the national interests of the United States, and under law or policy requires protection from unauthorized disclosure; typically this type of information is referred to as FOUO, "for official use only."

FOUO should be safeguarded and withheld from public release until approved by the originating agency. "Law enforcement sensitive," is another name for FOUO information. In either case, this type of information should only be shared with those with a need-to-know, and precautions taken to prevent the inadvertent release of the information.

What Intelligence Information Can and Cannot Do

Intelligence can improve decision making, while hindering our enemies decision making. It can warn of potential threats, provide insight into current events, provide better situational awareness, provide long-term assessments on issues of ongoing threats, provide pretravel security overview, and support and provide reports on specific topics based on need (Interagency Threat Assessment & Coordination Group, 2012). However, intelligence information cannot predict the future nor can it violate U.S. law (Interagency Threat Assessment & Coordination Group, 2012).

Aviation security practitioners may obtain unclassified intelligence products through the Homeland Security Information Network—Intelligence, Law Enforcement Online, Intelink-U, Regional Information Sharing Systems Network, DHS Technical Resources for Incident Prevention (TRIPwire), and Federal Protective Service Portal, Open Source Center. Secret-level information can be accessed through proper clearances at NCTC-Online-SECRET, the Office of Intelligence and Analysis, FBINet, and FBI intelink/SIPRNet (Interagency Threat Assessment & Coordination Group, 2012).

Raw intelligence reports, alerts, or notifications provide timely dissemination of unevaluated intelligence. It may be nonspecific, fragmentary, or just constitute suspicious activity. The key to evaluating such information is knowing how the information was obtained, its credibility and reliability, and the type of source. There are two types of access: direct and indirect. Direct access means the intelligence source has direct knowledge of the fact or appears to be in direct contact with those knowledgeable. Indirect access means there is some distance between the source and the origin of the information. Excellent access means the source may have learned about the fact or event from the decision maker or a source document. Good access suggests credibility but no direct access to the information (Interagency Threat Assessment & Coordination Group, 2012).

Also useful in evaluating intelligence is to understand the chain of acquisition, whether the information is credible and reliable, and the level of relationship the IC has with the source. The chain of acquisition speaks to the sources' relationships and the potential changes in the information as it passes from one individual to another. The longer the chain of acquisition, the more likely the information will have changed, which affects its accuracy. Credibility speaks to the extent to which something is believable. The credibility of a source is both time and context dependent. A person may be more credible at certain times and less at others. Reliability is a criterion of credibility and, as applied to the source, speaks to the likelihood that the report is an accurate reflection of the events reported. "Reliable," in this context, means that the established reporting record is judged to be accurate, while "uncertain" reflects that there is uncertainty in the report (and may indicate to what extent), and "unknown" means the information is from a previously unreported source (Interagency Threat Assessment & Coordination Group, 2012).

Sources are categorized into contact, collaborative, established, walk-in (or call, write), or sensitive (Interagency Threat Assessment & Coordination Group, 2012).

- *Contact* means that the information is from someone who has provided information but for whom a formal relationship has not been established.

- *Collaborative* means this likely comes from a newly established source and a formal relationship has been established.
- *Established* sources come from individuals with a previously established relationship and, if revealed, would probably endanger their status, reputation, or security.
- *Walk-in/call-in/write-in* sources are previously unknown individuals who make contact with an official (person, call, website, etc.).
- A *sensitive* source is a source whom, if compromised, might reduce the ability to use the source in the future.

While much threat information is available at the unclassified and confidential levels, sources and methods of gathering information are often classified at "secret," "top secret," or secure compartmentalized TS information.

In understanding estimative language, often the IC will use language such as "we estimate," "we assess," or "we indicate" when they are trying to convey an analytical measurement, based on incomplete fragments of information or information that is not factual. The IC does not use the word "unlikely" to imply that an event will not happen, and uses the words "probably" or "likely" to indicate that there is a greater than even chance something will happen. Words such as "maybe" or "suggest" are used to reflect situations in which the IC is unable to assess the likelihood because relevant information is unavailable, fragmented, or sketchy.

Sometimes the words "high," "moderate," and "low" confidence are used to convey degrees of likelihood. High confidence generally indicates the IC's judgments are based on high-quality information. Moderate confidence generally means that the information is interpreted in various ways or that the information is credible and plausible but not sufficiently corroborated to warrant a higher level of confidence. Low confidence generally means the information is scant and it is difficult to make solid analytic inferences or that the IC has significant concerns with the sources (Interagency Threat Assessment & Coordination Group, 2012).

Operated by the FBI, the National Terrorist Screening Center (NTSC) was established by Homeland Security Presidential Directive 6, which directed that a center be established to consolidate the government's approach to terrorism screening and to provide for the appropriate and lawful use of terrorist information in screening processes. The NTSC began operations on December 1, 2003. The vision of the NTSC is to be the global authority for watch listing and identifying known and suspected terrorists. Its mission is to consolidate and coordinate the U.S. government's approach to terrorism screening and facilitate the sharing of terrorism information that protects the nation and our foreign partners while safeguarding civil liberties.

The NTSC is a single database of identifying information about those known or reasonably suspected of being involved in terrorist activity. Prior to 9/11 various government agencies maintained nearly a dozen separate watch lists monitoring persons of interest to U.S. law enforcement and intelligence agencies. There was also not a central clearinghouse for all of the information that law enforcement and the IC could access about a potential person of interest. The terrorist watch list contains thousands of records that are updated daily with federal, state, local, territorial, and frontal law enforcement and members of the IC. The no-fly and selectee lists are two much smaller subsets of the terrorist watch list.

Although aviation security practitioners may be unable to prevent the activities of terrorists operating in a foreign country plotting to attack a component of the aviation system, awareness of those activities is still relevant to the protection of U.S. aircraft, airports, and the national aviation infrastructure. The TSA routinely notifies airport and airline security managers to increase security because of potential threats from situations in other areas of the world.[7] Aircraft operators must pay attention to worldwide terrorist threat alerts, because understanding the nature of these threats is essential to protecting flight crews, passengers, and aircraft.

Aviation security practitioners must also pay attention to intelligence related to non-criminal or nonterrorist threats that could jeopardize the aviation system. An example of this is the avian flu. At first glance, this may seem a public health threat, not an aviation security threat. It is, however, a threat to both. The expected fatality rate of the avian flu is 50%; a pandemic outbreak would cause widespread casualties, including among travelers and airport personnel. Even a local outbreak would certainly affect some airport and airline personnel, with some dying or becoming incapacitated for weeks or months, and others needing time off because of sick relatives. This would seriously jeopardize the ability of the local and national aviation systems to ensure security because of the reduced numbers of personnel essential to the airport and airline operations.

Airports are lifelines to the outside world, particularly during a disaster. For example, when Hurricane Katrina hit New Orleans in 2005, the airport became a triage location and then a staging area for rescue operations. When Hurricane Ivan hit Pensacola, FL, in 2004, the airport worked rapidly to recover so it could receive military and relief aircraft flying in to provide emergency services to the community.

Although the focus is often on intelligence related to international terrorist organizations, domestic terrorist activity must not be ignored. Domestic terrorists have been responsible for numerous attacks within the United States. While many domestic attacks have been minor as compared to 9/11 or WTC bombing in 1993, the vehicle bombing of the Alfred P. Murrah Federal Building in Oklahoma City, OK, in 1995 was a tragic example of the capabilities of domestic terrorism. It is just as important to monitor the activities of domestic terrorist organizations as it is to monitor the activities of international terrorist organizations. Both are threats to the aviation system.

Aviation security practitioners should become students of terrorist practices and strategies, be aware of local and world events, and have an understanding of current motives and methods of terrorist operations. Practitioners should develop excellent working relationships with local, state, and federal law enforcement agencies to effectively

---

[7]Usually, airport and aircraft operator personnel receive briefings in the form of TSA information circulars, or briefings from FBI or TSA federal security directors.

share intelligence and other mitigation strategies. They must constantly evaluate changing threat conditions and be prepared to implement higher levels of security.

Under the Aviation Safety and Security Act of 1996, airport operators are required to have a consortium of aviation security and law enforcement professionals who meet regularly to share information and develop strategies to mitigate or prepare for possible incidents. These committees are more effective if they are purpose-driven and key stakeholders have face-to-face meetings.[8]

After 9/11, Boston Logan International Airport officials began the practice of a daily 8:30 a.m. meeting. The airport director chairs this meeting, which is attended by the airport security coordinator, police and fire personnel, and representatives from the airlines that use the airport. This daily meeting is to share security information and to focus all parties on security concerns every day. That the airport director chairs the meeting and that division heads of various airport and airline entities attend demonstrates that the highest priority of the Boston Logan Airport is security. At many airports, security meetings are held rarely and even more rarely attended by decision makers. To emphasize the importance of security and to ensure that timely decisions concerning security are made, it is important that decision makers attend security meetings, not just "note takers" comprised of assistants and administrative personnel.

Airport security practitioners should maintain high levels of awareness with respect to global aviation security. Any attack on another airport, airplane, transportation, or infrastructure facility, in the United States or abroad, should be carefully studied. Early warning indicators should be identified, and similar vulnerabilities should be assessed and appropriate measures taken. There may be value to sending security personnel to a site to assist with recovery and to assess measures that could be used in their own facility under similar circumstances.[9]

Fusion Centers, TLOs, and InfraGard

Through the fusion centers and the Terrorism Liaison Officer Program, airport operators have new pathways to access relevant threat information and keep up on existing and potential threats. This can enable aviation security practitioners to be more aware of suspicious activities and patterns to watch for, possibly helping to turn information in to *intelligence*.

---

[8]Airline security managers and airport security personnel can receive security information on technologies and practices through organizations such as the American Association of Airport Executives. Airline security information on technologies and practices is available through certain aircraft operator–specific organizations such as the Air Transport Association (now known as Airlines for America, A4A). Airport law enforcement personnel can maintain an awareness of airport police practices and challenges through the Airport Law Enforcement Agencies Network (ALEAN).

[9]Various U.S. aviation security practitioners sought involvement in the recovery efforts after Hurricane Ivan in Pensacola, FL, in 2004, and at the Louis B. Armstrong New Orleans International Airport after Hurricane Katrina in 2005 as a way to assist and learn from these disasters.

The national network of fusion centers serves as a focal point for the analysis, gathering and sharing of threat related information throughout a state or region. With timely, accurate information on potential terrorist threats, fusion centers can directly contribute to and inform investigations initiated and conducted by federal entities. (DHS, 2012)

Located in states and major urban areas throughout the country, fusion centers are uniquely situated to empower front-line law enforcement, public safety, fire service, emergency response, public health, critical infrastructure protection, and private sector security personnel to understand local implications of national intelligence, thus enabling local officials to better protect their communities (Joint Regional Intelligence Center, 2012). Fusion centers provide interdisciplinary expertise and situational awareness to inform decision making at all levels of government. They conduct analysis and facilitate information sharing while assisting law enforcement and homeland security partners in preventing, protecting against, and responding to crime and terrorism (Joint Regional Intelligence Center, 2012).

Fusion centers are owned and operated by state and local entities with support from federal partners in the form of deployed personnel, training, technical assistance, exercise support, security clearances, connectivity to federal systems, technology, and grant funding.

The Colorado Information Analysis Center (CIAC) is one fusion center that was created in response to the 9/11 attacks, and was awarded Fusion Center of the Year in 2010.

The CIAC is designed to link all stakeholders in Colorado, from local and federal law enforcement officers, to bankers and schoolteachers. It emphasizes detection, prevention, and information-driven response to protect the citizens and critical infrastructure of Colorado. This counterterrorism effort is centralized in order to enhance interagency cooperation and expedite information flow. (CIAC, 2012)

CIAC has been successful in promoting the eight signs of terrorism, through online videos and public awareness campaigns:

1. Surveillance
2. Elicitation
3. Tests of security
4. Funding
5. Supplies
6. Impersonation
7. Rehearsal
8. Deployment

DHS is working with the Department of Justice in their nationwide Suspicious Activity Reporting (SAR) initiative, which provides law enforcement with another tool to help prevent terrorism and other related criminal activity, by establishing a national capacity for gathering, documenting, processing, analyzing, and sharing SAR information. The Nationwide SAR Initiative (NSI) is a standardized process—including stakeholder outreach, privacy protections, training, and facilitation of technology—for identifying and reporting suspicious activity in jurisdictions across the country, and also serves as the unified focal point for sharing SAR information (DOJ, 2012).

DHS has also launched the "If You See Something, Say Something™" public awareness campaign. "If You See Something, Say Something" originated with the Port Authority of New York and New Jersey (PANY&NJ), and through DHS now consists of a new series of public service announcements encouraging the public to contact local authorities if they see suspicious activity. Some videos have also been produced and can be viewed at DHS- and TSA-related websites.

A terrorism liaison officer (TLO) functions as the principle point of contact for a public safety agency in matters related to terrorism information (Joint Regional Intelligence Center, 2012). The TLO, though not necessarily an expert in terrorism, attends meetings and receives terrorism training and information from the local fusion center or other local entities engaged in terrorism intelligence or investigations. The TLO then educates others within his or her department or area of responsibility (Joint Regional Intelligence Center, 2012). Airport managers and airport security coordinators are eligible to become TLOs.

TLOs are a vital link in keeping those engaged in public safety professions aware of current terrorist tactics, techniques, and practices. Through the diligent performance of their duties, public safety personnel are alerted to terrorism indicators and warnings that might otherwise go unreported (Joint Regional Intelligence Center, 2012).

TLOs are typically contacted when suspicious activities are witnessed that could potentially be related to terrorism. They in turn forward the lead or suspicious activity report to their local police, fusion center, or the Joint Terrorism Task Force (JTTF).

InfraGard is an information sharing and analysis effort that combines the knowledge base members. InfraGard is a partnership between the FBI and the private sector and incorporates an association of businesses, academic institutions, state and local law enforcement agencies, and other participants dedicated to sharing information and intelligence to prevent hostile acts against the United States. InfraGard chapters are geographically linked with FBI field office territories (InfraGard, 2012). Aviation security practitioners can become members of InfraGard and are eligible to access some nonpublic intelligence and law enforcement information.

Domestic and Regional Aviation Security

Aircraft operators conduct flights throughout domestic and regional areas (those areas more than a mile beyond the geographical boundary of an airport). It is important that aircraft security practitioners understand that threat levels can vary within the United States, Mexico, and Canada, and that airport security measures differ slightly throughout the United States, affecting the airline personnel operations. For example, escort requirements at airports may differ. Some airports allow one *badged*[10] individual to escort only one other *unbadged* individual, whereas other airports allow a badged individual to escort multiple unbadged people.

The primary layers of government security organizations and agencies at the domestic and regional levels consist of the DHS, U.S. Northern Command, FBI, TSA, and state agencies. Individual states have developed their own law enforcement, intelligence gathering, and emergency response capabilities, and although the FAA was largely relieved of its aviation security responsibilities after 9/11, the organization still plays a role.

A key failure pointed out after the 9/11 attacks was the inability of the air traffic control system to effectively notify its radar and tower facilities, and thus the aircraft operators themselves, about the hijackings. Created after 9/11, the Domestic Events Network (DEN) is a 24/7 FAA-sponsored telephone-based conference-call network (recorded) that includes all of the air route traffic control centers (ARTCCs) in the United States, and includes other governmental agencies that monitor the DEN. Its purpose is to provide timely notification to the appropriate authority that there is an emerging air-related problem or incident.

North American Aerospace Defense Command

The North American Aerospace Defense Command (NORAD) is a U.S. and Canadian organization charged with the aerospace warning and aerospace control for North America. This includes monitoring fabricated objects in space and the detection, validation, and warning of attacks against North America, whether by aircraft, missiles, or space vehicles, through mutual support arrangements with other commands. The NORAD–U.S. Northern Command (USNORTHCOM) Integrated Command Center serves as a central collection and coordination facility for a worldwide system of sensors designed to provide the commander and the leadership of Canada and the United States with an accurate picture of any aerospace threat.

USNORTHCOM was established after 9/11 to provide command and control of DOD homeland defense efforts and to coordinate defense support of civil authorities. USNORTHCOM's civil support mission includes domestic disaster relief operations that occur during fires, hurricanes, floods, and earthquakes. Support also includes counterdrug operations and managing the consequences of a terrorist event employing a weapon of mass destruction. The command provides assistance to a lead agency when tasked by the DOD. Per the *Posse Comitatus* Act, military forces can provide civil support, but cannot become directly involved in law enforcement.

In providing civil support, USNORTHCOM generally operates through established joint task forces subordinate to the command. An emergency must exceed the capabilities of local, state, and federal agencies before USNORTHCOM becomes involved.

---

[10]*Badged* is a common industry term, which generally means an individual in possession of approved access/ID issued by the airport or aircraft operator.

Department of Homeland Security

The Department of Homeland Security, established in January 2003, was created through the Homeland Security Act of 2002. The primary mission of DHS is to help prevent terrorist attacks in the United States, reduce the country's vulnerability to terrorism, and assist in recovery after an attack (DHS, 2007).

DHS combines 22 separate government agencies including the U.S. Coast Guard, the Secret Service, the Federal Emergency Management Agency (FEMA), and the TSA. U.S. agencies such as the U.S. Border Patrol, U.S. Customs, and the U.S. Immigration and Naturalization Service (INS) were blended into DHS and reorganized under Immigration and Customs Enforcement (ICE) and Customs and Border Protection (CPB). Agencies with involvement in aviation and homeland security include the FBI, the Central Intelligence Agency (CIA), and the National Security Agency (NSA).

As with the TSA, DHS has undergone several reorganizations since its inception.[11] DHS is comprised of directorates responsible for the following:

1. Managing border and transportation security assets used to prevent terrorists from entering the United States.
2. Protecting air, land, and sea transportation systems.
3. Enforcing immigration laws.
4. Managing emergency preparedness and response.
5. Coordinating the federal government's response to terrorist attacks and major disasters.
6. Assisting in recovery efforts.
7. Employing science and technology personnel overseeing efforts to protect the United States from chemical, biological, radiological, and nuclear attacks.
8. Funding research related to homeland security.
9. Gathering and analyzing intelligence information from federal, state, and local agencies to detect terrorist threats or vulnerabilities in the country's infrastructure.
10. Enhancing nuclear detection efforts of federal, state, territorial, tribal, and local governments.
11. Aiding the private sector in developing coordinated responses to security threats.

An important department within DHS is the Domestic Nuclear Detection Office, responsible for developing and deploying domestic nuclear, fissile, or radiological detection systems, and for detecting and reporting attempts to import or transport these materials for illicit use (DHS, 2006). Aviation is a possible method of transport or delivery for nuclear or radiological materials to the United States.

Transportation Security Administration

Originally organized under the Directorate of Border and Transportation Security, the Transportation Security Administration's mission is to prevent terrorist attacks and to protect the U.S. transportation network. TSA is now a stand-alone division of DHS, with the

---

[11]The current organizational structure for DHS can be reviewed at *www.dhs.gov.*

administrator serving as an assistant secretary of DHS directly responsible to the DHS secretary/deputy secretary. The Aviation and Transportation Security Act of 2001 created the TSA to protect the public's security in all forms of transportation. However, the majority of changes to the nation's transportation network have been made in the area of aviation, and the majority of federal funding is still appropriated for aviation needs (Wodele, 2005). In the budget for fiscal year 2004, $4.22 billion (86%) of the TSA budget was allocated for aviation security (Bullock et al., 2006, p. 215).

The TSA's initial responsibility was to take over airline screening at the nation's 450+ commercial-service airports. The TSA met a demanding deadline by hiring 55,000 screeners in just one year. The rapid hiring created imbalances in the workforce, with some airports having too few screeners and others having too many (GAO, 2004, p. 9). TSA continues to adjust screener staffing models in an attempt to balance the number of screeners with the passenger activity levels at each airport.

In its rapid formation, the TSA experienced its share of public relations issues, missteps, and organizational challenges. Despite all of these, Paul C. Light, a public service professor at New York University and a Brookings Institution scholar who has studied the TSA's evolution, called it "one of the federal government's greatest successes of the past half-century" (Goo, 2005). In addition to screening and regulatory compliances, the TSA oversees numerous other programs, including:

- *Federal Air Marshal Program.* Oversees hiring, training, and operations of federal air marshals (FAMs).
- *National Explosives Detection Canine Team Program.* Started in 1970 by the FAA to train and certify canines and their handlers, and now the responsibility of the TSA.
- *Training and certification of federal flight deck officers.* The program to provide firearms and self-defense training to pilots of commercial air carriers.
- *Crew Member Self-Defense Training Program.* A basic self-defense awareness program combined with hands-on training at local community colleges.
- *Armed Security Officers Program.* Provides armed security officers for general aviation flights arriving and departing from Reagan National Airport.
- *Office of Training and Development.* Provides rapidly deployable, national-level resources regarding all aspects of chemical, biological, radiological, nuclear, and explosives (CBRNE).

The U.S. GAO and DHS have long promoted risk management strategies (GAO, 2005, p. 10). Risk management is based on conducting a risk analysis, then allocating funding and resources to those areas with the highest risk of attack or the areas where an attack would create catastrophic damage. It acknowledges that not all life and infrastructure can be completely protected all the time. TSA risk managers are aware that an attack could penetrate an area that has not been assessed to be of critical importance.

One area addressed by TSA Director Kip Hawley's administration is establishing flexible and random security procedures. This policy has resulted in numerous security measures being required of airports and air carriers. The TSA has expanded its workload substantially beyond screening and oversight of federal regulations. Since Hawley's tenure, the TSA has added behavior detection officers (BDOs) and bomb appraisal officers (BAOs) and has taken on employee screening through the Aviation Direct Access Screening Program (ADASP). According to Hawley, "If we follow the same procedures everywhere, every time, we make it easier for terrorists to break the security code" (Tablott, 2006). To this end, the TSA initiated a surge program (TSA, n.d.a.) as a method of increasing security focus and forces in key areas for short periods. The surge program is known as *Playbook,* and includes teams of FAMs, local law enforcement, screeners, and other federal, state, and local police personnel conducting random anti-terrorism measures throughout airports. Additionally, transportation security inspectors (TSIs) frequently are temporarily reassigned to another airport to conduct inspections or focus on security practices at specific areas of an airport or airline. The benefit of having outside personnel is that it provides an impartial look at security issues in specific situations or locations.

Federal Security Director

The TSA is represented at the local level by a federal security director (FSD). The FSD is employed by the TSA and reports to TSA headquarters. FSDs ensure that airport and aircraft operators within their jurisdiction follow regulations and oversee airport security screening operations. There were initially about 350 FSDs for the 450 commercial-service airports in the United States; that number is now estimated to around 125–150 as TSA figured out that an FSD was not needed at many of the airports that have low levels of commercial service. Most FSDs have jurisdiction over one airport, whereas a few have jurisdiction over several smaller airports. The FSD provides daily operational direction for federal security at airports and is the ranking TSA authority for the coordination of TSA security activities.

FSDs have operational authority over the security screening workforce. This is unusual as the regulator of an activity, such as screening, is not normally the operator of the same activity. Outside the United States, most airport security functions are carried out by one entity and regulated by another.

Other FSD duties (TSA, n.d.b.) include:

1. Organizing and implementing the Federal Security Crisis Management Response Plan.
2. Implementation, performance, and enhancement of security and screening standards for airport employees and passengers.
3. Oversight of passenger, baggage, and air cargo security screening.
4. Managing airport security risk assessments.
5. Implementing and maintaining security technology within established guidelines.
6. Providing crisis management to airports.
7. Protecting and recovering the data and communications network as it impacts federal security responsibilities.

8. Employee security awareness training.
9. Supervising federal law enforcement activities within the purview of the FSD and TSA.
10. Coordinating federal, state, and local emergency services and law enforcement.

Operationally, an FSD handles all incoming intelligence and disseminates it to the airport security coordinator (ASC). The FSD assists the airport operator with the interpretation of federally required procedures at the local airport level, particularly information circulars and security directives that are issued from the TSA. The FSD also has the authority to stop aircraft and airport operations (P. Ahlstrom, personal communication, 2006).

TSA employs area (i.e., regional) directors to oversee FSDs. Area directors have administrative oversight for the FSD workforce, conducting performance reviews and related activities.

An FSD does have the authority and responsibility to make decisions on behalf of the TSA. Certain actions, such as approving an ASP or an amendment or procedure within an approved security program, must be reviewed by the TSA's legal department. Although an FSD may have a TSA lawyer on hand, the lawyer receives direction from TSA headquarters, not the FSD. This is important to security coordinators attempting to get a security program or modified procedures approved as the process may take longer and be reviewed by individuals who are not onsite.

FSDs preside over three departments: compliance, operations, and business management. There is also an assistant federal security director (AFSD) for regulatory inspection who is responsible for compliance, and an AFSD for screening who is responsible for operations. FSDs have had their authority expanded to cover other modes of transportation, such as trucking, rail, and maritime activities (P. Ahlstrom, personal communication, 2006). FSDs also have some level of operational control over the federal air marshals assigned to their areas, but their reporting structure has changed frequently over the years.

Another AFSD is assigned to some airports, one for law enforcement. Besides the federal air marshals, this is one of the few actual law enforcement officers within the TSA's structure. Although transportation security officers are called "officers," they do not have police powers or law enforcement arrest authority.

The major responsibilities of the AFSD for law enforcement are to serve as the principle law enforcement specialist, providing advice and assistance to the FSD, maintaining contacts with local law enforcement agencies, and establishing working relationships with key airline and airport personnel. This AFSD also coordinates movements of dignitaries, VIPs, and other law enforcement personnel; serves as the primary contact for TSA for law enforcement in the state; and shares information with TSA personnel concerning criminal activities.

Transportation Security Inspectors

The compliance department, headed by the AFSD for regulatory inspection (sometimes referred to as "compliance"), consists of transportation security inspectors (TSIs) who conduct ongoing audits of airport and aircraft operator security programs and procedures. Although the AFSD for regulatory inspection reports to the FSD, much of the TSI's direction comes from TSA headquarters.

TSIs open cases and investigate alleged violations of security regulations. TSIs act as liaisons to the airport and aircraft operator security coordinators, providing briefings and guidance on industry issues and policy changes. TSIs undergo four weeks of classroom training and then are assigned to an airport working with senior TSIs. TSI training addresses the transportation regulation; the various security programs including the ASP, Aircraft Operator Standard Security Program, Indirect Air Carrier Standard Security Program, Model Airport Security Program (for foreign air carriers), Private Charter Standard Security Program, Twelve-Five Security Program, and the Alien Flight Training Program; compliance and enforcement techniques; and the methods of conducting investigations and inspections.

TSIs conduct reviews of the various security programs including audits and inspections of the actual security procedures as they occur in an airport or within an airline. Violations of such procedures can result in an Enforcement Investigative Report (EIR). If the infraction or procedures violation is not corrected, the EIR could eventually result in a fine to the violator and possible criminal charges in extreme cases. TSIs also conduct enforcement actions against individuals, such as when a passenger attempts to bring a prohibited item through a screening checkpoint (L. Hamler Goerold, personal communication, 2006).

TSA also employs TSIs who serve as air cargo security inspectors. Air cargo security inspectors are TSI agents specializing in security issues related to indirect air carrier programs and air cargo. TSIs may also go through additional training to become qualified as an international transportation security inspector. International TSIs are sent to foreign airports to inspect whether the airport's security procedures meet U.S. standards.

Although enforcement is a necessary part of their job, TSIs frequently work with airports and airlines to help them comply with regulations. Enforcement actions take a long time to adjudicate and often involve extensive administrative and legal work on the part of the violator and the TSA. TSA inspectors generally prefer to see actual compliance with the security regulations rather than to continuously process violations against those individuals and entities (L. Hamler Goerold, personal communication, 2006).

The AFSD for screening heads the operations department, which includes the security screening workforce, screeners (i.e., TSOs), lead TSOs, scheduling officers, supervisors, trainers, and managers.

Before 9/11, airlines were responsible for security screening. Following 9/11, the United States took the approach of hiring a federal screening workforce. At most of the world's airports, screening is not handled by the national government but by airport operators through contracts with private organizations. TSA directly employs nearly 43,000 personnel to conduct screening, which includes personnel training, performance standards, certification, and testing of screening equipment. With the exception of a handful of airports that retained or have elected to opt for private screeners, all of the nation's commercial-service airports now have TSA personnel conducting passenger, carry-on, and checked-baggage screening. TSOs conduct passenger, carry-on, and checked-baggage screening; conduct travel document checks; and are sometimes provided additional training and fulfill the role of behavior detection officers.

The third area of responsibility for AFSDs is the business management department, which includes standard business functions such as accounting, budgeting, and human resources. Other personnel, such as public relations officers, may be located in the FSD's main offices. Airport and aircraft operators may interface with the stakeholder liaison. Stakeholder liaison personnel work with airports, airlines, vendors, contractors, and others to help build and maintain working relationships between all parties. The stakeholder liaison works with the FSD to coordinate, implement, and maintain communication with stakeholders on TSA policies, programs, and directives.

TSA expert transportation security officer (ETSO) is a title that's been associated with TSA's bomb appraisal officers (BAOs) and other specialists within TSA's ranks. ESTOs–BAOs are charged with finding effective ways to share their expertise and real-world experience with the TSO workforce. BAOs build simulated explosive devices and run them through the screening process to show TSOs what terrorists are doing and what they are capable of. BAOs also conduct advanced alarm resolution when the conventional alarm resolution process has been exhausted and the alarm has not been resolved. At his point, BAOs are responsible for resolving the alarm, with zero margin for error. If a BAO has reason to believe that there is an improvised explosive device, he or she notifies airport law enforcement for resolution. BAOs are not explosive ordinance disposal personnel (although many come from the military EOD community); they are expected to assist in the resolution of suspect items but not disarm or remove them. BAOs also serve as the TSA subject matter expert liaison for law enforcement and bomb squad partners and act as on-call technical-assistance personnel for other law enforcement personnel.

TSA's Office of Intelligence

One of the criticisms of the intelligence community is that there was actionable intelligence available or known by certain agencies, but that the information wasn't shared. The FAA security office possessed limited intelligence analysis capabilities, relying largely on reports from other law enforcement and intelligence agencies, rather than developing their own intelligence products.

In response to this shortcoming, TSA developed an Office of Intelligence (OI) to provide threat information to the transportation community. OI was mandated by ATSA and further revised by the Homeland Security Act to receive, assess, and distribute intelligence information related to transportation security; assess threats to transportation; develop policies, strategies, and plans for dealing with threats to transportation security; and act as the primary liaison for transportation security to the intelligence and law enforcement communities.

OI coordinates and shares information with other DHS agencies, the intelligence and the law enforcement community, and other government departments and agencies such as DOT, FAA, and the transportation industry. OI has placed liaison officers with key IC and law enforcement agencies across the federal government to assist in their mission.

OI is focused on supporting TSA's risk-based security strategy. OI provide a threat framework to prioritize security resources, which is regularly used by federal air marshals, federal security directors, and the transportation industry. The office works in conjunction with the Transportation Security Operations Center (TSOC) to disseminate warnings and notifications of credible and imminent threats (TSA, 2006). OI provides and maintains the Remote Access Security Program (RASP) that provides the TSA field with access to classified information in a timely and secure manner.

The OI consists of the Intelligence Watch and the Outreach Division (IW&O), which functions as a 24-hour watch, and the Current Intelligence and Assessments Division, which functions as an analysis center tracking current and emerging threats across all modes of transportation (TSA, 2006).

IW&O maintains a full-time liaison officer presence at seven key IC and law enforcement locations including DHS's Office of Intelligence and Analysis, the Director of National Intelligence's National Counter Terrorism Center, the FBI's National Joint Terrorism Task Force, Customs and Border Protection's National Targeting Center, the National Security Agency, the DEA-administered El Paso Intelligence Center (EPIC) Air Watch, and the Terrorist Screening Center (TSC) (TSA, 2006).

Current Intelligence and Assessments (CI&A) produces *The Transportation Intelligence Gazette*, weekly field intelligence summaries, suspicious incidents reports, and specialized assessments on terrorist groups, weapons, explosives, CBRNE threats, modus operandi, tactics, and trends (TSA, 2006); CI&A also provides baseline threat assessments.

OI directs TSA's red cell activity (i.e. red teaming, discussed in Chapter 11) to identify potential vulnerabilities in the transportation system through the use of adversarial (terrorist) role-playing and scenario development (TSA, 2006).

TSA employs field intelligence officers (FIOs) to analyze incoming threat information, serve as the principal advisor to FSDs on intelligence matters, and develop and maintain working relationships with federal, state, local, and private entities responsible for transportation security (TSA, 2006). FIOs gather law enforcement and intelligence information and disseminate it throughout the national intelligence community. Law enforcement information is vetted, validated, and formatted as homeland intelligence reports (HIRs) by TSA's Office of Intelligence HIR program (TSA, 2006).

Federal Air Marshals

The mission of the FAMs program is to "promote confidence in our Nation's civil aviation system through the effective deployment of Federal Air Marshals to detect, deter, and defeat hostile acts targeting U.S. air carriers, airports, passengers and crews" (TSA, n.d.c.). FAMs detect, deter, and defeat hostile acts against U.S. air carriers, passengers, and crews. As armed federal law enforcement officers deployed on passenger flights worldwide, the mission of the FAMs is to protect airline passengers and crew against the risk of criminal and terrorist violence. FAMs perform investigative work and participate in multi-agency task forces and in land-based investigative assignments to proactively fight terrorism. FAMs promote public confidence in the safety of the nation's aviation system as a "quiet professional" in the skies (TSA, 2012).

FAM intelligence personnel collect, analyze, evaluate, and disseminate foreign and domestic intelligence and threat warning information (TSA, 2012). Air marshals dress as normal travelers to blend in with their surroundings and attempt to act as "normal" passengers. Air marshals carry firearms and are authorized to use lethal force in the protection of the flight deck from terrorist takeover. They have the same use-of-force restrictions as other federal law enforcement agents. The ammunition used by air marshals is designed to stop when it hits an individual rather than penetrate an aircraft's hull or continue through the cabin of a crowded commercial airliner.

The air marshal program was informally started in 1963 when President Kennedy ordered the federal government to deploy law enforcement officers to act as security officers on certain flights. The program was formalized in the early 1970s as the Sky Marshal Program, a name that was later changed to Federal Air Marshal Program. By the 1980s, there had not been a significant hijacking in the United States for many years, and the number of air marshals was reduced. By the 1990s, the air marshals were primarily used on international flights. On 9/11, there were only 33 FAA employees serving as air marshals.

The ATSA 2001 legislation called for the revitalization of the air marshal program. The actual number of air marshals is classified, but estimates put the number into the thousands. Although there still are not nearly enough to be on every U.S. flight, their presence can act as deterrence as they operate undercover. Not all passengers are aware if an air marshal is on a particular flight.

FAMs must meet the highest firearm standards of any federal agency. They train on aircraft donated by the airlines. Airline personnel also train FAMs on basic aircraft operations, such as how to open the doors and flight deck familiarization. When FAMs are not flying, they are usually training and honing their instincts to take immediate action to protect an aircraft. These instincts were used when a distraught man in Miami was shot and killed by the onboard FAMs when he threatened to blow up the aircraft. Federal air marshals have a unique operating environment and must be prepared to handle a variety of in-flight situations without the opportunity for backup. FAMs always work in numbers greater than one. They know that for an aircraft in flight, backup from other law enforcement agencies will not occur until the aircraft lands.

FAMs are assigned to certain high-risk flights based on a variety of intelligence information and other classified factors. Their assignments are coordinated out of the Transportation Security Operations Center (TSOC). Although there are not enough FAMs to be onboard every flight, TSOC helps create a force multiplier with other federal agencies.[12]

---

[12]A force multiplier is an increase in the strength or effectiveness of an entity. In the case of federal air marshals, knowing on which flights other federal agents are also flying allows the FAMs to cover more flights.

Every day throughout the United States, hundreds of federal agents also travel by commercial airlines as part of other duties or missions. TSOC keeps track of agents from other federal agencies and can make FAM flight assignments based on this information. Dave Adams, of the Director's Office of Federal Air Marshals, described this process by stating, "For example, if we know there are already several secret service agents on a flight going somewhere for other duties, then we can make better decisions about how many air marshals are needed on that flight, if any" (D. Adams, personal communication, 2006).

Although other federal agents have not been trained in the specifics of in-flight defense tactics, they do represent additional law enforcement presence on an aircraft. All federal agents are authorized to carry firearms onboard an aircraft at all times, regardless of whether they are traveling for business or pleasure. The same is not true of state and local police officers.

FAMs perform a variety of other duties including surveillance at airports. Using their issued personal digital assistants (PDAs), air marshals with wireless Internet access can send messages to the TSOC regarding suspicious individuals and receive photos and information from the TSOC on developing situations and new intelligence. FAMs are represented on the local Joint Terrorism Task Force and participate in local "surge" initiatives for airports and other transportation modalities. FAMs receive additional support from the Tactical Intelligence Branch of the TSA, which monitors and analyzes terrorist trends (D. Adams, personal communication, 2006).

Special Programs

The TSA also sponsors two programs directed at crewmember self-defense, the Federal Flight Deck Officer Program and the Crew Member Self-Defense Program.

Under the FFDO program, eligible flight crewmembers are authorized by TSA to use firearms to defend against acts of criminal violence or air piracy attempting to gain control of an aircraft. A flight crewmember may be a pilot, flight engineer, or navigator assigned to the flight, or cargo pilot.

FFDOs are trained by the Federal Air Marshal Service on the use of firearms, use of force, legal issues, defensive tactics, the psychology of survival, and program standard operating procedures. FFDOs are not paid for their duties as an FFDO. The FFDO program is covered more extensively in Chapter 8.

The Crew Member Self-Defense Training (CMSDT) Program is available to all actively employed or temporarily furloughed U.S. carrier crewmembers. The program teaches basic self-defense tactics that can be executed in the confines of an aircraft cockpit or cabin, and additional techniques to use "on the street." The additional training is useful as crewmembers travel extensively to a wide variety of destinations and may need to protect themselves against attack. CMSDT is provided at no cost and crewmembers are welcome to attend the training as often as desired.

Federal Bureau of Investigation

A critical role of the FBI, post-9/11, is to protect and defend the United States against terrorist and foreign intelligence threats. Each FBI office with an airport within its geographic jurisdiction has an agent assigned to handle aviation security. This agent is responsible for disseminating information to local airport and aircraft operators and law enforcement that may not have come through the FSD information channels. The agent also serves on the airport's security consortium and is the liaison to commercial-service and general aviation airports.

The FBI is called whenever there is destruction of a commercial aircraft. It is assumed that this is a terrorist or criminal act until determined otherwise. The FBI has been involved in several high-profile aviation incidents including the bombing of Pan Am Flight 103 over Lockerbie, Scotland, and the destruction of TWA Flight 800 (Holden, 2005). The FBI was part of the investigation into the cause of the accident, with the National Transportation Safety Board, until it was determined that the cause was unlikely to be a terrorist bomb or other type of attack.

The FBI has an extensive history of involvement in domestic and international terrorist actions. When there is a terrorist incident overseas, even if U.S. citizens are not involved, the FBI will often send agents to assist with the investigation and learn more about terrorist operational methods. After 9/11, the FBI sent special agents to interview more than 3,000 crop duster owners and operators regarding the potential use of agricultural aircraft for terrorist purposes.

The FBI's Civil Aviation Security Program provides key investigative resources to 56 FBI field offices including aerial surveillance and photography. The program also includes transportation of critical personnel, equipment, and evidence in crisis situations (e.g., hijackings and airport incidents) (FBI, 2006).

All TSA security–regulated airports have at least one FBI airport liaison agent (ALA) assigned to it for investigative purposes, and larger airports have slightly larger contingents. An ALA is the point-of-contact for the airport and aircraft operator security personnel to the FBI. The ALA investigates violations of federal criminal laws at airports, provides security intelligence and training to members of the airport community including local police, and speaks at airline ground security coordinator classes.

During security incidents at airports or on aircraft under the jurisdiction of the United States, the FBI has jurisdictional authority—not the TSA. Local law enforcement has authority until the FBI arrives and takes over primary responsibility for incident command. The FBI "special jurisdiction" onboard aircraft begins the moment all doors are closed after boarding and ends when they are reopened. If the aircraft is still on the physical ground and the doors are closed the FBI has concurrent jurisdiction with the local police. While in flight, in the actual air, only the FBI has jurisdiction. The FBI investigates federal crimes on aircraft including:

- Air piracy or hijacking
- Interference with flight crew and federal screeners
- Crimes aboard aircraft
- Destruction of aircraft
- Loaded weapons in checked baggage
- Weapons at checkpoints

- Bomb threats (often handled by the local police)
- Interstate transportation of stolen property
- Falsification of records (FAA regulates A/C, maintenance, pilot records; typically the FBI gets involved after an accident or incident regarding a logbook or other falsification)
- All terrorism matters
- Accident/crash investigations

The FBI also investigates a fair number of nonviolent cases, such as theft of electronics on an aircraft. Because they take place underneath the FBI's special jurisdiction, they are investigated as a federal crime.

Airport and aircraft operators should be familiar with the following resources that the FBI can bring to an incident:

1. *Joint Terrorism Task Force (JTTF).* The link between the FBI and state and local law enforcement officers. A specific JTTF exists for every major metropolitan city. They are staffed with FBI special agents, other federal agents, state and local law enforcement officers, and first responders (Holden, 2005). ASCs should be familiar with their local JTTF to access the latest threat and response information. Connected to JTTF is the National Joint Terrorism Task Force (NJTTF), based in Washington, DC. NJTTF is the fusion point for intelligence acquired about counterterrorism operations (Holden, 2005).

2. *Critical Incident Response Group (CIRG).* A team of 100–150 special agents and other personnel able to respond to high-profile events such as the Olympics, the World Economic Summit of the Eight, terrorist acts, major crimes, or natural disasters. The local FBI office contacts FBI headquarters for CIRG assistance. Resources consisting of crisis negotiators, behavioral analysts, computer specialists, and others back the CIRG. Depending on the nature of the incident, the CIRG can establish an incident command post in as little as two hours (Holden, 2005).

3. *Crisis Negotiations Unit.* This unit is responsible for hostage negotiations and are notified if there is a hijacking or a hostage situation at an airport or involving aviation.

4. *George Bush Strategic Information Operations Center (SIOC).* A central crisis management and special-event monitoring center supporting major cases, tactical operations, and exercises. The SIOC maintains a 24/7 watch operation observing terrorists and terrorist-related activities throughout the world. The SIOC also conducts analysis of potential terrorist activity and sends updates to agents in the field (Holden, 2005). It is likely that much of the intelligence information related to threats against aviation or transportation is generated at the SIOC.

5. *Hostage Rescue Team (HRT).* A special counterterrorist unit responding to terrorist incidents in the United States. The HRT provides a tactical option to a hostage or hijacking situation. However, its members are not U.S. military operatives, meaning they are able to make arrests, testify in court, and must operate within the confines of the U.S. Constitution. In contrast, the U.S. Army's Delta Force or the U.S. Navy's Seal

Team Six, both charged with duties of responding to hijackings and hostage takings, normally operate outside of the United States. Military operations within the United States are restricted under the *Posse Comitatus* Act of 1878.[13]

6. The FBI can bring additional resources in the form of special agents, a nationwide network of international liaison contacts, the FBI laboratory and disaster teams, evidence response teams, special agent bomb technicians (SABTs), hazardous material and WMD specialists, technical agents, electronics experts (ETs), computer analysis response team (CART), personnel electronics technicians, and victim witness coordinators. Agents also work with specific airlines to provide airline resources and information as needed.

Another capability of the FBI related to airport and aircraft operator security is the Terrorist Information System (TIS). TIS is an online database with the names of 300,000+ individuals and 3,000 organizations of interest. The FBI is also responsible for the investigation of crimes against interstate property, which brings air cargo into the purview of the FBI that may be conducting active investigations of personnel in an airport.

What should airport authorities do when the FBI arrives at an airport in response to a security incident? This is an important question for airport operators and airport security coordinators to address, because the FBI's presence creates a significant impact on airlines and passengers. Although it is not the purpose of this text to go into the specific tactics of the FBI, airport and aircraft operators can significantly assist the FBI in their response by understanding a few basic operational considerations (Price, 2006). The FBI will need to take the following steps:

- Access resources such as secure meeting rooms, secure phone lines, data links, CCTV access, and other resources normally associated with an incident command center.
- Likely have a few agents at the airport incident command center (ICC) to act as liaisons; however, the FBI prefers to coordinate much of their activities from another location to maintain operational security. This strategy depends on the working relationship between the airport's law enforcement agencies and other federal agencies such as the TSA, the FBI, and the security of the airport's ICC. Airport ICCs without tight controls on access are likely to be shut out of information coming in from other agencies because of the potential compromise of operational security. Although this may be a perceived hindrance to the effectiveness of the airport's incident command system, law enforcement agencies may feel it is an effective trade-off to maintain operational security.
- Arrange for a separate room to conduct hostage negotiations, if such a situation exists.
- Arrange for staging areas for their tactical response teams; escorts by airport operations or law enforcement officers to move around the airfield; equipment such as air stairs, aircraft tugs, and fuel trucks; and access to the air traffic control tower and/or other air traffic control facilities.

[13]*Posse Comitatus* Act (18 USC 1385). This law enables the use of U.S. military forces to "execute the laws" except where expressly authorized by the Constitution or Congress.

Customs, Immigrations, and Agricultural Enforcement Agencies

With the creation of the Department of Homeland Security (DHS), the Customs Service, Immigration and Naturalization Service (INS), and other related agencies were reorganized (as related to airport operations) into Immigration and Customs Enforcement (ICE) and Customs and Border Protection (CPB). ICE is the investigations arm of the DHS; CPB combines elements of the former inspection arms of U.S. Customs and U.S. Immigration, along with Animal and Plant Health Inspection and the former U.S. Border Patrol. Formerly, the Customs Service managed the INS and the Department of Agriculture was responsible for overseeing foods, plants, and animals brought into the United States.

CPB personnel staff customs and immigration checkpoints, watching as passengers and their belongings arrive in the United States, checking immigration and citizenship documents, and inspecting the persons and baggage of suspicious travelers. CPB staff also conduct animal and plant health inspections. ICE personnel are less in the public view, often in charge of investigations of personnel working at an airport who may be involved in illegal smuggling, or of passengers using aviation as a transit point to conduct illegal activities such as smuggling.

CPB personnel manage the Advance Passenger Information System (APIS), which requires aircraft operators to submit passenger manifests and other information that is subsequently checked by CPB personnel.

General aviation (GA) airport operators, particularly those along the U.S. borders, may also encounter CPB or ICE personnel. Some GA airports host their own CPB offices to facilitate business travelers arriving on corporate or private aircraft from outside the United States. GA airports are sometimes used as launch or recovery points for smuggling illicit goods and human cargo, so ICE (investigative) or CPB (active inspection and interdiction) personnel may be involved in operations at such airports. A relatively new phenomenon has drug smugglers using ultralight aircraft and landing at small airports or in open fields to deliver illegal narcotics to the United States. Unmanned aerial vehicles (UAVs) are also used to patrol the U.S. border and certain offshore areas and may operate from civilian airports.

CPB and ICE agents should be included in an airport's emergency incident plans. Airport operators must understand that the area where CPB personnel operate represents the border to the United States, thereby increasing security awareness and protections for those areas.

Other Federal Agencies

Airport and aircraft security personnel may encounter a number of individuals from a variety of federal agencies, including the U.S. Secret Service, the Drug Enforcement Administration, the U.S. Marshal Service, and the Diplomatic Security Service. The U.S. Secret Service, charged with the protection of the president and certain other dignitaries, requires special handling when Air Force One, Air Force Two, Marine One, or other aircraft under their protection arrive at an airport. Advanced teams will arrive weeks ahead to coordinate aircraft parking and site security for an event. Airport personnel can expect that access to the aircraft will be highly restricted; flight operations are shut down while the aircraft arrive and depart, and often detours are established on the airfield for other taxiing aircraft and ground vehicles.

The U.S. Marshal Service (USMS) conducts several significant missions that occasionally involve airports and airlines. The apprehension of fugitives, the protection of federal witnesses, the protection of federal judges, and the transportation of federal criminals are the primary missions of marshals. Marshals may track and arrest fugitives on airport property or conduct investigations on passengers or employees at an airport. Prisoners are frequently transported both on commercial aircraft and on aircraft owned by the USMS through an operation known as the Justice Prisoner and Alien Transportation System (JPATS).[14] JPATS flights facilitate the transfer of approximately 175,000 federal prisoners and some 3,000 other state and military prisoners from various prisons and jails every year. When an airport is used for a JPATS flight, the USMS requires a location on the airport to conduct the transfer as it is too dangerous and complicated to conduct transfers in the concourse and sterile areas. The selected site must have adequate space and pavement loading to handle the size and weight of the JPATS plane. JPATS operations use various types of aircraft, from small corporate aircraft to Boeing 737 s. The transfer site must allow marshals to provide security during the transfer. In some transfers, only a few USMS vehicles may be needed, whereas others require several busses of prisoners to be transferred.

USMS pilots work with air traffic controllers to keep aircraft from taxiing too close to the prisoner transfer operation. Airport operations and law enforcement personnel are usually called on to escort USMS personnel to and from the aircraft.[15]

The Drug Enforcement Administration (DEA) conducts frequent investigations on the transshipment of narcotics and narcotic traffickers using commercial and general aviation aircraft. The DEA also has an aviation division and may establish a base of operations at a hangar, often at a GA airport. DEA special agents work frequently with aircraft operator security coordinators on investigations where illicit drugs are being transported on commercial aircraft.

DEA agents conduct surveillance operations at GA airports because drug traffickers often use GA airports. Flight schools are sometimes used as recruiting grounds for drug traffickers looking for pilots. Airport operators must understand the secretive nature of such investigations and assist the DEA and other federal agencies in maintaining confidentiality.

---

[14]Sometimes informally referred to as "Con Air."

[15]Other local law enforcement agencies may also conduct prisoner transfers, which should be coordinated with the airport security and airport law enforcement personnel ahead of time. The carriage of weapons through the screening checkpoint and regulations regarding prisoner escorts on commercial aircraft are covered elsewhere in this text.

State Aeronautical Agencies

State aeronautical agencies exist in all 50 U.S. states. Often, the aeronautics agency is a subdepartment of each state's transportation department or division. State aeronautics agencies may serve in an advisory or regulatory capacity over the airports within their state and often have a role in advising or distributing financial grants to airport operators.

Each state aeronautical agency varies in organization structure and has various levels of power and influence over the operation and management of the state's airports. For example, Florida has approximately $129.90 million (fiscal year 2010) in annual state funding that is available to airports. That is in addition to the funding Florida airports receive from the federal government. By contrast, Colorado has approximately $20 million (fiscal year 2012) a year budgeted to airports. In both cases, each state has significant input over what federal monies an airport will receive through their agreements and relationships with the FAA's airport district office.

State aeronautical (or aviation) agencies provide other services for airports and pilots including the formulation of state aviation systems plans, installation of weather reporting facilities, and other improvements. In New York, the state's aeronautical agency published general aviation airport security guidelines. In other states, the aeronautical agencies are involved in legislation regarding airport security. These policies may sometimes infringe on the powers of the federal government to regulate air commerce. In 2002, a review of the website of the National Conference of State Legislators showed 31 separate state bills that would add to the federal aviation security requirements in the areas of flight school operations, airport security regulations, and airport personnel and law enforcement, including one bill that would make flight school training to non-U.S. citizens unlawful (NCSL, 2002). Many of the bills that have been sponsored since then were withdrawn, died in committee, or were never enacted because of conflict with the federal government's right to regulate the aviation industry. A few state measures did survive, particularly in the areas of general aviation security. Several states have also adopted the TSA's general aviation airport security guidance (TSA document A-001) as law for GA airports within their state.

Most state aeronautical agencies participate at some level in the annual capital improvement funding of airports and, therefore, participate in deciding what monies are used for aviation security purposes.

Local Law Enforcement

Local law enforcement agencies also receive and sometimes develop their own intelligence information. Some police departments like the New York and Los Angeles police forces have their own intelligence-gathering branches with personnel actively involved in intelligence gathering. Where local law enforcement agencies do not have these resources, they still have access to a wider intelligence network than an airport or aircraft operator. These local agencies can disseminate information to airport and airline operators and other federal agencies.

Smaller commercial-service and general aviation airports may not have the large and active law enforcement agencies of larger commercial-service airports, but that does not diminish the possibility that a smaller airport may be considered a target of opportunity by criminals or terrorists and should not be ignored by security agencies.

State and local law enforcement can sometimes be of greater assistance to airport and aircraft operators than federal agencies, as local law enforcement has a smaller purview and a more active interest in activities that directly affect the local community. Additionally, criminal activities and terrorist plots can be planned surreptitiously in rural parts of the country and in major metropolitan cities. The development of local intelligence and the sharing of that intelligence through the FBI's JTTF are important to the overall security of the community and the airports and airlines that serve it.

Investigations and arrests of individuals suspected to be engaged in terrorist or criminal activities can provide additional information contributing to antiterrorist efforts. For example, a local theft of a large amount of fertilizer may be larceny or it could be part of a terrorist plot to bomb the airport. Information of this nature may be actionable intelligence once it is brought up to other agencies including airport and aircraft operators. It is up to the agency making the arrest to investigate these possibilities and make the information known to state and federal law enforcement agencies and, when applicable, to local airport and airline operators. The individual arrested for the crime may provide information as to what his or her intent was in using the stolen substance. Other intelligence reports may also be able to shed light on similar thefts or intelligence information indicating that terrorists are actively plotting to attack a target using a fertilizer-based improvised explosive device. The 9/11 Commission was critical of the inability of government agencies to share intelligence information with one another.

Required Law Enforcement Support for Commercial-Service Airports

Under U.S. Regulation Title 49 CFR Parts 1542.215 and 1542.217 Law Enforcement Personnel and Support, commercial-service airports are required to maintain either a law enforcement presence or the ability for law enforcement personnel to respond in the case of a security incident. The exact number of law enforcement officers (LEOs) is determined on a case-by-case basis at each airport. Smaller commercial-service airports will likely have fewer LEO requirements because of the smaller number of flights. Airports with limited LEO requirements may only have a response time for LEOs to arrive on scene instead of officers actually assigned to the airport. The exact response times are considered sensitive security information. At larger commercial-service airports, numerous LEOs are often required both to be available to respond to incidents and to be able to patrol the landside and airside operational areas. Specifically, LEOs must be able to respond to incidents at the screening checkpoints and to any location on the airport where an aircraft incident is taking place. At the largest commercial-service airports, a LEO is often required to be present at the screening checkpoint anytime the checkpoint is in operation.

LEO presence at screening checkpoints is thought to deter potential criminal or terrorist activity. Much of the theft at an airport occurs at the screening checkpoints when travelers have to give up control of their personal belongings for screening. LEO presence is also designed to make criminals or terrorists nervous as they know the police are directly observing them. Overall, screening personnel are less subject to verbal abuse or upset passengers when police are present. If a screener detects a weapon or explosive on either a passenger or the passenger's carry-on baggage, the police are available instantly to respond.

Airports must also have enough LEO personnel either on duty or available to respond to foreseeable incidents, normally included in the contingencies section of the ASP. The total number of LEOs required at a particular commercial-service airport is stated in the ASP. The minimum law enforcement requirements for LEOs with airport responsibilities are as follows:

- Have arrest authority with or without a warrant while on duty at the airport for crimes committed in the presence of the LEO and for felonies the LEO has reason to believe that the suspect committed.
- Are identifiable by appropriate indicia of authority (uniforms, badges).
- Are armed with a firearm and authorized to use it.
- Have completed a training program that meets the requirements for law enforcement officers prescribed by either the state or local jurisdiction in which the airport is located.
- Airports are required to maintain LEO training records until 180 days after the departure of that particular LEO from the airport.[16]

Shortly after 9/11, the National Guard was called out to the nation's airports. The National Guard does not normally have arrest authority unless martial law has been declared, which it had not. The presence of the National Guard was a response by the government to help get the public flying again and to protect airport infrastructure from imminent attack until intelligence could be developed to determine if another attack on aviation was pending. The National Guard is an integral part of protecting the nation's infrastructure, including airports. Airport operators may consider conducting exercises and training that include National Guard forces to acclimate them to the airport environment.

Under U.S. Regulation Title 49 CFR Part 1542.219, when an airport operator cannot meet the minimum law enforcement staffing levels for their commercial-service airport, the operator can request the TSA to authorize staffing from either the TSA or another federal or authorized agency. The factors taken into consideration are the number of passengers enplaned at the airport during the preceding calendar year and the current calendar year as of the date of the request; the anticipated risk of criminal violence,

---

[16]Airports may also use privatized law enforcement personnel provided they meet these standards.

sabotage, aircraft piracy, and other unlawful interference to civil aviation operations; and the availability of law enforcement personnel who meet the requirements of Part 1542.217.

The TSA also requires airport operators to maintain security-related records for at least 180 days. However, many municipalities, airport authorities, and corporations have record retention schedules that far exceed the requirement. Aviation security practitioners should check with their governing agency to determine the exact requirements for their operation. The specific records that must be retained include bomb threats and actual incidents of weapons and explosives found within the airport and at screening checkpoints, attempted acts of air piracy (hijackings), and arrests of individuals arrested at the airport.

Conclusion

Identifying and understanding structures inherent to government and aviation security can be problematic because of flexibility and variations in the mission of each related government agency, the relationships between those being regulated and the regulating agencies, and the diverse nature of aviation systems in the United States and other nations. In this chapter, we identified and explored aviation security concerns related to various U.S. government agencies such as the Department of Homeland Security, the Transportation Security Administration, and the Federal Bureau of Investigation.

The transition of responsibility from the FAA to the TSA for managing aviation security within the United States continues to have important ramifications for aviation security practitioners. The TSA routinely issues TSRs affecting the national aviation system. TSRs and amendments to TSRs can affect systems such as access control, badging, security area boundaries, personnel changes, and incident management procedures.

To effectively plan and manage aviation security systems, practitioners must always have a current understanding of the role of government in aviation security. In addition, aviation security practitioners should become students of terrorist practices and strategies, be aware of local and world events, and understand current motives and methods of terrorist operations. Therefore, practitioners and government agencies must work together to find effective ways to disseminate information and knowledge related to aviation security. New consortiums of stakeholders to aviation security should meet regularly to develop strategies for the mitigation of and preparation against new threats to aviation security. In addition to regulatory responsibilities, government should help aviation practitioners reach these goals.

References

Bullock, J., Haddow, G., Coppola, D., Ergin, E., Westerman, L., Yeletaysi, S., 2006. Introduction to Homeland Security, second ed. Butterworth-Heinemann, Boston.

CIAC, Aug. 7, 2012. Colorado Information Analysis Center, retrieved Aug. 7, 2012, from Colorado Information Analysis Center, www.ciac.co.gov.

Department of Homeland Security (DHS), 2006. Fact Sheet: Domestic Nuclear Detection Office. Homeland Security, retrieved July 12, 2006, from www.dhs.gov/dhspublic/display?content=4474.

Department of Homeland Security (DHS), Oct. 2007. The National Strategy for Homeland Security, Homeland Security, retrieved July 21, 2008, from www.dhs.gov/xabout/history/gc_1193938363680.shtm.

Department of Homeland Security (DHS), Aug. 7, 2012. State and Major Urban Area Fusion Centers. Homeland Security, retrieved Aug. 7, 2012, from http://www.dhs.gov/files/programs/gc_1156877184684.shtm.

Department of Justice (DOJ), Aug. 7, 2012. NSI, The Nationwide SAR Inititative, retrieved Aug. 7, 2012, from http://nsi.ncirc.gov/default.aspx.

Federal Bureau of Investigation (FBI), 2006. FBI About Us: Quick Facts, retrieved July 13, 2006, from www.fbi.gov/quickfacts.htm.

Goo, S.K., 2005. Air Security Agency Faces Reduced Role. Washington Post, retrieved July 12, 2006, from www.washingtonpost.com/wp-dyn/articles/A35333-2005Apr7.html.

Government Accountability Office (GAO), 2004. Aviation Security: Improvement Still Needed in Federal Aviation Security Efforts (Publication No. GAO-04-592T), retrieved July 23, 2008, from www.gao.gov/new.items/d04592t.pdf.

Government Accountability Office (GAO), 2005. Strategic Budgeting: Risk Management Principles Can Help DHS Allocate Resources to Highest Priorities (Publication No. GAO-05-824T), retrieved July 23, 2008, from www.gao.gov/new.items/d05824t.pdf.

Holden, H., 2005. To Be an FBI Agent. Zenith Press, St. Paul, MN.

InfraGard, Aug. 7, 2012. InfraGard, retrieved Aug. 7, 2012, from www.infragard.net.

Interagency Threat Assessment & Coordination Group, 2012. Intelligence Guide for First Responders. ITACG.

Joint Regional Intelligence Center, Aug. 7, 2012. Terrorism Liaison Officer Information Network. retrieved Aug. 7, 2012, from http://tlo.org:www.tlo.org/what_is_tlo.html.

National Conference of State Legislatures (NCSL), 2002. Aviation Security State Legislation in 2002, NCSL Transportation Reviews, National Conference of State Legislatures, retrieved July 23, 2008, from www.ncsl.org/programs/transportation/aviationrev02.htm.

Price, J., 2006. Airport Security and Law Enforcement. Paper presented at Airport Certified Employee (ACE): Security. American Association of Airport Executives, Alexandria, VA.

Tablott, B., Apr. 4, 2006. TSA Official Says Airport Security Is Too Predictable. Government Executive, retrieved July 23, 2008, from ref=rellink">www.govexec.com/story_page.cfm?articleid=33756&ref=rellink.

Transportation Security Administration (TSA), n.d.a. Building Security Force Multipliers, retrieved July 23, 2008, from www.tsa.gov/what_we_do/tsnm/mass_transit/force_multipliers.shtm.

Transportation Security Administration (TSA), n.d.b. Federal Security Director Job Description, retrieved July 23, 2008, from www.tsa.gov/join/careers/careers_security_jobs_fsd.shtm.

Transportation Security Administration (TSA), n.d.c. Our People: Law Enforcement, retrieved July 23, 2008, from www.tsa.gov/lawenforcement/people/index.shtm.

Transportation Security Administration (TSA), Aug. 8, 2012. Federal Air Marshal Service Careers, retrieved Aug. 8, 2012, from http://www.tsa.gov/lawenforcement/people/fams_join.shtm.

Transportation Security Administration (TSA), June 14, 2006. Testimony by William Gaches, retrieved Aug. 7, 2012, from http://www.tsa.gov:http://www.tsa.gov/press/speeches/testimony_gaches_061406.shtm.

Wallis, R., 2003. How Safe Are Our Skies? Praeger, London.

Wodele, G., Aug. 10, 2005. Bush Administration's TSA Budget Request Faces Steep Cuts. Government Executive, retrieved July 12, 2006, from http://govexec.com/dailyfed/0805/081005tdpm1.htm.

Commercial Aviation Airport Security

Objectives

This chapter explores the more practical and routine responsibilities for airport operators and government in managing airport security. Included is a discussion of the requirements and applications of the airport security program, as well as an overview of general aviation airport security. The chapter assesses the current Transportation Security Administration's organizational model related to airport security and the relationship of that model to various positions held by security practitioners. Security threats and challenges between the landside, terminal, and public and nonpublic areas of an airport are analyzed. Applications for using secured areas, security identification display areas, and air operations areas to differentiate various levels of security are also presented.

Introduction

The business of securing a commercial airport is, ideally, a shared responsibility between the airport operator, the air carriers and tenants, and the Transportation Security Administration (TSA). Before 9/11, every aspect of airport security was the responsibility of the airport operator with the exception of passenger screening (an airline responsibility), and the Federal Aviation Administration (FAA) oversaw both airport and airlines compliance with aviation security regulations. Since 9/11, the boundaries for managing security among various regulators have become increasingly blurred. The TSA is clearly in charge of screening, but as specified in Title 49 CFR Part 1542, the responsibility for most other aspects of airport security remains with airport operators.

The TSA continues to migrate into many areas of what has traditionally been the responsibility of an airport operator. For example, the TSA has created the Behavior Detection Officer (BDO) Program, and hired transportation security specialists for explosives (formerly known as bomb appraisal officers), civilian K-9 officers, deployable law enforcement personnel, and visible intermodal protection and response teams, and has also taken over the travel document check process. TSA has also experimented with deploying a form of broadcast millimeter wave imaging technologies throughout the public areas of a terminal building. Adding to this complexity is that the TSA now regulates three general aviation (GA) airports and may develop regulations for additional GA airports.

Under Part 1542, airport operators assume certain security responsibilities. These include appointing an airport security coordinator (ASC), developing an airport security program (ASP), managing access control to protect the security areas of the airport, and the credentialing required for any employee requiring an airport access/identification media (ID badge). These also include ensuring the airport meets its law enforcement response requirements and developing airports' contingency and incident management measures.

Challenge of Commercial Airport Security in the United States

Fundamentally, airport security focuses on protecting the airfield and aircraft through controlling access of the general public and aviation employees, while still allowing those passengers and employees to efficiently move through the facility. Through leases and operational requirements, agencies such as the TSA, U.S. Customs and Border Protection (CBP), and domestic and foreign air carriers control many areas of an airport facility, further complicating the airport operator's ability to secure the facility. The expanding and changing authorities of the TSA also present challenges to airport operators. For example, the TSA can change oversight for a particular area of an airport depending on the time of day or staffing requirements. For example, in 2006, the TSA determined that the screening checkpoint exit lanes[1] were the responsibility of the airport operator—sometimes. Some in the airport community accused the TSA of trying to fix its own budget and staffing challenges under the guise of security. This led the TSA to modify its decision to add criteria as to when the exit lanes would be the responsibility of the airport and when they would be the TSA's responsibility.

Publicly, airport security is a team effort, with many individuals and organizations focusing on the objective of protecting the airport system and related infrastructure. The Government Accountability Office (GAO) summarized the role of the government and local airport operators in the following April 2006 report describing the TSA's progress in protecting airports:

> ATSA [Aviation and Transportation Security Act of 2001] also granted TSA the responsibility for overseeing U.S. airport operators' efforts to maintain and improve the security of commercial airport perimeters, access controls, and airport workers. While airport operators, not TSA, retain direct day-to-day operational responsibilities for these areas of security, ATSA directs TSA to improve the security of airport perimeters and the access controls leading to secured airport areas, as well as take measures to reduce the security risks posed by airport workers. Each airport's security program, which must be approved by TSA, outlines the security policies, procedures, and systems the airport intends to use in order to comply with TSA security requirements. FSDs [federal security directors] oversee the implementation of the security requirements at airports. (Government Accountability Office, 2006, p. 7)

---

[1]The area where inbound passengers exit the sterile area of an airport to the public area.

Although the TSA provides regulatory oversight over airport security practices, it is the airport operator who must develop and implement prescribed security practices. A misunderstanding resulting from this is that many in the public eye view the FSD, who is employed by the TSA, as being "in charge" of security at commercial-service airports.

With respect to an airport, the FSD clarifies the application of transportation security regulations and provides guidance on complying with the policies set forth by the TSA. The FSD provides guidance on systems, methods, and procedures by which airport and aircraft operators may comply with regulations and security directives, and approves the ASP. However, the airport operator conducts actual hands-on security functions, such as access control to the airfield, management of the computerized access control system, issuance of airport identification media, law enforcement and security patrol, and response to emergencies.

Title 49 CFR Part 1542 addresses airport security, and thus this section discusses the regulatory responsibilities and practical applications of airport operators, notwithstanding changing TSA authority and direction.

Airport Security Coordinator

An ASC is one of the most important people within the airport security system. Title 49 CFR Part 1542.3 describes the responsibilities of the ASC, which includes drafting and enforcing provisions of the ASP, implementing policies set forth by security directives, and overseeing credentialing and access control. The ASC is the primary liaison between the FSD and other agencies with aviation security responsibilities.

The ASC must be available 24 hours a day. To this extent, most airport operators designate an alternate ASC (or several) to be available when the primary is not. The ASC may have other airport management duties, a situation more common at smaller commercial-service airports, but the preference is for the ASC to be focused solely on security. The ASC must have a comprehensive understanding of the security regulations and policy requirements of the TSA to carry out key responsibilities.

Any ASC must be trained in curriculum areas defined in Title 49 CFR Parts 1540, 1542, 1544, 1546, and 1548. The TSA does not approve individual ASC training programs. Rather, the TSA specifies that any ASC training program must cover the required curriculum elements. Although any ASC can train any other individual to be an ASC, the best practice is for the individual to attend formal industry training or certification courses, such as those offered by the American Association of Airport Executives (AAAE) or other entities with expertise. AAAE's program provides a standardized curriculum with the latest industry trends and discussions that exceed regulatory requirements.

Although an ASC is not required to undergo retraining unless there has been a two-year break in serving as an ASC, as a best practice, many ASCs choose to continue their security education through a variety of commercially available industry training programs and conferences, such as the AAAE Airport Certified Employee-Security course (for more information, see *www.aaae.org*).

In accordance with an old FAA security amendment AP-97-01 (still in effect), ASCs must maintain a liaison with domestic and foreign air carriers' station managers on security issues. Airport security coordinators routinely work with the FSD, the assistant federal security director (AFSD)[2] for screening, and the AFSD for compliance, along with airport operations, law enforcement personnel, and major airport tenants. The ASC chairs the airport security consortium.[3]

Drafting and carrying out the provisions of the ASP is the ASC's primary duty. Airport security coordinators must draft and enforce the ASP, including correcting any actual conditions not in compliance with the ASP. The ASC must make adjustments to the security program in response to TSA security directives and to changing threats to aviation, and write amendments to the ASP to accommodate airport tenants when construction, maintenance, and expansion projects are in progress, or to conduct the occasional emergency exercise.

Airport security coordinators decide how to distribute the ASP and to which entities. It is then the ASC's responsibility to ensure that changes to the ASP and the provisions of security directives are disseminated to the authorized airport parties. A significant part of the ASC's duties is getting the TSA to approve the security program and then working with the TSA on implementing the provisions of the ASP, and address the expanding security roles of the TSA in the airport terminal building.

There is often debate between the ASC and TSA personnel about what constitutes compliance with the regulations as described in the ASP. When TSA was first created, many TSA compliance and FSD personnel had limited airport and airline management experience, whereas many airport security coordinators have multiple years of experience in either airport or airline operations. The TSA has recognized the need for a mix of backgrounds for its compliance inspectors, bringing in those with aviation backgrounds and those from other backgrounds, so the ranks have been infused with a mix of personnel over the past decade.

Other duties of the ASC include implementing and overseeing the access control and credentialing programs, establishing the security identification display area (SIDA) training program, providing for law enforcement coverage, and responding to security contingencies and incidents.

Airport Security Program

Part 1542.101 outlines the general requirements of an ASP, specifying that only those airports with a TSA-approved ASP in place may conduct scheduled service or large private charter operations. The ASP is the foundation for the entire airport security system.

---

[2]AFSDs are assigned to supervise areas such as screening and compliance.

[3]Aviation security legislation in 1996 mandated that airport operators develop an airport security consortium consisting of the airport security coordinator, airport law enforcement representation, FAA security representation, and other federal, state, and local agencies with aviation security responsibilities.

It establishes the security areas and details how access to these areas will be controlled, defines the process for obtaining an access/ID badge (i.e., the credentialing process), explains how the airport will comply with the law enforcement requirements of Title 49 CFR Part 1542 (including contingency and incident management programs), and specifies the actual practices for airport compliance with federal regulations. An ASP is specific to each airport, drafted by the ASC and approved by the TSA.

Certain components of the ASP must be published in an airport's rules and regulations, which regulate the conduct of employees and the general public at the airport. The sections of the ASP included in the rules and regulations (which are accessible by the public) result from discussion and consensus between the ASC and the FSD. Typical sections include procedures for new airport tenants to follow when setting up businesses at the airport and procedures for their personnel to obtain airport access badges; the requirements for an individual to enter the security areas of the airport and approved identification for personnel in security areas; penalties for disobeying the rules and regulations related to security; a warning not to disclose sensitive security information, to which every aviation employee who is issued a SIDA access badge is exposed; badge and access key fees; methods for escorting unbadged individuals and for challenging anyone without the appropriate identification in security areas; the requirement to submit to screening and proper procedures for entering/exiting airport access gates and doors; contraband or weapons prohibitions in the security areas; and general conduct in the security areas.

Since Part 1542 applies to all types of commercial-service airports, from Fairbanks, AL, to Atlanta Hartsfield International, in Georgia, and everything in between, means that each airport is unique in physical and operational characteristics; therefore, the ASP functions as the regulations for each airport. Violating a process within an ASP carries the full weight and force as if transportation security regulations (TSRs) had been violated, subject to civil and, in some cases, criminal penalties.

In addition to outlining the geographical boundaries of the security areas of the airport and describing the various measures applied in each area, the ASP also describes credentialing as related to the issuance of airport access media for aviation employees and how the airport will meet the law enforcement officer requirements under the security regulations. The ASP also describes how the airport will respond to contingencies and security incidents.

Applicability and Testing

This section covers Title 49 CFR Part 1542. Please note that the TSA regulations are subject to frequent change. Readers should review Title 49 CFR Part 1542 to ensure compliance with the current iteration of any regulation.[4]

The following interpretations of Title 49 CFR Part 1542 are derived from the authors' perspectives and experiences related to defining and examining working agreements

---

[4]Current copies of the transportation security regulations can be obtained at the TSA's website, *www.tsa.gov.*

among airport operators and the TSA. Official TSA interpretations may differ and would take precedence, and readers should be aware of this caveat.

Title 49 CFR Part 1542.5 gives the TSA authority to inspect commercial-service airports for regulatory compliance. Part 1542 does not currently apply to GA airports[5] if those airports conduct fewer than 2,500 annual passenger enplanements. Part 1542 does apply to those who receive security directives and information circulars. However, GA airport managers receiving such information need not implement the applicable commercial-service airport security measures.

Part 1542.5 establishes the authority of the TSA to conduct tests or inspections of an airport's security program, including the copying of records to determine if the airport is in compliance with federal regulations and the ASP. Airport operator (and sometimes aircraft operator) employees may also be designated within the ASP to conduct system tests and inspections. This regulation allows the TSA to be within the security areas of an airport for the purposes of conducting tests and inspections, but it extends this responsibility only to the TSA compliance inspectors and those authorized by the TSA (typically, transportation security officers, that is, screeners, are not allowed to conduct tests of the ASP). Part 1542.5 pertains specifically to TSA-compliance inspection personnel including transportation security inspectors (formerly aviation security inspectors and principal security inspectors) and air cargo security inspectors conducting validity and reliability tests of the security system. There have been occasions at some U.S. airports where TSA screening and other airport and airline personnel, having neglected to wear the proper airport credentials in the security areas of an airport, have claimed, when caught, to be conducting a test of the security system. This is not valid; Part 1542.5 allows the TSA-compliance inspectors and certain other personnel designated by the ASP to conduct valid inspections of the airport security systems. Additionally, this testing privilege *never* extends to the media. Airport operators are required to issue access credentials to TSA personnel if they have undergone the airport's site-specific SIDA training program. This helps ensure the TSA personnel are familiar with some of the unique security traits of the airport.

The ASP is classified as a sensitive security information (SSI) document and must be protected from public dissemination (discussed in Chapter 4). The ASP is not subject to individual state open records laws. Each page of the ASP must include a paragraph that notifies the reader that the contents are SSI. The exact language in the warning is obtained from the TSA.

When the TSA approves an ASP, it embodies the regulatory practices for that airport and enforces it with the same level of authority as the TSRs. If there is a violation of the airport-specific ASP, the TSA may take enforcement actions against the airport based on a violation of the airport's own security program.

---

[5]Over the past several years, the TSA has continued to develop various regulations that address general aviation flight operations and has discussed regulating general aviation airports. It is suspected that within a few years, general aviation airports will see some form of regulation.

Part 1542.103 specifies the content of the ASP. The requirements of a particular airport's security program are based on a variety of factors, such as the number of annual enplanements; the airport's location relative to other security-sensitive locations such as military bases, nuclear facilities, or centers of government; and other conditions usually identified through a risk management study.

For security purposes, ASPs classify commercial-service airports as Category X, Category I, Category II, Category III, or Category IV. Category X airports are the largest commercial-service airports in the United States. Category I, Category II, and Category III airports are large-, medium-, and small-hub commercial airports, respectively. Category IV airports are usually nonhub or GA airports with more than 2,500 enplanements per year. Some Category I airports, considered Category I based on the level of enplanements, are treated as Category X because of their proximity to other security-sensitive locations. Baltimore/Washington International Airport and Ronald Reagan National Airport are two examples. A smaller airport may have higher levels of security based on the geopolitical circumstances at the time, if a civilian airport is colocated with a military base, or other factors as determined by the TSA.

The TSA has established three levels of security programs:

- *Complete security programs.* Required for large-, medium-, and small-hub airports, these security programs are the most comprehensive and pertain to airports with scheduled passenger or public charter service using aircraft with passenger seating configurations of more than 60 seats.
- *Supporting security programs.* These security programs are required for many nonhub airports and pertain to airports with scheduled passenger or public charter service using aircraft with passenger seating configurations of fewer than 61 seats that enplane or deplane into any sterile area. They also pertain to any private charter operation that enplanes or deplanes into a sterile area.
- *Partial security programs.* These security programs are required for small commercial-service airports with infrequent or seasonal commercial service, with scheduled passenger or public charter operations using aircraft with 30–61 seats that do not enplane or deplane into a sterile area. They include scheduled passenger or public charter operations using aircraft with fewer than 61 seats that depart to, or come from, outside the United States and do not enplane or deplane into a sterile area. Typically, partial programs are seen in Alaska, and very few other locations in the United States.

There are 21 sections to the complete security program, as shown in Table 5.1 (and explained in detail throughout the text). Supporting security programs include only a few of the items required for a complete security program.

The complete ASP includes maps delineating security areas. These usually include charts showing the locations of airport access–controlled devices such as card readers and the locations of closed-circuit television (CCTV) cameras. The other ASPs, supporting and partial, are distinctly different in that they do not have a requirement to identify airfield security areas.

Table 5.1 Comparison of Requirements for Complete and Supporting Security Programs

Section	Complete Security Program	Supporting Security Program
1	The name, means of contact, duties, and training requirements of the airport security coordinator required under Part 1542.3.	The name, means of contact, duties, and training requirements of the airport security coordinator required under Part 1542.3.
2	Reserved for future regulations.	Reserved for future regulations.
3	Description of secured areas and associated descriptions of the access control methods and practices, access media, and perimeter signage.	
4	Description of air operations areas with associated descriptions of the access control methods and practices and perimeter signage.	
5	Description of the SIDAs including descriptions of the boundaries of the SIDAs.	
6	Description of the sterile areas including the boundaries and access control measures when the screening checkpoint is not active and other methods of controlling access to the sterile area.	
7	Procedures used to comply with Part 1542.209 regarding fingerprint-based criminal history records checks.	
8	A description of the personnel identification systems as described in Part 1542.211 (access/ID badge).	
9	Escort procedures in accordance with Part 1542.211(e) relating to escorting personnel that do not have unescorted access authority within the SIDAs.	
10	Challenge procedures in accordance with Part 1542.211(d) outlining the process whereby those with unescorted access authority are to challenge others who are not wearing an access/ID badge in the SIDAs.	
11	The content of the airport's SIDA training programs required under Parts 1542.213 and 1542.217(c)(2).	
12	A description of law enforcement support used to comply with Part 1542.215(a), specifically law enforcement personnel support requirements and training standards.	A description of law enforcement support used to comply with Part 1542.215(a), specifically law enforcement personnel support requirements and training standards.
13	A system for maintaining the records described in Part 1542.221, specifically security-related reports and forms such as law enforcement response to the screening checkpoint.	A system for maintaining the records described in Part 1542.221, specifically security-related reports and forms such as law enforcement response to the screening checkpoint.
14	The procedures and a description of facilities and equipment used to support TSA inspection of people and property and aircraft operator or foreign air carrier screening requirements of Parts 1544 and 1546. Addresses the facilities an airport must have to accept international traffic that has not been	

Table 5.1 Comparison of Requirements for Complete and Supporting Security Programs—cont'd

Section	Complete Security Program	Supporting Security Program
	screened to meet U.S. regulatory requirements, such that screening may be conducted before passengers and baggage are allowed entry into the airport.	
15	A contingency plan required under Part 1542.301 specifying how the airport operator will comply with increased security measures.	A contingency plan required under Part 1542.301 specifying how the airport operator will comply with increased security measures.
16	SSI procedures for the distribution, storage, and disposal of documented ASPs, security directives, information circulars, implementing instructions, and, as appropriate, classified information.	SSI procedures for the distribution, storage, and disposal of documented ASPs, security directives, information circulars, implementing instructions, and, as appropriate, classified information.
17	Procedures for posting public advisories as specified in Part 1542.305. Addresses the requirement of airports to post warning notices to passengers who may be traveling to certain countries that the Department of Homeland Security (DHS) deems noncompliant with U.S. aviation security standards or that the Department of State deems too dangerous for U.S. citizens to travel to.	Procedures for posting public advisories as specified in Part 1542.305. Addresses the requirement of airports to post warning notices to passengers who may be traveling to certain countries that the DHS deems noncompliant with U.S. aviation security standards or that the Department of State deems too dangerous for U.S. citizens to travel to.
18	Incident management procedures used to comply with Part 1542.307, including bomb threats, hijackings, and other unlawful acts of interference with aviation.	Incident management procedures used to comply with Part 1542.307, including bomb threats, hijackings, and other unlawful acts of interference with aviation.
19	Alternate security procedures, if any, that the airport operator intends to use in the event of natural disasters and other emergency or unusual conditions.	
20	Each exclusive area agreement as specified in Part 1542.111 governing the security responsibilities of regulated parties, such as aircraft operators.	
21	Each airport tenant security program as specified in Part 1542.113 governing the security responsibilities of airport parties not covered under Title 49 CFR Parts 1544, 1546, or 1548.	

The supporting and partial security programs are frequently used at small commercial-service airports serving only a small number of enplanements per year. Supporting and partial programs are also used at airports that are predominantly general aviation in nature but host periodic commercial-service flights and are still classified as "commercial service" by the FAA.

In some small airports, screening checkpoints are located beside ticketing and gate operations (Figure 5.1). After passengers go through the checkpoint, they exit into the airfield and proceed to board the aircraft. Airline personnel monitor the boarding to ensure

FIGURE 5.1 Salina Municipal Airport, Colorado. The airline ticket counter is at the left. The lobby/waiting area is at the lower left. The screening checkpoint is at the right. The exit to the small sterile area is beyond the checkpoint.

that only passengers who have successfully cleared security screening board the aircraft. These airports typically have supporting or partial security programs. Supporting security programs do not have requirements for identifying security areas (secured area, air operations area, or SIDA). Therefore, there is no need for requirements related to personnel identification (access media), addressing escorting individuals who do not have unescorted access authority, or challenging individuals not wearing access media.

Partial security programs include all of the items in a supporting program, with the exception of providing contingency plans that specify how the airport operator will comply with increased security measures.

Some airports with limited commercial service may be approved to have an alternate means of compliance (Part 1542.109) with the security program requirements. This is the designated ramp observer (DRO) program and is most common at remote airports in Alaska. The DRO is assigned the task of overseeing the security of the commercial-service operation for the duration of that operation. Before the arrival of commercial-service aircraft, the DRO inspects the commercial-service operation area for unauthorized individuals or contraband. The DRO then visually observes the entire commercial-service operation including the arrival of the aircraft, debarkation and boarding, any screening, and the departure of the flight. The DRO cannot have other duties during this time (such as ticketing or loading baggage).

Certain airport tenants, particularly airlines, prefer to control most of the security measures within their leasehold areas. This is possible using an exclusive area agreement (EAA), under Part 1542.111, or an airport tenant security program (ATSP), under Part 1542.113. An EAA permits a regulated party, such as an airline, to take responsibility for any component of the airport's security program including access control measures and personnel identification, within its leasehold areas. The airline, the airport operator, and the TSA must agree to the terms of the EAA, and the EAA must be included into the ASP. In no case may a tenant take over law enforcement responsibilities from an airport operator. EAAs are common for air carriers (including air cargo) that have large hubs or operations at particular airports.

Although the EAA may specify that a regulated party will conduct certain security functions, the airport operator must ensure the regulated party is complying with the terms of the agreement. For this purpose, the airport operations and security personnel still have the ability to be in and around EAA-specified boundaries.

ATSPs are similar to EAAs, but apply to nonregulated parties and require the airport operator to take on certain TSA functions, such as developing a compliance audit and inspection program. General aviation operators of a commercial-service airport, such as fixed-base operators (FBOs) and corporate flight operations, would be typical candidates for an ATSP, but they remain largely unpopular throughout airports in the United States due to the increased level of accountability and responsibility placed on the airport operator.

The same requirements for EAAs apply to ATSPs, including the specification of certain security functions and the responsibility of the airport to monitor and ensure the tenant is complying with the terms of the agreement.

Enforcing the Airport Security Program

Title 49 CFR Part 1540 establishes TSA's enforcement authority, but the airport operator is also required to enforce the airport's security program. Title 49 CFR Part 1542.3(4) directs the airport security coordinator to "immediately initiate corrective action for any instance of noncompliance with this part, its security program, and applicable Security Directives." The clause in Title 49 CFR Part 1542.3(4) defines the responsibility of the ASC to ensure the requirements of the ASP are met and, when not, to take corrective actions. When aviation personnel with security responsibilities violate the ASP, the ASC is responsible for ensuring that appropriate punitive action is taken. The ASP must outline the enforcement program and specify the penalties and the penalty procedures.

Some airport operators have converted their airport's rules and regulations into a local ordinance, making them enforceable under local laws—this enforcement extends to violations of the ASP. Other airports require user agreements with tenants or include an agreement within an access/ID badge application, which an individual signs when issued access media for a specific airport. The ASP must specify penalties for noncompliance. Normally these penalties include the confiscation of airport access media for a specified period or the issuance of an airport violation notice, going through the SIDA training program again, or have a hearing with the ASC to discuss the violation. Depending on the nature of the offense, criminal enforcement actions may be in addition to a TSA enforcement action or an airport enforcement action.

Common violations of the ASP include leaving a SIDA access door open and unattended, allowing other authorized employees into the SIDA by "piggybacking" (i.e., allowing a second employee with approved access to pass through an access control point without using his or her own access media to record the passage), failing to wear an access/ID badge in the SIDA, and failing to challenge an unbadged individual in the SIDA. Some of these breaches in security are considered minor violations, particularly if the violation was unintentional; ASCs must use discretion in weighing penalties for such violations. Other violations are more serious and should result in the immediate revocation or suspension of the access/ID badge and the immediate removal of the access/ID badge holder from the airport property. These include such things as loaning one's airport access media, intentionally breaching security by blocking a door open, jumping over the airport perimeter fence or allowing unauthorized individuals into the security areas, interfering with or assaulting security personnel, or falsifying or altering airport access media.

The goal of any airport violation notice program is to force compliance with the ASP and regulations. Airport operators who do not strictly enforce their security programs undermine their own efforts and show other airport personnel that security procedures need not be taken seriously. Some analysts have suggested that careless attitudes by airport employees toward maintaining security procedures may have been the reason certain airports were selected as the launch points for the terrorists on 9/11 (Airline Security, 2002).

Operations

Title 49 CFR Part 1542 Subpart C comprises the essence of the airport security system. It includes the regulatory directives for segregating the airport into four different levels of security and the necessary access control systems to provide the minimum levels of security for each, the procedure for the criminal history record check (CHRC), the access/ID systems and processes, the SIDA training requirements, and law enforcement support and response requirements. Whereas Subparts A and B primarily address the administrative construct of the airport security program, Subpart C addresses how the airport will meet its daily regulatory requirements.

A commercial-service airport has three areas: landside, terminal, and airside. The landside area has commercial and private vehicle areas; the terminal area has public, non-public, and sterile areas; and the airside area has secured areas, air operations areas, and security identification display areas. The sterile areas, secured areas, air operations areas, and security identification display areas are collectively known as the "security areas."

Landside operations include private and commercial passenger vehicle pickup and drop-off areas and, perhaps, rail or subway access. Landside operations do not have a specific security classification, but security practitioners must be aware that there are still highly important security considerations such as the threat to the airport of a vehicle-borne improvised explosive device, such as was used in Glasgow, Scotland, in 2008 when a car bomber attempted to drive into the terminal building.

The terminal area includes public areas such as ticket counters, nonpublic areas such as vendor storage areas and tenant administrative offices, and the sterile area. Most terminals divide the public areas and the sterile area with a security screening checkpoint. The TSA or a domestic or foreign aircraft operator primarily controls access to the sterile area. Sterile areas extend beyond the screening checkpoints into the concourses or piers of the airport. Some airports, such as Dallas/Fort Worth International Airport, have multiple separate terminals. Each terminal has its own associated concourses, and the concourses are connected by a series of bridges or a mobile people-mover system.

The airside area is divided into the secured area, SIDA, and air operations area (AOA). The secured area represents the highest level of security protection, and the AOA the lowest level. Normally, the traveling public is not allowed access to the airside near the commercial-service aircraft except in instances where the traveling public walks outside to the aircraft. This is typical at small airports and in the commuter airlines areas of large airports. The security areas can also include exclusive areas and areas under an airport tenant security program. The GA ramp commonly has passengers and flight crew walking in the airside areas to and from GA aircraft and is usually considered an AOA.

Terminal Area

The terminal area is normally associated with the building or group of buildings used by the public to park, check in at the ticket counter, and pass through screening checkpoints to connect with scheduled airlines. *Terminal* could also refer to fixed-base general aviation operators or private terminals for corporate flight or charter operations. For our purposes, terminal will be used in reference to scheduled airline service activities.

The public area includes areas accessible by the public without undergoing screening. The challenge of securing the public area is similar to that for securing any public venue, except the target (an airport) is typically a much higher-value target because of the potential impact and disruption on the national aviation system. Of special concern is that, in the United States, there are no specific access requirements to enter the public areas of an airport—anyone may enter.

Some airports have a central terminal connected by concourses or piers that extend outward into the airfield. This configuration is a *centralized terminal* and is an older airport design. Many newer large commercial-service airports have adopted a decentralized terminal design, which features one or more publicly accessible terminal buildings and concourses detached from the terminal and accessible by a people-mover system. Some airports feature a centralized terminal and attached concourses and have additional concourses located on the airfield and accessible by a people-mover system (known as a *hybrid* terminal design). Each airport design has advantages and disadvantages in the implementation of security measures.

Centralized terminals can consolidate screening checkpoints to a minimum number of locations, maximizing efficiency; however, a single screening breach in a centralized terminal can cause an entire airport to shut down. Decentralized terminals may be able to isolate the impact of a screening breach, but require substantially higher levels of screening staff and equipment due to multiple screening checkpoints.

The nonpublic area of the terminal includes those areas where the general public is generally not permitted, such as vendor storage rooms, airline and airport administrative offices, employee child daycare, employee fitness facilities, and airport loading docks. Although federal regulations do not restrict public access to these areas, the airport operator may want some level of access control just as a matter of loss prevention. Airport security coordinators often include the nonpublic areas in their ASPs and require personnel operating in those areas to wear airport-issued access media. Most nonpublic area access media do not allow access to security areas unless the individual also has an operational need to operate in the security areas. Access media helps distinguish those with a legitimate need to be in the nonpublic areas and security areas.

Sterile Area

The TSA offers the following description of the sterile area:

> *Sterile Area portion of an airport, specified in the Airport Security Program that provides passengers access to boarding aircraft and to which access generally is controlled by TSA, or by an aircraft operator under 49 CFR 1544 or a foreign air carrier under 49 CFR 1546, through the screening of persons and property. (TSA, 2006, p. 16)*

The sterile area is the area beyond the screening checkpoints and includes concourses or piers. Passengers and aviation employees who work in the sterile areas are normally required to go through screening to access the sterile area. Certain approved aviation employees with access/ID badges can also access the sterile area through one of the other security areas without having to undergo public screening.

Before 9/11, friends and family who were not traveling commonly passed through the screening checkpoints with departing passengers and were allowed to go with them to the gate. With the increase in screening times because of post-9/11 security procedures, access to the sterile area is now restricted to those with a boarding pass and aviation personnel working in the sterile area. Although some U.S. airports have asked for relief from this requirement, it is likely to remain in place for the foreseeable future. For an ASC, the sterile area is a security challenge as various security policies and procedures used in those areas are formulated and issued by different authorities.

Security Areas

The airside security areas help airport operators identify levels of security processes required throughout the airfield—such as access controls, personnel identification, signage, law enforcement response, and so on. Assigning security areas versus one security standard throughout the entire airfield is predicated on the perceived level of risk being greater at some parts of the airfield, such as around scheduled passenger service flight operations, and the perceived level of risk being lower at other areas, such as around general aviation operations. Lower security standards in lower-risk areas allow operators in those areas greater flexibility and potential cost savings by not having to meet the higher security requirements found around the scheduled or commercial-service (i.e., higher-risk) areas of the airfield. Ever since airport operators began treating commercial-service areas with a different standard than general aviation and lesser-risk areas,[6] the goal has been to prevent individuals operating in the lower-risk areas from accessing the higher-risk areas, using the runways, taxiways, and vehicle service roads.

The International Civil Aviation Organization (ICAO) recommends two types of security areas in an airport: the security identification display area and the air operations area. Before 9/11, the FAA, through Federal Aviation Regulation Part 107.14 Airport Security–Access Control, listed three primary airfield security areas: (1) the secured area, (2) the security identification display area, and (3) the air operations area. These definitions have survived the transition to the TSRs, but they can be confusing to new airport security practitioners. They also provide a great deal of flexibility if they are properly understood and applied.

The key points to understand when discussing security areas and access control are the following:

- There are three types of security areas, each having its own security requirements (Figure 5.2).
- There are different types of access control systems approved for separating security areas from the public areas, and security areas from each other. One type of access control system is for the secured area only and meets the highest security standards; another is for the AOA and meets lower security standards (Figure 5.3).
- Any security area can incorporate security measures inherent or required in another. For example, the airport operator can require individuals working within the AOA to wear personnel identification (not required under the strictest definition of the AOA).

The secured area includes areas where the highest level of security must be applied. It includes passenger enplaning and deplaning areas and baggage sorting, loading, and unloading areas. Secured areas are established wherever there are commercial-service (i.e. scheduled passenger service) operations. All secured areas are also considered to be SIDAs, as the requirements of the secured area also include all the SIDA requirements.

Aviation personnel operating within the secured area or the SIDA must undergo a criminal history record check and security threat assessment and must complete the SIDA training program before being allowed unescorted access to the SIDA. All personnel are required to wear access/ID badges at all times while operating in the SIDA.

---

[6]Other lesser-risk areas could include corporate flight hangars and flight operations, flight schools, aircraft maintenance operations, airport maintenance shops, caterers, and, up until 2006, air cargo.

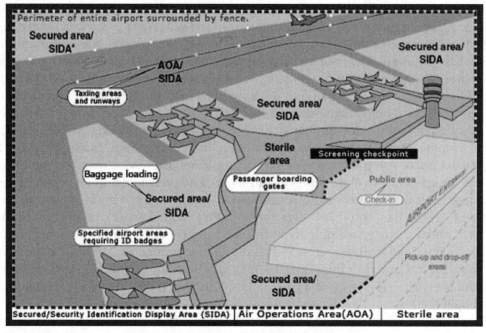

FIGURE 5.2 Example of various security area locations within the airport environment.

	PRODUCT	APPLICATION	SIZES	WT. / ROLL	MATERIAL	ATTACHMENT SPACING LENGTH	BREAK LOAD
	RAZOR RIBBON - Single Coil with Core Wire	Medium Security Fence Topping	18" 24" 30"	13 lbs. 17 lbs. 21 lbs.	AISI 430 Stainless Steel, .098 dia. high Tensile Wire	6" - 16.67' 9" - 25' 18" - 50'	2800 lbs.
	RAZOR RIBBON MAZE - Single Coil with Wire, Concertina Style	Ground Barrier Max. Security Fence Topping	24" 30" 36"	15 lbs. 19 lbs. 23 lbs.	AISI 430 Stainless Steel, .098 dia. high Tensile Wire	12" - 15' 16" - 20'	2800 lbs.
	RAZOR RIBBON MAZE - Concertina Style, Double Coil	Ground Barrier Max. Security Fence Topping	24" inside 30" outside	34 lbs.	AISI 430 Stainless Steel, .098 dia. high Tensile Wire	12" - 15' 16" - 20'	2800 lbs.
	MIL-B-52775 B Type II - Austenitic Double Coil	Ground Barrier Max. Security Fence Topping	24" inside 30" outside	35 lbs.	AISI 301 / 304 Stainless Steel .047 dia. Stainless Wire Rope	24" - 66'	2250 lbs.
	MIL-B-52775 B Type IV - Austenitic Double Coil	Ground Barrier Max. Security Fence Topping	24" inside 30" outside	35 lbs.	AISI 316 Stainless Steel .047 dia. Stainless Wire Rope	24" - 66'	2250 lbs.
	RAZOR RIBBON - Single Coil	Min. Security Fence Topping. Commercial Use	18" 24"	9 lbs. 12 lbs.	AISI 430 Stainless Steel	6" - 16.67' 9" - 25' 18" - 50'	1260 lbs.
	BAYONET BARB - Concertina	Ground Barrier	27 1/2" 37 1/2"	23 lbs. 34 lbs.	ASTM A 526 Zinc Galvanized .098 dia. high Tensile Wire	20" - 50'	1300 lbs.

FIGURE 5.3 Fence types and fabric (TSA Security Design Guidance).

Those within the AOA do not have to wear an access/ID badge. However, personnel who are based at the airport and operate in the AOA, but not in the SIDA, are required to apply for airport access media, undergo and pass a security threat assessment, and, under the strictest interpretation of the requirement, keep their access media on their person while operating in the AOA (but are not required to wear it). Individuals who are operating within an AOA on an airport they are not based at are not required to have an airport-issued access media, but should carry some form of photo identification.

Access to the secured area must be controlled with an approved access control system meeting Title 49 CFR Part 1542.207 requirements (Table 5.2). These include the ability of the system to authorize entry to those with authorized access, deny entry immediately if authorization is changed or withdrawn, and differentiate access between particular portions of the secured area.

Personnel within the secured area must challenge anyone they see not wearing access media; signage must also be posted warning of imminent entry into a secured area. The secured area extends to adjacent areas not separated by adequate access controls or barriers (e.g., the runway, airport fencing, a building, an electronic barrier, or natural barriers). The definition of *adequate access controls* depends on the ability of an access control system, measure, or procedure to detect, delay, or respond to an unauthorized individual attempting to gain access, and is largely determined by an evaluation by the TSA.

The SIDA may include airline administrative offices, airport fuel farms, and any other area in which the airport operator or federal regulator feels that identification should be worn. The SIDA requires a higher level of security than the security requirements of the air operations area. Within the SIDA, individuals must wear an airport access/ID badge, have passed a CHRC and security threat assessment, challenge those not wearing an access/ID badge, escort those without an access/ID badge, and have completed the SIDA training program. The SIDA does not specify access control requirements, but a SIDA could have access control requirements when combined with an AOA or a secured area (see Figure 5.2).

Table 5.2 Airside Security Areas

Security Areas	Secured Area	Security Identification Display Area	Air Operations Area
What areas are covered	The highest level of security. Passenger enplane and deplane areas, and baggage load and sort areas.	This could include areas where identification is necessary but access control may not be such as air carrier administrative areas, fuel farms, and cargo areas.	Aircraft movement and parking areas, loading ramps, and safety areas, which may also include general aviation areas and those areas near runways and taxiways.
What security measures are required	Access CHRC ID. Display/challenge training post signs.	Access CHRC ID. Display/challenge training post signs.	Access control but does not have to meet secured area access control standards. Orientation (provide security information but not necessarily training). Post signs.

The AOA includes runways, taxiways, and aircraft movement and nonmovement area, parking areas, usually general aviation areas, and, until recently, air cargo facilities.[7] The AOA represents the lowest level of security required at any commercial-service airport with a complete security program. AOAs are usually physically separated from the areas where commercial aircraft are handled (passenger loading and unloading, baggage, and cargo) and require a lower level of security. The AOA must be protected by an access control system, but that system does not have to meet the standards of a secured area. Airport access media are not required to be worn, and anyone operating in the AOA only needs to be provided with security information, not SIDA training.

Commercial-service airports are required to designate an AOA, unless the area is designated as a secured area (in which case the AOA label is dropped and it's just called the "secured area"). Some airport operators designate an entire airfield as a secured area and do not have lesser areas of security, such as AOAs or SIDAs; however, there is a high cost associated with this practice. AOAs are common in general aviation areas of an airport where issuing access media to all personnel operating airside is either too costly or logistically difficult. Before the air cargo regulatory update in 2006, many air cargo areas were also designated as AOAs. Regulations now require air cargo operations to be located within SIDAs.

Some airport security coordinators have a difficult time distinguishing between the requirements of the secured area, the security identification display area, and the air operations area. Many ASCs label these areas with their own terminology and then work with their federal security director to interpret the definitions to meet individual airport needs.

Access Control Systems

Once the security areas have been defined (see Table 5.2), the next requirement for the airport security coordinator is to determine the access controls that will be put into place to prevent access to the airfield by unauthorized personnel and to separate the security areas from each other (Table 5.3). The following section lists the regulatory requirements for each of the security areas, the access control system requirements, and provides an example of how to establish the security areas with appropriate access controls.

The underpinning of controlling access is to permit authorized personnel into only those areas they are authorized to enter and to prevent all other forms of unauthorized access. An access control system is more than just a system of computerized card readers or other measures to prevent unauthorized access. An airport access control system includes signage, personnel identification systems, law enforcement response, airfield security patrols, security training, and other measures designed to prevent unauthorized access to the airfield. Access points include gates, doors, guard stations, electronic access

---

[7]With the air cargo regulatory revision in 2006, air cargo areas must now be designated AOA/SIDA, at a minimum.

Table 5.3 Airport Personnel Identification Matrix

Person	Public Areas	Nonpublic Areas	Sterile Area	Security Areas (SIDA, AOA, etc.)
Visitors	Yes	No	No*	No
Passengers	Yes	No	Yes	No
Vendors, contractors, and tenants	Yes	Maybe	Maybe	Maybe
Airport or airline employees	Yes	Maybe	Maybe	Yes

Note: The level of required access for aviation employees depends on their job duties.

*Unless required as part of escorting a minor, assisting an elderly or disabled passenger, or authorized visitor.

points (e.g., doors with access controls, sensor line gates, automated portals), and vehicle inspection stations, which may incorporate crash barriers and blast protection.

Types of access may be categorized as doors or gates needed for a particular employee's job duties. For example, an employee of a fast-food restaurant within the airport would not need access to an airline flight planning room.

At most commercial-service airports, the computerized system is known to the TSA as the Access Control and Alarm Monitoring System (ACAMS) and is addressed under Part 1542.207(a); alternative access control measures are covered under Part 1542.207(b). The ACAMS is not intended to control the flow of passengers; rather, the purpose of ACAMS is to prevent passengers from accessing airside areas used by authorized personnel.

The RTCA 230[8] standard provides guidance for designing an access control and alarm monitoring system. RTCA 230 specifies technical standards for operating systems, the *Airport Security Policy and Guidance Handbook* (a TSA SSI document), and the TSA Guidance Package—Biometrics for Access Control (March 31, 2005).

Access Control: General Description

At a commercial-service airport, there is typically a centralized access control computer and a series of access control readers (card, proximity, PIN, or biometric)[9] located at airside access doors and gates; CCTV cameras watch over these locations. The access control computer system contains a database of aviation employees with approved access to various parts of the airfield. When an aviation employee uses an access/ID badge to access a door or gate, a microchip within the access reader is queried to determine if the individual has approved access at that location. If access is approved, the reader

[8]RTCA, Inc. is a private, nonprofit corporation. RTCA functions as a federal advisory committee. The FAA uses its recommendations as the basis for policy, program, and regulatory decisions. See *www.rtca.org/aboutrtca. asp#spec_comm*.

[9]Access media control devices include (1) card readers that detect a magnetized strip on an airport identification badge (similar to a credit card), (2) proximity readers that read a holographic image off of the airport identification badge, (3) 10-key PIN code device where individuals must enter the approved code to gain access, and (4) biometric readers used to scan and identify fingerprints, hands, facial features, voiceprints, or irises. There are also keys implanted with a computer chip that meet the access control requirements. These keys only work in special locks, but they can be useful in areas such as bag belt maintenance hatches and rarely used airfield perimeter gates.

triggers the door or gate to open. If the individual does not have access, the system temporarily issues an audible alarm to let the individual know he or she does not have access. The access control reader will send an alarm message to the central computer where an operator may elect to take additional action, such as dispatch a security guard or police officer to investigate the situation. Depending on the complexity of the system, many access control readers have an associated CCTV camera watching the access point, and the camera has intercom ability for the operator at the security control center to be able to converse with the person attempting access.

The ACAMS at a minimum is required to do the following:

1. Monitor access to the secure and sterile areas, to annunciate any security violations, and to record and log events.[10]
2. Verify that the holder of the access/ID badge is entitled to pass through the portal and either unlocks it to allow passage or denies passage and provides a local indication of this denial. It is not a requirement that the ACAMS monitor all access control points. Rarely used access points can be controlled with a padlock, with keys provided only to airport maintenance, emergency services, or airport operations and security personnel. Many airports use their ACAMS to control access to their air operations areas, but this is not required.
3. Automatically log all attempts to enter secure and sterile areas as appropriate, whenever successful or unsuccessful.
4. "Airport security systems should be high availability systems operating 24/7/365. System availability should meet or exceed 99.99%: higher performance requirements should be considered for higher risk airports" (TSA, 2006, p. 153).

There are numerous other methods of access control that must be implemented where a computerized access control system cannot be used. Examples of such a location are the runways and taxiways of an airport or where it is impractical to install an access control reader, such as roads used to service the AOA. Another example in use at many general aviation FBO at commercial-service airports is to have a customer service representative log in and out individuals who access the AOA, which is typical for corporate and private flight operations departing out of an FBO.

At many airports the general aviation ramp is located on the opposite side of the airport from the airline terminals, with the runways separating the two entities. Vehicle service roads wind around the approach end of runways and connect the two areas. In this situation, a checkpoint can be put on the vehicle service road with a security guard and an access control reader. This prevents unauthorized individuals from transiting from the general aviation area to the scheduled service area. It does not prevent an unauthorized person from walking or driving across runways and taxiways from the general

---

[10]This annunciation can be accomplished locally by means of a local alarm and remotely at a dispatch or control center monitoring the alarms and capable of dispatching appropriate personnel to the scene of the breach or attempted breach.

aviation area to the scheduled service area, but the concept of time and distance can be used—the distance between the lesser and greater security areas is far enough that an intruder will be spotted by an individual (e.g., an air traffic controller, security or ops officer, or police officer) and prevented from reaching the area of higher security. Putting an access control reader or security guard gate is not feasible on runways and taxiways, so an alternative form of access control must be implemented.

Alternate forms of access control covered under Title 49 CFR Part 1542.207(b) include any other system, measure, or procedure that meets or exceeds the Part 1542.207(a) standard. To get an alternate program approved, the airport operator must demonstrate that there is a need for the access point or that one naturally exists through the design of the airport, such as a runway separating the general aviation side of the airport from the commercial-service side. The airport operator must also explain why a computerized system cannot be used, and justify how the alternative system meets or exceeds the computerized standard and can accommodate current and projected throughput of the access point. Alternate systems must still be able to detect, delay, and provide notification for response to the presence of unauthorized personnel. Some examples of alternate systems include:

- Security guard continuously (or frequently in some cases) watching a particular area.
- CCTV coverage with continuous human monitoring or smart CCTV software, motion detectors, or other technology that sends an alarm when an intruder enters the area.
- Runways and taxiways that are watched by FAA air traffic control personnel during flight operations. SIDA training for personnel in the SIDAs who are required to report unauthorized personnel in the SIDAs.

For an alternative form of access control to be approved, the ASC must often implement a combination of procedures. For example, to separate an AOA from a secured area or SIDA, an airport operator may elect to paint a visible indicator, such as a red line, on the airfield delineating the border between the two areas. This, by itself, is normally not enough to satisfy the requirement to "detect and respond" to an unauthorized entry. Therefore, the ASC will incorporate the identification of the line and its meaning within the SIDA training program and combine that training with the requirement that all individuals who have been issued an access/ID badge challenge anyone they see who is not in possession of and displaying an approved access/ID badge. Airfield patrols by airport operations, security, and police personnel, along with CCTV monitoring by security operations center personnel, may be required for the TSA to approve this alternative form of access control.

Airfield Perimeter

FAA Regulation Title 14 CFR Part 139.335 Public Protection requires commercial-service airports to take measures to "prevent inadvertent entry to the movement area by unauthorized persons or vehicles ... and [install] fencing meeting the requirements of 49 CFR Part 1542." The most common method of preventing such access is with a perimeter fence, meeting regulatory minimums of 7 feet of fabric topped by 1–3 feet of barbed wire. The TSA offers the following description of airfield perimeter fencing or barriers:

> *To delineate and adequately protect the AOA, SIDA, and other Security Areas from unauthorized access, it is important to consider boundary measures such as fencing, walls, or other physical barriers, electronic boundaries (e.g., sensor lines, alarms), and natural barriers in the planning and design process of an airport. Access points for personnel and vehicles through the boundary lines, such as gates, doors, guard stations, and electronically controlled or monitored portals, must also be considered. In addition, there are other security measures which should be part of the design that enhance these boundaries and access points such as clear zones on both sides of fences, security lighting, locks, monitoring systems such as CCTV, and signage. (TSA, 2006, p. 22)*

From 2010 to 2012, there were a high number of breaches into the security area by vehicles driving through airfield access gates and onto the aircraft areas, and in July 2012 an airline pilot who had been placed on administrative leave and was under investigation for murder, jumped the fence at the St. George Airport in Utah and accessed an unattended regional jet. He drove the aircraft through the fence and into the parking lot, where he committed suicide. These incidents have called into question whether the airport perimeter security requirements are adequate. Simple fencing, or in the case of natural barriers such as bodies of water, essentially keep the "honest people honest," and will not stop a determined individual with the means and motive to overcome the barriers. A consideration must be made, however, whether the threat of individuals attacking the airport or accessing or attacking aircraft on an airport is significant enough to change the requirements.

Barriers should prevent unauthorized access and ideally differentiate between an authorized and unauthorized user. There are four types of barriers: physical barriers, electronic boundaries, natural barriers, and access points.

Physical barriers, such as fencing, should be selected by conducting a vulnerability assessment. Some airports may be at higher risk of perimeter intrusion and therefore should install fencing that exceeds the standard 7 feet of chain-link fabric plus 1–3 feet of barbed or razor wire (see Figures 5.3–5.6).

Depending on the threat level at a location, fencing can include motion, tension, or other electronic sensing means. These features are tied into an intrusion detection system and monitored in the airport's security operations control center.

Buildings may be incorporated into the physical barrier of the airport perimeter, abutting the airfield perimeter fence or security gates. Interior walls of a building and interior doors are often tied into the airport's ACAMS, regulating airside access for authorized personnel. Interior walls used in such fashion should reach to the ceiling, not just the suspended ceiling. Exterior walls on an airport perimeter (versus fencing) are less common in

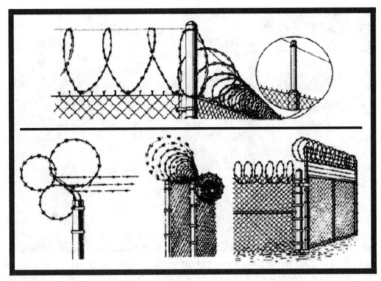

FIGURE 5.4 Chain-link fence barbed-wire configurations (TSA Security Design Guidance).

FIGURE 5.5 Vertical bar fence (TSA Security Design Guidance).

the United States because of their cost, but they can provide the advantage of reducing the visibility of storage or security areas and be less scalable because of the lack of handholds. They are also in use at some airports as jet blast deflectors, and serve the dual purpose of being a physical barrier and a safety barrier.

U.S. airports are in the early stages of evaluating electronic boundaries. One concept incorporates airport surface detection equipment (ASDE), which is ground radar used to

FIGURE 5.6 This type of reinforced fencing has been has successfully tested to the U.S. Department of State's K8 Anti-Ram rating, stopping a 15,000-pound truck traveling at 40 miles per hour within 3 feet of the fence line (TSA Security Design Guidance).

track the movement of vehicles and aircraft on the airfield in low visibility. The TSA offers the following description of ASDE systems:

> *Tests by TSA's Transportation Security Laboratory are designed to demonstrate that modified ASDE radar could differentiate between "approved" and "unauthorized" targets, including persons and ground vehicles as well as marine craft approaching a waterside perimeter. The radar is intended to determine the origin and track the paths of movement of these targets. With further development, the system is expected to classify an object, to predict its likely next movements or directions, and to assist the operator in providing an appropriate level of response. Some of these functions can also be automated and applied to pre-programmed zones of priority to enhance security decision-making. (TSA, 2006, p. 28)*

A second electronic boundary uses radio frequency identification (RFID) technology. The implementation of RFID technology includes active RFID tags, stationary RFID readers, and mobile RFID readers. All authorized airport personnel and vehicles have RFID tags in the access/ID badges to facilitate real-time tracking of vehicles and personnel. Stationary RFID readers are located around airport perimeters or within security areas and work on the basis of signal triangulation to observe the real-time motion of a tagged object or person. If an unauthorized signal is detected, then the system sends a notification to the security operations center for law enforcement or security response.

The third type of airport perimeter barrier is a natural barrier (e.g., bodies of water, expanses of trees, swampland, or cliffs). Some natural barriers may border an airport in such a way as to make physical barriers or fencing impractical. In other instances, fencing or physical barriers would conflict with aircraft navigation, communications, or runway clear areas beneath approach paths. Two other forms of natural barriers are known as *time and distance* and *detect, delay, and respond* (DDR).

The time and distance barrier depends on the idea that the extensive time it would take an intruder to reach a critical facility, such as a concourse or the secured area, combined with the high visibility from a variety of employees in and around the airplanes, effectively reduces the likelihood of the intruder successfully reaching the protected location. The concept of detect, delay, and respond is that remote areas are "sufficiently removed from the primary security-related areas to allow the airport to detect an intrusion, and delay its progress until an appropriate security response can be implemented" (TSA, 2006, p. 29).

While not considered a "barrier" SIDA training and challenge programs may enhance access control measures within a security area.

Access points are protected by three types of gates that allow vehicular access to the airfield for routine maintenance and emergency operations. Vendors, contractors, airport operations and maintenance personnel, aircraft operator personnel, cargo delivery vehicles, and just about any vehicle that needs authorized access to the airfield to conduct business all use routine operation gates. These high-throughput gates should be designed for long life and to minimize delays to users. Because of their high throughput, routine operation gates also represent a higher risk of allowing unauthorized individuals or vehicles access to the airfield. These gates are typically electrically operated, connected to ACAMSs, staffed with security personnel, and may include secondary barriers (in addition to the primary sliding or swing-style security gate) such as retractable bollards, drum barriers, or cable barriers.

Maintenance gates are typically seldom used, often padlocked, and used by airport and FAA personnel to conduct maintenance activities on airfield equipment, utilities, or grounds maintenance. Emergency gates are for use by on- and off-airport emergency response personnel. Remote control units within aircraft rescue and firefighting (ARFF) trucks enable some gates to be operated electrically. These gates should be on frangible mounts to allow emergency vehicles to crash through in the event of rapid response to an aircraft accident or incident. Emergency gates are sometimes padlocked or equipped with a Knox-Box®, which provides emergency rapid nondestructive entry for emergency personnel.

Security gates should have a 10-foot clearway at the surface and ideally be recessed 3 feet into the ground to prevent tunneling by both humans and animals (there must still be 7 feet of visible fencing above ground). Some airport operators elect to cement the entire fence line at the surface to prevent tunneling and digging. Security fencing must have no more than a 4-inch clearance between the ground and the bottom of the fence.

Within buildings also serving as a physical barrier between the landside or sterile areas and airside, the number of doors must be kept to an operational minimum and, ideally, equipped with an alarm and CCTV in the case of emergency ingress. Many doors within an airport also allow emergency ingress in case of a fire or other emergency, which presents a challenge to airport security. Usually, these doors will open after the activation of a crash bar on the door or after the alarm system is activated. If the system is so equipped, an alarm will sound in the security operations center and a CCTV camera will trigger at the door and begin recording the activities around that door, enabling security operations personnel to make the appropriate response.

At large airports, CCTV monitors are used to view access doors and are tied electronically into the ACAMS. Unfortunately, because of costs, the CCTV is usually located on the interior side of a door. This is for a practical purpose; most door alarms are not the result of intentional unauthorized access but of an aviation employee having functional problems with the door. Either he or she does not have access at that particular access point and does not know it, requiring the security operations personnel to explain it, or he or she does have access but something else (e.g., a hardware or software problem, weather, or power) is preventing authorized access. The CCTV system is therefore located on the inside so that security operations personnel can converse with the person at the door through the intercom. This system requires intervention by the security operations center to assist with problems. Cameras on both sides of an ACAMS controlled door would provide security personnel a higher level of security and a frontal view of an offender, but this is still rare, although some airports are implementing this technology.

Another security measure to prevent or deter unauthorized entry into security areas is the integration of a door opening delay. Because the majority of access control doors double as fire exits, a conflict exists between security and public safety. The ICAO recommends a 15-second delay before a fire door will open for anyone who activates its crash bar. Some local fire codes and the approval authority (the fire marshal) have different perspectives, so in some cases there is only a 5-second delay and at some airports there is no delay. Another important consideration is a power failure or fire that triggers the fire doors to automatically open. When this occurs, airport security personnel closely monitor the situation and likely will dispatch personnel to help contain people entering security areas. A security guard, a law enforcement officer, or an airport operations employee most often initiates intruder response. In some cases, such as passenger boarding gates, there is a mantrap door, with the secondary door serving as an adequate intruder response (Figure 5.7). Subsequent to an intruder alert, security personnel will follow up to determine the identity and intent of the intruder.

Staffed guard stations are used in some locations where higher levels of security are required or desired. These are usually at vehicle entry points, but they can also be used at employee turnstiles to reduce or prevent piggybacking or tailgating. Guard stations should be sheltered and guards should be able to communicate with the security operations center. Guards should be trained in how to search a vehicle for prohibited items, suspicious awareness, and actions to take should an unauthorized entry occur. Guards should receive the latest intelligence information relevant to their responsibilities, and briefings of any special occurrences on the airfield, such as a VIP or prisoner escorts that would cause additional or unusual traffic through the checkpoint. Guards should receive a "stop list," which contains the names of people who no longer have access authority but may still be in possession of a valid access/ID badge. Some guard stations include a law enforcement officer with access to a vehicle to provide an armed response or pursuit to an intrusion.

FIGURE 5.7 A mantrap door in place at Ben Gurion International Airport in Israel.

Some gates include a vehicle inspection station, which provides an area outside the blast envelope (Figures 5.8 and 5.9) in which to inspect suspicious vehicles or conduct inspections of all vehicles during high threat levels. The station should be far enough away from the entry gate so as not to impede the movement of other vehicles not subject to search.

Other security measures can include fence clear zones,[11] security lighting, locks, CCTV coverage, and signage. Higher-tech sensor line gates are being tested. A sensor line uses microwaves, infrared, or other electronic sensor technology other than a hard physical barrier (Figure 5.10).

Authorized Signatory

Virtually all personnel working at a commercial-service airport require some sort of credential (i.e., ID badge) that accesses their work areas and areas of the airport where they are required to go to do their job. ASCs need some way to verify that individuals applying for airport access/ID badges actually work for a company at the airport, and to determine what access (doors and gates) are required for their job functions. Therefore, every company that works at the airport and that requires badges and access is required

---

[11]In these zones there are no climbable objects such as trees or adjacent buildings abutting the fence.

Type of Explosive	Explosive Capacity in TNT Equivalents	Lethal Air Blast Range
Pipe Bomb	5 lbs. (2.3 kg)	
Briefcase, Backpack, or Suitcase Bomb	50 lbs. (23 kg)	
Compact Sedan (in trunk)	500 lbs. (227 kg)	100 ft. (30 m)
Full Size Sedan (in trunk)	1,000 lbs. (454 kg)	125 ft. (38 m)
Passenger or Cargo Van	4,000 lbs. (1,814 kg)	200 ft. (61 m)
Small Box Van (14th ft box)	10,000 lbs. (4,536 kg)	300 ft. (91 m)
Box Van or Water/Fuel Truck	30,000 lbs. (13,608 kg)	450 ft. (137 m)
Semi-trailer	60,000 lbs. (27,216 kg)	600 ft. (183 m)

FIGURE 5.8 Blast distance radius.

to designate an authorized signatory (AS). The AS is responsible for authorizing their employees to receive a badge and approves their level of access.

Smaller companies with few employees usually only have one or two authorized signatories, while air carriers and other tenants with large operations at the airport may have several individuals designated as authorized signatories. For tenants with hundreds of employees, badge access is often broken down by department or job title, or in some cases, both. For example, an airline gate agent has different access requirements to do his or her job than a ramp worker loading bags, or a pilot based at that airport. Many ASCs divide up the businesses that operate at the airport into air carriers, tenants, contractors, vendors, and government employees (including the airport staff). Also, in many cases, contractors and vendors must be sponsored by an air carrier, tenant, or government agency to prevent companies from freelancing services or accessing the airport when they do not have a valid reason.

The AS also provides a way for the ASC to hold the company and their employees accountable to the rules and regulations of the airport. Many times it is the AS who is required to respond to violation notices issued to their personnel, or be accountable for badge suspensions, the return of access/ID badges that are no longer required (employee termination, retirement, or resignations), and other issues that relate to the company's compliance with the regulations and the ASP.

Anyone designated an AS is required to be trained in their responsibilities by the airport security office. The training includes the proper method to fill out the paperwork so that employees can be issued access/ID badges and the method to fill

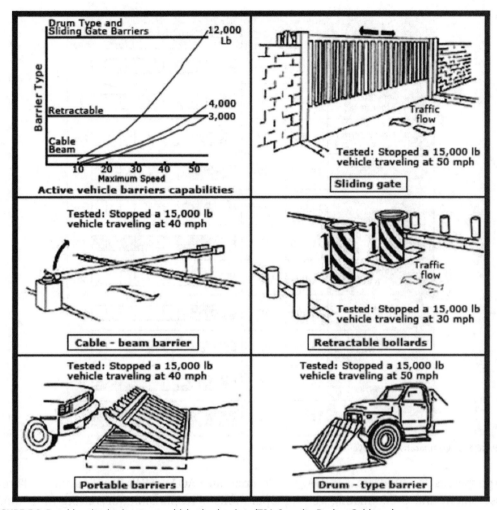

FIGURE 5.9 Road barricades in use at vehicle checkpoints (TSA Security Design Guidance).

out paperwork to request access changes (adding, subtracting, or changing access to doors and gates).

Credentialing

In accordance with Part 1542 and TSA Security Directive 1542-04-08 G[12] (OIG, 2011), applicants are required to undergo a CHRC and have an approved security threat assessment (STA) from TSA before receiving a badge and obtaining unescorted access to certain

---

[12]While we are citing a security directive, normally an SSI-protected document, unfortunately this SD was posted on the Internet. In this particular citation, this information was published in a publicly available OIG report.

		Vehicle		Protection Level (0-10)	
		Weight	Speed		10
Passive Barrier Test Results	Concrete Filled Steel Bollards	4,500	30		1
	Jersey Barrier	4,000	50		2.6
	Straight Retaining Wall	15,000	30		3.6
	Sloped Back Retaining Wall	15,000	40		6.4
	Concrete Planter Retaining Wall	15,000	50		10
Active Barrier Test Results	Cable - Beam Barrier	10,000	15		.6
	Rectractable Bollards	15,000	30		3.6
	Portable Barriers	15,000	40		6.4
	Drum Type Barriers	15,000	50		10
	Sliding Gate	15,000	50		10

FIGURE 5.10 Comparative effectiveness of barrier types.

airport areas (secured areas, SIDA, sterile area, and AOA). TSA's Transportation Threat Assessment and Credentialing Vetting Operations is responsible for vetting individuals with unescorted access to secure areas. This is accomplished by comparing the applicant's information against data sets to discern whether the applicant is a threat to transportation or national security. Approximately a half a million individuals are vetted in this process. This is collectively known as the *credentialing process*.[13] While undergoing credentialing, individuals may be escorted by those having properly issued access media, if allowed by the ASP.

A CHRC is a listing of certain information taken from fingerprint submissions retained by the FBI in connection with arrests and, in some instances, federal employment, naturalization, or military service. A STA is a check conducted by the TSA of databases,

[13]Before 2008, this was commonly called the CHRC, or criminal history record check process. However, in 2007 and 2008, with the continuous addition of processes that must be completed before an individual can receive an airport access badge, the term *credentialing process*, or simply *credentialing*, is being used.

including terrorist watch lists, to confirm that an individual does not pose a security threat and possesses lawful status in the United States, and to verify an individual's identity. To pass the CHRC, a person must not have been found guilty, or not guilty due to insanity, within the past 10 years of a number of disqualifying offenses (Table 5.4).

TSA's Threat Assessment and Credentialing (TTAC) adjudication service completes the STAs for applicants; the FBI conducts the CHRC while TSA inspector personnel provide oversight of the airport badging process. A regulated party, such as the airport operator or air carrier, conducts the badging process itself.

Designated airport operator employees, known as trusted agents, perform the essential functions of the badging process, including collecting, verifying, and inputting applicant data used for the STA process, and fingerprinting applicants for the CHRC. Airport operator personnel ensure badge applications are complete with the required biographical and

Table 5.4   Disqualifying Offenses

Criminal History Record Check	Disqualifying Offenses
Forgery of certificates, false marking of aircraft, and other aircraft registration violations	Assault with intent to murder
Interference with air navigation	Espionage
Improper transportation of a hazardous material	Sedition*
Aircraft piracy	Kidnapping or hostage taking
Interference with flight crewmembers or flight attendants	Treason
Commission of certain crimes aboard aircraft in flight	Rape or aggravated sexual abuse
Carrying a weapon or explosive aboard an aircraft	Unlawful possession, use, sale, distribution, or manufacture of an explosive or weapon
Conveying false information and threats	Extortion
Aircraft piracy outside the special aircraft jurisdiction of the United States	Armed or felony unarmed robbery
Lighting violations involving transporting controlled substances**	Distribution of, or intent to distribute, a controlled substance
Unlawful entry into an aircraft or airport area that serves air carriers or foreign air carriers contrary to established security requirements	Felony arson
Destruction of an aircraft or aircraft facility	Felony involving a threat
Murder	Felony involving willful destruction of property, importation or manufacture of a controlled substance, burglary, theft, dishonesty, fraud or misrepresentation, possession or distribution of stolen property, aggravated assault, bribery, illegal possession of a controlled substance punishable by a maximum term of imprisonment of more than one year
Assault with intent to murder	Violence at international airports

*Conduct that is directed against a government and that tends toward insurrection but does not amount to treason. Treasonous conduct consists of levying war against the United States or of adhering to its enemies, giving them aid and comfort. See *www.lectlaw. com/def2/s020.htm*.

**Knowingly and willfully operating an aircraft in violation of an FAA regulation related to the display of navigation or anti-collision lights, or knowingly transporting a controlled substance by aircraft or aiding or facilitating a controlled substance offense.

fingerprint data for the STA and CHRC. Critical data processed from the application includes full legal name, date of birth, place of birth, passport number, and alien registration number. Social security numbers are not required but generally the STA cannot be completed without it.

Airport operator employees receive the applicant's clearance status from AAAE's Transportation Security Clearinghouse, or another approved clearinghouse. If the STA and CHRC results are favorable, the airport operator employees will issue the badge with access to secured airport areas. Air carriers and flight schools use similar methods to vet their personnel.

Clearinghouses also handle fingerprints submitted for the Alien Flight Training Program and the public and private charter programs, and it assists with inked-card submissions if a live-scan device is not available. The clearinghouse will also assist in the reconciliation of fingerprints that are missing in the submission of difficult-to-classify fingerprints (construction workers, cement workers, and others whose occupations have made their fingerprints difficult or impossible to classify) and in handling issues such as amputations, scars, and deformed fingers. Prior to April 2012, the clearinghouse handled fingerprint records, but now must submit fingerprints and STA information as a package to the TSA and FBI. The clearinghouse does not see the actual results or the details of the criminal records.[14]

Every applicant for access media must produce two forms of ID; at least one must be a government-issued photo ID and one must be proof of citizenship. Enhanced background screening services can provide more in-depth identity checks to help ensure the person presenting the identifying information is in fact that person and not using stolen ID information. In a 2011 investigation by the DHS Office of the Inspector General, they found that in many cases airport operator employees were not using available tools to assist in the identification of fraudulent documents, such as scanners, ultraviolet lights, and loupes (magnifying lenses), even when the tools were available at their workstations (OIG, 2011), and only one airport had a formalized training program focused on airport operator employees' duties and responsibilities. Part 1542 requires each airport operator to ensure that individuals performing security-related functions are briefed on specific requirements or guidance as they relate to the performance of their duties. Briefings must cover the provisions of Title 49 CFR Part 1542: security directives, information circulars, and airport security programs (OIG, 2011).

---

[14]Before 1996, background checks were not specifically regulated. Responsibility for conducting the checks fell mainly on the employer. The employer decided who needed access media to the SIDA. In 1996, the Aviation Security and Anti-Terrorism Act required aviation employees to submit 10 years of employment history. The authority issuing access media verified the most recent 5 years of employment history. If there were significant gaps within the entire 10-year period, then the individual was subjected to a fingerprint-based CHRC. Individuals without employment gaps were eligible to be issued access media. This process was known as the *access investigation* and many airports and aircraft operators are still conducting the checks, even though it is no longer required.

It is a common misconception that the CHRC identifies an individual. It does not. The CHRC only compares the fingerprints it has been given to the fingerprints of convicted criminals. If an individual does not have a criminal record, it is possible to use a fake ID to obtain airport access media—a crime under Title 49 CFR Part 1540.103 Fraud. Operation Rampcheck, conducted by the Immigrations and Customs Enforcement (ICE) service, has arrested thousands of individuals who received airport access media using fake identification. Many illegal aliens have been arrested at commercial-service airports in the United States for obtaining airport access/ID badges using fake identification.

Airports are not required to conduct recurring CHRCs to ensure that badge holders maintain their reputable status (OIG, 2011), but under SD-1542-08 G badges must be renewed every two years, whereupon the employee must bring the same identifying information to be issued a new badge. According to the Inspector General's report, passing an initial CHRC does not preclude employees from engaging in subsequent criminal activity and presenting an insider threat at airports (OIG, 2011).

Individuals who have been issued access/ID badges are required to self-report if they are convicted of a disqualifying offense after they have been issued the ID. However, the OIG report states that the self-reporting policy is ineffective because most employees would not report themselves for fear of losing their job (OIG, 2011). The past actions of some employees reinforce this notion. For example, in 2007, a customer service agent with no prior record was found guilty of repeatedly accepting bribes totaling $21,500 from an undercover agent, and agreed to smuggle $396,000 in cash, plus illegally export weapons, military night-vision goggles, and a cellular phone "jammer" to a foreign country. Also in 2007, two workers at another airport were arrested after bringing guns and drugs on a flight. One worker was able to stow the guns and drugs near the departure gate ramp after using his airline uniform and badge to bypass TSA security.

Some airports have taken additional measures, such as having names run through local and state criminal databases, using data mining software to investigate connections and associations, requiring badges to be reissued annually, and in one case, hiring a private detective to check 100 names per month for new or outstanding warrants that have not been reported. In the OIG report, both the TSA and the airports agree that recurrent CHRCs are necessary. In response, the TSA is developing rulemaking that will create a standard duration for all STAs and CHRCs.

Another issue for security practitioners is the acceptance of a CHRC conducted by another regulated agency, such as another airport or an airline. Airports can decide whether to accept the CHRC conducted by an aircraft operator. Provided the air carrier has a record of providing honest information to the airport, the airport operator usually accepts the air carrier's CHRC. This reduces by thousands the number of people the airport must process through the credentialing process. When an employee of one airport transfers to another airport, it is up to the ASC at the new airport to decide whether to accept the CHRC from the previous airport. Thirty days between employment dates is considered the limit to accept another airport's or aircraft operator's CHRC.

The ASC can also accept a U.S. government employee's CHRC, provided the CHRC was required by the government agency. This is a common occurrence for TSA, FBI, FAA, CBP, and Department of Agriculture personnel needing SIDA access. ASCs should consult with the TSA regarding special circumstances such as furloughed employees and extended leaves of absence.

Personnel Identification Systems

Personnel identification for unescorted access to an airport is required for access to the SIDA. The ID badge is known by a variety of names, including airport ID badge, access/ID, SIDA credentials, SIDA badge, and so on. Identification must be large enough to be seen easily on the ramp by others working nearby; 2 × 3 inches is standard. Access/ID badges must include a full-face photograph, the employee's name, the employee's employer, a badge expiration date (annually), a unique sequenced number, and the scope of access privileges. The scope is usually shown by the color of the ID badge. Access/ID badges usually denote airfield access privileges using color or icons placed on the badge. Individuals cannot have their faces covered on their ID photo, but whether an individual is allowed to wear a hat for their photo is at the discretion of the ASC.

The TSA periodically requires the reissuance of all airport access media and a change in the appearance of the badges.[15] When this occurs, airport operators will often attempt to maintain some semblance of continuity from the old badges to the new, while still making them distinguishable from previous designs. For example, airports may use a red badge to denote movement area access privileges and, upon reissue, the airport may switch to a red striped badge or a half-red/half-white badge. This reduces confusion on the part of airport workers, while still distinguishing the badge from previous patterns.

In some cases, the airport operator may allow airport tenants and air carriers to use their own forms of identification, provided they meet the standards of Part 1542.211 and receive TSA approval. Under Title 49 CFR Part 1542.5 Inspection Authority, all airports must allow TSA inspectors access to the airport. TSA-compliance inspectors carry their own identification cards, which must be accepted as valid access/ID at any commercial-service airport. FAA aviation safety inspectors have similar privileges and their own FAA 110-A identification cards. Other accepted forms of identification include flight crew access/ID badges issued by the employer (an air carrier regulated under Parts 1544, 1546, or 1548) that meet the requirements of this section.

Other generally acceptable forms of access/ID media at airports include, for airports with military operations or facilities, U.S. military–issued identification cards; for offsite emergency services personnel, photo identification and access/ID proving employment as an emergency worker; and for airports with general aviation facilities, an FAA pilot certificate and a government-issued photo ID for pilots transiting directly to or from

---

[15]There is no set time schedule for conducting the reissuance process. It is a decision made by the TSA based on a particular set of circumstances, such as outside intelligence indicating a system has been compromised or the overall number of lost access/ID badges.

aircraft or conducting preflight inspections and passenger and baggage loading. Whether airport operators accept these alternate forms of identification depends on the individual ASP.

Although FAA Regulation Part 1542 did not specifically require airport operators to provide access/ID badges to personnel working at the airport not requiring SIDA access, subsequent federal directives require that everyone working at the airport, regardless of location, apply for and receive an access/ID badge. This does not generally include taxi and limo drivers, bus drivers, or delivery personnel but some airports have required that landside personnel undergo an STA and receive an identification badge that does not provide security area access. In fact, it was this process that helped identify the movement and activities of Najibullah Zazi, an airport shuttle bus driver at Denver International Airport who pled guilty to charges that included conspiracies to use weapons of mass destruction, commit murder in a foreign country, and for providing material support for a terrorist organization, for his role in plotting to bomb the NYC subway system.

Access/ID badges must be worn on the outside of the outermost garment whenever the individual is in the SIDA. Airport operators can make exceptions to this requirement for activities where wearing the badge on the outside of a garment may endanger the badge wearer, such as during aircraft maintenance activities, but any such exception must be articulated in the ASP.

Lost or stolen badges must be immediately rendered inactive from the security system and the badge placed on a stop list. Although a deactivated access/ID badge should alarm a computerized security system, there is the possibility that someone in possession of a deactivated access/ID badge may jump the perimeter fence or enter the airside through a fire alarm door and then place the illegal access/ID badge on himself or herself. Putting the deactivated access/ID badge onto a stop list provides security and law enforcement personnel with a tool to help determine that a person is not in possession of valid access/ID media. Only one badge should be issued to each person unless that individual also is employed with another airport tenant, in which case two badges—or more depending on the number of jobs the individual has—are generally acceptable.

Badges have an expiration date within two years of issuance. Some airports require annual reissuance or revalidation. Temporary badges can be issued at the discretion of the ASC. Individuals receiving temporary badges must still pass the credentialing process, and the airport must retrieve the badge when the user no longer needs it. Temporary badges work well in situations where an individual already has passed a CHRC at another airport, airline, or government organization and will be temporarily assigned to the airport (e.g., airline mechanics brought in for specialized maintenance or government inspectors and agents brought in for special projects or to cover employees on leave), or for construction personnel who are on projects for less than two years, or less than what is dictated in the ASP.

Airport security coordinators must conduct routine audits of issued badges and maintain a high percentage of accounted-for badges, or the TSA may require all airport badges to be reissued or revalidated. Revalidation generally involves everybody who has been issued a badge returning to the badge issuance office to have his or her identity verified and to ensure he or she is in fact in possession of a badge.

Training

Individuals with SIDA access must be trained by the airport in SIDA-specific policy and procedures. This training must be completed before an airport ID badge is issued for SIDA access. Subjects that must be included in the training are as follows:

- Policies on unescorted access authorities
- Control, use, and display of access media
- Escort and challenge procedures
- Security responsibilities as identified in Title 49 CFR Part 1540.105
- Protection of SSI

Those who only need access to the air operations area must be provided with security information, and not necessarily trained, although it is common practice to provide security training to those needing only AOA access for better overall security. Airports provide security information in a variety of ways; smaller airports may simply brief new personnel on their security responsibilities; medium and large airports may use computer-based training programs such as that available from AAAE.

Challenge Program

Under the security regulations, anyone issued an airport identification badge must not only wear the badge whenever in the SIDA but must verbally challenge anyone seen without the appropriate badge. Should the person challenged refuse to show, or cannot produce, the appropriate media, the challenger must immediately report that individual to airport security or law enforcement personnel (or other entity per the ASP). Airport security programs often include incentive-based testing programs to encourage the challenge program. At times, a person authorized by the ASC purposely does not wear a badge in the SIDA. If challenged, a prize is awarded; if unchallenged by a passerby, a warning is issued. An airport and the TSA can issue violation notices and take enforcement actions against anyone failing to challenge someone without a badge. This type of testing must be specified in the ASP to be approved by the TSA.

Escorting Programs

The process of escorting involves individuals with unescorted access authority to the SIDA escorting individuals without such authority. Title 49 CFR Part 1542 states that escorted individuals must be continuously monitored or accompanied by the escorting individual. Each airport has discretion to interpret how this regulation should be applied in its facility. Some airports allow only one escorted party per one individual (1:1 ratio) with an airport ID badge. Airports typically have anywhere from a 1:1 ratio to a 1:15 ratio. This definition is also subject to wide interpretation. Some airports use a distance qualification rather than the number of escorted parties (e.g., within 5 feet, within 10 feet, etc.). The actual determiner is completely up to the airport operator, subject to TSA approval through the ASP. Aviation security practitioners should keep in mind that the airfield can be a large and noisy environment; the ability to control an escorted party can change depending on aircraft movements, weather and visibility, and whether it is day or night.

Construction on airports can push the boundaries of the escort policy. Construction activities often bring many unbadged workers into the SIDA and frequently with only a few badged individuals to watch over them. Some airport operators include within their ASPs the provision that construction personnel can be escorted by fewer authorized individuals provided that the construction area is demarcated by some type of physical barrier such as a snow fence or a temporary chain-link fence.

Transportation Worker Identification Credential

The Transportation Worker Identification Credential (TWIC™) program is a TSA and U.S. Coast Guard initiative. The TWIC™ program provides a tamper-resistant biometric credential to maritime workers requiring unescorted access to secure areas of port facilities, outer continental shelf facilities, and vessels regulated under the Maritime Transportation Security Act (MTSA), and all U.S. Coast Guard credentialed merchant mariners. An estimated 750,000 individuals will require a TWIC™. To obtain a TWIC™, an individual must provide biographic and biometric information such as fingerprints, sit for a digital photograph, and successfully pass a security threat assessment conducted by the TSA.

While TWIC™ may be implemented across other transportation modes in the future, the TWIC™ Final Rule, published in the *Federal Register* January 25, 2007, sets forth regulatory requirements to implement this program in the maritime mode first.

Some TWIC™ cardholders have insisted that their card provides them access to an airfield SIDA, but this is only the case if an ASC has authorized the TWIC™ as an acceptable alternate form of ID within the ASP, which many have not.

Before 9/11, there was concern from the pilot community about the feasibility of being issued a universal access/ID badge that would work at any airport. The concept became known as *universal access control*, and the TWIC™ card seems to represent the post-9/11 manifestation of that. The primary challenge to creating a universal access control system was that many large and medium commercial-service airports have proprietary access control systems. Because of the security-sensitive nature of these systems, many are not designed to be accessed from outside their own airport network. This is part of the basic cyber-security protections of the system. An access/ID badge at one airport is not compatible with an access/ID badge at another. Further, ASCs are responsible for the access/ID badges that are issued at their airport, or of which the usage is allowed by a second party (such as an airline) as identification in their SIDA. Therefore, ASCs are understandably concerned about the quality and integrity of the background check of any individual who has received an access/ID badge but did not go through the normal CHRC process. These two realities will make implementation of the TWIC™ card into the aviation industry very difficult.

Requiring all airports to redesign their ACAMSs to read a new form of access/ID media would cost millions, if not billions, of dollars. Airport security coordinators argued that the problem was already solved by allowing airlines to issue their own access/ID badges, in compliance with Part 1544.211 and as defined within each airport's ASP. In these cases, the ASP would establish the flight crew access/ID badge as an acceptable form of access/ID for the airport. Flight crews domiciled at a particular airport can receive personal identification numbers and other systems to be stored and recognized at other airports.

The universal access control card is more a matter of convenience than security. A universal access control access/ID badge actually creates additional security concerns; if an access/ID badge is stolen, it could be used to access numerous facilities. If an airport access/ID badge is stolen, the access/ID badge will only work at a single airport. The addition of biometrics to the TWIC™ is an added layer of security intended to prevent or mitigate this issue, but does not solve the access/ID background check issue.

TWIC™ cardholders must undergo a security threat assessment conducted by the TSA and may not have been found guilty of a list of felonies, similar to the Title 49 CFR Part 1542.209 CHRC process. When the Maritime Transportation Security Act introduced TWIC™, airport operators feared being required to spend money on a new security ID system when an effective system was already in place. However, the maritime, trucking, and rail industries lack the formalized and time-tested access/ID systems of commercial-service airports, and there has been a great deal of effort to bring TWIC™ to those industries. TWIC™ cards are being tested at a few U.S. airports.

Contingency Plans and Incident Management

Airports are required to have contingency plans to respond to increases in the National Terrorism Advisory System (NTAS), which replaced the previous Department of Homeland Security Alert System (the color-coded alert levels), and to specific incidents of unlawful interference with aeronautical activity.

Airports with complete or supporting security programs must have contingency plans and must implement them when directed by the TSA. These airport operators must also exercise their contingency plans as specified within their security program (usually one tabletop exercise a year with one full-scale exercise every other year). All parties involved in the contingency plans must know their responsibilities, which mean periodic meetings with those entities involved in the plans and ensuring point-of-contact information is continually updated. The TSA can approve alternative contingency measures, which still provide an overall level of security equal or greater than that of the measures in the security program. The TSA can also approve alternatives to exercises in the contingency plan, such as when an airport has actually implemented a contingency plan because of a real-world event.

NTAS communicates information about terrorist threats by providing detailed information to the public, government agencies, first responders, airports and other transportation hubs, and the private sector, rather than vague references to a potential threat and a color-coded system that was roundly criticized as being ineffective. NTAS alerts are issued only when there is credible information of a threat, and the alerts contain a sunset provision, a specific date they are automatically cancelled, thus eliminating "blanket" warnings.

Prior to 9/11, airports used a system known as AVSEC, aviation security contingency, to know what measures and equipment are necessary when the FAA increased the security measures at an airport. There were four basic levels, plus additional special-circumstances levels (measures) that could be added in. When the FAA would increase security from normal to AVSEC 1, then each ASP had within its contingency plan section certain measures to take, such as increased law enforcement patrols or random vehicle inspections (most actions are SSI).

After 9/11, DHS implemented the color-coded Homeland Security Alert System, which used five colors to designate the alert level for the United States. Like with AVSEC, each level called for certain measures to be taken by airport operators and air carriers. While there were five levels, the country never went below yellow and airports almost immediately went to orange and stayed there for several years. The colors served the same purpose as the AVSEC levels previously served.

Under the new NTAS, airports and air carriers must convert their previous AVSEC plans to match the new threat levels. However, NTAS only issues two types of alerts: imminent and elevated. Imminent warns of a credible, specific, and impending terrorist threat against the United States. Elevated warns of a credible terrorist threat against the United States.

Since there are fewer levels of alerts than there were color-coded levels, ASCs work with their FSD to modify existing contingency plans to meet the new NTAS standards. Additionally, under NTAS some alerts may be sent directly to law enforcement, certain government areas, or affected areas of the private sector, while some alerts will be issued more broadly to the American people through official and media channels.

The AVSEC program allows airport operators to better prepare for increased security measures.[16] By outlining what the airport operator is required to do when the NTAS changes and having the TSA previously approve those actions, the airport operator can better project and acquire needed resources and plan for increased funding and staffing levels. An additional benefit of the AVSEC system is that the airport public relations department is part of the AVSEC process, so that whenever the next AVSEC level calls for additional procedures, some of which may impact passenger flow, the public can be briefed on what to expect in terms of adjustments.

Incident Management

Airport operators must establish procedures to evaluate bomb threats, threats of sabotage and aircraft piracy (hijackings), and other security incidents as specified by the TSA. When an airport operator receives such a threat, it must evaluate the threat based on the

---

[16]Aircraft operators have similar contingency programs, which also increase with increases in the NTAS.

information presented and any other protocols in the ASP. The airport must then initiate the appropriate response as outlined in the airport emergency plan under Title 14 CFR Part 139.325. The TSA must immediately be notified of any act or threat to aviation. Every year, airport operators must review the responsibilities outlined in the incident plans with all entities that are part of the plan.

The difference between what constitutes a contingency plan and an incident management plan can be confusing. However, in the TSA's vernacular, it is helpful to think of contingency plans as those plans that are required to be developed and implemented when the DHS increases the NTAS, while incident management plans are those plans the airport implements when an actual threat or attack has occurred.

When discussing contingency planning and incident command, it is important to remember that neither is to be done in a vacuum. Contingency planning is the responsibility of both the airport operator under Title 14 CFR Part 139 and the ASC under Title 49 CFR Part 1542. The airport operator and the ASC must work with other agencies, onsite and offsite (fire department, paramedics, law enforcement, adjacent agencies, FBI, etc.), to develop and successfully implement their contingency plans and incident management plans, taking into account mutual aid responses, nongovernmental organizations, and the National Incident Management System protocols.

Security Tools and Considerations

Public Advisories

Airports must post signs advising the public when the secretary of Department of Transportation has determined that a foreign airport does not meet requirements for administering effective aviation security. These signs are usually posted at the security screening checkpoints. This is part of the State Department's program to protect U.S. citizens flying to foreign destinations. The program was initiated in 1978 when the United States, in the interest of protecting its citizens, desired more control over the aviation security programs of foreign airports.

TSA Programs for Airport Security

TSA provides the Airport Security Self-Evaluation Tool (ASSET) on their secure Web board for airport operators. ASSET is the third and final product associated with the TSA's Innovative Airport Security Measures Initiative, which sought to identify, develop, and publicize to airport operators best security practices. ASSET gives airport operators the ability to evaluate their airport's current level of security compared to that of the most innovative security measures in place across the U.S. commercial airport network. The tool is part of a larger effort, which sought to identify innovative security measures currently in place at U.S. commercial airports and to examine the risk assessment and resource allocation decisions that led airport operators to implement these measures.

To collect the data set of innovative security measures, the TSA surveyed commercial airport operators in 2009 and visited 24 of the 108 responding airports throughout 2010. Over 700 innovative security measures from commercial airports of all sizes were identified as either "exceeded existing security requirements," established within Title 49 CFR Part 1542, or "meeting the requirements in a unique way."

Conducting a self-evaluation via ASSET allows airport operators to:

1. Gain visibility into the most innovative practices in place across the industry by accessing the roughly 700 measures reported in the Commercial Airport Innovative Security Measures report.
2. Determine the full range of innovative measures that apply to their particular airports.
3. Create a report summarizing the innovative measures in place at other airports that may also be relevant to their own facilities.

Security Operations Center

The security operations center (SOC), known by various names including "operations center" and "communications center," is the location of the primary access control computer. The primary access control computer receives door and gate alarms as they occur and logs transactions as aviation personnel move through the facility. The SOC also includes the CCTV monitoring stations, video or digital recording equipment, and communications dispatch equipment to communicate with airport security, operations, and law enforcement personnel (Figure 5.11). The SOC is generally staffed 24/7. At some airports, the SOC is combined with other functions such as police, fire and emergency medical services (EMS) dispatch, and weather and airfield operations monitoring and emergency communications.

Closed-Circuit Television

CCTV has numerous security applications. It primarily provides a low-cost (compared to human monitoring) surveillance option that is also a visual record of events. Within aviation, CCTV has a wide variety of applications including surveillance over the following:

- Terminal and public areas
- Landside passenger pickup and drop-off areas
- Cargo and loading dock areas and perimeter gates
- Baggage-handling areas[17]
- Perimeter fence areas
- Fuel farm and remote storage areas
- Passenger and employee parking areas
- Concourse and gate area access points

[17]This form of surveillance is frequently used for employee theft prevention.

FIGURE 5.11 The SOC portion of Denver International Airport's communications center (left). The center consoles include airport operations communications, and weather and airfield condition monitoring stations with police, fire, and EMS dispatch stations are shown on the right.

Although CCTV can be used in a wide variety of applications, its operation is affected by a number of performance factors:

> *The performance of a surveillance system depends on a number of factors including the characteristics of the object to be observed (e.g., its size, reflectance, and contrast); local environmental conditions (e.g., atmospheric clarity and scene illumination); camera characteristics (e.g., detector size, sensitivity, and resolution); characteristics of the camera objective lens (e.g., focal length and relative aperture); characteristics of a display or monitor (e.g., resolution and contrast); and the ability of the human eye to resolve target details (which also depends on whether this is done in daylight or at night). (TSA, 2006, p. 162)*

Of these criteria, one of the most important in aviation security is the overall resolution of the image along with its field of view. The resolution capability depends on how a particular CCTV camera is being used. Fixed cameras over an access control door or a sterile area checkpoint should provide enough resolution to identify the individual who has breached security. CCTV cameras in use at a baggage screening checkpoint, where the purpose of the camera is to deter theft of items within a passenger's bag, may need a high enough resolution to identify objects within a piece of luggage for it to be admissible as evidence in a court of law. Digital cameras are becoming the industry standard for their resolution and the ability to store tons of data on hard drives, rather than videotapes.

CCTV placement will depend on factors including lighting and the purpose of the camera: a fixed camera watches over a fixed area such as an access control point; a pan-tilt-zoom (PTZ) camera offers a wider field of view, the ability to move the camera, and a zoom capability. Lighting is particularly important in placing a CCTV camera. Some cameras can have filters installed that allow better resolution in low-light conditions. Lighting conditions often change throughout the day, requiring the use of additional cameras or variable aperture lenses that can adjust to accommodate the lighting level. In some situations, such as at night or in dimly lit interior areas of the airport, it may be advantageous to use CCTV cameras with infrared capability.

Ideally, all access control points and sterile area screening checkpoints should be fitted with CCTV cameras. At larger airports, this means the purchase, installation, and integration of hundreds or thousands of CCTV cameras. PTZ cameras should be used throughout the public and sterile areas to monitor passenger movements. PTZ cameras should be placed on the exterior building areas of the airfield to monitor the airfield for security purposes and to provide on-scene imagery of aircraft incidents. CCTV cameras over fixed access points are often tied directly to the ACAMS. When a door goes into alarm at an access control point, the CCTV camera associated with the point automatically activates and begins recording. The TSA's Recommended Security Guidelines for Airport Planning, Design, and Construction provide additional guidance to airport operators on how to measure the resolution and overall effectiveness of CCTV systems.

Control of CCTV cameras and storage of recording information is often a function of the SOC. Storage of images from analog recorders is usually on videotape. More modern systems use digital cameras and store the information on digital video recorders. In some systems, video analog imagery is received from analog video recorders and then digitized. The type of system—analog/tape, digital, or tape-to-digital—depends on the funding available for the CCTV system and other requirements, such as the desire to meet a higher security standard. The costs of CCTV systems can increase significantly depending on the use of the system, such as incorporating motion detection or the ability to discriminate between human and vehicle, animals, trees, or other nonthreatening object movement.

Intelligent video started with early motion detecting abilities programmed into the software that interprets CCTV imagery. Then it advanced to possess discrimination abilities:

Intelligent video is also able to analyze an object and make a determination whether there is a possible threat, based on "rules" established by airport security. A basic application of this is the monitoring of passenger traffic jetway—if persons exiting an aircraft reverse course, the camera monitoring that jetway will see the change in course and the object tracking software can "decide" to notify an operator in the Security Operations Center. More sophisticated behavioral "rules" are under development and will appear on the market as digital CCTV equipment continues to improve.

Intelligent video can also "associate" behavior or events, including events detected by other sensors such as infrared (thermal) imaging cameras and ground surveillance radars, to further aid security operations. DHS is pursuing a research project, known as the Automated Scene Understanding Program (ASUP), to develop intelligent surveillance systems capable of correlating and interpreting fragments of information derived from video, radar, seismic, acoustic and other monitoring technologies. The intent is to reduce thousands of objects, tracks, events, situations, behaviors and scenarios to the few that matter—so that security teams can respond before an attack occurs. (TSA, 2006, pp. 164–165)

Intelligent video has numerous security applications, particularly on airport perimeters where CCTV cameras can observe large areas without constant human monitoring. Some systems are in development that can detect nonmovement and thus alert airport security if an object has been left unattended.

Perimeter Security Considerations

In 2012, when a Skywest pilot jumped a perimeter fence at the St. George Airport in Utah, the public could no longer overlook the issue of airport perimeter security. Leading up to the Skywest incident were two notable incidents of individuals driving through the perimeter gates of an airport and leading police on high-speed chases across the tarmac. In one incident, an elderly man sped through a security gate at a Florida airport and drove onto one of the main runways, which highlights that perimeter security is not just a security issue, but also a safety issue. Prior to all of these incidents, the issue was previously addressed by the GAO in a 2006 report, which noted that access controls to airports continue to improve but that more improvement is needed:

ATSA contains provisions to improve perimeter access security at the nation's airports and strengthen background checks for employees working in secure airport areas and TSA has made some progress in this area. For example, TSA issued several security directives to strengthen airport perimeter security by limiting the number of airport access points, and they require random screening of individuals, vehicles, and property before entry at the remaining perimeter access points. Further, TSA made criminal history checks mandatory for employees with access to secure or sterile airport areas. To date, TSA has conducted approximately 1 million of these checks. TSA plans to review security technologies in the areas of biometrics access control identification systems (i.e., fingerprints or iris scans), anti-piggybacking technologies (to prevent more than one employee from entering a secure area at a time), and video monitoring systems for perimeter security. Further, TSA plans to solicit commercial airport participation in a pilot Airport Security Program and is currently reviewing information from interested airports. (p. 16)

It is widely acknowledged by aviation security practitioners that the perimeter fences around U.S. airports largely serve to "keep the honest people, honest" and are no real protection against attack. However, the minimum standard for fencing in the United States remains 7 feet of chain-link fabric topped by 1–3 feet of barbed or razor wire. To provide real security around an airport perimeter, there must be some form of detection system to alert the airport operator when a breach has taken place or is being attempted.

A challenge to airport operators is protecting the airport perimeter while considering the surrounding environments, which vary widely at U.S. airports as shown by the following examples.

Newark Liberty International Airport is bordered by maritime shipping and trucking facilities, which increases the overall security sensitivity as several transportation nodes, and thus targets of opportunity, are in close proximity. An attack on a port or shipping facility at Newark is likely to affect the airport as well. San Francisco International and Boston Logan International airports are predominantly bordered by water, which requires the airports to consider perimeter defenses other than the standard perimeter fence. Denver International Airport is located on a massive parcel of land, and it can take well over an hour for a security guard to drive the perimeter road, which is a challenge to the airport security department. San Diego's Lindberg Field is next to the U.S. Marines recruitment depot. It, along with other airports located on or near military airfields, is in a symbiotic relationship with the military airport; should something detrimental happen to one, it may affect the other. Houston Bush Intercontinental Airport is surrounded by trees, which may conceal attackers approaching the airport.

Two airports in particular, Boston Logan International Airport and Houston Bush Intercontinental Airport, are notable in the field of aviation security because both airport operators have taken unique approaches to perimeter security. Much of Boston Logan International Airport is surrounded by water, but one of the primary challenges that the airport faced after 9/11 is that the water that surrounds the airport is also home to a thriving clam fishing industry. For many generations, clam fishers (known as "clammers") have been fishing the shores near what is now the Boston Logan International Airport. After 9/11, government officials suggested that clammers be removed from the proximity of the airport because their business activity interfered with the airport's ability to maintain security. To the Massachusetts Port Authority's credit, officials determined that who better to know who should and should not be allowed into those shores than those who have spent careers in the area. The airport could have prohibited clamming operations but chose to makes the clammers part of the security solution by providing security awareness training with phone numbers to call if they spotted suspicious activity. This is an example of a mitigation strategy that created allies instead of enemies.

Houston Bush Intercontinental Airport faces the potential challenge of a criminal or terrorist using the heavily wooded areas around the 11,000-acre airport for concealment to fire a rocket or attempt to penetrate the airfield perimeter. The airport has taken a variety of measures to mitigate or prevent these types of attacks. One of the first measures was to commission "airport rangers," a group of citizen volunteers who ride horses in areas

FIGURE 5.12 D-Fence product in place at Ben Gurion International Airport in Israel.

surrounding the airport and report suspicious activity. Each ranger is given a background check and provided with security awareness training. This is another example of working with the community rather than against it. The airport also is experimenting with a variety of perimeter security measures including ground surveillance radar and smart CCTV video imaging.

Israel's Ben Gurion International Airport, in addition to being heavily protected with a barbed-wire primary perimeter fence encircling the entire airport property (including landside and the terminal), has a delaying fence similar to that in place at many U.S. prisons. The final defense layer for the airport is a perimeter fence built by D-Fence, which manufactures wrought-iron fencing designed for airport security concerns (Figure 5.12). The fence is equipped with numerous motion sensors and smart software to distinguish between nuisance alarms (animals, weeds, and other nonthreatening items blowing against the fence). The alarm monitoring station is in the security control center. When a motion alarm is triggered, a CCTV camera provides an image of the location of the alarm, and security units can be dispatched if necessary to investigate the disturbance.

Vehicle Access Control Programs

Airports are required to have vehicle access control programs in addition to personnel access control systems. These often serve dual roles by allowing the airport to comply with the security requirement to have fencing and certain sections of Federal Aviation Regulation Part 139, which requires fencing to protect the public and to prevent wildlife from entering the airfield. In some cases, the airfield driver training offices are colocated with the security offices to maximize airport resources and reduce the processing time for new airport workers to get security and the airfield driver training required before an individual is allowed to drive on the airfield.

Vehicles on the airfield must be clearly marked by their business affiliation. All vehicles should have consecutively numbered permits, issued and reissued annually. Permits are color coded to indicate access and restrictions and have expiration dates that are easy to see from other vehicles. Requiring all airfield vehicles to be registered with the airport ensures that airport access control personnel can determine if the required insurance levels are met.

Other issues that must be considered when constructing a vehicle access control program are protocols for identifying and stopping a vehicle in the air operations area. At some airports, security guards and airport operations personnel can stop vehicles in the air operations area. However, there may be local ordinance or legal issues that restrict the right to stop vehicles to law enforcement only. All vehicle access control programs must be described in the ASP.

Conclusion

Airports serve as transit points for millions of passengers every day. Therefore, airport security is the central concern within the aviation security system. ASPs and related systems protect access to aircraft on the ground. The airport serves as a critical line of defense, ensuring that passengers and baggage do not possess or contain prohibited items capable of destroying aircraft.

The ASP is an essential document used to describe how the airport will protect its infrastructure, the aircraft within its airfield, and the traveling public. The airport security coordinator is the individual responsible for ensuring that the airport complies with TSA regulations and the ASP, and for protecting the airport from unlawful acts of interference.

Airport security centers on controlling access to sterile areas and airside areas. Various components are involved in access control systems, including physical barriers, computerized control systems, employee security training, and other methods that all contribute to providing a layered airport security system.

References

Airline Security, 2002. Investigative Reports: Airline Security (DVD). Arts and Entertainment Television Network, New York.

Government Accountability Office (GAO), 2006. Aviation Security: Transportation Security Administration Has Made Progress in Managing a Federal Security Workforce and Ensuring Security at U.S. Airports, but Challenges Remain (Publication No. GAO-06-597 T), retrieved July 31, 2008, from www.gao.gov/new.items/d06597t.pdf.

OIG, 2011. TSA's Oversight of the Airport Badging Process Needs Improvement. OIG, Washington, DC.

Transportation Security Administration (TSA), 2006. Recommended Airport Security Guidelines for Airport Planning, Design and Construction, retrieved July 31, 2008, from www.tsa.gov/assets/pdf/airport_security_design_guidelines.pdf.

6

Introduction to Screening

Objectives

This chapter will give you further insight into the functions and processes of screening in the aviation security system. The fundamental design elements inherent with screening checkpoints and emerging issues such as checked-baggage screening, cargo screening, and inspection and aviation employee screening are included. This chapter also describes the roles of the Transportation Security Administration (TSA) and airport and aircraft operators as related to screening.

Introduction

Screening has the highest visibility and is the most scrutinized component of the aviation security system. Of all the layers within the system, screening is the most personally intrusive but also one of the most effective at deterring attacks. Passenger and baggage screening has been the cornerstone of the aviation security system for nearly four decades. The term *screening* can encompass many different meanings. The physical inspection of individuals and property using technology is the most commonly understood form of screening; however, other forms of screening are routinely used. These may include (depending on which agency's definition is used) criminal history record checks (CHRCs) of airport personnel, positive passenger/baggage matching, and even the known-shipper cargo security program. Security screening extends beyond the inspection of passengers and their baggage. Cargo screening, computerized prescreening of travelers, and profiling are all included in the screening process.

Security screening is intended to prevent hijackings, bombings, weapons, incendiaries, or other dangerous objects from being brought on board a commercial aircraft. Screening is carried out on ticketed passengers, visitors in an airport's sterile area, certain airport and airline employees, concession employees, and vendors.

Screening passengers and carry-on baggage started in the early 1970s as a way to deter the numerous hijackings occurring at the time. Through legislation, airlines were assigned the responsibility for the screening function. This responsibility remained with the airlines until shortly after 9/11, when it was transferred to the TSA.

From a regulatory perspective, aircraft operators remain responsible for ensuring that no person or piece of property (bag or cargo) is allowed onto their aircraft without proper screening. Federal transportation security regulations specify that it is the aircraft operator's responsibility to ensure that the screening has been done. Therefore, in the United States the TSA or a contract company under the Screening Partnership Program (SPP; also known as Opt-Out) provides the screening service, while foreign governments and foreign contractors provide screening for U.S. flights overseas.

There are four major screening areas: passengers, carry-on baggage, checked (or hold) baggage, and cargo. Checked-baggage screening did not start until the mid-1980s and only for Canada, and by the late 1980s, for the United Kingdom. The International Civil Aviation Organization (ICAO) holds checked-baggage screening as a fundamental part of any security program (Shanks and Bradley, 2004).

Canada initiated their checked-baggage screening program after the 1985 Air India bombings; Great Britain after the bombing of Pan Am Flight 103. The United States implemented 100% checked-baggage screening after the passage of the Aviation and Transportation Security Act of 2001 (ATSA, 2001). Some checked bags were screened prior to 9/11 in the U.S., mainly those of passengers who were selected for additional scrutiny under the Computer-Assisted Passenger Prescreening (CAPPS) program, however, the number of screened checked bags was still less than 5% of all checked bags in the United States (Shanks and Bradley, 2004).

Cargo screening is one area that has historically received little attention, domestically or worldwide. Before 9/11, the United States and much of the world did not screen cargo in the way that passenger baggage is screened (e.g., X-ray, EDS, ETD). Items accepted as cargo are those to be carried onboard a commercial aircraft without being accompanied by the individual or a representative of the entity shipping the item. Most cargo is accepted in a separate air cargo facility for each air carrier; however, some cargo is accepted at the ticket counter. Airlines use a shipping document, an air waybill, for shipping cargo. This is a contract between the shipper and airline stating the terms and conditions of transportation, shipping instructions, a description of the commodity, and transportation charges. Items accepted and treated as checked baggage accompany the owner on the same flight and do not have an associated air waybill. According to Shanks and Bradley, "the lack of cargo screening is a global dilemma" (Shanks and Bradley, 2004, p. 41). Until recently, the primary air cargo security program has been the known-shipper program. The passage of ATSA 2001 brought the industry's attention to the issue of cargo screening. In 2007, the Implementing the Recommendations of the 9/11 Commission Act required 100% screening of all air cargo that is shipped on commercial airliners in the United States. Cargo must by physically inspected, which includes the use of X-ray and canine and explosive trace detecting technologies.

Evolution of Screening in the United States

Initial passenger and carry-on baggage screening programs were designed to detect items commonly used by hijackers. The first passenger and baggage screening systems were designed to detect guns and hand grenades, carried either on an individual or in carry-on baggage. In the early 1970s, airline bombings were so infrequent they were not considered a significant threat, and thus, no programs were implemented to prevent or deter such attacks.[1]

---

[1]During the early 1970s, bombs were more commonly brought onboard aircraft as checked baggage rather than carried onboard by passengers.

In 1973, walk-through metal detectors (WTMDs) were installed at airports to detect whether an individual was carrying a gun or hand grenade.[2] Conventional X-ray machines, which X-ray an object from only one angle, were used to detect weapons and explosives in carry-on bags; neither device actually "detected" weapons or explosives. WTMDs sounded an alarm when metal was detected within its magnetic field. A security screener then investigated whether the individual possessed a weapon or had a nonprohibited object such as a belt buckle, steel-shoed boots, car keys, and loose change. X-ray systems only provided an image of the objects within a bag leaving the interpretation of the imagery to the X-ray operator.

Often, screeners would use metal detecting hand-wands to conduct a secondary screening. During secondary screening, passengers were asked to roll up shirtsleeves (to reveal a watch) or to remove a belt. These secondary screenings were not foolproof. Additionally, WTMDs do not detect plastic explosives. In the early 1970s, plastic explosives were available only to the military and had rarely been used in any aircraft bombing; dynamite was the preferred explosive. Regardless, it was assumed that if an individual attempted to conceal a bomb on his or her person, he or she would be deterred by a WTMD, assuming the metal detectors would detect the metal components within a bomb (timer, wiring, blasting caps, etc.). It was thought that if an individual tried to sneak in a bomb in baggage, it would be detected by the X-ray operator. Conventional X-ray machines, however, could not detect explosives or weapons; they showed only the density of objects in the carry-on bags. Screeners had to judge whether the density of the object posed a threat or whether a physical search of the bag was warranted.

As mentioned, security screening was an airline responsibility. Airlines usually subcontracted screening to an approved vendor, often by selecting the lowest bid. Before 9/11, contract screening companies were notorious for high turnover rates of staff and for hiring individuals who were less than qualified for the task. Because of these factors, numerous failures of the system occurred when under airline control. From 1973 through the 1980s, airline-contracted screening companies using WTMDs and conventional X-ray machines for passenger screening and carry-on baggage represented the traditional security screening system. During that period several aircraft were bombed using checked baggage as the means of introducing the bomb; undetectable plastic explosives were becoming more available and being used. Hijackers were also using aviation employees to help bypass screening to hide weapons on targeted aircraft. Despite new strategies being used by terrorists, screening processes and technologies remained largely unchanged.

The challenges of passenger and carry-on baggage screening are not new. In a 1987 U.S. General Accounting Office (GAO) report, the agency pointed out glaring inadequacies including a lack of performance standards set by the Federal Aviation Administration

---

[2]Despite Hollywood movies, such as *In the Line of Fire*, where a character fashioned a plastic gun undetectable by metal detectors, a working plastic gun has yet to be made. The closest plastic gun is the Glock 17, an Austrian-made semiautomatic weapon that still contains 83% metallic parts. The CIA is rumored to have a ceramic pistol, although this cannot be verified.

(FAA) and a lack of FAA enforcement authority related to aviation security screening. This assessment concluded that evaluation of the effectiveness of the screening programs was inadequate, and wide variations existed in the frequency with which test weapons were detected at screening checkpoints throughout the United States (GAO, 1987). In the GAO study, at 34 U.S. airports, the detection rate ranged from 48% to 99%. The average detection rate of the test weapons was 80%.

In 1990, Congress passed the Aviation Security Improvement Act (ASIA 1990), which required the deployment of better screening technology by November 1993. However, by 1997 the FAA still had not deployed such technology:

> *This delay was due primarily to the technical problems slowing the development and approval of the explosives-detection devices. But we also found that FAA did not develop an implementation strategy to set milestones and realistic expectations or to identify the resources to guide the implementation efforts.*

ASIA 1990 also established a requirement for employee background checks for anyone with security identification display area (SIDA) access and for security screeners. However, little guidance was provided on the nature of the employment record check, and few employees underwent the check.

Throughout the 1990s, airline lobbyists opposed new security measures or policies that could have increased security but also slowed passenger flow. Failure of the screening process was not entirely the fault of the airlines. The system was regulated and overseen by the FAA, which had the power to fine airlines for failing to pass screening system tests but did not. According to Kip Hawley, former director of the TSA:

> *A more subtle shortcoming of pre-9/11 security was how it discouraged responsibility in the face of failure. When something did go terribly wrong, the FAA could blame the airlines for providing insufficient security, while the airlines blamed the contractors for faulty execution. The contractors, in turn, claimed to have met the government's "specific requirements" and pass the blame back to the FAA for not setting appropriate measures. The punch line was that all three of them could be right. But this was cold comfort for the families of dead passengers. (Hawley and Means, 2012, p. 62)*

The FAA's minimum performance standard for screeners was to detect 100% of the test weapons (GAO, 1987) whenever FAA inspectors conducted such tests. The punishment for a screener who failed a test was usually immediate dismissal. This, rather than being an incentive to not miss a prohibited item, was an incentive not to miss a *test object* and conditioned screeners to look more for the FAA test weapons than for actual weapons and explosives.

The policy of immediate termination also affected employee morale. Rather than setting reasonable expectations and assessing the totality of the circumstances surrounding a screener's failure to detect a test weapon, screeners were simply fired. This policy did not develop better screeners but did hinder the ability of the screening companies to hire personnel and drove down morale for the remaining employees. Additionally, screeners were trained using the same limited number and types of FAA test weapons and explosives, which were obvious representations of guns or hand grenades encased in plastic and difficult to miss, rather than a variety of realistic-looking weapons and devices. The TSA has corrected this issue by using computer-imaging technology capable of simulating images from a variety of weapons, explosives, and bomb configurations.

In the 1990s, the FAA did not have a minimum training standard for screening personnel or performance or qualification guidelines for screening companies. As a result, screener training was inconsistent from one company to the next. The FAA fined airlines based only on the failure of screeners to find test weapons and explosives. The fines were transferred directly to the security subcontractor, which reduced the financial impact on airlines and the effectiveness of the fines. The fines the FAA did levy ended up being more a cost of doing business than an incentive to improve the performance of screeners. The FAA also did not allow the investigations conducted by their own red teams[3] to be used in civil actions against airlines or airports.

In 1996, the Gore Commission recommended raising the professionalism of screening personnel. The commission recommended "development of uniform performance standards for the selection, training, certification, and recertification of screening companies and their employees" (Gore, 1997). Commissioner Victoria Diaz drafted the minority report on behalf of several commission members including family members of victims of Pan Am Flight 103, TWA Flight 800, ValuJet Flight 592, and others. One family member, the widow of John Cummock, a victim of the Pan Am Flight 103 bombing, wrote a letter deriding the Gore Commission's recommendations as being too vague, unenforceable, or useless and without appropriate funding mechanisms. Writing specifically on the recommendations to improve screening, Mrs. Cummock stated:

> *This recommendation contains a number of admirable objectives but it, like its predecessor recommendation in President Bush's Commission on Aviation Security and Terrorism, lacks teeth. Following President Bush's Commission of Aviation Security and Terrorism and the follow-on Aviation Security Improvement Act in 1990, the FAA established standards for the selection and training of aviation security personnel. Those standards were, and still are, totally inadequate. There is nothing to prevent the same inadequate actions by the FAA to this recommendation. The Commission should specifically recommend that the FAA mandate 80 hours of intensive classroom/laboratory and 40 hours of On-the-Job training before performance certification for all airline security screening personnel. (Gore, 1997, 3.10 Recommendation)*

---

[3]*Red teams* are a common security and military term for inspection teams that use nonstandard methods to attempt to breach a facility or bypass some security measure, system, or procedure.

Another recommendation of the Gore Commission concerned the deployment and use of explosive detection system (EDS) technology, using computerized tomography[4] and explosive trace detection (ETD) technology, capable of detecting trace amounts of explosives on people or their belongings. A handful of EDS and ETD machines were deployed to the nation's commercial-service airports with the intent of using them for travelers selected under the CAPPS program or selected for secondary screening. The machines were rarely used as few individuals were flagged under the CAPPS program for additional screening. When the 1996 legislation was passed, neither the Gore Commission's recommendations nor the concerns of the minority report to the Gore Commission were addressed.

The Aviation Security and Antiterrorism Act of 1996 required that security screeners, and all other aviation employees with SIDA access, undergo a 10-year employment history check. If an individual could not satisfy the requirements of this check, then he or she would be required to undergo a fingerprint-based CHRC before being allowed unescorted access to SIDAs or to serve as a security screener. Reasons for being unable to pass the 10-year check were most often due to gaps in employment. Essentially, these were the same requirements embodied in the Aviation Security Improvement Act of 1990, but the 1996 legislation formalized the process and provided more guidance on what an employment history check entailed. Since the background check, known as an *access investigation*, was relatively easy to pass, less than 3% of the total number of employees working at U.S. airports had to undergo the fingerprint-based check.

The Airport Security Improvement Act of 2000 (ASIA 2000) finally mandated screener training standards. These required at least 40 hours of classroom instruction and 40 hours of on-the-job training for screening personnel. ASIA 2000 also introduced computer-based training[5] for threat image projection (TIP) technology to security screener testing procedures. These were better simulations of how a weapon or explosive might look to a security screener viewing an X-ray monitor.

Before 9/11, FAA policy allowed passengers to carry knives that were shorter than 4 inches onboard, which means that poor screening was not necessarily a factor in the 9/11 attacks. The screening system and related policies were largely inadequate. In a GAO report published nine days after 9/11, the GAO found that screeners were still missing 20% of prohibited items. Screeners missed even more items when GAO inspectors used more realistic tactics and test weapons than the FAA used in its inspections:

> *As we reported in June 2000, tests of screeners revealed significant weaknesses as measured in their ability to detect threat objects located on passengers or contained in their carry-on luggage. In 1987, screeners missed 20 percent of the potentially*

---

[4]A CT scan uses a combination of X-rays and computer technology to produce cross-sectional images commonly referred to as "slices."

[5]Training or instruction using a computer program to provide instruction and feedback in place of a live instructor.

dangerous objects used by FAA in its tests. At that time, FAA characterized this level of performance as unsatisfactory. More recent results have shown that as testing gets more realistic—that is, as tests more closely approximate how a terrorist might attempt to penetrate a checkpoint—screeners' performance declines significantly. A principal cause of screeners' performance problems is the rapid turnover among screeners. Turnover exceeded over 100 percent a year at most large airports, leaving few skilled and experienced screeners, primarily because of the low wages, limited benefits, and repetitive, monotonous nature of their work. Additionally, too little attention has been given to factors such as the sufficiency of the training given to screeners. FAA's efforts to address these problems have been slow. (GAO, 2001, p. 2)

Despite 14 years and three major legislative actions (Aviation Security Improvement Act of 1990, Aviation Security and Anti-Terrorism Act of 1996, and Airport Security Improvement Act of 2000), the performance of screeners was no better than 1987 levels. The GAO report went on to note that part of the problem rested with the FAA and its inability to promulgate regulations to certify screening companies, noting that it had been two and a half years since the FAA had originally planned to implement certification standards.

The Aviation and Transportation Security Act of 2001 (ATSA 2001) transferred screening to the TSA and added checked-baggage screening to the process.

Screening under the Transportation Security Administration

The takeover of screening by the U.S. federal government from the airlines has not been without problems and challenges. In addition to passenger and baggage screening, ATSA 2001 also set deadlines for the TSA to complete key benchmarks, including:

1. The complete takeover of the screening by November 19, 2002. This included more than 400 commercial-service airports.
2. Establishing the basic screener training minimums at 40 hours of classroom instruction and 60 hours of on-the-job training.
3. Annual proficiency reviews and testing of screeners, including remedial training of any screener who fails a security test.
4. The screening of all checked bags using EDS technology by December 31, 2002.

The TSA met its first goal for taking over passenger and carry-on baggage screening when it hired more than 55,000 new federal employees within a year and met the November deadline. Meeting the checked-baggage screening deadline was hindered in 2002 because only two types of EDS machines were certified by the FAA for use in airports and the manufacturers of EDS equipment could not produce the hundreds of needed units within the time constraints. Airports were considered in compliance with the deadline by using a combination of positive passenger/baggage matching, inspections by K-9 explosive detection bomb dogs, and ETD technology. In his book *Permanent Emergency*, former TSA administrator Kip Hawley noted that the TSA met the 100% checked-baggage screening deadline. However, the deadline was essentially met because the policy of positive passenger/baggage matching was accepted as checked-baggage screening; the actual physical inspection or X-ray of checked baggage at all U.S. commercial-service airports would not be complete for several more years (Hawley and Means, 2012, p. 55).

Although the TSA met the first deadline, the hiring of so many workers quickly created an imbalance across the nation; some airports had too many screeners, and others had too few (GAO, 2004a). Since 2004, the TSA has proffered several staffing models in an attempt to fix workforce imbalances.

Today, in an attempt to compensate for screener shortages, the TSA established a mobile National Screening Force consisting of 700+ screeners who can be reassigned to understaffed airports.

Although staffing and training issues are important and contribute to the ability of the screener to detect prohibited items, the most important issue in screening is the ability of the screener or the technology to detect weapons and explosives. There is little public information on this issue, as the federal government has classified the performance of both its screening workforce and its detection equipment. However, some documents were made public that indicate that the ability to detect weapons and explosives has been less than satisfactory. In 2003, Department of Homeland Security (DHS) inspectors were able to sneak weapons past screeners at 15 airports. In screener performance tests conducted by the GAO and revealed publicly in 2004, the agency determined that TSA screeners and private screeners tested about the same and that there was no significant difference in the success of detecting prohibited items between the groups (GAO, 2004b). The less than desirable detection rates pushed the TSA to continue the testing and deployment of other systems, such as portal-trace detectors and body imaging devices, particularly after two Chechen female suicide bombers snuck bombs past the same type of metal detectors in the Russian Federation that were in use in the United States at the time.

Another challenge affecting effective and efficient airport security is the retention of government security screeners. The overall lack of promotional opportunities and career advancement for screeners has resulted in an attrition rate unacceptable to the TSA. In response to these concerns, TSA administrator Kip Hawley announced in December 2005 that he was changing the job title of screeners to transportation security officers (TSOs) in an effort to expand the identity and job duties of screening personnel (Hawley, 2005). Hawley also stated he would take steps to provide TSOs a career path into the Federal Air Marshal Program, the TSO Expert Program, and the Behavior Detection Officer Program.

Sterile Area

Once an individual and their belongings have undergone the screening process it is important to ensure that they nor their belongings mix with individuals who have not been screened. This is done through the establishment of a *sterile area*. The TSA offers the following descriptions of the sterile area:

> *A Sterile Area is a portion of an airport, specified in the airport security program, that provides passengers access to boarding aircraft and to which access generally is controlled by TSA, or by an aircraft operator under 49 CFR 1544 or a foreign air carrier under 49 CFR 1546, through the screening of persons and property.*
>
> *The primary objective of a sterile area is to provide a passenger holding and containment area, preventing persons in it from gaining access to weapons or contraband after having passed through the security screening checkpoint and prior to boarding an aircraft. (TSA, 2006b, p. 65)*

Sterile areas are those areas within the terminal where passengers wait after screening (Figure 6.1). From a regulatory perspective, access to these areas is the responsibility of the aircraft operator and the airport operator. However, from a practical stance, the TSA, through the performance of the screening function, "controls" much of the access to the sterile area at airports in the United States. Unauthorized personnel must be prevented from entering the sterile area without undergoing screening.

For aviation personnel entering a sterile area through a Part 1542.207–regulated airport access-controlled door, their CHRC is considered the required screening. This concept often creates confusion to neophyte security practitioners, particularly when there is a common belief that all personnel must undergo the screening process before entering

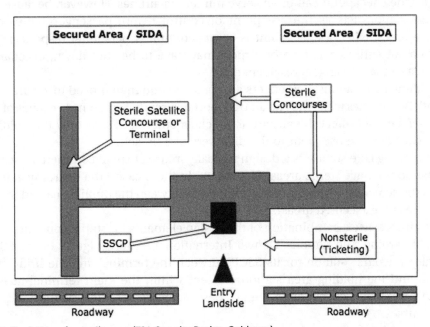

FIGURE 6.1 Depiction of a sterile area (TSA Security Design Guidance).

a sterile area. Personnel entering the sterile area by way of a security screening checkpoint (SSCP) and not a Part 1542.207–regulated access-controlled door must undergo passenger and carry-on baggage screening. The issue of requiring all aviation personnel with SIDA badges to undergo the passenger screening process continues to be debated in the United States. Some U.S. airports have adopted the practice, establishing screening checkpoints at airfield access gates and other locations, but private contractors working for the airport conduct the actual employee screening, not the TSA. An employee's job responsibilities also determine whether he or she can access the sterile area through an airport access-controlled door or through the screening checkpoint. An employee possessing an authorized access/ID badge may, depending on job title and access authority, access the sterile area through an airport door, as long as he or she does not board an aircraft and fly out. Almost without exception, any aviation employee who boards an aircraft must be screened at the SSCP, even if the individual has approved access. Among the limited exceptions are flight crew personnel at their home airport who are on duty and will be serving in a flight crew capacity once they board the aircraft.

Entry into sterile areas is through SSCPs, of which there are three types: the sterile concourse station, the holding area station, and the boarding gate station. Sterile concourse stations are placed at points between public-use terminal areas and sterile areas within the airport. This configuration is the most desirable as it provides the highest level of passenger security at the best cost, and is in use at most every U.S. airport.

The benefits to security offered by the sterile concourse station include the consolidation of screening staff, closed-circuit television (CCTV) monitoring, and law enforcement coverage. One checkpoint can often serve numerous airlines. However, because a sterile concourse station acts as the last point of control before passengers enter multiple concourse areas, when a checkpoint is breached, unless personnel respond rapidly to the breach, an entire concourse or airport may have to be shut down, evacuated, and searched, and the passengers rescreened.

In holding area stations, passengers are screened and then moved to secure rooms or areas with boundaries while they await transport to aircraft. Screening breaches of holding area stations do not affect the airport as much as other SSCPs, as only the holding area must be shut down subsequent to the discovery of a breach.

The boarding gate station is a design typically in use at small commercial-service airports that do not have sterile areas (Figure 6.2). In these cases, the boarding gate station serves as the screening checkpoint and boundary between the public area and the airfield security areas (i.e., secured areas).

Some airports use a combination of the design elements for sterile concourse, holding area, and boarding gate stations. Denver International Airport (Figures 6.3 and 6.4) uses the sterile concourse station for its SSCP between the terminal and the bridge to Concourse A, and the holding area station concept within the main terminal at its north and south screening checkpoints, before passengers access the underground train to the concourses.

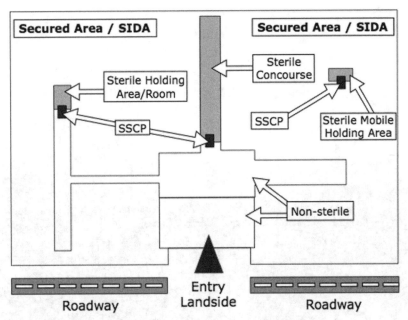

FIGURE 6.2 A boarding area station design (TSA Security Design Guidance).

FIGURE 6.3 Denver International Airport's north screening checkpoint, August 2001.

FIGURE 6.4 Denver International Airport's north screening checkpoint, post-9/11.

The queue area is where passengers wait and line up for the screening process at the SSCP. After 9/11, the space needed for passenger queuing and the additional space needed for more screening equipment and personnel significantly increased the overall space required. Significant terminal redesigns were done in a short period to accommodate the new, larger checkpoint demands.

The space allotted to passenger queuing is based on the airport's peak-hour enplanements (based on the highest amount of passengers measured on the busiest days of the year, averaged over the four busiest months of the year). The TSA has updated the specific design specifications and resources, such as passenger throughput, personnel, or space requirements (TSA, 2006b). The TSA provides this data to airport planners and engineers for design and building purposes. The TSA does admit that the average throughput for rough computational estimates is 175–250 passengers per hour per screening lane, but the agency cautions that this estimate can vary considerably at higher levels of security or with the development of new technologies at an SSCP. These numbers will also continue to change with the implementation of precheck and other risk-based screening procedures, and the integration of advanced imaging technology (AIT) machines.

During the initial days of the TSA, a Disney engineer helped to design the queuing system with ropes or bands attached to poles to form passenger movement lanes. The cutbacks or zigzag method increase passenger throughput and reduce waiting times and maximizes space within the terminal.

Sterile Area Breaches

The sterile area represents a high threat level to the airport operator, as it enables direct access to aircraft and the airfield. Unauthorized individuals in a sterile area (i.e., those who have not undergone the proper screening) must be identified immediately, sequestered, and questioned by law enforcement personnel as to their intentions and whereabouts in the sterile area. A breach of a sterile area often causes an entire concourse, or sometimes an entire airport, to shut down all flight operations until the affected areas can be evacuated and searched for prohibited items. This represents a loss of millions to the airlines and disrupts the entire national airspace system. Ensuring that unauthorized individuals do not access the sterile area is a priority for the TSA, the airport operator, and the aircraft operators.

Protecting sterile areas from unauthorized personnel and items involves more than just screening all personnel and their belongings. There are businesses and aviation employees who require access to prohibited items in the sterile area to do their jobs. For example, airline and airport maintenance personnel often carry a variety of tools that are normally prohibited at a screening checkpoint. Catering, vendor, and restaurant personnel have access to knives and box cutters within their shops in the sterile area—again, items that are prohibited through a screening checkpoint. The airport operator must maintain a strict inventory of which businesses have prohibited items and in what quantity. Business operators must train employees on the proper use and security of these items. Airline and airport personnel authorized to carry prohibited items in sterile areas must be trained in the proper security of such items, and the airport operator must maintain a database that includes the names and job titles of these individuals.

In the 2000s, the TSA mandated the physical inspection of catered goods by the airlines and the physical inspection of goods delivered to the sterile area by vendors.

Screening Checkpoint Operations and Design

The SSCP is a dynamic and busy environment and one of the most important components of the aviation security system. Many SSCP-related factors, such as staffing and the layout or number of checkpoints, affect security and ultimately reduce or increase risk in the airport security system. In the early 2010s, the implementation of AIT machines into the SSCP created additional complications for airport managers who had to find room to accommodate the larger machines. Additionally, flooring had to be strengthened and the AIT machines have higher power requirements than metal detectors.

Inadequately designed checkpoints can result in more passengers waiting in public areas, which can be potential targets for terrorists. Inadequately designed checkpoints can also reduce the performance of screeners who may perceive the need to work faster to reduce the wait for passengers. Longer passenger lines often result in low levels of customer service and can cause passengers to miss their flights. Inadequate staffing of a checkpoint can reduce the effectiveness of detecting prohibited items. Poorly designed or managed checkpoints can result in more breaches of screening if passengers become disoriented in the SSCP area.

Design

According to the TSA (2006b, pp. 91–92), SSCPs should:

1. Prevent persons with prohibited items from boarding a commercial aircraft.
2. Prevent exit lane breaches.
3. Secure exit lanes for arriving passengers during both operational and nonoperational hours of the SSCP.
4. Accommodate persons with disabilities requiring wheelchair accessibility or allowances for other assistive devices.
5. Provide for minimal interruption or delay to the flow of passengers and others being screened.
6. Provide effective and secure handling of tenant goods that cross from the nonsterile area to the sterile area (i.e., vendor deliveries).
7. Allow adequate equipment maintenance and interference spacing requirements.
8. Provide for operational flexibility in response to changes in passenger loads, equipment, operational processes, and security levels.
9. Be flexible enough to accommodate new technology and processes.
10. Be an efficient and effective use of terminal space.
11. Provide acceptable and comfortable environmental factors, such as air temperature, humidity, air quality, lighting, and noise.
12. Be of safe and ergonomic design.
13. Ensure adequate power, data, and CCTV equipment requirements.

According to ICAO, detailed planning and consideration of security, human factors, and operational requirements can help optimize the passenger screening process. The size of screening checkpoints is somewhat based on how many ticket counters are available to process passengers. *The International Air Transport Association Airport Development Reference Manual* set recommended parameters for certain passenger processing functions: (a) slightly over 2 minutes for ticketing/check-in, (b) 15 seconds per passenger for passport control, and (c) 12 seconds per passenger for security screening. These times do not take into account the queuing time before each process. With advanced security X-ray machines and AIT (body imagers) the average screening processing time is closer to 1–2 minutes per passenger, after waiting in the line.

The number of ticket counters is one method used to help measure passenger throughput, however, the expanded use of airline common-use equipment has accelerated check-in times. With more self-service kiosks in the terminal and as passengers use the Internet and smartphones to check in for flights, bypassing the ticket counters altogether, benchmarks such as peak hourly enplaning passenger and annual passenger enplanements must also be used to determine how many screening checkpoints are needed for a particular airport.

The "goal" time for an individual to wait in the screening queue is 10 minutes. This objective was started after a comment made by former Secretary of Transportation Norman Mineta in response to a media question (Hawley and Means, 2012, p. 43). Former TSA administrator Kip Hawley and TSA personnel were using 10 minutes as the basis for checkpoint design, however, once the media and Congress grabbed onto the sound bite, the goal for processing of an individual through a screening checkpoint became 10 minutes and that is the current expectation of passengers and airlines. A security screening checkpoint should be designed to use automation whenever possible.

Checkpoints should have explosive and weapons detecting technologies that do not pose a threat to passengers or screening personnel, minimize the footprint and construction costs, maintain revenue-generating areas within the terminal wherever possible, increase throughput rates, and preserve passenger privacy.

Airports should limit the number of screening checkpoints to as few as possible, while retaining acceptable passenger throughput levels (as established from the normal level of passenger enplanements at each airport). SSCPs should be designed to accommodate airport concessionaires and maintenance and operation personnel who move equipment and supplies through the SSCP. SSCPs must also allow the passage of emergency services personnel, their vehicles, and equipment (e.g., electronic paramedic carts and mobile explosives detection equipment), and provide discrete areas where undercover air marshals, federal agents, law enforcement personnel flying armed, and armed pilots can be processed without being in plain view by the traveling public. Concessionaires represent a particular challenge to airport security, as deliveries occur throughout the day and delivery personnel change frequently. It is best if delivery personnel can use nonpublic-area loading docks to deliver their products. Concessionaire employees working at the airport who have access/ID badges can then transport the items to their stores and in or out of the sterile area (TSA, 2006b).

Security screening checkpoint areas should, when possible, be located away from public view and prevent the unauthorized passage of articles from the nonsterile area to the sterile area. Walls or partitions should be placed to minimize surveillance by criminals. Such sheltering also helps X-ray operators maintain focus. When it is not possible to shield the SSCP areas from public view, the airport security coordinator (ASC) can deploy law enforcement and security patrols to watch for individuals who might be conducting covert surveillance or to thwart an attack on a crowd of passengers in a screening area.

SSCP design considerations must also include the space available for expanding to accommodate larger passenger loads and new equipment. In 2004, portal trace detectors were added as additional equipment at U.S. SSCPs. While these were experimental at first, and later removed from the SSCPs, they were soon replaced with AIT machines, which took up even more space.

Another important component of SSCP design is closed-circuit television coverage. Video cameras can provide surveillance for law enforcement in the deterrence of theft at SSCPs. CCTVs can be very valuable in the event of a screening breach. CCTVs should be positioned to capture the images of the faces of passengers. CCTVs can also be used to help monitor closed and unstaffed SSCPs (TSA, 2006b). They should be of high digital quality as lesser-quality imagery is not often useful in determining the unique characteristics of individuals passing through SSCPs.

Other design considerations include secondary screening areas, walls and partitions to keep passengers separated from other passengers and their baggage during screening, and a holding area for passengers waiting for secondary screening. Break rooms, locker areas, and administrative areas are also needed for security screening personnel.

The use of extended tables leading up to X-ray machines helps facilitate and expedite passenger and baggage screening. Chairs and tables located at the exit of a screening area help maintain passenger flow without interfering with screening by giving passengers room to put their shoes back on, put laptops back into carry-on bags, and so forth. At Vancouver International Airport in Canada, screening checkpoints are equipped with circular countertops located near the exit areas of SSCPs. This provides passengers with a location to retrieve and reorganize their belongings and gives screeners a convenient location to conduct secondary screenings (Price, 2004). A simple rule of checkpoint design to improve passenger flow is to put in more divestiture and recomposure tables.

Operations

Entry Lane Protection

The TSA, airport tenants, law enforcement, and local building inspectors and fire inspectors should be consulted as part of the SSCP design and follow the standards in the TSA's *Checkpoint Design Guide Standards* (CDG), published in 2006 (TSA, 2006a) and revised in 2009. The TSA's publication, *Recommended Security Guidelines for Airport Planning, Design, and Construction* (TSA, 2006b), also contains detailed information on SSCP, as well as *TSA Security Checkpoint Layout Design* (2006c), both produced by ACI.

The following elements are characteristic of an SSCP:

- Prescreening preparation instruction zone, located in front of the SSCP and usually includes signage, instructional videos, "ambassador" staff, and sometimes posters discussing prohibited items, the effects of screening technology on film, private screening options, TSA warning signs, and airline carry-on and pet restrictions. Some SSCPs feature pictures and diagrams to help passengers with the process.
- The queuing space, which includes queuing stanchions to define the lines, the travel document check stations, and tables and bins for passengers to begin the divesture process, removing metallic items and wallets, cell phones, and other devices that may trigger the screening devices.
- Walk-through metal detector (the TSA has upgraded all earlier WTMDs to an "enhanced" status) or other weapon or explosive detecting technology.
- The nonmetallic barrier is used to prevent individuals bypassing the WTMD and to stabilize the WTMD units themselves. As stand-alone units the WTMDs are rather unstable.

- The nonmetallic American's with Disabilities Act (ADA) gate is used to accommodate passengers with wheelchairs or otherwise disabled and not able to use the WTMD. The use of nonmetallic materials prevents disruption of the performance of the metal detecting devices.
- Carry-on baggage X-ray machine, which is a single-angle X-ray device for screening passengers' carry-on baggage. In some cases, EDS machines may be implemented in the screening checkpoint.
- Divest and composure tables are used for passengers to retrieve their belongings and to improve the throughput rate.
- Adjacent walls and barriers separate sterile from nonsterile areas and are at least 8 feet high. If the walls are adjacent to the exit lane, they should be transparent, allowing security personnel to view the exit lanes.
- Passenger containment and inspection, which is the holding station (sometimes combined with a wanding station) that is used to hold passengers waiting to undergo secondary screening. This station must be positioned such that affected passengers can be diverted without obstructing the flow of passengers not requiring secondary screening. Additionally, this area should include a secure door to prevent passengers from walking away without the knowledge or consent of the screener.
- Wanding stations are not enclosed areas, but rather specified points near the SSCP where hand-wands or pat-downs are used to perform additional screening.
- Explosive detection machines for use in secondary screening. ETD machines provide considerably more imagery and interpretation than conventional X-ray machines. They are used to protect the safety of the screener and to minimize the intrusion on a passenger's privacy.
- Egress seating areas are where passengers can put their shoes and coats back on, and get their personal belongings back in order.
- The law enforcement officer (LEO) station is where LEOs may be positioned to enable them to view the entire screening operation, providing as unobstructed view as possible.
- Private screening areas should provide enough space (and privacy) to accommodate one passenger, two TSOs, a chair, and a search table. If space is available, this location should ideally accommodate passengers with disabilities, passengers under escort (e.g., a prisoner who requires an LEO to stand by), or who have interpreters.

The prescreening preparation and instruction area is the primary area where airport personnel often serve as instructional guides, helping to inform passengers of standard security requirements, such as reminding them to remove their laptops and not to attempt to take prohibited items through the checkpoint. Videos may also be shown in these areas to provide the same types of instructions.

Queuing spaces are where passengers await their turn for screening. Queue lines are designed to increase the speed at which passengers move through the lines. Security agents also conduct document checks in queuing spaces. Passenger document checks are performed to ensure that each individual has a boarding pass and government-issued photo identification. At many foreign airports, the document check is also used to conduct security questioning.

After 9/11, access to sterile areas was primarily restricted to those with boarding passes. The majority of people passing through a screening checkpoint are airline passengers. However, other personnel who have business within sterile areas but are not boarding an aircraft may need access. For example, there may be individuals visiting airport personnel for business meetings or training. Others may be accompanying passengers who require assistance in getting to their gate. The airport and the TSA work together on procedures to allow these individuals sterile area access. Access is usually granted along with a visitor's pass (also referred to as a concourse pass or pier pass).

Other personnel requiring access through checkpoints include members of law enforcement. Law enforcement officers traveling armed on aircraft require special handling through a checkpoint, particularly if operating undercover. Law enforcement officers who are conducting investigations not related to aviation crimes but operating undercover or are nonuniformed and not flying out on an aircraft also require special handling through a screening checkpoint. Emergency response personnel such as paramedics, emergency medical technicians, firefighters, and airport operations personnel require expedited access through the checkpoints.

In 2006, a computer student at Indiana University constructed a website that printed fake boarding passes. These passes supposedly would allow the user to bypass airport security checkpoints. The student claimed that the website was not intended to be used for criminal or terrorist purposes but to show "security theater," the student's term for showing that security is deeply flawed (Silverstein, 2006). This attempt to show a gap in the aviation security system is typical of assumptions made by those who do not understand the layered security system. Although there are gaps and there is always room for improvement, aviation security practitioners should not jump to the latest perceived security hole without first carefully considering if there is an actual security risk. Security practitioners must make decisions on what actions to take and what monies to spend as part of an overall risk assessment, rather than jumping to fix what may only be a problem in the public's mind. Although a fake boarding pass could get an individual through a sterile area checkpoint (as could a fake driver's license), he or she still needs a real boarding pass to get on the aircraft, and the individual still must undergo screening, thus reducing the likelihood that he or she could conceal a prohibited item within the sterile area. The document check serves as an additional layer of the security system.

After passing the document check, passengers select their screening lane and begin divestiture. This often includes removing laptops from their cases and removing shoes, belts, and outerwear. The TSA also requires that any liquids be presented in a one-quart transparent bag, separate from other baggage. Divestiture procedures change as security levels increase and decrease. These changes should be incorporated into the prescreening preparation and instruction zone briefing. The TSA implemented a "diamond" line system for a period of time in the hopes of speeding up the screening process, but the results of this remain a subject of debate. Based on the system used by ski areas to denote beginner, intermediate, and expert skier, the same concept was attempted by the TSA—a green circle indicated an infrequent traveler, a blue square indicated a somewhat more frequent traveler, while a black diamond indicated the line was for frequent flyers who completely understand the screening system and are good at getting through it. While some declared it a success, passengers tended to drift to the shortest line, ignoring the system altogether.

Following the queue line, passengers must submit their carry-on bags for screening and personally pass through the WTMD. After 9/11, all WTMDs were enhanced for better metal detection. WTMDs should be located next to the explosive detection equipment or X-ray machine. Nonmetallic gates should be next to the WTMD to allow wheelchair access.

A holding station to detain passengers requiring secondary screening should be located immediately beyond the WTMD. Shortly after 9/11, there were several incidents at U.S. airports where individuals had been asked to wait until a screener could conduct a secondary screening. For a variety of reasons, the individuals did not wait and proceeded to the sterile area, creating a screening breach. Soon thereafter, airports incorporated holding stations into their SSCPs (Figure 6.5) to prevent individuals waiting for secondary screening from going into the sterile area.

The wanding station is the location used for secondary screening. If a passenger sets off the WTMD alarm, a security screener must pass a metal detecting hand-wand over the individual's body. The screener may also conduct a pat-down search and may ask the

FIGURE 6.5 Close-up view of an SSCP.

passenger to step into a private area to remove clothing if warranted. These searches are conducted under supervision by screeners of the same sex as the passenger.

ETD equipment may also be located at the SSCP. The ETD machine is used when the X-ray operator notes a suspicious item within a bag. In these cases, another screener will investigate the contents using the ETD machine. The ETD equipment is often used as part of this secondary screening along with a physical search of the suspect bag.

Other elements of SSCPs include portal-trace detectors, a law enforcement officer section where police are stationed to watch over the screening, a supervisor administrative workstation, private search areas, CCTV coverage, data port connections, and exit travel lanes. The TSA's *Recommended Security Guidelines for Airport Planning, Design, and Construction* show variants of multiple SSCP lane design.

Exit Lane Protection

Passengers and others use exit lanes within sterile areas to move back to the public areas. Exit lane protection is an important consideration in the design of SSCPs. Specific concerns related to exit lanes include breach alarms, physical barriers, personnel coverage, CCTV coverage, and the implementation of response corridors (long hallways that provide a time and distance barrier to breach attempts).

A breach occurs when an individual enters a sterile area without being successfully screened. The most common form of security breach is when an exit lane is compromised. Typically, exit lanes are protected by security personnel who physically monitor the exits. When individuals attempt to enter the sterile area through the exit lane, security personnel redirect them to the screening entry lanes. If a person refuses and continues to enter the sterile area, security notifies additional resources to respond while attempting to keep a visual on the offender. If two security guards are protecting the exit lane, one can pursue the offender while the other remains in position to prevent additional unauthorized access. If the security guard loses sight of the offender, then the entire concourse or airport (depending on the configuration of the facility) may have to close until the person can be located and the entire sterile area is resterilized. Many security breaches occur when individuals exit the sterile area, realize they forgot something, or believe they are in the wrong area, so they turn around and re-enter the sterile area through the exit lane. While these are seemingly innocent actions, it is still considered a security breach and must be handled as such.

Some airports have incorporated revolving doors or turnstiles capable of blocking entry from public areas while permitting egress for those departing sterile areas. Such designs must also allow sufficient space for the passage of the person with baggage and accommodate the disabled (TSA, 2006b).

Many pre-9/11 airport designs located the exit lanes next to the SSCPs. Where possible, the TSA (TSA, 2006b) recommends locating exit lanes away from SSCPs to physically separate those entering sterile areas from those exiting. If exit lanes are located next to entry lanes, a see-through barrier between the SSCP and the exit gates, such as a substantial plastic or laminated glass wall, should be installed. This allows law enforcement and other security personnel at the SSCP to observe exit lanes and immediately respond to a breach attempt.

Some new exit lane designs have incorporated a long response corridor to enhance detection of anyone attempting to breach the exit lane. Located just beyond the SSCP, a response corridor can give security time to catch and detain individuals who have not been screened.

Additional measures, such as the use of CCTV equipment and alarms, should be taken at both the exit and entry lanes to prevent breaches. CCTV equipment can be used to record the breach attempt, making it easier to identify and locate offenders. CCTV combined with motion sensor and alarm technology can be installed to monitor exit lanes.

Alarm systems that can be activated by security personnel are another important tool to prevent breaches. If a breach occurs, the screener can activate an audible alarm (i.e. panic button) that alerts all personnel at the SSCP including law enforcement officers. The alarm should be tied into the airport's access control alarm monitoring system (ACAMS), thus immediately notifying the security operations center, which can take additional actions to prevent unauthorized access.

Until December 2005, the TSA was responsible for protecting exit lanes, but that responsibility is presently a topic for industry debate. In 2006, the TSA issued a policy that attempted to transfer exit lane staffing to the airport operator when the SSCP closes, or if the SSCP and exit lanes are visually separated. The airport industry has not universally agreed with this policy and the issue remains in debate.

Screening: Terminal Operations and Profiling

In addition to the passenger and carry-on baggage screening at the SSCP, other screening methods are used before entering SSCPs and during screening. Document verification and passenger profiling are two such additional layers of security. Both require special training before security agents can be effective and efficient at either. Like all other security measures, neither documentation nor passenger profiling is 100% effective in preventing threats to security. Nevertheless, each is a critical component of the layered aviation security system.

Travel Document Verification

Document verification is an important component of screening to ensure the ticketed passenger is the same passenger who boards the aircraft. At some foreign airports, the document check is conducted at the ticket counter, at the beginning of the screening queue line, again at the WTMD, and once more before boarding. In the United States, the documents are checked at the ticket counter and at the beginning of the screening queue. At some U.S. airports, depending on the existing threat level, other document checks occur at the WTMD and occasionally at the boarding gate.

It is important to have individuals properly trained in the verification of authorized identification but the TSA is also implementing the Credential Authentication Technology/Boarding Pass Scanning System (CAT/BPSS), also known as the travel document scanner. The CAT/BPSS will scan a passenger's boarding pass and photo ID and automatically verity the name provided on both documents, then match and authenticate the boarding pass. The technology also identifies altered or fraudulent photo IDs by analyzing and comparing security features in the IDs.

Special-category passengers must also be accounted for at screening checkpoints. Passengers with diplomatic privileges and their protective personnel, law enforcement officers, deportees, and prisoners and their law enforcement escorts must also be processed at the screening checkpoints. Diplomats with protected document holders and their armed law enforcement escorts are considerations for the screening operator. The proper handling of such individuals and their armed escorts must be precoordinated through the U.S. Department of State before travel. Diplomatic pouches[6] are exempt from screening provided the proper procedures as set forth from the U.S. Department of State are followed (2004).

Armed law enforcement officers who must show identification to supervisory screening personnel must be able to do so out of sight of the public. It may also be advantageous to screen prisoners separately from the public screening areas for the protection of the public and the anonymity of the armed law enforcement escorts.

Profiling

Martin Aggar is the commercial director of Group 4 Securicor U.K. Security. In an article in *Aviation Security International,* Aggar described passenger profiling as a rapid risk analysis based on suspicious signs in a passenger's documentation, itinerary, appearance, or behavior (Aggar, 2005). Profiling has been successfully used to identify and deter terrorist attacks and to apprehend drug smugglers, money launderers, petty thieves, and illegal immigrants. The first objective in profiling is to determine the absence of the normal and the presence of the abnormal (through document inspections, behavior observations, etc.). Aggar stated, "It is a complete waste of time and money, and plays into the terrorists' hands to subject all passengers to a detailed bag and body search when clearly, if they have not been duped, 99.99 percent pose no threat to the flight" (Aggar, 2005, p. 21).

Profiling is a sensitive topic, particularly in the United States, where the term is often interpreted to imply racial profiling. Racial profiling has been largely dismissed by security professionals throughout the world as being ineffective and a naive and insensitive approach to maintaining aviation security. In fact, racial profiling can increase risk when it causes security personnel to focus on only individuals of a certain race or religious background.

---

[6]A *diplomatic pouch* (container of some sort, not necessarily an actual pouch) has diplomatic immunity from search or seizure.

Racial profiling would not have identified Timothy McVeigh or Terry Nichols, perpetrators of the 1995 vehicle bomb attack on the Oklahoma City Murrah Federal Building, or Richard Reid, the "shoe bomber" who attempted to blow up an American Airlines flight in 2001. Nor would it have identified John Walker Lindh, the "American Taliban." Because of 9/11 and the rise of Islamic fundamentalist terrorism, there is a tendency among some to believe that terrorists are only young Middle Eastern males. However, the face of terrorism changes over time, as Louise Richardson noted in her 2006 book, *What Terrorists Want*: "Today, the term 'terrorist' connotes the image of a radical Islamic fundamentalist from the Middle East. Thirty years ago, the term conjured up images of atheistic young European Communists" (Richardson, 2006, p. 10).

There is a tendency to focus on a particular demographic, primarily when an individual of a specific demographic was responsible for a recent terrorist attack. This is not so much an issue of race as of nationality. Rafi Ron, former security director at Ben Gurion International Airport in Tel Aviv, Israel, in an interview for *Aviation Security International* in 2002, noted that most of the terrorists acting against the United States are from extreme Muslim groups:

> *I know people can say there are always the likes of Timothy McVeigh so why pick on the Muslims. And I say, don't pick on the Muslims, but at the same time, don't ignore the fact that most of the terrorists right now are coming from extreme Muslim groups. Now, that doesn't mean that you should automatically make every Muslim a selectee in your security process, but, at the same time, don't ignore it. I was invited to testify before a House committee in Washington on profiling. I received tremendous support from the committee members. There were about 30 of them, and I think that about 28 were extremely supportive of the use of Behavior Pattern Recognition[7] because it excluded the racial element. One of the committee members asked me specifically whether nationality should be an element in security profiling. And my answer was "if somebody's home address is a cave in Afghanistan, it would be ridiculous to ignore it." There's nothing unconstitutional to use nationality as part of your security profiling when it comes to protecting the U.S. against its enemies. The people who wrote the Constitution never intended to blur the view of the government in this country identifying who is their enemy and who is not. So, it may be politically correct to go and pick on a 72-year-old Norwegian, but it's not very practical.*

When it was determined that the Oklahoma City Murrah Federal Building was bombed by individuals who held extremist views and had connections to certain militia groups in the United States, security and law enforcement began to focus on others participating in such groups.

---

[7]*Behavior pattern recognition* is the study and process of observing an individual's behavior, and interpreting that behavior as an indication of potential risks or threats.

Although the debate continues as to whether nationality should be used in profiling, the trend has been to keep profiling focused on document checks and the observation of behaviors— passive and active (through observation or the use of security questioning). Behavioral profiling is the process of understanding how an individual will behave when telling a lie or attempting to deceive or when acting in a manner consistent with preparing to commit a crime. Profiling, according to Dan Korem, author of *The Art of Profiling*, is the ability to assess a comprehensive amount of information about a person's personality (Korem, 1997). Profiling can help a security or law enforcement officer better predict criminal strategies and intent and obtain useful information during interviews (Korem, 1997). In the aviation security field, the definition of profiling extends to identifying enough information about an individual's intent to determine if he or she is a threat to aviation security. This can be accomplished by noting behaviors, body language, conversation, and other factors.

Humans have learned to profile as a survival mechanism, as noted by Malcolm Gladwell, author of *Blink: The Power of Thinking Without Thinking*. Gladwell remarked, "The only way that human beings could ever have survived as a species for as long as we have is that we've developed another kind of decision making that's capable of making very quick judgments based on little information" (Gladwell, 2005, p. 11). Gladwell's comment relates directly to profiling methodology, such as the ability to interpret body language.

Body language experts Allan and Barbara Pease and noted that "the ability to read a person's attitudes and thoughts by their behavior was the original communication system used by humans before spoken language evolved" (Pease and Pease, 2004, p. 7). As an example, subsequent to the advent of broadcast media, today's politicians understand that politics is about image and appearance and will often hire personal body language consultants to help get their message across (Pease and Pease, 2004).

Profiling begins the moment an individual encounters another, human or animal. Korem calls the first step in profiling a *read* and offers the following description of the read process:

> *Reading people is a natural reaction that none of us can avoid. We all spend time reading and interpreting people's actions, trying to predict how they will act in the future. Each time an observation of an individual's behavior is made, it is called a read. When several reads are compiled together, we identify a profile which provides a more complete picture of a person.* (Korem, 1997, p. 3)

When someone walks down a street late at night and spots a group of people approaching, profiling begins. In this case, profiling is initiated by asking various questions: Are they looking at me? How are they looking at me? Is it intently, the way a predator studies its prey, or casually, as if to wonder if I am a threat? Are they taking any actions that I need to be aware of? Are they attempting to conceal something? Do I know them? What is their reputation? How are they dressed? How many are there? What are they doing here? Our example profiler continues the internal questioning while observing the group's behavior and making further evaluations. The individual then decides on a course of action based on these evaluations.

Successful police officers and criminal investigators rely on intuition, often characterizing it as *a funny feeling that something is just not right*. Gladwell's (2005) research showed that intuition may be related to experience and training in what patterns and nonverbal clues are present when a crime is about to be, or has been, committed. Although some people are better at noticing nonverbal clues, the skill of profiling can be taught.

The first step in profiling is to establish a baseline. This is how a person typically behaves under normal circumstances. Korem (1997) noted that people typically act in consistent, similar ways, called *traits*. When two or more traits are combined, we establish *types*. Combine two or more types, and you develop a *profile* (Korem, 1997).

Establishing a baseline within an airport environment can be difficult as travelers often behave differently at an airport and when traveling by air than they do when going about day-to-day business and personal activities. Many people are anxious about flying and display signs of nervousness. Additionally, an airport is a busy and often overwhelming environment for many infrequent fliers, which increases their stress levels, throwing off what would be their normal behavioral baseline. In fact, the stress felt by passengers and the manner in which they exhibit that stress is frequently argued as the primary weakness against the effectiveness of profiling in an airport environment. However, proper profiling should consider these factors when establishing the behavioral baseline. The general public and the media have expressed concern that the general nervousness associated with flying, or the anger occasionally associated with the process of air travel (long lines, invasion of privacy, etc.), could be indicators of suspicious activity. However, the emotions of anxiety and of anger exhibit different signs than does that of deception.

When establishing a behavioral baseline, profilers should be aware that individuals can be trained to deceive. For example, professional actors spend years perfecting their art. In this example, believability is derived from the actor's talent at modifying behavior coupled with the audience's lack of training at identifying deceptive body language and behavior.

Profiling focuses on a variety of factors, but body language is central. Albert Mehrabian, one of the original body language researchers, found that the impact of a message is about 7% verbal (words), 38% vocal (tone, inflections), and 55% nonverbal (Pease and Pease, 2004). However, we tend to rely on the spoken word to communicate and tend to ignore body language. Research subsequent to Mehrabian's studies has shown that between 60% and 80% of communication, particularly in a business setting, is nonverbal (Pease and Pease, 2004). Since the 1990s, several researchers have explored the meaning of body language including Eckman (1992), Gladwell (2005), and Pease and Pease (2004). Much of this research has focused on the ability to read body language to identify behaviors, emotions, or truthfulness that may contradict the spoken word.

Although body language is not admissible in court in determining whether an individual is innocent or guilty, this inadmissibility does not mean that a person's body language should be dismissed. Body language is similar to reasonable suspicion in that it may indicate that further investigation is needed.

When an individual lies, the body tends to give away clues, often called *tells*, that the person intends to deceive—for instance, raising the pitch of the voice, looking away from the person one is talking to, or flying into a tirade when questioned. Sometimes, tells may be normal responses, particularly in an airport. Although many emotional responses are not absolute evidence of deception, they are worthy of additional attention while the person conducting the profile seeks more information to make a threat assessment. Although we may communicate primarily with our voices, our bodies can project many different messages. Deception clues may be in the form of a microfacial expression, swallowing, a change in breathing patterns, sweating, body and leg positioning, a slip of the tongue, or a voice inflection (Eckman, 1992).

In *Telling Lies: Clues to Deceit in the Marketplace, Politics, and Marriage*, Paul Eckman (1992) explained why it is difficult to get away with a deception: "The question is, why can't liars prevent these behavioral betrayals? Sometimes they do. Some lies are performed beautifully; nothing in what the liar says or does betrays the lie." Liars do not always anticipate when they will need to lie, so there is not time to prepare a response and its delivery. Even when there has been time, someone may not be clever enough to anticipate all the questions that may be asked (Eckman, 1992, p. 43). This is why good security questioning often involves a series of questions, some of them seemingly unrelated to security.

Eckman explained that the second reason liars cannot always prevent body language from revealing deception is because of the difficulty in concealing or falsely portraying emotion. When liars choose to lie, they generally prefer to conceal rather than falsify. When concealing, nothing has to be contrived and there is less chance of being suspected. Concealment is passive, not active (Eckman, 1992). When a lie is told, the liar must remember the same version of the lie as originally told. In law enforcement, this challenge is often characterized as "getting your stories straight."

Security questioning is considered more effective than passively observing behaviors. A liar loses the choice of whether to conceal or falsify once challenged (Eckman, 1992). When being questioned, a liar must create stories either in advance or on the spot. This stress causes the body to exhibit certain noticeable characteristics. A person who is being observed without being questioned can more carefully select and rehearse how to behave so that the behavior more closely matches the norm.

Questioning can also create emotions that may not be exhibited during passive observation. Eckman stated that "people do not actively select when they will feel an emotion, and negative emotions may even happen despite themselves" (Eckman, 1992, p. 47). Questioning can also cause an emotion to appear suddenly, rather than gradually. When emotions gradually appear, it is harder to detect the deceit as the behavioral changes are less noticeable. Strong emotions brought on quickly are harder to control (Eckman, 1992).

However, there are situations where criminals have rehearsed their "performance" so carefully as to not show many of the visual or auditory signs of deception. In these cases, a professionally trained profiler can still spot the deceit, as liars usually monitor and try to control their words and face more than their voice and body (Eckman, 1992). Eckman stated that the body is a good source of leakage and deception of clues as it is not tied directly to areas of the brain involved in emotion, as the face and voice are. Eckman explained that, "The body leaks because it is ignored. Everyone is too busy watching the face and evaluating the words" (1992, p. 85).

Additionally, Eckman (1992) pointed out that if a liar has not worked out the lies in advance, he or she will have to be cautious, carefully considering each word before it is spoken. And even if the liar has worked out the lie in advance, the body language, specifically *illustrators*,[8] or emphasis on words or gestures, may actually decrease (Eckman, 1992). This is another clue for profilers that a deception may be in the works.

Israel is known for its profiling practices and its ability to ferret out criminals and terrorists through profiling. It was profiling and security questioning that prevented Anne Marie Murphy from boarding an El Al flight in 1986. Murphy had a bomb in her suitcase that had been planted by her fiancé, unbeknownst to her, who was a Syrian intelligence agent.

In Israel, profiling starts when a traveler books his or her flight. Background checks are run to ensure a ticket is not being issued to a known terrorist or criminal. Travelers arriving at the airport usually first experience profiling at the vehicle checkpoints, before entering airport property. Security questioning begins when the traveler enters the terminal building. Before a passenger checks in for the flight, security personnel question him or her about travel plans, agenda, luggage, purpose of the trip, and so on.

Challenges to implementing the Israeli model of profiling in the United States include the number of passengers that U.S. airports enplane daily, versus the relatively small number enplaned in Israel, and the civil rights issues surrounding profiling in the United States, which are not problematic in Israel. In spite of this and shortly after 9/11, Boston Logan International Airport hired Rafi Ron, a former security director at Ben Gurion International Airport in Israel, to teach airport and airline personnel methods of profiling that were scalable to their operations.

Started in 2004 as a pilot program, Ron's patented behavior pattern recognition (BPR) method teaches personnel working at an airport, such as air marshals, screeners, and state police, how to identify suspicious behaviors that could indicate a terrorist plot and then to report such behaviors to law enforcement officers. Uniformed and undercover officers watch people as they move through terminals, looking for odd or suspicious behavior such as heavy clothes on a hot day, loiterers without luggage, and anyone observing security methods. Screening supervisors use score sheets containing a list of suspicious behaviors. If a passenger hits a certain number, a law enforcement officer is notified to question the person (AP, 2004).

Air marshals have also been trained to watch airport crowds as they await flights, looking for things outside the normal range of behavior. In an article for the *Toronto Star*, Jack Shea, special agent in charge of the federal air marshals in Boston, MA, made the following

---

[8]Eckman classified illustrators as emphasis that can be given to a word or phrase or gestures that repeat or amplify what is being said. Brow and upper eyelid movements or the trunk of the body can also be illustrators, serving to emphasize an expression.

remark regarding BPR: "What I like about it, it's very basic, it's common sense, it's effective, it works" (AP, 2004, p. H10). Although the BPR program has caught several individuals with outstanding warrants, critics of the program, including the American Civil Liberties Union, are concerned that BPR could become a pretext for racial profiling. Barry Steinhardt, the director of the American Civil Liberties Union's technology and privacy program, qualified this concern by stating, "Not every police or security officer who's going to be heading up a local operation is going to be sensitive to the racial implications of the project, especially as you roll it out nationwide" (AP, 2004, p. H10). Other privacy advocates prefer the BPR program to the government's plans to conduct screening via computer databases to try to identify potential terrorists. In this regard, David Sobel, general counsel for the Electronic Privacy Information Center, a leading critic of the background checks, stated that "Targeted interviewing is certainly preferable [to screening via computer databases]" (AP, 2004, p. H10).

TSA's Behavior Recognition Programs

In 2004, the TSA began implementing a program known as SPOT (Screening Passengers by Observation Techniques). SPOT involves specially trained security officers scrutinizing people in security lines and elsewhere in the airport. By assessing a person's body language and travel details, screeners can make a quick judgment on the threat level (Meckler and Michaels, 2006). Suspicious individuals can then be questioned and referred for either secondary screening or to law enforcement personnel. SPOT has led to the use of behavior detection officers (BDOs). BDOs are transportation security officers who receive special training in identifying suspicious behaviors. Initially, BDOs were deployed to passively observe passenger behavior, which is not as effective as actually questioning individuals. In 2010, the GAO concluded that there was not a scientifically validated basis for using behavior detection for counterterrorism purposes in the airport environment, and further that there was not a means of recording and tracking data on suspicious individuals (GAO, 2010, p. 16). Unfortunately, many in the media and the general public took the GAO report to mean that behavior detection is ineffective, which is not what the report concluded. Additionally, it is difficult to measure the effectiveness of behavior detection when part of the objective of the program is to deter criminal and terrorist activity. Deterrence, measuring the number of times something did not happen, particularly when it is not known how often a criminal act was attempted but then deterred, is extraordinarily difficult. In nine months and with about 7 million people flying out of Washington Dulles International Airport, several hundred people had been designated for intense screening through the BDO program, and about 50 were turned over to the police for follow-up questioning. Several resulted in arrests, mostly for immigration violations (Lipton, 2006).

In 2011, the TSA started the assessor program at Boston Logan International Airport, which added the security questioning process to the BDO process. The program rapidly was nicknamed "chat-down," a word-play on the pat-downs conducted by TSOs at the checkpoint.

One problem with the BDO program is that TSOs do not have law enforcement powers, which means if they see a suspicious person, they can request that the individual talk with them, but they do not have the legal right to detain them. They can require the passenger to go through secondary screening if the individual still desires to board a commercial aircraft, or they can refer the individual to the local police. But, even police officers must fulfill requirements called for in the Fourth Amendment to the U.S. Constitution before they can detain an individual. Therefore, when a BDO requests law enforcement assistance on an individual the BDO finds suspicious, the LEO can still elect to not question the individual, and unless there is a reasonable suspicion that a crime has been committed or is about to be committed by the individual, then the individual has the right to leave the area without further LEO interference.

The SPOT/BDO programs raised another issue directly relevant to airport security coordinators, who are charged with overall security including law enforcement and security in the terminal building. While some suspected the program is an example of TSA "mission creep," another real logistical issue with TSOs going into the public areas, expanding their security purview into the terminal where airport police and security personnel are already on patrol, are the coordination and control issues among essentially two or three security groups, all of which report to different agencies and have different requirements under the airport security program (ASP). Questions are raised, such as how will a police officer prioritize two conflicting security issues? For example, a TSA behavioral officer spotting a suspicious individual at the same time the airport needs that same police officer to respond to an issue that he or she is required to respond to under the ASP.

Aviation Watch and See Something, Say Something

Aviation Watch is a program initially developed at Boston Logan International Airport and in use at some airports. It is based on the Neighborhood Watch program used in residential communities throughout the United States. Aviation Watch was adopted and formalized into a training video distributed by the American Association of Airport Executives.

See Something, Say Something is a program similar to Aviation Watch and is promoted by the TSA. Initially developed by the Port Authority of New York and New Jersey, the program has spread even beyond the aviation community to other transportation industries and even American society in general.

Aviation Watch teaches airport workers how to spot suspicious activities, actions to watch for, and how to respond when they notice something odd. Aviation Watch focuses less on questioning and more on observation in an airport environment. It established the concept of understanding norms, such as distinguishing the difference between a group of passengers on the shuttle between Washington, DC, and Boston versus a group of passengers traveling to Orlando, FL. If it is a weekday, the shuttle flight is usually filled with passengers in business attire, carrying laptops and small briefcases, whereas the flight to Orlando is more likely to be filled with families going on vacation. Although the business traveler would not be out of place on the Orlando flight, a family of four might be out of place on the shuttle. Although not necessarily a cause for alarm, this activity is enough to warrant some additional attention by airport personnel.

The Aviation Watch program focuses on other suspicious behavior, such as individuals photographing or videotaping security procedures, emergency exit doors, and other areas not normally the subject of photography at an airport. Aviation Watch is typically not affected by civil liberties issues, as it does not involve active security questioning, just observation of activity and law enforcement notification.

Several airports, like Salt Lake City International Airport in Utah, have instituted their own versions of Airport Watch programs. In Salt Lake City, aviation employees are trained to report suspicious activities to law enforcement. The training is reinforced with signage in both English and Spanish around the airport to remind employees and passengers to be watchful for suspicious activity, along with a phone number to call to report such activity. It is useful to include the emergency contact phone number, as it cannot be assumed that calling 911 will reach the airport police. Some cities do not have the 911 system in place, and in many other cities, calling 911 from the airport will put the caller in touch with the local police department, not the airport police. This causes the local police dispatcher to route the call back to the airport, thus delaying the law enforcement response.

Salt Lake City has also embraced a population of aviation hobbyists who enjoy plane watching (plane spotting). These individuals are commonly referred to as *spotters*. These groups exist at many airports across the country, with individuals gathering around the airport perimeters to watch aircraft take off and land. Often they photograph aircraft, note tail numbers when trying to spot specific planes, and are equipped with VHF scanners to listen to pilot-ATC communications, binoculars, and cameras. Such activity is legal provided the spotters stay outside the airport perimeter fence and protected air navigation zones.

Shortly after 9/11, spotters came under greater scrutiny from federal officials and police officers who did not understand their motives. Spotters can be a tremendous security asset. They are often a tight-knit group and are keenly aware when there is a newcomer in their midst, particularly when the newcomer engages in activities not normally associated with plane spotting. In Toronto, Canada, one group of plane spotters, calling itself YYZ Airport Watch, even has its own website (see *www.canairradio.com/airportwatch.html*) and includes as part of its mission "to observe, record, and report suspicious activity" while engaged in plane-spotting activities.

Plane spotters are often very familiar with normal airport operations and activity in and around an airport. Airport operators should create a formal program to include such individuals as part of any suspicious activity reporting program. The first step is to verify that none of the spotters is wanted or has a criminal history, particularly involving interference with aviation operations. If possible, a CHRC should be conducted and, for the spotters who are at the airport frequently, the airport should consider issuing non-SIDA identification badges, similar to those issued to aviation employees who work only in the landside areas of the airport. These identification badges can reduce the time police and airport security and operations personnel spend when conducting checks of the spotting groups. Any identification issued should have a photo, an expiration date, a distinctive appearance so the badge cannot be mistaken for a SIDA access/ID badge, and rules that the badges cannot be loaned or used improperly. Approved plane spotting areas should also be delineated, with proper signage to ensure individuals know where they can and cannot be. This facilitates safety and security, as airport operators do not want individuals within the runway critical areas and safety zones.

There are many examples of this sort of program in the United States and Canada. The Royal Canadian Mounted Police have Airport/Coastal Watch whereby citizens can report suspicious vessels and aircraft. The Aircraft Owners and Pilots Association (AOPA) has Airport Watch, featuring airport signage, a toll-free number (1-866-GASECURE), and an instructional video.

Secure Flight

One of the first passenger profiling programs in the United States was CAPPS. Started in 1996, whenever an individual purchased an airline ticket with certain peculiarities, such as a one-way ticket with cash bought on the day of travel, the computer would flag that individual for additional scrutiny upon arrival at the airport. However, CAPPS was geared at preventing bombings, not hijackings. Several of the 9/11 hijackers were flagged under CAPPS. However, the only requirement was additional scrutiny going through the screening checkpoint, and that the individual's bags would not be loaded onto the plane until it was confirmed that the individual had already boarded. CAPPS remains in place and is controlled by the airline that checks the passenger's reservation information contained in the air carrier's passenger name record (PNR) against a set of established system rules, referred to as the CAPPS I rules (GAO, 2005a).

Added to screening are checks of passenger names against the no-fly list and the selectee list. The no-fly list contains the names of individuals who are not allowed to fly on U.S. carriers or must first be cleared by law enforcement officers. The selectee list contains the names of individuals who require additional scrutiny such as secondary screening or law enforcement officer notification. Both lists are shared with airlines, because carriers must check passenger names against the list. The program is now known as Secure Flight, which is a behind-the-scenes program that enhances the security of domestic and international commercial air travel through the use of improved watch list matching. When passengers travel, they are required to provide the following Secure Flight passenger data (SFPD) to the airline:

- Name (as it appears on government-issued ID the passenger plans to use when traveling)
- Date of birth
- Gender
- Redress number (if applicable)

The airline submits this information to Secure Flight, which uses it to perform watch list matching. This serves to prevent individuals on the no-fly list from boarding an aircraft and to identify individuals on the selectee list for enhanced screening. After matching passenger information against government watch lists, Secure Flight transmits the matching results back to airlines so they can issue passenger boarding passes.

Consumers initially complained that it was difficult to be removed from the selectee list, even after proving the name is a sound-alike of a real name on the list and that the individual is, in fact, not the person the TSA is looking for. In one case, the list nearly prevented a 4-year-old boy from boarding because of a sound-alike conflict. The TSA has since told airlines not to deny boarding to anyone under 12. Even Senator Edward Kennedy had been hindered from traveling commercially and had a difficult time getting his name separated from a sound-alike on the list.

The selectee list was scrubbed in 2005 to ensure its accuracy (Burbank, 2005) and additional improvements have corrected many of the previous identity confusion problems. The Secure Flight database is being tied into the Transportation Vetting Platform (TVP), which the TSA is developing to ensure that transportation workers, including flight crews, aviation workers with access to secure areas, alien flight school candidates, commercial hazmat truck drivers, screeners, and others with access to transportation infrastructure, are properly screened. The TSA does not currently plan for Secure Flight to include checking for criminals, performing intelligence-based searches, or using alert lists (GAO, 2005a).

The TSA did make progress though coordinating with stakeholders such as Customs and Border Protection and the Terrorist Screening Center (TSC).

Intelligence-Driven Risk-Based Screening

With the appointment of John Pistole to the position of TSA administrator, the long sought ideal of screening personnel and baggage based on risk, rather than a one-size-fits-all security model, started becoming a reality. In fact, risk-based screening (RBS) may fulfill the original intent of the Trusted Traveler Program called for in the Aviation and Transportation Security Act of 2001, but quickly relegated to second-class status by the TSA (becoming the defanged Registered Traveler Program). RBS is based on the fact that the majority of airline passengers present little to no risk. RBS takes into account that the TSA may have more information about some travelers, that screening detection technology continues to improve, and that certain types of people naturally present a lower risk than others. The overriding principle, however, for acceptance of a risk-based model of security is that risk-based assumes some level of risk is present and accepted by the traveling public. While the concept of accepting a level of risk may seem counterintuitive to aviation security there is already an acceptance of risk in the one-size-fits-all screening model. The heavy reliance on regimented processes and well-known technology allows certain holes to develop and be exploited. Risk-based security accepts that no security system is perfect but that it can be better by focusing on higher-risk populations.

TSA's Pre✓™ is an expedited screening initiative that places more focus on prescreening individuals who volunteer to participate to expedite the travel experience and provides them their own screening lane at the SSCP. The program started with airline frequent flyers and is expected to expand to other populations. Travelers enroll in TSA's Pre✓™ through either being preselected (as an airline frequent flyer with known travel habits), or they can enroll through the Customs and Border Protection Global Entry program. TSA's Pre✓™ may not have to remove their shoes or laptops, and in most cases are only subjected to the WTMD, rather than the AIT machines.

According to the TSA they will always incorporate random and unpredictable security measures throughout the airport and no individual will be guaranteed expedited screening, to retain a certain element of randomness to prevent terrorists from gaming the system. In addition, individuals who commit certain offenses, such as violations involving firearms, explosives, and fraudulent documents at an airport or on board an aircraft, are likely to be denied expedited screening for a period of time. The duration of disqualification for expedited screening will depend on the seriousness of the offense. TSA's Pre✓™ is one method that allows TSOs to focus their efforts on other passengers who are more likely to pose a risk to transportation.

Screening for active-duty U.S. service members is another RBS program that allows U.S. service members who possess a valid common access card (CAC) at participating airports to receive expedited screening and no longer have to remove the following: shoes, 3-1-1-compliant bag from carry-on[9], laptop from bag, light outerwear/jacket, belt.

While the TSA is required to screen all passengers, they have enacted risk-based checkpoint screening procedures for passengers 12 and under that include allowing children 12 and under to leave their shoes on, allowing multiple passes through the WTMD and AIT to clear any alarms on children, and using explosives trace detection technology on a wider basis (rather than immediately resorting to a pat-down) to resolve alarms on children. While some critics have speculated that terrorists could decide to use children to conceal explosives or other prohibited items in their shoes, there is still the chance for metal detectors, AIT, or ETD machines to pick up on the substance or item, and so far, terrorists have not attempted to use children in this capacity. Further, the amount of explosives that can be concealed in the shoes of an individual under 12, due to their shoe size, is not as significant as what could be concealed in an adult shoe size.

Similar processes now exist for individuals 75 years of age and older since the TSA has determined that individuals in this age range present a lower risk.

In 2006, the liquid restrictions became effective, however, as part of RBS, medically necessary liquids and gels, including medications, baby formula and food, breast milk, and juice, are exempt from the 3-1-1 rules, and are allowed in reasonable quantities exceeding 3.4 ounces (100 ml). They are not required to be in a zip-top bag but TSOs may ask travelers to open these items to conduct additional screening and passengers should declare them for inspection at the checkpoint.

Another RBS initiative is a combination of risk-based security with the behavior detection program. At certain travel document check locations, specialized behavioral analysis

---

[9]The 3-1-1 security policy allows for liquids to be contained in no larger than three-ounce containers that must fit in a one-quart baggie and only one bag per passenger.

techniques are used to determine if a traveler should be referred for additional screening at the checkpoint. The vast majority of passengers at the pilot checkpoints experience a "casual greeting" conversation with a behavior detection officer (BDO) as they go through identity verification.

Conclusion

The discussion of airport screening is continued in the Chapter 7. Please see the Conclusion in Chapter 7 for a summary of the topics presented in both chapters on screening.

References

Aggar, M., 2005. Passenger Profiling: Dispelling the Myths Surrounding the Controversy. Aviation Security International 11, 20–22.

AP, Apr. 24, 2004. Airport Keeps Watchful Rye; Logan Airport Staff Specially Trained to Spot Strange Behaviors in a Crowd. Daily Mercury (Toronto Star Newspapers), H10.

ASIA, 1990. Aviation Security Improvement Act of 1990, P.L. 101–604, Sec. 318.

ASIA, 2000. Airport Security Improvement Act of 2000, P.L. 106–528 Sec. 3.

ATSA, 2001. Aviation and Transportation Security Act of 2001, P.L. 107–71 Sec. 110.

Burbank, L., Dec. 13, 2005. How to Avoid TSA No-fly Name Games. USA Today, retrieved July 8, 2005, from www.usatoday.com/travel/columnist/burbank/2005-12-06-burbank_x.htm.

Eckman, P., 1992. Telling Lies: Clues to Deceit in the Marketplace, Politics, and Marriage. W.W. Norton, New York.

General Accounting Office (GAO), 1987. FAA's Implementation of a Performance Standard for Passanger Screening Process (Publication No. GAO/T-RECD-97-90), retrieved July 9, 2008, from http://archive.gao.gov/d39t12/134217.pdf.

GAO, 2001. Aviation Security: Terrorist Acts Demonstrate Urgent Need to Improve Security at the Nation's Airports (Publication No. GAO-01-1162T), retrieved July 9, 2008, from www.gao.gov/new.items/d011162t.pdf.

GAO, 2004a. Aviation Security: Improvement Still Needed in Federal Aviation Security Efforts (Publication No. GAO-04-592T), retrieved July 9, 2008, from www.gao.gov/new.items/d04592t.pdf.

GAO, 2004b. Aviation Security: Private Screening Contractors Have Little Flexibility to Implement Innovative Approaches (Publication No. GAO-04-505T), retrieved July 9, 2008, from www.gao.gov/new.items/d04505t.pdf.

GAO, 2005a. GAO 2005 Aviation security: secure flight development and testing under way, but risks should be managed as system is further developed (Publication No. GAO-04-356), retrieved July 9, 2008, from www.gao.gov/new.items/d06374t.pdf.

GAO, 2010. Efforts to Validate TSA's Passenger Screening Behavior Detection Program Underway, but Opportunities Exist to Strengthen Validation and Address Operational Challenges. Government Accountability Office, Washington DC.

Gladwell, M., 2005. Blink: The Power of Thinking Without Thinking. Little, Brown and Company, New York.

Gore, A., 1997. White House Commission on Aviation Safety and Security, Final Report to President Clinton, retrieved July 8, 2008, from www.fas.org/threat/212fin∼1.html.

Hawley, K., Dec. 12, 2005. AAAE/DHS/TSA Annual Security Summit. Washington, DC.

Hawley, K., Means, N., 2012. Permanent Emergency: Inside the TSA and the Fight for the Future of American Security. Palgrave Macmillan, New York.

Korem, D., 1997. The Art of Profiling: Reading People Right the First Time. International Focus Press, Richardson, TX.

Lipton, E., Aug. 17, 2006. Faces, Too, Are Searched at U.S. Airports. The New York Times, retrieved July 8, 2008, from www.nytimes.com/2006/08/17/washington/17screeners.html.

Meckler, L., Michaels, D., 2006. Aircraft-Security Focus Swings to People. The Wall Street Journal, retrieved July 8, 2008, from http://online.wsj.com/public/article/SB115535025189634122-aLFdB7oNQYQNq15VKhKL7kyvQDK_20070814.html?mod=tff_main_tff_top.

Pease, A., Pease, B., 2004. The Definitive Book of Body Language. Orion Books, London, U.K.

Price, J., 2004. Airport Certified Employee: International Aviation Security, Module 1. American Association of Airport Executives, Alexandria, VA.

Richardson, L., 2006. What Terrorists Want. Random House, New York.

Shanks, N., Bradley, A., 2004. Handbook of Checked Baggage Screening: Advanced Airport Security Operation. Professional Engineering, Suffolk, U.K.

Silverstein, J., Oct. 27, 2006. Web Site Lets Anyone Create Fake Boarding Passes. ABC News, retrieved July 8, 2008, from page=1">http://abcnews.go.com/Technology/story?id=2611432&page=1#.UMJgn6Wo_Sc.

TSA, 2006a. Checkpoint Design Guide Standards. TSA, Washington, DC. revised November 7, 2006, Version 1. http://www.aci-na.org/static/entransit/Checkpoint_Layout_Design_Guide_v1r0-0.pdf.

TSA, 2006b. Recommended Airport Security Guidelines for Airport Planning, Design, and Construction. U.S. Department of Homeland Security, Washington, DC. TSA document, Revised in 2011. http://www.tsa.gov/sites/default/files/assets/pdf/airport_security_design_guidelines.pdf.

TSA, 2006c. TSA Security Checkpoint Layout Design. http://www.aci-na.org/static/entransit/Checkpoint_Layout_Design_Guide_v1r0-0.pdf.

U.S. Department of State, 2004. Addendum to New Pouch Procedures, Diplomatic Note 04-181, retrieved July 8, 2008, from www.state.gov/documents/organization/35403.pdf.

7 ⣿

Passenger and Baggage Screening

Objectives

In this chapter, we focus more closely on processes and policies related to passenger and baggage screening. Methods and techniques for screening passengers and baggage are fundamental to sustaining effective aviation security. Within the layered aviation security system, screening passengers and baggage are the most important processes that security practitioners use to mitigate potential threats to the airport and aircraft environments. You will learn about these fundamental concepts and special issues in screening that are vital to securing the national airspace system.

Screening Passenger and Carry-On Baggage

ICAO Annex 17 to the Convention on International Civil Aviation—Safeguarding International Civil Aviation Against Acts of Unlawful Interference calls for procedures to ensure passengers do not bring a firearm, explosive, or dangerous device into a passenger cabin (International Civil Aviation Organization [ICAO], 2006). Unfortunately, while threats to aviation security have changed significantly, up until 9/11 the technologies and processes developed in the 1970s and 1980s were largely unchanged.

In August 2004, Chechnyan terrorists (known as the "Black Widows") brought down two Russian commercial airliners by concealing explosives beneath their braziers, highlighting the weakness of the walk-through metal detectors (WTMDs), which do not detect explosives. A passenger could wear plastic explosives and conceal the detonator in their carry-on bag and not set off any security alarm. A passenger could pass through the type of metal detector currently in use with a suicide belt made of plastic and fabric while concealing the detonator elsewhere and not set off any alarm (Hughes, 2005).

This weakness was again exposed in August 2006, when U.K. officials discovered a plot to blow up commercial airliners departing the United Kingdom on intercontinental flights. The unfortunate reaction in the United States and the European Union was to prohibit or severely restrict passengers from carrying liquids onboard aircraft. Had technology and funding priorities kept up with the existing and known threats, technology or strategies may have existed to prevent or detect the presence of liquid-based explosives. Such technology is now being developed, years after the initial threats were known. Newer passenger screening technology involves graphic body imaging X-ray machines, portal-trace detectors, advanced metal detection equipment, and multiview X-ray equipment.

Passenger and carry-on baggage screening deters a number of threats, specifically bombings and hijackings, thus it is one of the most important security layers. The basic screening process is explained in Shanks and Bradley's *Handbook of Checked Baggage Screening* (2004):

1. *The divestiture process.* The passenger or employee is called by security staff to remove outer attire and anything that may set off the metal detector, or be detected by the advanced imaging technology, such as belt buckles, watches, jewelry, coins, mobile phones, or PDAs. These items are placed in a polymer container and sent through the X-ray machine. Laptops and large electronics are often removed from their containers and placed in separate bins. In the United States and many other countries, individuals must also remove their shoes, but this policy changes both from airport to airport and country to country. The individual also loads their baggage onto the bag belt for analysis by the X-ray machine or explosive detection system (EDS) technology. Security staff may sometimes realign the bag to randomize its position and to allow proper separation of bag images enabling a static image.

2. *Passenger screening.* The passenger is requested to move through the metal detector or AIT machine. Passengers who set off the detector's alarm or may have a suspicious item as indicated by the advanced imaging technology (AIT), are asked to step aside until a secondary search can be conducted. The secondary search is usually a physical pat-down or hand-wand metal detector. In some instances, passengers may be allowed to go back through the metal detector after divesting themselves of additional items that may have triggered the alarm, rather than going to secondary screening.

3. *Carry-on baggage screening.* As the passenger is being screened, security staff members analyze the contents of the passenger's bag using conventional X-ray technology[1], multi-angle X-ray machine, or EDS technology. Baggage that contains questionable items or threat items is often checked physically through a bag search, analysis by an explosive trace detection (ETD) machine, or, in some cases, both. If a bag contains an apparent bomb, then the screener will likely keep the suspect item within the X-ray machine, stop the belt to prevent the bag from advancing out of the machine, and hinder attempts by the owner to pull the bag away from the security staff. The screener will notify law enforcement and supervisory personnel for further assessment of the X-ray image. If the item appears to be a bomb, then an immediate evacuation of the security screening checkpoint (SSCP) and surrounding area may be required. Explosive demolition teams or K-9 explosive detection teams are called to further verify the threat.

4. *Exit process.* Provided passenger belongings have been cleared through the X-ray analysis, passengers are reunited with their belongings and allowed to proceed into the sterile area.

5. *Special circumstances.* Disabled passengers and those in wheelchairs often must be hand searched.

---

[1]Conventional X-ray technology is the most common form of technology to check carry-on baggage. Some countries have implemented EDS technologies in lieu of the conventional X-ray, and in the United States conventional X-ray machines are being replaced with multi-angle X-ray machines that offer better imagery displays.

6. *Newer technology.* New technology is being developed such as liquid explosive detectors and document scanning devices. These technologies are being integrated into the passenger and carry-on baggage screening process.

Passengers

Screeners must be aware of several dynamics in screening, including the ability to ensure that individuals do not pass through a screening checkpoint with a prohibited item, individuals do not bypass primary or secondary screening, and wait times are kept to a minimum to maintain the advantages of air travel and prevent massive crowds from gathering, thus creating a vulnerable target for a suicide bomber or active shooter attack.

There are methods that criminals and terrorists use to get prohibited items, including explosives and weapons, through the SSCP. Individuals might carry only a portion of an explosive through a screening checkpoint, such as plastic explosives that are undetectable with metal detector technology, while concealing detonators in carry-on bags. Screeners should watch for bulky clothing and passengers with odd-shaped bulks and sharp angles underneath their clothing. Passengers should always remove coats and excessive outerwear, which could conceal explosives or other prohibited items that may escape the detection abilities of the passenger screening technology and security screeners. Screening personnel must also be aware of body language that is consistent with attempts to deceive, and be trained in interpreting verbal and nonverbal cues.

Customer service should not be sacrificed in screening. Screeners must always consider that the majority of passengers are not terrorists or criminals. They must ensure that passengers selected for secondary screening can keep in view of their own belongings, increasing customer service and reducing the possibility of theft. This also reduces stress on the individual who has been selected for secondary screening. In the United States the Transportation Security Administration (TSA) has attempted to quiet the SSCP area with the theory that a quiet and calm environment allows those who are attempting to deceive to be more easily detected. Some of these methods have included signage, but one of the most effective methods of bringing calm to the SSCP is the use of earpieces and personal microphones by transportation security officers (TSOs). Rather than yelling for assistance, which just adds to the noise and confusion inherent in a screening area, TSOs can communicate quietly.

Passenger screeners must ensure that individuals requiring secondary screening are not allowed to leave the SSCP area until the secondary screening is complete, preferably by a second screener to avoid disrupting the passenger flow. Passenger screeners who staff entry gates should strictly control the passage of individuals through the metal detector (or ETD) by allowing only one individual at a time to pass through.

The standard method of screening in the United States and abroad has been using the WTMD, often called the *magnetometer*. The WTMD creates a magnetic field, which is disrupted with the presence of metal. This disruption, if high enough, causes an alarm to sound. WTMDs can be set to specific levels depending on the sensitivity desired by the operator. WTMDs quickly process passengers, requiring only a second or two for the passenger to pass through.

WTMDs cannot detect plastic explosives or other types of explosive substances, and it can be assumed that other bomb-making devices (blasting caps, timers, etc.) that do have metal components would be concealed in carry-on baggage and assembled after the individual has passed through the checkpoint. Individuals who set off the alarm, or who for any reason cannot pass through the detector (e.g., disability), are referred to secondary screening. The secondary screening generally consists of using a hand-wand. If the hand-wand cannot determine the problem, then a physical pat-down search or even a strip search, done privately by a member of the same sex and with supervision, may be needed. In some cases, a handheld explosives detection device may also be used.

More advanced metal detectors can pinpoint where a metallic object is on an individual (Figure 7.1). The Garrett PD 6500i™ divides the magnetic field into 33 distinct zones

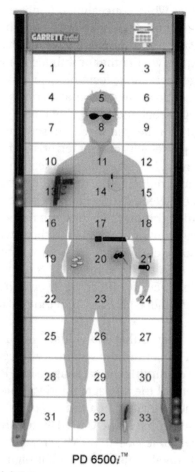

PD 6500i™

FIGURE 7.1 Garrett PD 6500i™ metal detector.

and has uniform coverage from head to toe. Lighting visible on the exit side of the detector notes the zone where a particular metallic object is located.

To make up for the weaknesses in the metal detection process, the TSA has widely deployed advanced imaging technology devices. This body imaging technology comes in two forms: backscatter X-ray or millimeter wave imaging. Backscatter technology projects low-level X-ray beams over the body to create a reflection of the body displayed on the monitor. Millimeter wave technology bounces harmless electromagnetic waves off the body to create the same generic image for all passengers.

Before going through this technology, passengers must remove all items from pockets and accessories, including wallets, belts, bulky jewelry, money, keys, and cell phones. Once inside the imaging portal, passengers stand in defined position with their arms raised over their head, and remain still for a few seconds while the technology creates an image of the passenger in real time (Figure 7.2).

For machines using backscatter technology, a remotely located officer views the image to identify any irregularity that appears on the screen. After review and resolution of any anomalies, the image is immediately deleted. All millimeter wave technology units are equipped with automated target recognition (ATR) software that detects any metallic and nonmetallic threats concealed under a passenger's clothing by displaying a generic outline of a person on a monitor attached to the AIT unit, which highlights any areas that

FIGURE 7.2 A passenger undergoes secondary screening at Denver International Airport.

may require additional screening. The generic outline of a person is identical for all passengers. If no anomalies are detected, an "OK" appears on the screen. The passenger then exists at the opposite side of the portal and collects his or her belongings. The entire process takes less than one minute.

Another device that could be implemented at screening checkpoints is a document scanner. The document scanner is based on the same trace detection technology as the ETD system. The premise of the document scanner's utility is that explosive particles are usually sticky, and anyone handling such materials is likely to transfer those materials to any travel documents he or she is carrying. Placing the individual's passport or other identification in a document trace scanner allows quicker screening, reduces the space needed to conduct the explosives trace search, and is available at a lower cost.

The TSA briefly experimented with explosives trace portal (ETP) systems, but the technology proved to be too fragile to handle an airport environment and was eventually replaced with the AIT machines.

Some passengers require or request to either opt-out of using the AIT systems. In these circumstances, they are subjected to a physical pat-down, similar to the search conducted by a police officer during an arrest. Other passengers may request to be screened privately, or there may be issues with the screening of a particular individual that necessitates the removal of clothing; in either case they are escorted by TSOs of the same sex into a private location, usually near the screening checkpoint, out of public view, and are patted down.

Pat-downs are used to resolve alarms at the checkpoint, including those triggered by metal detectors and AIT units. Pat-downs are also used when a person opts out of AIT screening to detect potentially dangerous and prohibited items. Because pat-downs are specifically used to resolve alarms and prevent dangerous items from going on a plane, the vast majority of passengers will not receive a pat-down at the checkpoint.

The TSA notes that the majority of pat-downs occur when a passenger alarms either the metal detector or the AIT unit. Passengers have rights during a pat-down including the right to request the pat-down be conducted in a private room and to have the pat-down witnessed by a person of their choice. All pat-downs are only conducted by same-gender officers. The officer will explain the pat-down process before and during the pat-down. The TSA initially was criticized for its pat-downs of children, but these processes have been modified to reduce some of the issues. TSOs have been given some discretion with children and will work with parents to resolve any alarms at the checkpoint. If required, a child may receive a modified pat-down, witnessed by the parent.

For passengers who are in a cast or have prosthetics, CastScope technology provides TSOs with a means to ensure that a cast or prosthetic does not contain a concealed threat. The TSA began deploying CastScope machines to several major airports nationwide in 2008, initially to airports based on the airport's proximity to military hospitals or large rehabilitation facilities that serve amputees, sporting events for disability groups, vacation destinations utilized by amputees, and airports that see large volumes of military severely injured.

The TSA worked closely with special-interest groups, such as the Amputee Coalition of America, to determine best practices, operational suitability, and modify the technology to meet the needs of the traveling public.

Carry-On Baggage

For more than 30 years, carry-on baggage screening has been conducted with conventional X-ray machines. These were discarded by the federal government for protecting buildings such as the U.S. Supreme Court and the U.S. Capitol, but they remained in use at U.S. airports. Upgraded X-ray equipment, including color filters to highlight suspicious items, was sent to U.S. airports in the mid-1990s and again after 9/11.

In 2006, the TSA announced a plan to implement new X-ray machines that present a three-dimensional view, supposedly better at spotting liquid explosives, guns, and other weapons (Frank, 2006). The TSA continues to analyze the use of EDS machines with smaller designs, enabling them to more easily fit into SSCP areas. Early-model EDS machines were the size of a small car (which takes considerable space), have high power requirements, and need strong flooring. As the industry continues to mature, machines are getting smaller, lighter, and require less power, but still take up more space than most conventional X-ray systems.

EDS machines using computerized tomography (CT) imaging are the best imaging technology presently available to detect weapons and explosives but costs about $850,000 per unit, multi-view x-ray costs approximately $200,000–$300,000 per unit, depending on functionality and other variables.

Multiview X-ray machines scan from several angles to create three-dimensional images of items in a bag. Proponents of the multiview technology say that the machines give screeners more detailed images and increase efficiency by reducing the number of bags that have to be searched by hand (Frank, 2006).

One of the most significant challenges to carry-on baggage screening is prohibited items. Since 9/11, the list of prohibited items has grown significantly. Passengers are prohibited from taking on board an aircraft nearly any weapon, explosive, or other substances that has been classified as hazardous. All knives are prohibited, along with tools and other items that could be converted into a weapon. Although pens, pencils, laptops, car keys, and other allowable items could be used as weapons, their effectiveness as such is considered less than that of a knife or gun. Also, with more air marshals, reinforced cockpit doors, and passengers unwilling to comply with hijacker demands, there is some consideration that even small pocket knives could once again be allowed to be carried on board, but public perception is likely to prevent this from happening in the near term.

After a liquid bomb plot was discovered in London in 2006, the list of prohibited items in the United States expanded. X-ray and magnetometer systems in place at U.S. and U.K. airports at the time of the attack were not capable of detecting liquid-based explosives, so the initial response was to ban all liquids from being carried onto a commercial aircraft. However, this is not a sustainable solution, so the TSA set out to determine just how much explosives it would take to bring down an airplane.

The TSA briefly considered continuing the complete ban, but this policy, while popular within the TSA at the time (Hawley and Means, 2012, p. 188), completely ignores the needs of a vast majority of the traveling public and particularly demonstrates an ignorance of the travel habits of the business traveler who finances much of the air transportation system. Business travelers predominantly only have carry-on baggage. Forcing this population of travelers to check their bags so they can bring essential liquids increases their travel time, decreases the space available in the cargo hold for revenue-generating cargo, and forces business travelers to spend hundreds or thousands dollars every year in buying liquids at every destination. While not a huge expense for the leisure traveler, for business travelers, some of whom travel several times a week with numerous overnight stays, the costs would add up.

The use of some explosives such as dynamite is detectable by X-ray technology, and plastic explosives such as C-4 are not easily acquired. The acquisition of such materials to a terrorist group brings unwanted attention to the group, increasing the chances they will be trailed and possibly stopped by law enforcement (Hawley and Means, 2012, p. 141). Ramzi Yousef used a bomb with a nitroglycerine base in the 1993 bombing of the World Trade Center and nitroglycerine is virtually undetectable by X-ray, magnetometer, and EDS technology; however, it is relatively unstable, requiring tamping materials to prevent it from exploding inadvertently if it's dropped or mishandled.

However, in 2005, hydrogen peroxide–based explosives became popular. Hydrogen peroxide is a common element in everything from hair dye to pool cleaner, and hydrogen peroxide–based bombs had already proven their worth in the 2005 London subway and bus bombings. It requires more bomb-making skill to create a hydrogen peroxide–based bomb, but their increased stability, blast power, and invisibility to security makes them an attractive option for an airplane bomb (Hawley and Means, p. 141). The liquid bomb represented a shift away from the previous types of bombs al-Qaeda had been using, which where predominantly commercially available explosive materials. The new liquid-based devices gave al-Qaeda a high-powered, reliable, stable, and repeatable explosive. In 2006, intelligence from the U.K.'s MI-5 indicated that operatives had created such devices using normal energy-drink bottles, undetectable by airport screening methods, and that a plot was in motion to bring down several commercial flights from the United Kingdom to the United States. If successful, the death toll in the bombing of nearly a dozen commercial flights could have exceeded the deaths suffered on 9/11.

The TSA developed the 3-1-1 program in response to the liquid problem. The current security policy allows for liquids to be contained in no larger than three-ounce containers that must fit in a one-quart baggie and only one bag per passenger. The TSA took a public beating for their new policy, and even a skit on the TV show *Saturday Night Live* mockingly questioned whether three or four ounces could bring down an aircraft, but what was not known at the time is that the type of liquid (or gel or aerosol) was not the critical issue—the size of the container was. TSA explosives experts determined that the hydrogen peroxide–based explosives could only be a serious threat to an aircraft if detonated in a container with sufficient diameter. The TV skit also questioned whether two individuals could bring on enough quantities of explosive materials and then meet on the plane to combine them. While the possibility exists, the nature of the hydrogen peroxide–based explosive device makes a successful marrying of these binary materials, unlikely to be successful.

Bottled liquids scanner (BLS) screening systems are used across the nation by TSOs to detect potential liquid or gel threats that may be contained in carry-on baggage. The technology differentiates liquid explosives from common, benign liquids and is used primarily to screen medically necessary liquids in quantities larger than three ounces.

Next-generation bottled liquids scanner systems have the ability to detect a wider range of explosive materials and use light waves to screen sealed containers for explosive liquids. The TSA has deployed more than 1,000 next-generation BLS units to airports nationwide. The TSA is currently testing new liquid screening systems with enhanced detection capabilities that use light waves to screen sealed containers for explosive liquids.

Liquid-based explosive detection devices have been in use in some airports throughout the world for several years, but were slow to receive certification in the United States. One ironical impact of the wars in Iraq and Afghanistan is that while they have allowed terrorists to improve their bomb-making skills, they have also driven technological advances in explosives detecting technology. Better liquid explosive detecting devices continue to make their way from the battlefield to TSA testing and eventually to airport screening checkpoints.

Overseas, all liquids were banned from carry-on baggage for a period, as were all laptop computers and other electronic devices. The rationale for restricting the electronic devices was that they could be used as a triggering mechanism for a bomb. When laptops were again allowed, the size of the laptop case was restricted, although no clear reason was given as to why the laptop case must be a certain size. Security screeners were provided with rulers but not with the ability to make commonsense interpretations on the risk presented by a laptop case only slightly over the allowable limit.[2] An ideal security operation includes flexibility and allowing security personnel to make commonsense decisions and interpretations of rules and standard operating procedures.

Checkpoint of the Future

In 2011, the International Air Transport Association (IATA) rolled out its idea of the future of airport screening, which features three lanes: one for known travelers, one for normal, and one for enhanced security. As a passenger approaches the checkpoint his or her identity is determined using biometrics and the individual is directed to the appropriate lane. Passengers who have preregistered with the appropriate government agency and undergone a background check are known travelers who will be subjected to lesser

---

[2]Musical instruments were also temporarily prohibited in the weeks after the discovery of the liquid bomb plot. Many musicians lost or had their high-value musical instruments damaged as a result of checking these items into the cargo hold. These policies did not make sense as there has never been an incident in the history of aviation security where an individual attempted to hijack an aircraft with a musical instrument. Violins, guitars, or any woodwind instrument do not make an effective weapon. Decision making leading to policies such as these must be avoided if the traveling public is to develop a trust in its security agencies.

security measures, similar to TSA's Pre✓™ program (see Chapter 6). Enhanced security would be for travelers on selectee lists, or that intelligence or other information dictates that they may present a higher security risk to aviation. Individuals in this lane would likely receive the equivalent of secondary screening. Normal screening is designed for the majority of travelers who neither have put their background information up for review by the government, nor are regarded as suspicious.

Additionally, IATA notes that screening technology is being developed that does not require passengers to remove any item of clothing or even have to be separated from their belongings. The ultimate goal is a seamless screening process where a passenger walks through a corridor, where security systems are integrated and hidden within the walls, and out the other side without ever skipping a step.

Screening Checked Baggage

An aircraft hold contains a variety of items including passenger bags, cargo, U.S. mail, and airline company mail and materials. Inside passenger baggage and cargo are any number of items including liquids, gels, compressed canisters, unloaded weapons, ammunition, human organs and tissue, and hazardous materials.[3] Many of these and other items are listed on the carry-on baggage prohibited items list but are allowed in checked baggage and in cargo, where they cannot be accessed, complicating screening and inspection.

The screening of checked baggage and cargo is a layer in the security system that focuses on preventing one form of attack—the bombing of an aircraft by the placement of an improvised-explosive device (IED) in the baggage hold. Most airline bombings have been carried out this way. Checked-baggage screening could have prevented several significant aviation tragedies including the bombings of Air India Flight 105, Pan Am Flight 103, and nearly 60 other in-flight bombing incidents. However, the screening of checked baggage and air cargo is costly. Costs include equipment purchase, hiring and training of personnel, and modifications to airport facilities to accommodate equipment. Modifications to an airport facility also include secondary search areas for suspect items, containment areas, and administrative and break rooms for screening personnel.

This section provides an overview of procedures for screening checked baggage and cargo. Detailed guidance for constructing a checked-baggage inspection system is covered in the TSA's *Recommended Security Guidelines for Airport Planning, Design, and Construction*.

Checked Baggage

As well as the fundamental purpose of preventing the placement of explosives onboard an aircraft, checked-baggage screening can detect threats such as the illegal transport of hazardous materials, narcotics, and weapons. Checked-baggage screening is complex

---

[3]Hazardous materials carriage is regulated by the Federal Aviation Administration (FAA).

and expensive. Whereas the United States transports between 600 million and 800 million passengers per year, carriers transport nearly three times that number in checked bags.

The fundamental tool in checked-baggage screening is the EDS machine, an X-ray machine that uses computerized tomography to analyze the contents of a bag. Although other systems are used internationally, the standard in the United States is that all baggage be checked by an EDS.

After 9/11, airports purchased EDS machines used to inspect checked baggage at the cost of approximately $1 million per unit. Some major airports had to purchase numerous EDS machines to meet baggage screening demands. Terminal facilities had to be modified and reconstructed to make room for the new systems.

Canada implemented its checked-baggage screening program after the bombing of Air India Flight 105, the United Kingdom implemented its checkedbaggage screening program after Pan Am Flight 103, and the United States implemented its checked-baggage screening system after 9/11. Russia implemented checked-baggage screening after the downing of two of its airliners in 2004 by suicide bombers, although the individuals smuggled the bombs on their persons, rather than in checked baggage.

There are several options in screening checked bags including the use of K-9 explosive detection dogs, ETD technology, conventional X-ray equipment, manual search, and explosive detection CT equipment. Canada and the United Kingdom use a five-tier system of checked-baggage screening.

Level 1 screening is an inspection by a conventional X-ray machine. The computer software within the X-ray machine makes a threat determination and either sends the bag to the Level 2 inspection for additional scrutiny or routes it to the aircraft if there is no threat. Approximately 70% of all checked bags are cleared at Level 1 (Shanks and Bradley, 2004, pp. 10–11).

Bags not cleared at Level 1 are sent to Level 2 where they are inspected by an automated EDS system. Another 5% are cleared at this level and proceed to the aircraft. Suspect bags are sent to Level 3. Level 3 consists of another inspection by EDS using CT (the same systems certified by the TSA in the United States) and a human operator interpretation or a check using ETD or quadrapole resonance (QR) technology. Another 2% of bags are cleared at this level (Shanks and Bradley, 2004).

Level 4 consists of reconciling the bag with the passenger or removing it from the inspection area to be remotely detonated (Shanks and Bradley, 2004). This raises the question of whether a suspect bag should be moved. The argument for moving a bag that may possibly have an IED is that most IEDs smuggled through the checked-baggage process are triggered by barometric pressure, a timing device, or a radio frequency trigger. They are not set off by motion, as the bomber would probably not be able to arm the device and then hand it to a ticket agent without the bomb detonating. The argument against moving the bag is that the IED may be so unstable, regardless of the triggering mechanism, that it may detonate. Whether the bag is moved or not is a decision for law enforcement responding to the incident and is based on the totality of the circumstances and internal department practices.

A Level 5 inspection occurs on about one or two bags in 50 million (Shanks and Bradley, 2004). A bag reaches Level 5 classification when the passenger cannot be found or refuses to open the bag. At this point, either the area is evacuated or the bag is moved via a mobile blast containment system where it is then inspected further or destroyed.

Because the United States requires all bags to be checked by EDS using CT, international airports using the five-tier system but conducting flights to the United States can meet the TSA's EDS screening standards by turning off (or bypassing) their Level 1 and Level 2 systems, which allows the bags to go immediately to the Level 3 EDS machines (provided they are TSA certified) (Shanks and Bradley, 2004).

The use of canines in explosive detection is highly effective. However, it takes a large number of dogs to scan hundreds of bags for explosives. According to the canine explosive detection trainers at Israel's Ashdod seaport, it takes three dogs to maintain continuous detection capabilities, but this is on a much smaller scale than at a large commercial-service U.S. airport. Each dog works a 20-minute shift, and then that dog rests for 40 minutes while another begins its work. Canines must also be exercised and they must have housing and play areas and facilities for trainers to store clothing, training equipment, and dog food. Canines are best used in specific circumstances such as random patrols in terminal or cargo areas or to search bags and aircraft.

Explosive trace detection technology and physical searches of checked baggage are effective strategies, but they are impractical on a large scale. ETD inspection of checked baggage was implemented at many U.S. airports that could not meet the 100% EDS deadline (December 31, 2002) as a stopgap measure until EDS systems could be deployed—this process was very slow and created long lines in ticketing areas.

In the United States, the Aviation and Transportation Security Act of 2001 (ATSA 2001) mandated that all checked bags would be screened using EDS technology certified by the TSA. At the time ATSA 2001 was passed, only two companies—InVision's 9000DSi and L3's eXaminer—produced EDS machines, making their availability to the U.S. airport community extremely limited. By December 31, 2002, the deadline for all U.S. airports to have full EDS screening equipment implemented, only a few airports were able to meet the deadline. The deadline was extended by one year, and the TSA decided that a combined use of K-9, ETD, and the process of positive passenger/baggage matching could substitute for EDS until the airports were fully EDS equipped.

An important lesson for airport and aircraft security coordinators (ASCs) can be learned regarding security mandates. Shortly after 9/11, the U.S. government promised to help airports pay for the 1,200 plus EDS machines that it now required at airports. However, by 2006, the government had reimbursed about 8 airports (out of 429) for EDS upgrades, and additional reimbursements have been slow. The lesson to U.S. airports and the aviation industry is that additional federal requirements are not likely to be funded or, if so, not funded long term. It is important for aviation security practitioners to consider that each operator (airport, airline, etc.) will have to account for additional security expenses and not rely on federal funding.

There are essentially six options for checked-baggage screening in the United States. The options are fully integrated EDS, inline EDS, ticket counter mounted, simple inline, stand-alone EDS, stand-alone ETD, and high-speed fully integrated inline. Each option is dependent on the configuration of the airport and its overall size. Approved alternatives included canine screening, physical inspections of baggage, and positive passenger/baggage matching (Government Accountability Office [GAO], 2006). EDS is TSA's preferred method. Most EDS systems installed after 9/11 were either EDS or ETD, hastily put into airport lobbies as stopgap measures to meet federal deadlines. Airports continue to construct inline systems to recover needed lobby space and improve the speed and efficiency of baggage movement and screening. Airport system requirements depend on the throughput rate (the number of bags being processed hourly), available space, and available funds. Inline systems are preferred over lobby screening options, as they improve efficiencies, increase throughput volume, and open terminal space in the lobby areas. Airport operators must consider the layout of their terminal facilities when deciding which form of checked-baggage screening to incorporate. A fully integrated inline EDS system is the fastest and most effective. However, inline EDS systems are costly, normally require the purchase of many scanning units, and require remodeling of bag sorting areas to accommodate the larger operation and additional personnel.

At some airports, it is not feasible to install an inline EDS system because of financial or physical layout restrictions. Large inline systems work better at large commercial-service airports with ample bag sorting facilities. Smaller airports may elect to place stand-alone EDS machines in their lobby areas or install a ticket counter EDS operation. The smallest commercial-service airports may go with an all-ETD checked-baggage search process, located in the lobby of the terminal.

At the smallest level of checked-baggage screening (without using alternative procedures such as canine, bag matching, or physical inspection), airports usually implement stand-alone ETD inspection stations located in the lobby of the terminal building (Figure 7.3).

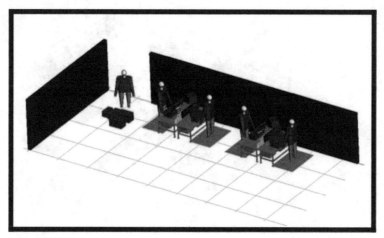

FIGURE 7.3 Stand-alone ETD system.

Stand-alone ETD machines are the most labor intensive and require several ETD stations and additional personnel to handle the desired baggage throughput. A 2006 GAO report showed that ETD machines could screen an average of 36 bags per hour, whereas even the lowest-performing EDS machines can screen at an average rate of 120–180 bags in the same time. Despite the efficiencies of inline, many small commercial-service airports do not have bag sorting facilities large enough to use inline EDS systems.

Stand-alone lobby EDS machines consist of single machines placed in an airport lobby area near the ticket counters. When a passenger completes checking in a bag, it is moved to a screener staffing the EDS machine. The screener then places the bag through the EDS and into the baggage system. Many airports initially went with this option to meet the ATSA 2001 deadline, as it does not require substantial modifications to the airport, as inline systems do. Airports with limited-sized baggage sorting areas or limited funds continue to use this method.

A ticket counter–mounted system consists of EDS machines placed at airline check-in ticket counters (Figure 7.4). When an airline representative accepts a bag, it immediately goes into an EDS machine. If the bag is cleared, it proceeds to the baggage sorting room where it is directed to the appropriate flight. If the bag requires further inspection, it will be routed to a screener (out of public view), who then conducts a secondary inspection. These systems are best used for airports with low baggage volumes. They may also be used for self-service check-in, curbside check-in, or international recheck-in areas.

The advantages of stand-alone or ticket counter–mounted systems is that bags that may have suspect items can be immediately reconciled as the passenger is often still in

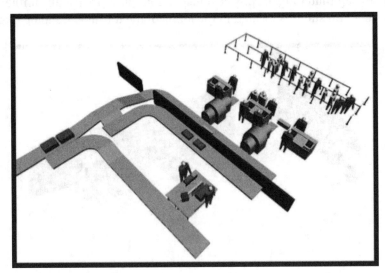

FIGURE 7.4 Ticket counter–mounted EDS station.

the vicinity. However, these systems can consume up to 20% of the existing check-in area (Shanks and Bradley, 2004, p. 9), creating both inconvenience to passengers and increased congestion in public areas.

For larger airports with higher throughput rates and more available facilities, an inline EDS system is considered the most effective and efficient form of baggage screening. Inline EDS systems are integrated with the airport's baggage system and are able to scan a bag quickly, and the system either selects the bag for additional screening or sends it off to the aircraft (Figures 7.5 and 7.6). Inline systems are the quickest method and the least labor intensive as the software does much of the scanning and threat interpretation. Inline systems can process up to 40% more bags in the same time as a stand-alone EDS system. Fully integrated EDS systems feature control rooms where images can be scrutinized, bags can be physically inspected, and bags reintegrated back into the baggage sorting system once they are cleared.

Baggage throughput for an EDS system depends on the type of machine used. Lower EDS rates average 80 bags per hour, whereas faster machines average 180 bags per hour. The TSA is working on future systems that have a throughput rate of up to 900 bags per hour. Baggage throughput is also affected by the amount of oversize bags that are put into the system. This is a particular problem for airports located near tourist facilities (golfing, skiing, etc.), and depending on the geographic location of the airport, anywhere from 5% to 20% of total checked baggage could be oversized (Shanks and Bradley, 2004, p. 53).

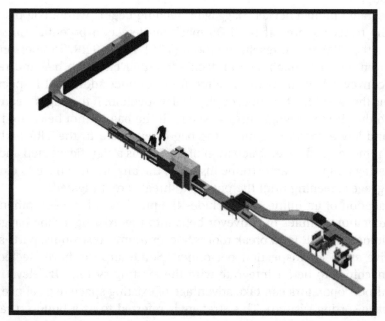

FIGURE 7.5 Simple inline EDS setup.

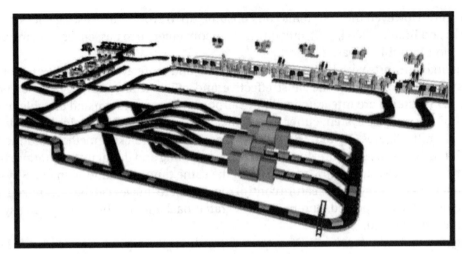

FIGURE 7.6 Fully integrated inline baggage screening system.

Building a Checked-Baggage EDS Screening System

There are numerous factors to consider when designing a checked-baggage EDS screening system. These factors include the size of the airport, the number of enplanements, integration with the airline's existing baggage handling system, costs, available space, and power supply.

Fully integrated inline checked-baggage screening begins with a bag placed into the system at the ticket counter. If the EDS machine detects a potential threat item, the bag is sent to the TSA's threat resolution room (TRR). In the TRR, TSA personnel analyze the image from the EDS machine and decide to either clear the bag and reintegrate it (through a conveyor belt) back to the aircraft or conduct additional inspections of the bag including the use of ETD, canine, or physical inspection. If the bag is cleared, it is reintegrated into the airline baggage sorting system. If the bag cannot be cleared, inspection personnel and law enforcement move the bag or evacuate to the TRR and attempt to detonate or remove the device. Evacuation of the TRR is a significant step and may mean baggage missing a flight; in an extreme instance, the airport may have to close down all checked-baggage screening until the potential threats are mitigated.

The installation of an inline EDS includes the purchase of the equipment, the addition and integration of miles of conveyor belts into the existing airline baggage system, and the addition of a TRR (and break rooms, locker rooms, restrooms, parts storage, and administrative areas for inspection personnel). Some airports have elected to elevate their EDS machines to better integrate with the existing system. By elevating the EDS machines, airport operators can take advantage of existing space (e.g., above the current airline baggage sorting system). This strategy is referred to as a build-up rather than a build-out.

Another key factor in designing the inline system is ensuring that bags selected for additional scrutiny are not accidentally mixed back in with bags that have been cleared. Many airports use barcode tags with laser readers. However, the most accurate baggage identification system is radio frequency identification (RFID) (Shanks and Bradley, 2004). RFID misread rates are in the tenths of a percent, whereas laser/barcode systems have misread rates in the tens of a percent. Additionally, RFID uses economical read stations, which enable multiple verifications along the conveyor route. This feature increases reliability and reduces the likelihood of a suspect bag being routed in with cleared baggage (Shanks and Bradley, 2004).

Airport operators must determine the required layout for a fully integrated, 100% checked-baggage EDS. The first step in this process considers the airport's current total throughput of bags per hour and the peak throughput of bags per hour. From these computations, the total number of EDS machines needed and the space required can be determined. If growth is projected, airport operators may want to use projected bag throughput to design the system to meet future demands.

Before 9/11, most existing outbound baggage systems were straight-line conveyors running directly from the ticket counters to the baggage makeup areas (where bags are sorted). Because airlines generally lease both ticket counter and baggage makeup areas and because any inline system is likely to affect these areas, airport operators may have to modify existing airline leases to account for the additional space required for security (and thus less available space for the airline or tenant).

One method of improving throughput rates and integrating into an existing system is to keep baggage conveyor routes as straight as possible or with curves that have a 6-foot centerline and a 12- to 18-degree slope (Shanks and Bradley, 2004) to better facilitate the movement of oversized bags.

As an example of some of these concerns, design criteria used at Jacksonville International Airport (JIA) in Florida were based on accommodating 8 million passengers. The machine count (10 EDS machines based on 8 million enplanements) established throughput at less than 300 bags per hour at JIA. JIA integrated several previously separate airline baggage handling systems and increased throughput to 500 bags per hour. Throughput is reduced at certain times because of bag jams and machine fault resets (Shanks and Bradley, 2004). In addition, 5% of JIA's baggage is made up of golf clubs, so the inline system was slightly overbuilt to handle golf bags. This improved throughput rates because employees did not have to physically move golf clubs along with other oversized bags (Shanks and Bradley, 2004).

Contingency planning for failure of the checked-baggage system should consider having an extra inline conveyor/EDS system to use as a backup or "hot standby" (i.e., if you need six machines to fulfill the throughput rate, buy seven). Another option is to reduce the workload on the main systems (Shanks and Bradley, 2004).

Failure of an EDS machine is usually the result of poor maintenance or a failure of the conveyor system rather than of the EDS machine itself. As opposed to EDS, when an airport uses ETD machines for checked-baggage screening, generally downtime is less than it is for EDS systems, as ETDs are solid-state and normally only require the replacement of consumables.[4]

Checked-Baggage Screening Approved Alternatives

Canine search and physical inspection are alternatives to the EDS/ETD systems. However, each carries inefficiencies when used as a sole screening mechanism. Positive passenger/ baggage matching (i.e., matching checked baggage with passengers on the flight) does not prevent suicide bombings and is no longer used in the United States for domestic flights. Also, this process creates operational inefficiencies for airlines, because when a bag is found not to have a matching passenger on board, the airlines often take a delay while baggage handlers go digging through the cargo hold for the unmatched bag (which usually results in the airline taking a risk and not bothering to even look for the bag). The use of canines requires extra floor space to spread out the baggage for inspection, and we have already mentioned how many dogs are required to conduct such inspections.

Physical inspection requires screening personnel to open each bag and inspect the contents. This is time consuming and potentially dangerous, should a bomb be accidentally triggered during inspection. The likelihood of detecting an IED is related to the skills of the individual screener to find secreted devices. Additionally, a physical inspection may miss hidden explosives such as C-4 formed into a sheet and sewn into the lining of the bag. Physical inspection is usually used in a secondary process, when a bag has already been identified as suspect by other means (EDS, ETD, canine, etc.). Good judgment must be applied when deciding whether to inspect a bag physically or to call an explosives ordinance disposal team to handle the suspect item.

■ ■ ■ ▬▬▬▬▬▬▬▬▬▬▬▬▬▬▬▬▬▬▬▬▬▬▬▬▬▬▬▬▬▬▬▬

Case Study Boston Logan International Airport

Logan International Airport, located in Boston, MA, is the largest airport in the New England area. Logan handles more than 1.3 million passengers per year and has five passenger terminals, each with its own ticketing, baggage claim, and ground transportation facilities. Nearly 30,000 flights are conducted out of Logan each year. Logan was also the launch point for two of the hijacked flights on 9/11.

Logan met the ATSA 2001 mandate of 100% hold baggage screening by December 31, 2002. The undertaking required the coordination of Massport[5], personnel at Logan International Airport, the TSA, and L-3 Communications, the selected provider of the EDS machine. Many other individuals assisted in making sure the mandate was completed successfully and on time. These stakeholders included management and line personnel, union contractors, construction workers, and the airport's airlines and tenants.

---

[4]Items such as oils, paint, adhesives, and cleaning materials that are consumed during their use.

[5]Massport, the Massachusetts Port Authority, is the governing authority for Logan International Airport.

Logan moves more than 5,000 bags per hour at peak times. The installation of the baggage screening system could not interrupt service to the airlines, passenger flow, or sacrifice baggage screening throughput. EDS equipment was placed at several locations throughout the facility, requiring the expansion of 9 of its existing 16 luggage rooms and seven major building additions.

The L-3 Communications' eXaminer 3DX 6000 was the selected EDS machine because of its certified detecting capability, throughput capability, low false alarm rates, ease of use, and size. The eXaminer 3DX 6000 also has a smaller footprint, allowing units to be placed in the lobby if necessary. L-3 engineers designed a new conveyor system that addressed baggage flow issues and met the TSA's goal of clearing 95% of baggage within 10 minutes of meeting an EDS machine. The system included routing for three levels of screening. At Level 1, the eXaminer 3DX 6000 screens each bag for explosives then either clears it or rejects it and sends it to Level 2 screening. Level 2 screening follows for rejected bags. If cleared at Level 2, then baggage is placed back into the system. Rejected baggage is then searched by ETD or physical inspection. Note that these levels are distinctly different from the international model of five levels. The international Level 3 check—EDS—is actually the Level 1 check in the U.S. system.

The initial installation of EDS machines at Logan included 30 integrated eXaminer units at Level 1 and operator workstations at Level 2. The eXaminer clears nearly 85% of all baggage at Level 1, with any images from noncleared bags being transmitted from the EDS machine to operator workstations.

Installation occurred mostly at night when the airlines were at lower passenger flows or idle. Electricians converted Logan's power environment to a clean and uninterrupted power supply while L-3 technicians worked continuously to ensure the systems were installed correctly and operated properly.

As systems went online, TSA screeners began testing and training on the machines. With all entities working in concert, Logan was able to meet the ASTA 2001 mandate of 100% checked-baggage screening by December 31, 2002—one of the very few airports that met the deadline.

The following example is from a case study, where inline EDS was installed after 9/11 at Boston Logan International Airport. This case was featured in Shanks and Bradley (2004).

■ ■ ■

Screening Aviation Employees

One of the most controversial issues in aviation security is the screening of aviation employees. Throughout the history of aviation security, airport and airline workers have been involved in numerous security incidents, yet many airport workers remain able to access the security areas of an airport, and even aircraft, without undergoing the same screening process as passengers. Some of these cases included the following attacks:

- Libyan Airline employees placing the bomb that brought down Pan Am Flight 103.
- The shooting of the flight crew on PSA Flight 1771 by an employee of USAir.
- The attempted takeover of FedEx Flight 175 by a second officer.
- The hijacking of TWA Flight 847, where catering personnel planted guns and grenades in the aircraft lavatory for later use by the hijackers.

Despite these and other cases, there are still serious concerns regarding the value of screening employees in the same way passengers are screened. Stimulating this debate was an incident that took place shortly after 9/11. In this incident, the FBI arrested three food workers with security identification display area (SIDA) access at the Detroit Metropolitan Airport in Michigan on suspicion of plotting terrorist activities. These individuals possessed airport diagrams and phony immigration documents and may have been planning a terrorist attack or conducting surveillance to learn more about airport operational and security protocols. In February 2010 an airport shuttle bus driver pleaded guilty to three terrorism charges after he was accused of conspiring to detonate an explosive in the New York City subway system. In 2007, an individual who may have been employed by Evergreen Airlines was one of the men convicted in a failed plot to bomb JFK airport in New York. In 2011, a former British Airways employee was arrested in England on terror-related charges after prosecutors say he plotted with Anwar al-Awlaki in 2010 to carry out both cyber and physical attacks against British Airlines passengers, including potentially bringing down a trans-Atlantic flight.

The issue of employee screening is primarily concerned with the ability of an aviation employee with SIDA access to bypass routinely the majority of the aviation security processes.

Depending on the level of access a particular aviation employee has, he or she may be able to bypass all of the security measures, right up until he or she accesses the flight deck of an aircraft. This level of access is necessary for pilots, mechanics, crewmembers, and others, but the debate continues about whether these individuals should be subjected to the same level of scrutiny as passengers.

Currently, passenger screening checkpoints are designed based on the demands of a certain number of passenger embarkments plus airport personnel who do not having SIDA access and therefore must use passenger checkpoints to gain access to their worksite. To require all aviation personnel to go through the passenger screening checkpoints would require significant increases in space, equipment, and personnel. Some airports in the United States employ more than 40,000 workers (at a single airport) who possess access/ID badges that allow them to enter secured areas. Requiring all of these individuals to use passenger checkpoints as they are currently designed and staffed would cause undue delay for both passengers and employees. However, the idea of employee screening continues to gain popularity because of its use at many foreign airports and a public and political outcry that the lack of employee screening represents a significant gap in the aviation security system. It should be noted that foreign airports feature physical design characteristics to reduce the number of individuals who have access to aircraft.

Rather than have all current aviation personnel begin using the passenger checkpoints, an alternative strategy is to construct employee screening checkpoints with X-ray machines and magnetometers. Many airport workers already use an access gate or door to get from parking areas to their worksites within secured areas, so checkpoint facilities could be installed at these locations, depending on available adjacent space. Even if the physical facilities were expanded, airports would be required to purchase millions of dollars' worth of additional X-ray and magnetometer equipment to use at these checkpoints, and the TSA would have to hire more personnel to staff these locations. A few airports in the United States, notably Miami and Orlando international airports in Florida, have initiated their own employee screening, employing their own private contractors to conduct the screening at airport access gates and other locations.

If the screening of airport employees is eventually required by law, there is a possibility that the airport operator could be given the responsibility to conduct the screening function of airport and airline employees, which would then shift the financial obligation from the TSA to the airport operator, leaving the airport operator to hire screening personnel and ensure they are trained and certified to TSA standards. Notwithstanding the startup costs to implement such programs, yearly operational costs have ranged up to $8 million at some large-hub airports—costs that the airport operator would have to absorb. Otherwise, if the TSA is given the responsibility, it would require the hiring of tens of thousands more TSOs, which is unlikely due to public pressure to temper TSA's expansion.

An additional challenge to employee screening is that many employees require the use of tools and other items that are considered prohibited if brought through the screening checkpoint. This raises the question, why screen someone for a knife or hammer when the person needs the item for the job? The intent of employee screening should focus on ensuring that explosives or guns are not being brought into secure areas, with less focus on tools and knives. Individuals required to be in possession of such items should be registered with the airport operator and required to maintain control of those items at all times while in secured areas.

When discussing employee screening, generally issues are raised that lead to a distinction between the terms *screening* and *inspection*. ATSA 2001 called for the TSA to develop systems to screen all employees. So far, the aviation industry has met this condition by requiring that all personnel with SIDA access undergo a criminal history record check (CHRC) before being allowed unescorted access. There are several loopholes within this type of screening program, such as the lack of a requirement for employees to undergo additional CHRCs in the future (once passed, employees are not required to go through the check again, but the TSA says they do random periodic checks of personnel who have already undergone the process).

Currently, it is the responsibility of an employee to advise the issuer of his or her access/ID badge if he or she has been convicted of a disqualifying offense used in CHRCs—that is, if the offense occurs after he or she has already been issued an access/ID badge. If an aviation employee is found guilty of a disqualifying offense, and considering that all disqualifying offenses are felonies, it is likely that he or she will be serving prison time and will not be available for work.

An additional problem with existing regulations is that they do not address individuals who may have been found guilty of a crime not listed as a disqualifying offense. These individuals may still work, but they remain a security risk. Each airport operator can address these offenses within its airport security program (ASP) and create additional restrictions on the requirements to receive an airport access/ID badge.

There continues to be a trend within the TSA to move forward with additional employee screening requirements, focused on using physical inspection processes. Since 9/11, the TSA has required airport operators to reduce the number of employees with SIDA access to the minimum number possible and the TSA conducts random security inspections at employee access points using magnetometers, X-ray machines, or physical inspections such as pat-down searches and bag dumps.

The TSA also requires that security threat assessments be conducted on all SIDA access/ID applicants and that all SIDA access/ID applicants are checked against the no-fly and selectee lists.

The TSA has piloted several programs at some airports to enhance employee security. These programs have included suspicious awareness training for employees, using biometric-based identification credentials and random screening of employees in the security areas. Employee awareness training about violence in the workplace is a good alternative as it's likely that an employee intending to cause harm, whether they are a terrorist or unstable individual or criminal, will exhibit unusual behaviors or pre-incident indicators. Employees should be trained to watch for these behaviors and should have a method for reporting such suspicious activity to their supervisors or airport security.

Special Issues in Screening

Canine

The National Explosives Detection Canine Team Program originated in 1972. A canine was first used for screening in 1972 for a flight from New York to Los Angeles that returned to the gate after receiving a bomb threat. After evacuating the passengers, a bomb-sniffing dog was brought onboard. The dog discovered an explosive device just 12 minutes before it was set to detonate (TSA, 2006). Shortly thereafter, the FAA Explosives Detection Canine Team Program was created. The program has operated continuously, with dogs and handlers providing an essential layer of the aviation security system.

The initial purpose of the FAA Explosives Detection Canine Team Program was to provide certified canine teams at select airports so that aircraft receiving bomb threats could quickly divert to one of these airports. The program started with 40 canine teams at 20 airports. Another 47 teams and 7 more airports were added in 1987, and by 1996, virtually all Category X airports and many Category I airports hosted canine explosive detection teams. Today, canine teams are used in greater capacities than just inspecting aircraft. Canine teams are actively involved in patrolling the airport's public areas, particularly the ticket counters where passengers are checking in their bags, in cargo areas, and at the screening checkpoints. Canine teams are also effective deterrents for would-be criminals or terrorists. With all existing technology, a canine's sense of smell is still the most reliable method of detecting an explosive!

Canine teams are also part of the airport's emergency response plan to address unattended and suspicious items. A canine can often detect the presence of an explosive device within a bag without physically examining the bag. This helps bomb squad members to better assess the potential for an IED. Dogs can quickly exclude the presence of explosives in aircraft, at the airport, or in vehicles, thus allowing a quick return to service and minimal disruption of commercial activity. Canine teams are also highly versatile and are able to adapt to new environments with little to no additional training. In 2005, when the TSA conducted pilot programs using dogs to search air cargo break bulk, containers, and airline ground support equipment for explosives, the GAO issued the following report:

> *Overall, the data suggest that even without prior training, the canine teams performed reasonably well, illustrating the ability of the teams to adapt to new and different environments and tasks. According to TSA officials, the results of both phases of the pilot program demonstrated that canines offer promising alternatives to inspecting air cargo. (GAO, 2005, p. 80)*

The TSA Canine Program trains and certifies dogs and handlers at Lackland Air Force Base in Texas. Traditionally, handlers were local city or county law enforcement officers (LEOs) based at commercial-service airports, but in the late 2000s the TSA initiated their own civilian canine explosive detection program, sending hundreds of TSA non-LEO personnel through the training and deploying them to airports in the United States. Dogs are acquired and trained by the TSA, then a handler is assigned and the dog and handler train together. The handler generally remains the handler until either the dog retires or until the handler retires or takes a new assignment.

Once a handler and dog are paired up as a team, the TSA provides the initial training and certification, plus partial funding for the handler's salary and care and feeding of the dog, along with veterinary and other costs associated with the dog once the team returns to the home airport. Dogs are trained using B. F. Skinner's operant conditioning reward-based system whereby the dog receives a reward for every success. Dogs are taught to identify the fundamental components of explosive material and then indicate to his or her handler that an odor has been detected. The dog is then given a reward, along with physical and verbal praise from the handler. The association of detecting an explosive odor and being rewarded is repeated hundreds of times until the dogs "learn the game."

The dogs are usually labrador retrievers, German shepherds, Belgian malanoises, or vizslas. They must meet higher standards than those used for other types of service dogs (Price, 1995). Once the handler and dog are teamed up, the TSA puts the team through a 10-week course on how to locate and identify dangerous materials using search techniques unique to aircraft, baggage, vehicles, and airports.

Human handlers also must undergo a rigorous military style of training before selecting their dogs; it is just as important to have an effective handler as it is to have an effective dog.

Once teamed up, handlers and dogs spend nearly every moment together. During the detection process, the handler must totally concentrate on the dog, watching for any changes in behavior that may indicate the dog smelled something questionable. The dog is also sensing the handler's actions. This intense concentration is a trademark of the bomb dog teams. Handlers must "read" the dog or they could inadvertently pull the dog off an odor (Price, 1995).

To this day, dogs are considered the gold standard of the explosive detection industry. Although some manufacturers of explosive trace detection technology claim to have products that are as good as a dog's sense of smell, there have been no significant studies to support such claims. The sensitivity of a dog's nose is virtually unrivaled.[6] Another benefit is that dogs are highly portable and can quickly search large areas. In 1995, during a recertification test, a bomb dog team at Denver International Airport cleared the west end of the B Concourse (about a half mile in length) in just over 1 hour and 15 minutes and found all eight test explosives.

The canine programs have been one of the most consistently successful explosive detection programs in the history of aviation security. Airport operators should place the addition of more canine teams at their airports as one of their highest funding priorities.

Another method of screening passengers that is being developed by the TSA is passenger screening canines (PSCs), which are dogs that are specially trained to detect explosive vapors coming off a general area and able to track the vapor back to its source, or close to it. This allows TSOs to conduct additional screening of a particular individual or group. Dogs also have the benefit of being passenger-friendly and provide an element of randomness to the security process.

Opt-Out: The Screening Partnership Program

With the passage of the Aviation Transportation Security Act of 2001, Congress mandated that five airports would continue to use privately contracted screeners but under federal direction and under federal acquisition. The program was initially a pilot program, but is now in effect and is known as Opt-Out or the Screening Partnership Program (SPP). By 2012, the number of opt-out airports reached 17, however, the program remains controversial. In 2011, the TSA attempted to reject any further airport opt-out applications unless the airport could prove that the private screeners could provide a security advantage over TSA personnel, a virtually impossible task. In 2012, Congress, a long-time supporter of opt-out in general, reversed the burden of proof forcing the TSA to approve opt-out applications unless they could prove that the private screeners would provide a lesser level of security than TSA—again, a nearly impossible task. The TSA's own studies on the SPP found that screening at SPP airports costs approximately 17.4% more to operate than at airports with federal screeners, and that SPP airports fell within the "average performer" category for the performance measures included in its analysis. Other studies have put the

[6]The human nose contains 5 million scent receptor nerves, whereas a dog's nose contains 225 million.

costs at 9% to 17% higher than at non-SPP airports, and private screeners performed at a level that was equal to or greater than that of federal TSOs. A 2009 GAO report that essentially studied the results of the TSA study and another study conducted by an independent contractor concluded that some of the increased costs resulted from redundancies between TSA and contractor personnel at the SPP airports (GAO, 2010).

Before 9/11, airlines subcontracted screening to private companies. The United States mandated in 1973 that aircraft operators were responsible for the screening and inspection of its passengers and their carry-on baggage. The aforementioned ineffectiveness of these private companies has greatly affected the SPP.

When ATSA 2001 was passed, it called for a pilot program to study the effectiveness of private companies conducting passenger and baggage screening and inspection, a model used throughout the world. In this evaluation, the U.S. federal government contracted private companies and held them to the same performance and training standards as federal screeners.

Five airports were selected for the pilot program, with each airport representing different levels of commercial service: San Francisco International Airport, CA (Category X); Kansas City International Airport, KS (Category I); Greater Rochester International Airport, Monroe, NY (Category II); Tupelo Regional Airport, MI (Category III); and Jackson Hole Airport, W (Category IV). All of the airports with the exception of Jackson Hole were provided with contract screeners who worked for a private company that had won a bid from the TSA to provide such services. The Jackson Hole airport was allowed to use its own airport employees as screeners but still had to meet federally mandated performance and training standards.

Before 9/11, employees of contracting screening companies were paid poorly and met only minimal qualifications. These companies had sought to keep expenses to a minimum to maximize profit from their airline contracts. With the backing of minimum hiring, training, and performance standards through legislation, the post-9/11 private companies have been more successful at increasing their standards for screening. After five years, some of the private screening companies were reporting workers' compensation rates of lower than 5%, as compared to the 20% or more workers' compensation rates of the federal workforce. In addition, some of the private screening companies were reporting fewer security breaches, fewer customer service complaints, and, in some cases, were able to provide the same level of service with 20% less personnel than similar federally staffed facilities. Contributing to the lower costs was that the private companies made effective use of part-time personnel, whereas the federal workforce was not set up to use such an advantage. The federal government has recently allowed part-time personnel.

Subsequent GAO studies have not identified a significant difference in the breaches or number of prohibited items identified between TSA and the private companies. It is reasonable to conclude that if the private companies were not achieving equal to or better detection rates than TSA personnel, the TSA would be telling this to Congress as a way to eliminate the contract program altogether.

In the 2004 study, private screeners exceeded the federal workforce in most every other measurable area (customer service, lower workers' compensation rates, lower staffing, etc.). This level of performance was not lost on the aviation industry. Because of this evaluation, any U.S. airport was allowed to apply to the SPP, effectively opting-out of using a federally provided screening workforce. However, there was another significant barrier to airports switching to private screeners—liability. There is wide speculation in the industry as to why the SPP had not caught on at first. Many airport operators perceive that the airport will place itself at higher risk in terms of public perception if the airport chooses to opt-out, whereas with federal screeners, if there is a security breach, the airport operator can always point out that screening is not an airport (nor an airline operator) function. However, the number of opt-out airports eventually did rise and the debate continues within the industry and the halls of Congress.

With ATSA 2001, the opt-out process called for the airport operator to initiate the decision to opt-out of the federal workforce. That is the only control the airport has over the process. Once an airport decides to opt-out, the federal government issues a request for proposal to qualified parties that want to provide the private screening service. The federal government makes the selection, negotiates the contract, then signs the contract and acts as the contract manger. Although the airport operator may be consulted on which screening company it desires to employ, the decision is ultimately up to the federal government. Once in place, the federal government continues to provide operational oversight over the contract and the private screening workforce.

As mentioned, the federal government is responsible for selecting the contractor, signing the agreement, and maintaining operational oversight for any airport deciding to opt-out. Despite these responsibilities by the federal government, the airport is still exposed to risk should a security incident occur resulting from actions taken by the subcontracting screening provider. The federal government several times at various industry conferences and through personal communications sought to assure airport operators that they were not liable for risks associated with using subcontracted screening providers. Regardless of this consultation, many airport operators were advised by their legal counsel that the airport could be exposed to lawsuits if such an incident occurred and it could be proven in a court of law that if the airport had not made the decision to opt-out that the incident may not have occurred.

In 2005, the U.S. Congress addressed the concerns of airport operators regarding liability when using subcontracted screening services. The government passed the Support Antiterrorism by Fostering Effective Technologies Act of 2002, also known as the Safety Act. The Safety Act (see *www.safetyact.gov*) provided a method whereby private companies supplying homeland security technology (including screening services) could apply for liability protection. The Safety Act sets minimal insurance limits for companies providing antiterrorism technologies and services. The Safety Act also ensures that the company will only be liable for its insured amount if the company is sued because of a terrorist act.

Another advantage of the Safety Act is that any airport or aircraft operator that sub-contracts a company that has Safety Act coverage is also, by extension, covered from related liability. In 2006, numerous companies applied for and began receiving Safety Act liability coverage.

One challenge still facing opt-out proponents and the companies that employ private screeners is reluctance by airport operators to take on any additional responsibility that already belongs to the federal government. Although federal screeners do not work as well as the private screeners in the previously mentioned areas, the anecdotal opinion of many airport operators is that the performance of the TSA is adequate. Although this qualification has not been officially evaluated, some conclusions can be drawn from the fact that only one airport has elected to opt-out since the option became available. Presently, it seems that airport operators do not perceive the performance of federal screeners as poor enough to employ opt-out subcontractors and thereby incur additional liability.

Although airport operators are presently reluctant to switch to private screeners, the U.S. Congress took action attempting to sway airports to seek the SPP. In 2005, Congress looked at legislation sponsored by Senator John Mica and Congressman Dan Lundgren that, in its original form, would have rewarded airport operators with federal capital improvement funds if an airport could demonstrate cost reductions in its screening operation. Private screening companies can provide the same level of service as the federal workforce, but with a 20% smaller staff and with 15% fewer workers' compensation claims. In this regard, the airport operator could potentially realize a significant cost savings by seeking the opt-out. Under the legislation proposed by Mica and Lundgren, the airport using opt-out would then be eligible for a percentage of the savings to be returned to the airport as a federal grant. Although the bill never made it to a final vote, it does show a continuing trend in the U.S. Congress to move away from using the federal workforce for airport screening purposes. The 2010 GAO study concluded that many airports would consider opting out if they could select the private screening contractor that would be awarded the passenger and baggage screening contract at their respective airports and share in any cost savings resulting from their participation in the program (GAO, 2010). Further, the reason many airports stated they had not applied to opt-out is (1) the airport was satisfied with TSA screening services, (2) screening is a federal government responsibility, and (3) the SPP does not allow airports to have managerial control (GAO, 2010).

Registered Traveler

ASTA 2001 created another pilot program designed to help frequent travelers to move through the system more efficiently. Originally, the program was called Trusted Traveler, but the name was quickly changed to Registered Traveler (RT), as the TSA did not want to give the impression that individuals were trusted and thus not subjected to all of the levels of screening and inspection as other travelers.

The initial concept for RT was that low-risk passengers could submit to a background check process. After passing the background check, the applicant for RT status would be issued biometrically encoded identification allowing for a more expeditious path through the screening process. Once he or she has cleared the CHRC process, the applicant for RT status will have a biometric identification card created by a qualified RT provider. The traveler would then have the ability to use a designated line for RT passengers at airport screening checkpoints. The checkpoint, while not subjecting RT participants to any lesser level of security, would be designated solely for registered travelers. The advantage to an RT program participant is that he or she will be in line with other RT participants, who are most likely business travelers familiar with the security process and can move through the line quickly. Additionally, RT participants would have their own security line and not have to stand in line with hundreds of others. This potentially allows RT participants the flexibility to arrive later at the airport. With the switch to risk-based screening, this philosophy has changed, however, the RT program remains in place but today is primarily a front-of-the-line pass.

Aircraft Charter Screening

Under CFR Part 1544.101, aircraft operators conducting private charter operations with aircraft having a maximum certificated takeoff weight greater than 45,500 kilograms (100,309.3 pounds) or with a passenger seating configuration of 61 or more must ensure that all passengers and accessible baggage are screened (using metal detection devices and X-ray equipment) before boarding the aircraft. This rule applies to aircraft operators regardless of the nature of the airport used. Therefore, it applies to private charter operators working out of commercial-service and general aviation airports. Most private charters conducted under this rule are commercial airline Part 121 operators. These operators can take advantage of normal commercial-service security screening checkpoints. In some cases, where the private charter operation is conducted either from the general aviation area of a commercial-service airport or from a general aviation airport itself, the aircraft operator must provide for the necessary equipment and trained and certified screening staff. The TSA has a modular training program whereby non-TSA screeners who have been trained and certified by the TSA through the TSA-approved Basic Screener Training Course are allowed to perform this screening.

Screening Conducted by the Domestic or Foreign Aircraft Operator

In some cases, the aircraft operator, particularly at airports that have infrequent commercial service, may conduct screening. In such cases, CFR Part 1544 Subpart E is used to ensure that airline employees and the equipment used for screening are certified by the TSA to perform inspection operations. In addition, in the case of foreign aircraft operators conducting screening outside the United States for flights departing for U.S. airports, the aircraft operator must meet TSA screening standards as related to passenger, carry-on, checked, and cargo screening. If a foreign operator does not meet these requirements, then the passengers on the flight will have to be screened by TSA-certified personnel and equipment before being allowed into sterile areas and before baggage and cargo are transferred to other domestic flights.

Conclusion

The screening of passengers and their carry-on bags has been the most visible airport security measure since its inception in the late 1960s. Screening is also one of the most important layers in the aviation security system. When performed effectively, with the proper equipment and personnel, screening can deter two of the most common forms of attack on aviation—airline bombings and airline hijackings.

When screening requirements were first established, the U.S. government placed the responsibility for conducting the screening on the airlines, with the FAA providing regulatory oversight of the airline screening programs. The main objective of screening was to ensure that passengers were prevented or deterred from bringing prohibited items, such as guns and explosives, through screening checkpoints and into sterile areas of an airport and onto commercial aircraft.

Over the years, regulatory oversight became scarce, personnel attrition at the private screening companies that had been subcontracted by the airlines to conduct the actual screening function reached 100% annually in some areas, and technologies did not keep up with emerging threats. Although several pieces of legislation required tighter screening oversight, little was done to improve the system before 9/11.

After 9/11, the TSA took over the screening responsibility from the airlines and subsequently found itself regulating its own activities. Manufacturers began providing better equipment to detect weapons and explosives, along with expanding their research and development programs to meet existing threats and counter emerging threats. Additionally, for the first time in the history of U.S. aviation security, checked baggage was required to be inspected. Cargo and aviation employee screening programs were also mandated.

Screening is a term with broad applications and definitions. Screening can consist of verifying a passenger's identity and boarding pass information to passive and active forms of profiling. Screening is also a term used interchangeably in the TSA with the term *inspection*, which infers the actual physical inspection of an individual, package, bag, or other cargo, through the use of metal detectors, explosive detection equipment, patdowns, or other approved methods. Research and development efforts continue to generate better methods of detecting prohibited items, yet one of the best detecting "technologies" continues to be one that has been in use since 1972—canine explosive detection teams.

The post-9/11 changes have also brought cargo screening to the forefront, along with new industry issues such as the Screening Partnership Program, the Registered Traveler Program, and the integrity of foreign aircraft operator screening programs.

References

ATSA, 2001. Aviation and Transportation Security Act of 2001, P.L. 107–71 Sec. 110.

Frank, T., Oct. 17, 2006. TSA Plan: X-ray for Liquid Bombs. USA Today, retrieved July 8, 2008, from www.usatoday.com/travel/news/2006-10-17-tsa-xray_x.htm2006.

Government Accountability Office (GAO), 2005. Aviation Security: Federal Action Needed to Strengthen Domestic Air Cargo Security (Publication No. GAO-06-76), retrieved July 9, 2008, from www.gao.gov/new.items/d0676.pdf2005.

GAO, 2006. Aviation Security: TSA Oversight of Checked Baggage Screening Procedures Could Be Strengthened (Publication No. GAO-06-869), retrieved July 9, 2008, from www.gao.gov/new.items/d06869.pdf2006.

GAO, 2010. Efforts to Validate TSA's Passenger Screening Behavior Detection Program Underway, But Opportunities Exist to Strengthen Validation and Address Operational Challenges. GAO, Washington DC.

Hawley, K., Means, N., 2012. Permanent Emergency: Inside the TSA and the Fight for the Future of American Security. Palgrave Macmillan, New York.

Hughes, D., Aug. 2005. Explosives Detection Gaps. Aviation Week & Space Technology 41–42.

International Civil Aviation Organization (ICAO), 2006. Annex 17 to the Convention on International Civil Aviation—Safeguarding International Civil Aviation Against Acts of Unlawful Interference, retrieved July 8, 2008, from www.icao.int/cgi/goto_m.pl?icaonet/anx/info/annexes_booklet_en.pdf2006.

Price, J., Aug. 5, 1995. These Noses Know. Rocky Mountain News 2D.

Shanks, N., Bradley, A., 2004. Handbook of Checked Baggage Screening. Professional Engineering Publishing, Wiltshire, U.K.

Transportation Security Administration (TSA), 2006. National Explosives Detection Canine Team, retrieved Nov. 29, 2006, from www.tsa.gov/lawenforcement/programs/editorial_multi_image_0002.shtm2006.

8

Commercial Aviation Aircraft Operator Security

Objectives

This chapter explains the roles of aircraft operators and the federal government in aircraft operator security. We discuss the issues airlines must address in their security programs, including crewmember safety and security training. We introduce you to the requirements and approval processes of the full and partial aircraft operator security programs. Characteristics that differentiate special security programs, such as the private charter and 12-5 security programs, are provided. We investigate the functions and responsibilities related to stakeholders, such as flight crewmembers, aircraft operator security coordinators, ground security coordinators, and in-flight security coordinators. Salient to these discussions is an introduction to the psychology of confrontation that may be encountered in aviation. The chapter concludes with generalized comparisons of security requirements for foreign air carriers to those of domestic air carriers.

Introduction

Aircraft operations offer unique challenges to security. The necessity of the airlines to move across national borders, carrying millions of passengers every year and billions of tons of cargo and baggage, creates the opportunity for significant security loopholes. This vast scope creates many security challenges to airline operators in terms of processes and regulatory enforcement. Communication lines are critical, particularly when transmitting security directives and threat information from airline headquarters' offices to hundreds of airline stations across the world.

This large span of control makes airline security operations different from airport security management in a number of ways. When a security incident occurs at an airport, the airport security coordinator (ASC) can go to the location in a matter of minutes and see the circumstances first hand. Those involved in the incident can be interviewed directly, and the ASC can address the problems. When there is a problem at an airline station far away from the airline's security manager's office, the security manager must fly to the location involved. Airline security managers at the corporate level are geographically farther away from the personnel actually implementing and carrying out day-to-day security measures. With potentially dozens of security incidents occurring daily, flying to the location of each is impractical and, in most cases, impossible.

Airline Security: Historical Context

Airlines first embraced a formalized role of security management in the early 1970s in response to numerous hijackings. During this time, airlines voluntarily implemented passenger and carry-on baggage screening. By 1973, screening was mandatory and regulated for all air carriers. Air carriers continued with screening throughout the 1980s and 1990s. In 1988, in response to the bombing of Pan Am Flight 103 (Lockerbie, Scotland), positive passenger/baggage matching became a regulatory responsibility of U.S. airlines conducting international flight operations. In 1996, in response to a ValuJet crash, the carriage of hazardous materials also brought cargo security to the forefront of aviation security. Also in 1996, the crash of TWA Flight 800 and subsequent legislation added limited screening of checked baggage to air carrier security responsibilities.

Throughout the 1980s and 1990s, those conducting screening were required to meet few federal standards and had minimum training requirements. With airlines operating as businesses, security was a loss prevention (i.e., expense) line item rather than a revenue-generating function. As with any business seeking to minimize expenses and maximize profits, funding the contract screener workforce was often at the bottom of funding priorities.[1] The screening workforce reflected this lack of focus and, before 9/11, was fraught with turnover and performance problems. In some areas, the screener turnover rate exceeded 100% per year, causing gaps in security as positions were left unfilled. Screeners were typically lured away from the stress and monotony of the poorly paid and boring job to higher-paying positions often at airport fast-food outlets.

Companies such as Argenbright, the nation's largest provider of contract screening personnel in the United States before 9/11, provided contracted screening services for numerous overseas air carriers and airports. This training required security personnel to be held to a much higher standard than their U.S. counterparts—highly trained in security profiling and questioning techniques in addition to their screening responsibilities. The lower performance of U.S. screeners was more a result of the low expectations from U.S. aviation security regulations rather than the performance of U.S. screening companies. Internationally, screening has traditionally been viewed a great deal more seriously than in the United States. Terrorist attacks on commercial aviation continued in foreign countries at a rate high enough to warrant continued vigilance and a continuing series of improvements to the security system.

In a 1989 report on Federal Aviation Administration (FAA) security programs, the General Accounting Office (GAO, 1987) stated that the quality of screener training varied widely among the airlines and that aircraft operators had different approaches for carrying out procedures, such as additional questioning of passengers, profile application, and

---

[1]Although it could be argued that loss prevention is a form of revenue protection or even a revenue-generating function (if too many aircraft of a particular airline are lost as a result of security incidents, passengers will quit flying that airline), the level of risk in the United States from terrorist or criminal attacks did not justify the additional security costs. In the risk management model, an analysis of attacks on U.S. aviation in the 1990s demonstrated that although the impact could be high, the rate of occurrence was quite low.

detection of plastic explosives, thus creating few consistencies in procedures or policies. The FAA did not evaluate formal aircraft operator security training at high-risk foreign airports. In 1996 and 2000, the U.S. government addressed minimum security standards for screeners—however, these regulations had not been completed by 9/11. The Aviation and Transportation Security Act of 2001 (ATSA 2001) transferred responsibility for screening to the federal government and reduced, but did not eliminate, the air carrier security role.

In 1996, the Gore Commission concluded that aviation security was a federal responsibility. The Air Transport Association (ATA), now known as Airlines for America (A4A), an organization representing the airline industry, quickly agreed. In 2001, Congress decided that screening and in-flight air marshal functions were federal responsibilities.[2]

Airline security focuses on protecting the aircraft, its passengers, and crew. To that end, air carrier security managers focus on a variety of aspects, not just terrorism. Airline security is similar to security functions that might be found at any major corporation (e.g., background checks, loss prevention, and employee safety). From the perspective of protecting corporate assets and personnel, airline security officers conduct their own inspections of airports that their aircraft use to verify that the airports serviced meet minimum security standards. No airline manager wants to lose an airplane because of a bombing or hijacking. Such events usually result in the loss of life, the loss of public confidence in the airline attacked, and the loss of corporate assets, but as with any other business venture, airlines must always balance how much time, energy, and effort is spent on security, with other areas such as safety and managing aircraft operations.

Aircraft Operator Standard Security Program

Airlines use security standards set forth in the Aircraft Operator Standard Security Program (AOSSP). The AOSSP describes how an aircraft operator must manage a security program to be approved for U.S. commercial aircraft operations. Unlike the airport security program (ASP), which is unique to each airport, the AOSSP is a standard document applicable to all airlines. Like the ASP, each airline creates its own internal security program that specifies how it will comply with the requirements of the AOSSP. The Transportation Security Administration (TSA) compliance inspectors enforce and provide oversight for the security programs. Specific policies and procedures of the AOSSP are classified as security sensitive information (SSI), so this section will only cover generalities regarding the AOSSP.

Before 9/11, the majority of information in the AOSSP focused on passenger and carry-on baggage screening. Although the revisions to the AOSSP still include those processes, they only apply where the TSA does not perform screening. The airline is still responsible to ensure that its passengers and baggage have been screened before allowing them onto

---

[2]This transfer of screening responsibilities is specified in ATSA 2001.

the aircraft. However, the responsibility of actually conducting the screening usually falls under the control of the TSA or foreign airport screening personnel. Other areas of the AOSSP focus on the *common strategy*, which addresses how an airline should handle hijackings, bomb threats, and bomb discovery procedures. The common strategy also addresses special flight procedures related to security and procedures against other threats to commercial aviation.

Title 49 CFR Part 1544—Aircraft Operator Security: Air Carriers and Commercial Operators

Before the passage of ATSA 2001, aircraft operator security was described under Federal Aviation Regulation Part 108. ATSA 2001 transferred Part 108 to Part 1544. Title 49 CFR Part 1544 generally addresses commercial flight operations. However, any aircraft operator may request to be regulated under Part 1544. This section addresses the requirements of aircraft operators as related to the airline's security program; screening of passengers, baggage, and cargo; the use of law enforcement personnel including the carriage of armed law enforcement officers on an aircraft; criminal history record checks (CHRCs); flight deck privileges; the Known Shipper program; and threat and response contingencies. Much of Part 1544 is similar in content and scope to Part 1542, particularly the areas that CHRCs, personnel identification systems, the security program amendment process, law enforcement response, and exclusive area agreement processes.

Applicability of the Regulations

Title 49 CFR Part 1544 applies to any aircraft operator with a certificate issued by the Federal Aviation Association (FAA) under CFR Part 119 Certification: Air Carriers and Commercial Operators. Part 119 specifies the requirements an aircraft operator must adhere to in order to conduct commercial operations in the United States. Specifically, Part 1544 applies to scheduled passenger operations, public charter passenger operations, private charter passenger operations, and the operations of aircraft operators holding operating certificates under Part 119 and operating aircraft with a maximum certificated takeoff weight of <u>more</u> than 12,500 pounds. Subpart A also provides the TSA with the inspection authority and the rights to conduct tests and copy records of the aircraft operator to establish compliance with Part 1544.

Subpart B: Security Program

This section covers the security program format, amendment, and approval process and addresses the six aircraft operator security programs: security, full security, partial security, twelve-five (12-5), private charter, and all cargo. In October 2008, the TSA issued a Notice of Proposed Rulemaking, which would add the Large Aircraft Security Program (LASP) affecting private aircraft operators in aircraft above 12,500 pounds, however, over

10,000 comments were received, most of them against the program, and the TSA went back to the drawing board. As of 2012, the TSA expects the newly revised LASP to be available for review in 2013 (and possibly renamed). Each security program necessitates a set of requirements based on the threat presented by the aircraft type in terms of passenger capacity and weight. This section also addresses the format of an aircraft operator security program and the amendment process.

The type of aircraft operator will generally determine the type of aircraft operator security program required. Scheduled passenger or public charter operators operating aircraft with a passenger seating configuration that allows 61 or more seats and scheduled passenger or public charter passenger operations with an aircraft with 60 or fewer seats when passengers are enplaned from or deplaned into a sterile area are required to have a full security program. Aircraft operators using aircraft with a passenger seating configuration of 31–60 seats not enplaning or deplaning from a sterile area or aircraft operators with passenger seating configurations of fewer than 60 seats but operating internationally (to, from, or outside the United States and as a registered U.S. air carrier) adhere to partial security program requirements. Aircraft with a maximum gross takeoff weight that exceeds 12,500 pounds used in scheduled or charter service, carrying passengers or cargo or both, and not already under a full or partial program must adhere to the 12-5 program. Aircraft with a takeoff weight in excess of 45,500 kilograms (100,309.3 pounds) with a passenger seating configuration exceeding 61 seats or any aircraft enplaning from or deplaning into a sterile area while conducting a private charter operation must adhere to the private charter program.

The Private Charter Standard Security Program (PCSSP) is primarily for commercial airline operators that offer their aircraft for private charter. Typical customers are sports teams or entertainers. Screening is limited to passengers and carry-on bags only and, usually, all of the passengers on a chartered flight know one another. The PCSSP is required to be followed even at general aviation airports. The PCSSP can also be used for noncommercial airline–type aircraft. Price (2008) described PCSSP as follows:

> *The PCSSP requires aircraft operators to ensure all passengers and accessible baggage are screened prior to boarding the aircraft. To comply with this requirement, the private charter rule allows "non-TSA" screeners who have completed TSA-approved private charter screener training to perform the screening. TSA screeners who have completed the TSA-approved basic screener training course may also perform screening at TSA checkpoints for private charter operations.*
>
> *Parts 121 and 135 operators using their aircraft in private charter operations with a maximum certificated takeoff weight greater than 45,500 kg (100,309.3 pounds), or with a passenger seating configuration of 61 or more, must ensure that all passengers and accessible baggage are screened prior to boarding the aircraft. Additionally, these operators must have a security program that establishes the required security components for private charter operations. The program must include use of metal detection devices, X-ray systems, security coordinators, law enforcement personnel, accessible*

weapons, criminal history records checks, training for security coordinators and crewmembers, training for individuals with security-related duties, training and procedures for bomb or air piracy threats, security directives, and all of Subpart E of 49 CFR Part 1544 concerning screener qualifications when the aircraft operator performs screening. (pp. 24–25)

Aircraft operators conducting cargo operations where the aircraft's maximum certificated takeoff weight exceeds 45,500 kg (100,309.3 pounds) carrying cargo and authorized personnel but no passengers operate under the all-cargo program. Although each security program contains a variety of requirements, it is best to remember that the TSA may add requirements if the TSA administrator feels that additional requirements are necessary to address heightened security concerns. The full security program requires commercial-service aircraft operators to comply with the regulations listed in Table 8.1.

The requirements for screener performance and training are also covered (Subpart E of Part 1544) within the full security program. These requirements only apply when the personnel employed by the aircraft operator or a private screening company (not TSA personnel) are used. Table 8.2 displays the measures called for by each security program.

Aircraft Operator Security Programs

Each aircraft operator must have an approved security program describing the facilities and equipment under its control and the specific security responsibilities. This includes geographic boundaries of exclusive areas, equipment used to screen passengers, carry-on baggage, checked baggage and cargo, and other factors. Security programs are approved and amended in the same manner as those for airport operators.

An initial security program for a new aircraft operator must be submitted for approval at least 90 days before start of passenger operations. The TSA then has 30 days to approve or give the aircraft operator written notice to modify the program. The aircraft operator can then either comply with the request to modify the program or appeal to the TSA administrator within 30 days of receiving the request. Amendments to existing security programs requested by the aircraft operator must be filed with the TSA at least 45 days before the effective date of the change. If the TSA denies a proposed amendment, the aircraft operator has 30 days to appeal the decision to the TSA administrator. Amendments to a security program requested by the TSA provide for 30 days advance notice to an aircraft operator.[3,4]

[3]The TSA largely conducts the screening of individuals, carry-on baggage, checked baggage, and cargo in the United States. However, it is still the responsibility of the aircraft operator not to let any individual, bag, or cargo item onboard its aircraft unless the proper screening process has been conducted.

[4]This regulation applies when the TSA does not manage screening and the airport operator does not meet LEO requirements as prescribed in CFR Part 1542.215.

Table 8.1   Full Security Program Requirements

Full Security Program Requirement	Regulation
Ensure individuals and carry-on baggage are screened	1544.201
Ensure checked baggage is properly screened	1544.203[2]
Screen cargo	1544.205
Ensure that the TSA, the aircraft operator itself in lieu of the TSA, or a foreign government has conducted screening of accessible property, checked baggage, and cargo	1544.207[4]
Use approved metal detection devices to screen persons	1544.209
Use an approved X-ray system to screen carry-on and checked baggage where applicable	1544.211
Use an approved explosive detection system to screen checked baggage on international flights	1544.213
Designate a security coordinator, known as the aircraft operator security coordinator (AOSC); duties of the AOSC are similar to those of the airport security coordinator; also addresses ground security coordinators and in-flight security coordinators	1544.215
Provide for law enforcement officer (LEO) support to respond to security issues onboard a flight or at a screening checkpoint	1544.217[5]
Ensure that processes for the carriage of accessible weapons for federal LEOs and state and local police officers and sheriff personnel are adhered to	1544.219
Carry prisoners	1544.221
Transport federal air marshals	1544.223
Prevent unauthorized access to exclusive areas and aircraft	1544.225
Carry out the provisions of any exclusive area agreements the operator has entered into	1544.227
Conduct a security threat assessment on individuals with access to cargo	1544.228
Conduct CHRCs on all aircraft operator personnel with access to checked baggage and cargo and any personnel with unescorted access authority	1544.229
Conduct CHRCs on flight crewmembers	1544.230
Use airport-approved and exclusive area–approved personnel identification systems	1544.231
Provide security training for ground security coordinators and flight crewmembers	1544.233
Provide security training for other employees with security-related duties such as security identification display area (SIDA) training for ramp workers, airline administrative personnel, and others who are not required to be a ground security coordinator or in-flight security coordinator	1544.235
Restrict access to the flight deck to only those the air carrier has specifically authorized in its TSA-approved security program	1544.237
Comply with the requirements of the Known Shipper program as related to the acceptance of cargo	1544.239
Have and implement contingency plans when directed by the TSA and participate in airport-sponsored emergency exercises	1544.301
Evaluate threats and have plans to handle bomb and air piracy threats	1544.303
Comply with security directives and receive information circulars	1544.305

The TSA may issue an emergency amendment with which an aircraft operator must immediately comply. The aircraft operator may appeal the conditions of the emergency amendment but must comply with the conditions during the appeal. Emergency amendments are usually used with foreign air carriers as they are not allowed to receive sensitive security information (security directives issued to domestic airport operators and air carriers are always classified SSI).

Table 8.2 Aircraft Operator Security Program Requirements (per CFR Part 1544 or 1550)

Security Program	Full Security Program (1544)	Partial Security Program (1544)	Private Charter Security Program (1544)	All-Cargo Security Program 1544	12-5 Security Program (1544 and 1550)
Type of aircraft operation requirements	Scheduled passengers or public charter with more than 61 seats or scheduled passengers or public charter when enplaned or deplaned into a sterile area	Scheduled passengers or public charter with 30–60 seats that do not enplane or deplane from a sterile area or scheduled passengers or public charter with fewer than 60 seats flying to, from, or outside the United States that do not enplane or deplane into a sterile area	Private charter above 100,309 pounds (45,500 kilograms) or more than 61 passengers seats (regardless of enplaning or deplaning into a sterile area)	Aircraft operators engaged in the carriage of air cargo in aircraft above 100,309 pounds (45,500 kilograms), configured for all-cargo operations	Any aircraft in scheduled operations or public or private charter carrying passengers, cargo, or both weighing more than 12,500 pounds maximum gross takeoff weight
Screen individuals and carry-on bags (1544.201)	Yes	No*	Yes	No	No (unless enplaning into a sterile area)
Prevent unauthorized explosives, weapons, and incendiaries from loading onto aircraft (1544.202)	Yes	Yes	Yes	Yes	Yes
Screen checked baggage (1544.203)	Yes	No	No	No	No
Screen cargo (1544.205)	Yes	No	No	Yes (prevent explosives, control cargo access, accept only from foreign authorized cargo operator)	No

Requirement				
Ensure that the TSA has conducted screening of persons, carry-ons, checked baggage, and cargo (1544.207)	Yes	No	Yes	No
Use metal detectors to screen persons (1544.209)	Yes	No	Yes	No
Use X-ray system for checked baggage if authorized by the TSA (1544.211)	Yes	No	Yes	No
Use explosive detection system for checked baggage (1544.213)	Yes	No	No	No
Designate a security coordinator (1544.215)	Yes	Yes	Yes	Yes
Provide for LEO support (1544.217)	Yes	Yes	Yes	Yes
Carry armed LEOs (1544.219)	Yes	Yes	Yes	Yes
Carry prisoners (1544.221)	Yes	No	No	No
Transport federal air marshals (1544.223)	Yes	Yes	No	Yes
Prevent unauthorized access to exclusive area and aircraft (1544.225)	Yes	No (exclusive area)	Yes (prevent access to aircraft)	Yes (prevent access to aircraft)

Continued

Table 8.2 Aircraft Operator Security Program Requirements (per CFR Part 1544 or 1550)—cont'd

Security Program	Full Security Program (1544)	Partial Security Program (1544)	Private Charter Security Program (1544)	All-Cargo Security Program 1544	12-5 Security Program (1544 and 1550)
Conduct security inspection of air cargo before placing in passengers operations (1544.225)	Yes	Yes	Yes	No	Yes
Carry out provisions of exclusive area agreement (1544.227)	Yes	No	No	Yes	No
Conduct a security threat assessment for personnel with access to air cargo (1544.228)	Yes (unless CHRC is done)	No	No	Yes	No
Conduct a search of the air cargo before departure (1550.7(b))	Yes	No	Yes (if flight is not already covered by a full or partial security program)	Yes	Yes (if flight is not already covered by a full or partial security program)
CHRCs for those with unescorted access, screeners, and those with access to checked baggage or cargo (1544.229)	Yes	No	Yes	Yes	No
CHRC of flight crewmembers (1544.230)	Yes	Yes	Yes	Yes	Yes
Airport-approved ID system (1544.231)	Yes	No	No	Yes	No

	Column 1	Column 2	Column 3	Column 4
Training for security coordinator and crewmembers (1544.233)	Yes	No	Yes	No
Training for individuals with security duties (1544.235)	Yes	Yes	Yes	Yes
Restrict access to flight deck (1544.237)	Yes	Yes	No	Yes (if door installed)
Contingency plans (1544.301)	Yes	Yes	No	Yes (but does not have to participate in airport-sponsored exercise of plan)
Bomb and air piracy threat procedures (1544.303)	Yes	Yes	Yes	Yes
Comply with and protect security of security directives and information circulars (1544.305)	Yes	Yes	Yes	Yes
Approved screeners and screening program (Subpart E)	Yes	No, unless the TSA has approved a request	No	No, unless the TSA has approved a request

Note: This matrix is for training purposes only. Each aircraft operator is responsible for contacting its TSA representative to get a list of requirements for its particular program.

*Unless required by the TSA. 1. Part 1550.5: Any operation enplaning or deplaning into a sterile area must conduct an aircraft search, screen passengers, crewmembers, and carry-on baggage before boarding. 2. Passenger or charter operations already operating under Part 1544 or 1546 (Foreign Air Carrier) must adhere to their regulatory programs. 3. For any flight operation (of any number of seats or weight) into or out of a sterile area for which there is no security program under Part 1544 or 1546 in place, the aircraft must be searched before departure, and passengers, crewmembers, and other individuals and their accessible property (carry-on items) must be screened before boarding in accordance with procedures approved by the TSA.

Aircraft Operator Security Operations and Screening Responsibilities

Subpart C of Title 49 CFR Part 1544 Operations details the day-to-day requirements for aircraft operators. It includes sections on personnel identification systems, personal and property screening, law enforcement personnel, exclusive areas, flight deck privileges, and the Known Shipper program (related to air cargo). Many of the requirements reviewed in this section can be met by other entities, such as the TSA (screening) and airport management (law enforcement support).

Either the TSA or an authorized private screening firm within the United States generally conducts screening. Nevertheless, it is the regulatory responsibility of the aircraft operator to ensure that it uses measures to prevent or deter the carriage of any weapon, explosive, or incendiary device on a person or carry-on baggage before the person or bag boards an aircraft or enters a sterile area. The aircraft operator has little involvement with physical screening when the TSA or an authorized U.S.-owned company conducts screening. When a U.S. air carrier is operating in another nation or at certain small air carrier stations within the United States, the aircraft operator must still ensure that all applicable measures are taken to properly screen individuals and carry-on baggage. Aircraft operators are obliged to deny transportation to anyone who refuses to submit to screening.

All-cargo aircraft operators who carry passengers (called supernumaries) must ensure that only authorized passengers and their carry-on baggage are allowed to board and that they do not bring any weapons, explosives, incendiaries, or other destructive devices, items, or substances onboard. Although all-cargo operators primarily carry cargo, there are occasions when they carry passengers, airline employees, animal handlers, hazardous materials supervisors, and those needed to accompany the carriage of human organs.

Aircraft operators under a full security program, an all-cargo program, or a 12-5 program must ensure that cargo placed within the holds of their aircraft are properly screened and inspected to prevent or deter the carriage of unauthorized persons, unauthorized explosives, incendiaries, and other destructive substances or items in cargo onboard an aircraft. Aircraft operators are obliged to maintain the integrity of any cargo from the time it is accepted until it is delivered to a proper entity. Operators must ensure that unauthorized explosives, incendiaries, and other destructive substances are not introduced to the cargo container, and they must refuse to transport any cargo that has not been subjected to the proper screening protocols. Further, cargo may only be accepted from a shipper with a security program, such as an indirect air carrier, similar to the aircraft operator's security program.

The TSA distinguishes between the terms *screened* and *inspected*. Screened means some form of vetting that is not necessarily a physical inspection (i.e., the Known Shipper program), and inspected means a physical inspection of a parcel by explosive trace detection (ETD), K-9, or other approved means. The regulation requires that parcels be screened and inspected. It is up to the TSA and the aircraft operator to determine what measures an aircraft operator must undertake to be in compliance with this section.[5]

---

[5]As of January 2007, the U.S. Congress has been assessing whether to require the physical inspection of all cargo that is carried by a commercial aircraft through the use of explosive detection system (EDS), ETD, K-9, or some other approved method.

Subpart E of Title 49 CFR Part 1544 addresses screener qualifications when the aircraft operator performs the screening. Aircraft operators/screeners must meet similar qualifications as TSA screening personnel. They must also complete at least 40 hours of classroom instruction, pass a screener readiness test, complete 60 hours of on-the-job training and testing, and pass an on-the-job exam before they are allowed to conduct screening independently. Screeners cannot conduct screening duties while under an impairment such as illegal drugs, sleep deprivation, medication, or alcohol.

Airline Security Coordinators

Each aircraft operator must appoint an aircraft operator security coordinator who is the primary point of contact to the TSA. This position is similar to the airport security coordinator in terms of responsibilities, but the AOSC's span of control is much larger. AOSCs must manage security programs for operations with dozens of remote locations across the United States and, for larger airlines, the world. The AOSC receives and disseminates security directives and information circulars and works with the ASCs at each airport to ensure compliance with security regulations and procedures. The AOSC also audits the hub and station security programs to ensure uniformity with the airline program for passenger, baggage, and cargo operations. At large airlines, the primary AOSC is often assisted by other security managers at the airlines' major hubs. Airline station managers are responsible for the operation of an airline at each hub or station and assume certain security responsibilities, which vary from airline to airline.

The AOSC oversees the airline's security program and works with the TSA to ensure compliance with the program and the regulations. AOSCs may also work with the U.S. Secret Service and the Department of State for the transport of dignitaries and with any other government agencies such as the Drug Enforcement Agency (DEA) and the Federal Bureau of Investigation (FBI).

Each flight has a ground security coordinator (GSC) responsible for the security of that flight. GSCs must review all security-related functions for which the aircraft operator is responsible, including the Aircraft Operator Standard Security Program and applicable security directives. GSCs correct each instance of noncompliance as determined by the TSA. At foreign airports where the government or a contractor provides security, the GSC must notify the TSA for assistance in resolving noncompliance problems.

The GSC can be a distinct job (more common at large airlines) or a collateral duty for a gate agent or cargo agent at low-cost and smaller airlines. GSCs are trained in many areas including the transportation of hazardous materials and the identification of prohibited weapons and explosives and in bomb threat and hijacking management procedures. Their primary responsibilities are to resolve security-related conflicts between gate agents and passengers, liaison with the airport operator during security incidents, intervene in disruptive passenger situations, oversee baggage- and cargo-handling acceptance procedures, and act as the primary contact for the airline in a security incident until relieved by a higher authority.

GSCs are required to make a pilot-in-command (PIC) aware of any security issues related to a flight, and they often work with a PIC on security problems (e.g., a passenger leaving bullets on an aircraft that are subsequently found during a cabin search). It may seem as though the TSA would intervene in such a situation. However, more than 600 million passengers travel every year. The average rate of prohibited items identified by screeners is about 80%, meaning that 20% of prohibited items get through screening. Sometimes passengers do not realize they still have prohibited items until after boarding a plane. Not wanting to be caught, nor wanting to confess, they leave the item somewhere in secured areas or onboard the aircraft. Because of the relatively large frequency of these cases, the GSC and PIC handle many of these *minor* security issues.

Before 9/11, GSCs were present at all screening operations. Today, GSCs are occasionally called to screening checkpoints to determine if a passenger carrying a prohibited item may check an item into checked baggage. GSCs in most locations no longer staff positions at security checkpoints because the TSA is responsible for activities at a checkpoint.

GSCs and crewmembers must complete annual security training as outlined in the airline's security program. The content of such training is considered SSI. In-flight security coordinators and crewmembers must complete an advanced qualifications program approved under Special Federal Aviation Regulation (SFAR) 58 in 14 CFR Part 121, which includes security training required by 14 CFR 121.417(b)(3)(v) or 135.331(b)(3)(v)—covering hijackings or other unusual situations.

Law enforcement officers traveling armed and special procedures for flights departing for Israel or into Reagan National Airport are examples of other situations that GSCs are trained to handle. GSCs watch to ensure that passenger boarding doors are secured when not in use and that all security procedures are being followed.

The PIC has final responsibility for the safety and security of a flight and its passengers. The PIC acts in the capacity of the in-flight security coordinator (ISC) on each flight.[6] The ISC receives a security briefing before each flight, both from the GSC and through preflight briefings. For practical purposes and to the extent possible, the PIC is "in charge" of security-related incidents onboard the aircraft during flight, including hijackings. During a hijacking, the PIC may also be a hostage and his or her ability to command, lead, and coordinate security activities will be severely affected. In such situations, law enforcement will typically treat the PIC and other flight crew personnel as hostages and make decisions on their behalf. The ability of the PIC to handle in-flight security issues is also hampered in the post-9/11 era. Previously, it was common for the PIC to exit the cockpit to attempt to handle a security situation in the cabin. Since 9/11, the PIC and all required flight crew personnel are now expected to stay in the cockpit, particularly during the time of a security incident in the cabin.

In some cases, a member of the flight crew may be trained as a federal flight deck officer (FFDO). However, FFDOs are not allowed to exit the cockpit to deal with a situation in the

---

[6]This responsibility kept Captain John Testrake and his crewmembers from attempting to escape into the streets of Lebanon during the hijacking of their aircraft in 1985.

cabin. The FFDO jurisdiction ends at the cockpit door, although there have been some attempts by groups to extend FFDOs' jurisdiction into the cabin and even into the airport and beyond. FFDOs only undergo one week of training in a specific function, not the 14 weeks of training that actual federal agents undergo at the Federal Law Enforcement Training Academy (FLETC), or 18–20 weeks for DEA or FBI agent training. The events of 9/11 make it clear that the integrity of the cockpit door is an essential layer of the security system. Therefore, the flight crew remains in the cockpit with the door locked as much as possible while in flight, only exiting in an emergency (not a security situation) or to use the restroom—in which case secondary flight deck barrier measures are used. The PIC has the final authorization as to whether a passenger will be allowed onboard an aircraft and whether to land the aircraft and deboard a passenger he or she deems a threat to the safety of the flight or other passengers and flight crew. The airline security manager, ground security coordinators, and pilots form a team for each flight and often have overlapping duties to ensure the security of each flight.

Law Enforcement Operations Related to Airline Security

Law Enforcement Personnel

Law enforcement requirements prescribed in Parts 1542.215 and 1542.217 must still be met where an aircraft operator conducts passenger operations at domestic airports not required to have a security program under Part 1542. This includes certain aircraft operators with full or partial security programs, 12-5 programs, all-cargo programs, private charter programs, and passenger operations at foreign airports. These requirements address the availability and training for LEOs. Specific response times and numbers of available LEOs are addressed in the aircraft operator security program and are considered SSI. Where LEO support is not available, the aircraft operator must coordinate with the TSA to provide such support. The aircraft operator must ensure that its employees, including crewmembers, have the training and resources to contact LEO support when needed.

Carriage of Firearms in Checked Baggage

Federal regulations require that firearms must be unloaded and carried in a hard-sided locked container. Only the person checking in the firearm can have the combination or key. Furthermore, the firearm must be declared at the ticket counter and stored on the aircraft such that it is not accessible to passengers.

Aircraft operators may require additional policies regarding the transport of firearms in checked baggage. Some aircraft operators allow ammunition to be stored in the same container but not loaded into the firearm. Some aircraft operators require advance notice upon booking a flight on which a passenger will be bringing a firearm. The transport of ammunition is covered in Title 49 CFR Part 175, which addresses the carriage of hazardous materials onboard aircraft. Ammunition must be kept separate from any flammable liquids or solids.

As a practical matter, when a passenger declares that he or she is placing a firearm into checked baggage, a qualified screener reports to the ticket counter and immediately inspects the firearm, and then the container is locked and placed in checked baggage. This prevents personnel conducting checked-baggage screening (which may or may not be located in the ticket counter or lobby area) from opening the container without the owner present.

Carriage of Accessible Weapons

This section addresses the carriage of accessible weapons (by passengers) onboard a commercial-service aircraft. Generally, the only individuals authorized to carry a loaded firearm onboard are LEOs.

For flights requiring screening, an LEO must be a federal law enforcement officer or a full-time municipal, county, or state law enforcement officer who is a direct employee of a government agency. This individual must be sworn and commissioned to enforce criminal statutes or immigration statutes. The individual must be authorized by the employing agency to have the weapon in connection with assigned duties and must have completed the training program Law Enforcement Officers Flying Armed (see *http:/www.tsa.gov/ lawenforcement/programs/traveling_with_guns.shtm*). A federal agent (whether or not on official travel) carrying the proper credentials is not required to show any other reason for carrying a weapon onboard, provided he or she is armed in accordance with an agency-wide policy governing that type of travel established by the employing agency by directive or policy statement.

An LEO must have a need to have access to the weapon during flight to carry it on the flight. This could include being on protective duty assigned to a principal or advance team, on travel requiring preparation to engage in a protective function, conducting a hazardous surveillance operation, on official travel to report to another location, armed and prepared for duty, controlling a prisoner, in accordance with other stipulations in Part 1544.221, or on a round-trip ticket returning from escorting or traveling to pick up a prisoner.

Armed law enforcement officers must notify aircraft operators of the flights on which an accessible weapon is needed at least one hour, or in an emergency as soon as practicable, before departure. LEOs present identification to the ticket agent, generally a law enforcement officer's agency ID card that must include a clear full-face picture, the armed LEO's signature, and the "letter of authority." State, local, territorial, tribal, and approved railroad LEOs flying armed must submit a National Law Enforcement Telecommunications System (NLETS) message prior to travel to receive the letter of authority. The NLETS message replaces the original letter of authority, commonly referred to as the "chief's letter." Presentation of an LEO badge or shield cannot be used as the sole means of identification. If the armed LEO is an escort for a foreign official, then the U.S. State Department provides such documentation.

Aircraft operators must inspect the documentation presented and advise the LEO of procedures unique to the airline. The aircraft operator must ask the LEO to confirm completion of the Law Enforcement Officers Flying Armed training program as required by the TSA, unless specifically exempted by the TSA. The Law Enforcement Officers Flying Armed training is a 1.5- to 2-hour block of instruction that is comprised of a structured lesson plan, slide presentation, FAQs, NLETS procedures, and applicable codes of federal regulation. The aircraft operator must ensure that the GSC knows the identity of the armed LEO and must notify the PIC and other appropriate crewmembers of the location of each armed LEO aboard the aircraft. All armed LEOs, including federal air marshals, must be notified of each other's seating on a flight.

For flights where screening is not conducted, LEOs may carry weapons onboard provided they are federal law enforcement officers or full-time municipal, county, or state law enforcement officers who are direct employees of a government agency. They must also provide the proper notification to the aircraft operator and present proper credentials.

LEOs traveling armed must not consume any alcohol while on the flight or board the flight armed having consumed alcohol within eight hours before the flight. When traveling armed, LEOs not in uniform must keep the firearm concealed on their person or within reach; if in uniform, they must keep the firearm on their person at all times.

The practical application of this section varies with each airport's security program and each aircraft operator's security program. An LEO intending to fly armed will declare such intent at the ticket counter. Some airlines require that LEOs flying armed notify the airline when booking the flight and present the proper credentials and documentation. The LEO does this again at the screening checkpoint, to a TSA supervisor. At some airports, an LEO assigned to the airport will provide the vetting at the screening checkpoint for another LEO traveling armed. LEOs traveling armed are not subject to screening, provided all other conditions of this section have been met.

The TSA will issue a specific alpha-numeric unique federal agency number (UFAN) for each federal law enforcement agency or entity or other organization, including those performing personnel security detail (PSD) missions, as approved by the TSA. This number will be known only to the respective agency and the TSA. This identifier will be verified at the airport LEO checkpoint prior to granting the LEOs access to the sterile area for the purpose of flying armed.

Prisoner Transport

Prisoners are classified as high risk or low risk. A high-risk prisoner presents an exceptional escape risk, as determined by the law enforcement agency, and has been charged with, or convicted of, a violent crime. A low-risk prisoner is any prisoner who is not classified as high risk. An aircraft operator may not carry a prisoner unless he or she has been classified as either high risk or low risk by the proper agency. For low-risk prisoners, at least one armed LEO must accompany the prisoner for flights that are shorter than four hours. One armed LEO can escort no more than two low-risk prisoners. A minimum of two armed LEOs is required for flights longer than four hours for one or two low-risk prisoners.

Only one high-risk prisoner may be carried on any particular flight and must be escorted by two armed LEOs. The LEOs must have no other prisoners under their charge during this time. The TSA can approve more than one high-risk prisoner on a specific flight provided that there is a minimum of one armed LEO for each prisoner and one additional armed LEO (Table 8.3).

Law enforcement officers escorting a prisoner must notify the aircraft operator at least 24 hours before departure, or as far in advance as practicable, and advise the airline of the identity of the prisoner and whether he or she is high risk or low risk. The LEO must arrive at least one hour before the departure time of the flight and ensure that the prisoner has been thoroughly searched. The LEO must be seated between the prisoner and the aisle and must accompany the prisoner at all times during the flight. Ideally, prisoners should be boarded first and deplane last, and they should be seated in the farthest rear section of the plane possible, but not near any exits. LEOs must ensure the prisoner is properly restrained from using his or her hands, but the use of leg irons is not permitted as the prisoner must still be able to egress the aircraft in an emergency. Leg irons would hamper egress of the prisoner and, potentially, other evacuating passengers and flight crew. Prisoners must not be provided with eating utensils (unless approved by the escorting LEO) or be provided with alcohol.

Transportation of Federal Air Marshals

Aircraft operators are required to carry federal air marshals on any scheduled passenger or public charter flight on a first-priority basis and without charge while on duty, including the repositioning of flights and in the seat requested by the air marshal. The TSA determines the number of air marshals required on a flight. Federal air marshals are required to identify themselves to an aircraft operator using their credentials (photo ID card)—a badge or shield is not adequate identification. Aircraft operators are required to keep confidential the location and identity of all federal air marshals onboard a flight. Federal air marshals directly contact and coordinate with other armed LEOs on each flight.

Security of the Aircraft on the Ground and in Flight

Aircraft operators must ensure that unauthorized personnel are unable to access the air carrier's aircraft, associated facilities (hangars, administrative areas that may have access to aircraft, etc.), and exclusive areas. Air carriers must conduct a security inspection of

Table 8.3 Ratio of LEOs Required to Prisoners Transported

Number and Type of Prisoners	Required Number of LEOs
One or two low-risk prisoners; flight shorter than four hours	1
One or two low-risk prisoners; flight longer than four hours	2
One high-risk prisoner	2
Two high-risk prisoners (only with TSA approval)	3

each aircraft if the aircraft has not been previously protected in accordance with the aircraft operator security program. Air carriers conduct a daily inspection of each aircraft that has been out of service before returning it to passenger operations.

Aircraft searching is an effective method of deterring hijackings and bombings. Aircraft searches are done by a combination of flight crew and cabin cleaners before each scheduled departure. Crewmembers conducting a thorough search of an aircraft may discover hidden weapons or explosives between flights. For example, in 1995, Ramzi Yousef's plan to destroy 12 commercial U.S. airliners on over-the-water international routes assumed such searches would not be conducted. Yousef and his operatives intended to hide explosives beneath passenger seats and then deboard at a stopover. Their expectation was that the aircraft would then take off again, with the explosives detonating while the aircraft were over the Pacific Ocean. The loss of life and 12 aircraft would have been catastrophic. Yousef had successfully tested the concept on an earlier flight, killing a Japanese executive, but in this instance, a cabin inspection prevented the catastrophe.

Protecting the flight deck is critically important! Aircraft operators are required to restrict access to the flight deck for all aircraft having a cockpit door, in accordance with the operator's security program. This restriction does not apply to federal air marshals, certain FAA and National Transportation Safety Board personnel such as check pilots and investigators, and U.S. Secret Service personnel.

In-Flight Security

There are several dynamics that must be accounted for with respect to aircraft in flight that experience an unlawful interference, including whether the pilot should attempt to use violent maneuvers to knock a hijacker or assailant off balance, whether the passengers should attempt to retake a hijacked aircraft, and how to relocate a device that is suspected to be an explosive device to a "safer" area of the aircraft.

In May 2003, an individual attempted to overtake an aircraft using sharpened wooden stakes, with the intent of crashing the aircraft into the Walls of Jerusalem National Park to deliver his souls and the souls of the passengers to the devil, to bring about Armageddon (Baum, 2011). Had he been successful, several individuals would have lost their lives and nothing related to al-Qaeda or any other terrorist organization would have had anything to do with it. It demonstrates that the attackers' perspective need not make sense, or that he or she is even thinking clearly. The hijacker demonstrated that using a low-tech tactic on a low-risk domestic route still presents a risk to aviation security. The individual had also tried to rush the cockpit on another flight four months earlier—in both cases flight attendants and flight crew prevented the takeover of the plane (Baum, 2011).

To address the threat of an aircraft being bombed, every commercial airliner has an identified least-risk bomb location (LRBL), which is the safest part of the aircraft for an explosion to take place. This location is usually near a natural opening, such as the rear doors. The use of the LRBL is upon discovery or notification of an improvised explosive device (IED) on board the flight. Unlike a suspicious device found in the airport, it is often considered better to relocate a suspicious item on an aircraft with the assumption that it is not triggered by motion (otherwise, how was it brought onto the flight, armed, and not detonated through the aircrafts' movement), and that it is better that an explosion take place at an area that is already designed to have a hole in it. Another consideration is that very rarely will an individual make a bomb, put it on a commercial flight, and then call in a bomb threat. If one goes to so much trouble to make the device, why would they then notify the aircraft operator of its presence? However, as we've seen earlier, the motives of criminals, terrorists, and others don't have to make sense.

In addition to relocating the device, flight attendants will use luggage and other items to create a blanket or buffer around the suspected device to mitigate the blast effect should it detonate. Bomb blankets are available commercially for just such an incident, but are not always found on a commercial flight.

Other considerations for flight crew include whether passengers or crew should attempt to retake a hijacked flight. Many in the United States after 9/11 believed that this was a foregone conclusion, but as many hijackings subsequent to 9/11 have shown, this is not always the case. There are too many variables in a hijacking scenario for any hard-and-fast standard operating procedures to dictate a course of action. The PIC will have to make a best judgment as to how to handle a hijacking, and also how to react to the actions of passengers. If it does not appear to be a hijacking to use the aircraft as a weapon of mass destruction (WMD), there are other dynamics to watch for such as the London syndrome, the Stockholm syndrome, the Lima syndrome, and the John Wayne syndrome (Baum, 2011). The London syndrome occurs when a passenger's words or actions brings the wrath of the hijackers down on the individual, usually resulting in the beating or death of the passenger. This should be avoided. The Stockholm syndrome occurs when the hostages feel sympathy for the hostage-takers. First responders must take this syndrome into account when dealing with hostages, particularly during negotiations and during any rescue attempt. The Lima syndrome, generally desired, occurs when the hostage-taker becomes sympathetic to the hostages. Uli Derickson cultivated this during the hijacking of TWA Flight 847 (Baum, 2011). The John Wayne syndrome occurs when individuals feel helpless in a situation they believe they should be doing something about (usually this occurs to males during a hostage crisis). Flight crew personnel particularly should watch for signs of this syndrome and remember that they have an important role to play, and that while active or physical intervention may at some point be necessary, they should do so based on opportunity and, ideally, preplanning and coordination with others, not go "flying off the handle."

Flight crewmembers have other options to handle an in-flight security emergency, such as a hijacking. One possibility is aggressively maneuvering the aircraft to knock a hijacker off balance. This has been done several times as a method to allow passengers to get the upper hand on a hijacker, most notably during the hijack attempt in 1970 of an El Al flight by Leilia Khaled, but also during the hijacking of a Brazilian Airlines flight in 1988, during the attempted hijacking of a FedEx flight in 1994, and during a 2007 hijacking of a Mauritanie flight. Most aircraft flight manuals and airline policies prohibit these types of maneuvers as a response to a hijacking, and pilots must also weigh the possibility of losing control of the aircraft or having the structure overstress and come apart during the process.

Other Aircraft Operator Security Requirements

Aircraft operators must conduct fingerprint-based CHRCs for any employee who conducts a *covered function*, which includes unescorted access authority, the authority to perform screening, or the authority to check baggage or cargo. These requirements are essentially the same as Part 1542.209 of the airport security regulations requiring those operating within the SIDA to complete a fingerprint-based CHRC to ensure he or she has not been found guilty (or not guilty by reason of insanity) of 28 listed disqualifying offenses.

Aircraft operators must also designate an individual to be responsible for conducting the CHRC, including receiving results on investigations from the FBI and conducting internal audits of CHRCs. Airport operators may accept the results of a CHRC conducted by an aircraft operator in the issuance of airport access/ID media, but this is up to the individual airport operator. Aircraft operators are restricted from revealing the results of a CHRC to anyone but the individual whose background was investigated, the TSA, or an airport operator with a "need to know" (i.e., to receive an airport access/ID badge). Flight crewmembers must undergo a fingerprint-based CHRC. As a practical matter, aircraft operators frequently conduct a CHRC on their personnel and then attest to each airport operator where that employee is required to obtain a SIDA access/ID badge that the individual has passed the CHRC. Some airport operators will still require the aircraft operator employee to undergo another CHRC to ensure that the correct procedure has been completed before issuing an airport access/ID badge.

Within an aircraft operator exclusive area, aircraft operators must establish and carry out a personnel identification system that is approved for use in the exclusive area. The personnel identification requirements are the same as in the airport security regulations under Title 49 CFR Part 1542.211. Aircraft operators must conduct yearly audits and include measures to retrieve expired ID, report lost or stolen ID, and secure unissued identification media stock. Temporary IDs may also be issued.

Training
Aircraft operators must provide SIDA training for their personnel and any additional security training that may be required under the AOSSP. Any employee with security duties must also be provided with information regarding applicable security directives and information circulars.

Contingency and Incident Management Planning
Aircraft operators must have contingency plans in the event of increased security levels required by a higher level of the color-coded Department of Homeland Security (DHS) alert system. Regulations require aircraft operators to participate in airport-sponsored contingency plan exercises; however, as a practical matter, aircraft operators at major

airports often alternate their participation in airport-sponsored exercises with other aircraft operators at the airport. Aircraft operators are required to have programs in place to handle threats that may be received, such as a bomb threat or the location of a suspicious object onboard an aircraft, and to have procedures in place to handle hijackings.

Receipt of Air Cargo

Aircraft operators are prohibited from carrying cargo unless that cargo is received from a known shipper meeting the requirements of the AOSSP. Any cargo received from an unknown shipper must be separated from cargo received by a known shipper and screened before being loaded onto an aircraft.

Airline Security Issues

Just as the airport security department has its own unique challenges and characteristics, so do airline security departments. General differences between airport and airline security departments include personnel (airlines generally employ more people than an individual airport), financial (airports are generally nonprofit and airlines are usually for-profit), and customer base (airports serve a variety of stakeholders including the airlines, vendors, passengers, contractors, and others, and airlines generally serve just passengers and cargo customers). Airports also have a smaller span of control than airlines in relation to security concerns. Airlines may have stations and bases all over the world that require security management and response. An ASC is usually concerned with security at one airport and can respond to situations firsthand. Airline employees usually have a higher level of contact with passenger bags, and cargo security managers are faced with issues such as baggage and cargo theft and narcotics trafficking.

Airlines have the responsibility and challenges of maintaining a high level of security while carrying thousands of passengers in a highly competitive industry. Passengers may choose other airlines or other modes of transportation—vehicle, rail, bus—or may not travel at all and use teleconferencing. With thousands of passengers every day, airlines must be prepared to handle issues such as air rage and on-board medical emergencies. These require that flight and cabin crews be trained in handling such events.

Positive Passenger/Baggage Matching

The Aviation and Transportation Security Act of 2001 mandated that positive passenger/baggage matching (PPBM) be used for all domestic and international flights. The PPBM process is no longer required in the United States as all checked baggage is screened either through EDS, ETD, physical inspection, or canine inspection.

Catering and Vendor Services

Another avenue to hide prohibited items onboard an aircraft is through the catering stores or others who have access to the aircraft to maintain and restock service items. Catering provides an opportunity for an individual to conceal an IED within the catering stock for loading onto an aircraft.

Security of catering begins with the employment of food services personnel and management. Extensive background checks are needed to ensure individuals have no criminal history, are legal residents, and do not have past affiliations with terrorist or organized crime groups. For catering personnel with access to the SIDA or the aircraft, a CHRC must be completed before the person is allowed unescorted access to either.

Catering and vendor facility security is also important. Visitors to flight kitchens and catering facilities should be verified and authorized, and then issued visitor's identification and be accompanied by a staff member throughout the time they are in the facility. Additional facility security should be considered, such as closed-circuit television (CCTV) monitoring of the food preparation and packaging areas and an access control system that requires authorized personnel to present an access/ID badge to access food prep and storage areas. Outside and in loading areas, trucks and vans used for food delivery should be monitored for tampering and theft when they are not in use.

Food and utensils need to be prepared in a secure area within the flight kitchen, then packaged, sealed, and transported to the aircraft. Seals must not be tampered with and the integrity of the transport trucks must be maintained throughout delivery. Within the flight kitchen, security, supervisory, and management personnel must be alert for employees attempting to introduce unauthorized substances or items into the packaging. Other food preparation personnel need to be aware of the potential of other employees smuggling unauthorized items within the packaging areas and know how to report such behavior.

Once packaged at the flight kitchen, security personnel should monitor the loading of the catered goods onto trucks. Trucks and personnel responsible for moving catered goods should be inspected and searched before accepting the goods each time they enter SIDAs. Random inspections should also be conducted using ETD, EDS, or physical inspections.

A particular challenge in securing catering goods is that catering personnel, such as loaders and drivers, often carry items such as box cutters that are normally prohibited in SIDAs and other sterile areas. Catering personnel should be required to register prohibited items that are essential for the job and be responsible for keeping track of those items. Periodic audits can be conducted to ensure that catering personnel are only carrying those items they are authorized to carry. The aircraft operator should also monitor catering personnel while they stock catered goods onboard an aircraft. A flight attendant often does this.

Airline Security Staff Operations and Issues

Airlines are major corporations with thousands of employees; they must handle general security and loss prevention issues similar to those of any large corporation. The corporate security responsibilities of an airline include dealing with disgruntled employees, protection of airline facilities, employees with addictions, employees with mental illness challenges, baggage and cargo theft, pilfering of airline assets, abuse of flying privileges, assault, intimidation, cyber-security, and general loss prevention. With virtually any business there is a level of theft—by employees, by customers, or by others who interact with the company or its employees. The airline industry has additional challenges to loss prevention measures as aircraft travel across state and country boundaries. Airline stations are often far removed from their headquarter offices and operate more independently and without much oversight. Narcotics smuggling and baggage theft rings are important issues for airline security managers.

Millions of dollars of personal belongings are entrusted to airlines every day, and unfortunately the reputation airlines have of losing bags works in the favor of those attempting to steal from a bag. When a bag is checked, the owner of the bag gives up control for the duration of the process. With so many bags being out of sight of the owners, the opportunity for theft is significant. Organized bag theft rings routinely move around the country. Airline security personnel find themselves working frequently with the FBI and other law enforcement agencies in an attempt to stave off these bag theft rings. Many rings consist of individuals who can pass a CHRC and are hired by the airline to work as a baggage handler. These rings attempt to move on after a short time and after many goods have been stolen, and before the police and federal agents can catch up with them. Good internal security is fundamental to reducing the impact of baggage theft. An employee reporting system for reporting suspicious activities together with CCTV to monitor baggage sort and cargo storage areas, along with working with police and federal agencies, can reduce the loss an airline experiences from theft.

The advent of the Internet and online sales websites have provided another means to anonymously offload stolen goods, Unfortunately, bag theft involves airline and airport employees and TSA personnel, who then attempt to sell stolen property on eBay or other websites:

- In Chicago in 2005, an airline employee was charged with stealing electronic recording equipment off an aircraft and selling it on eBay.[7]
- More than 60 TSA screeners at 30 airports were arrested as of November 2004 on suspicion of bag theft. The TSA has settled 15,000 passenger claims and paid out more than $1.5 million in damages (ABC News, 2004).
  In Washington state, camera surveillance caught a TSA screener rummaging through a passenger's bag, then taking the passenger's prescription medication, literally opening the bottle and popping a pill in her mouth (ABC ActionNews, 2005).
- In New York City, four screeners were convicted on theft charges, including the theft of cameras, laptops, and cell phones (ABC ActionNews, 2005).
- In Los Angeles in 2012, more TSA personnel were arrested on drug and bribery charges and according to a *Los Angeles Times* article in 2011, 500 TSA personnel have been arrested on suspicion of theft (Forgione, 2011). With a screener staff of over 50,000, that's less than 0.01%, but many passengers expect a higher standard from a federalized security workforce.

[7]After the theft, friends chided the rightful owner that he was going to see his belongings on eBay. The owner logged on to eBay later that night, saw his own belongings up for bid, then notified the police.

Baggage theft of this nature is particularly difficult to investigate as several individuals including airline employees and TSA screeners handle a passenger's baggage after it is checked as luggage. Although passengers can lock their bags with a lock, it must be a TSA-approved lock and only TSA screeners are supposed to have a key. Breaking the lock or the zipper it is attached to can snap many locks, and they can also be picked. The construction of inline baggage systems, which greatly reduces the handling of baggage by human hands, is one method for reducing baggage theft.

Another form of baggage theft occurs at the baggage claim areas as bags are returned to their owners. Airports and airlines often do not require that each bag claimed be matched with the passenger's baggage claim tag as the level of theft in this area is relatively low compared to the cost of establishing secure areas around baggage claim and hiring more personnel to check the bags and claim tags. Often, a security presence is enough to deter theft at baggage claim carousels. Common sense works in our favor here as well. Although it is easy for a criminal to walk into a baggage claim area and grab a bag, unless the criminal has prior knowledge of a particular bag, the criminal has no idea whether what he or she has stolen has any real value. The thief would probably end up with clothing, rather than laptop computers, jewelry, or anything of higher value. For bag theft to be beneficial to the thief, he or she must have time to go through at least a few bags and see if there are valuables. It is hard to do this discreetly in the middle of a baggage claim area, where the owner is likely searching for his or her bag. The baggage claim area is a public venue and open to robbers who do know what they are looking for, so it is more likely that bag theft in the public areas of an airport will be of laptop bags and briefcases, which are more likely to contain electronics and other valuables.

The liquid bomb plot discovered in London in 2006, which caused passengers being required to place nearly every item, including laptops, MP3 players, DVD players, cameras, camcorders, and other high-value items that passengers usually take onboard, into checked baggage represented a huge economic risk to the airline industry. For a short period when passengers were forced to check such items, the financial impact to the airlines was significant with passengers reporting lost or stolen laptops and other electronic devices. This reinforces the need to assess the total impact of security decisions before implementing new prohibitions or solutions. With business passengers fearing the loss of their laptops, which often include proprietary or classified business materials, they sought other travel methods such as corporate or charter business travel.

Drug smuggling is another area of concern, particularly with international flights. Airline security personnel often work with the DEA to help stop narcotics trafficking operations on their airlines. With the borders of the United States being more carefully watched, particularly for general aviation air traffic not on a flight plan, commercial airlines have become a widely used method of transporting illegal narcotics. Colombian drug trafficking organizations continue to smuggle marijuana into New York and Florida using commercial airlines (National Drug Intelligence Center [NDIC], 2002). Heroin is smuggled by commercial or private aircraft from South America and Mexico into the United States. In a 2003 report before the House Judiciary Committee Subcommittee on Crime, Terrorism, and Homeland Security, Rogelio E. Guevara, chief of operations for the DEA, noted that couriers traveling on commercial airlines are the primary smugglers of Colombian heroin to the United States. Their primary entry points are Miami, FL, and New York City (Drug Enforcement Administration [DEA], 2003).

In some cases, even federal air marshals can be involved in the illicit yet highly profitable drug trade. In April 2006, two air marshals pled guilty to accepting $15,000 to smuggle cocaine onboard a flight out of Las Vegas, NV, using their positions to bypass airport screening (Associated Press, 2006). Again, this is an area where employee vigilance can prevent or deter this type of activity.

Aircraft Security Requirements

Aircraft manufacturers spent, and continue to spend, a great deal of money and time after 9/11 making modifications to cockpit doors to meet new federal standards. Previously, cockpit doors were no more difficult to get into than the doors to an airplane's lavatory. Airlines had resisted strengthening the doors because of added weight and potential egress safety concerns in case of an aircraft emergency.

Arguments were made that the cockpit door should remain unlocked to prevent a hijacker from threatening a flight attendant as leverage to get the door open, as occurred in TWA Flight 847. Conversely, a locked and reinforced cockpit door may have prevented David Burke from shooting the pilots on PSA Flight 1771 and downing that airplane.

ATSA 2001 mandated that cockpit doors be strengthened and that access to the cockpit be strictly controlled. By *strengthening*, the government means making it harder to physically knock in the door or to punch or kick through it. Some doors have been strengthened to the point where they can withstand a small explosion. Bulletproofing the cockpit doors was not mandated because of the additional weight.

With the doors strengthened, limiting the ability of individuals to force their way into the cockpit, there remain a few other methods for getting into the cockpit, including waiting until a crewmember opens the cockpit door, coercion, or bypassing the locking method. Aircraft operators must implement procedures to ensure the cockpit is not invaded during the critical times when the cockpit door is opened for a short time, such as for a crewmember to use the bathroom, to swap crews on longer flights, or for food to be served. Many airlines use a simple method of bringing flight attendants to the front of the aircraft and then positioning the food/drink cart sideways across the aisle; an individual attempting to jump over the food cart should be slowed down and this action also demonstrates intent to access the flight deck or do harm to crewmembers. During this process, a flight attendant will standby in the cockpit and open the door for the pilot when he or she needs to reenter, while allowing the other pilot to stay at the controls. For this method to be effective at preventing an intruder from entering the cockpit, the flight crew and flight attendants must still be able to overpower an intruder if an attempt is made.

The food/drink cart is an obstacle, but its effectiveness is limited in preventing someone from entering the cockpit.

At the request of some airlines, some aircraft manufacturers have installed reinforced curtains, which are pulled across the forward section of the cabin seating area and locked. Steel cabling similar to a child-safety gate and that can be pulled across the front bulkhead area also can be used to protect the cockpit when the door is open. A secondary flight deck barrier provides a higher level of protection than a beverage cart.

Coercion is commonly used in hijackings to access the cockpit. Intruders will take a flight attendant or passenger hostage and then demand access to the cockpit, often beating the hostage until access is granted. No amount of strengthening of the cockpit door will be able to stop an intruder who can successfully leverage the sympathy of the flight crew to gain access. Flight crewmembers are aware that they may have to make a decision whether to allow the hostage to be beaten and possibly killed, along with other hostages, versus allowing access to the cockpit where the aircraft may be used as a weapon on a ground target.

Another method of accessing the cockpit is through a popular airline benefit known as *jump-seating*. General employees (not pilots or mechanics or dispatchers) of a particular airline can only occupy the flight deck jump-seat on a case-by-case basis, with the approval of the airline, the PIC, and the FAA through the use of FAA form 8430 (Title 14 CFR Part 121.547). Pilots of that airline can occupy the flight deck jump-seat with PIC approval. Offline airline employees may not ride in the flight deck jump-seat, except for offline pilots, who must first be approved through the Cockpit Access Security System (CASS), and also have PIC approval.

The jump-seat pulls down from the front bulkhead at the back of the cockpit. It does not allow the occupant direct access to the flight controls. The 9/11 attacks and subsequent legislation placed new restrictions on the practice of jump-seating, but it does highlight the importance of conducting thorough background checks of airline employees and of monitoring the behaviors of such personnel. Flight crewmembers not comfortable with a particular individual in their cockpit have the right to not permit that individual to occupy a jump-seat.

Airline Employee Safety

Although terrorism is dominating the news, employee safety is often the concern of the day at airlines. An aircraft in flight is in a very remote location; often 35,000 feet or more above the ground, traveling more than 400 miles per hour, and usually without law enforcement personnel onboard. This environment has a high-risk potential. Drugs, alcohol, and mentally disturbed individuals traveling alone exacerbate the potential. Not only are airline security personnel concerned with the safety of passengers during flight, but also the safety of their flight crews both onboard and while staying in cities around the globe. In some countries, U.S. citizens are particularly high-value targets, and airline flight crews in uniform are easy to spot.

Air Rage

Air rage has been around nearly as long as commercial aviation. Instances of air rage increased after airline industry deregulation in 1978. With deregulation, thousands of travelers who before could not afford air travel suddenly filled the skies and brought on a wave of new air carriers with various business models. Some airlines allowed first come–first served seating or did not have first class or business class, which meant the vacationing family would sit next to the business executive. The 1980s also saw dozens of airlines quickly go out of business, often leaving passengers with useless tickets or, in extreme cases, stranded away from home.

Before deregulation, airline travelers had become accustomed to decorum about air travel—albeit more expensive, it did seem more "civilized." Deregulation brought new problems and new travelers as suddenly air travel was as available and nearly as affordable as other forms of transportation.

Air travel tends to be stressful. Travelers must often arrive hours early (particularly post-9/11) to clear ticket counter and security screening lines, then wait for an hour or more for their flight to board, and often pay high prices for airport food and drinks. Security regulations change frequently and are subject to numerous interpretations. An item deemed acceptable at a passenger's departure airport may be confiscated at another airport, increasing stress to the traveler. Overbooking, long waits at ticket counters, long waits and expensive food at airport restaurants, and being completely outside of their normal day-to-day environments all contribute to passenger stresses, and all occur before passengers board the plane.

Onboard, travelers are subjected to cramped quarters, limited eating and entertainment options, and the perception of a lack of control. The combination of these factors makes the cabin of an aircraft a high-stress environment. Passengers also bring their own stresses onboard—for example, the business traveler trying to get work done, the parent traveling with a child who must go to the bathroom while the fasten-seat-belt sign is on, the military officer who has just spent a few weeks of liberty at home for the birth of a child and is heading back to duty in a volatile country, or the infrequent traveler who jumps every time the plane hits turbulence. Sometimes, the combination of these types of factors creates *air rage*. There is no official definition for the term air rage, but a commonly accepted definition is that it involves angry or violent passengers onboard an aircraft. Factors contributing to air rage include:

- Alcohol and drugs (both legal and illegal)
- Being "stuck" on an aircraft
- Poor service and limited food and drink options
- Cramped quarters and an invasion of personal space
- Gender and sexual preferences
- Weight and size
- The feeling of losing control over your life
- Being forced to check carry-on baggage and a lack of space in the overhead bins

In many air rage cases, a flight attendant or another passenger is assaulted. Some have included fights among intoxicated passengers, child molestation of an unaccompanied minor, sexual harassment, vandalizing, and refusal to stop smoking or drinking. In addition to the threat of violence to passengers and crewmembers, air rage incidents can end up costing airlines thousands of dollars in lost time and fuel. In September 2006, a passenger on a Canadian Zoom flight had to be subdued during a trans-Atlantic flight and was allegedly drinking vodka that he had smuggled onboard. The passenger became unruly, and the pilots diverted to Scotland's Glasgow Airport where the passenger was taken into custody. The cost of the unplanned landing fees and additional fuel was $188,000. The man was found guilty and sentenced to 240 hours of community service.

Before 9/11, the FAA did not take air rage very seriously, nor did the agency learn about other vulnerabilities in the aviation system resulting from air rage (Morrison, 2001). In 2001, *USA Today* conducted a study on air rage and found the FAA rarely opened an investigation despite hundreds of onboard incidents responded to by airport police, nor did the FAA send inspectors to interview witnesses. When investigations did occur, the FAA either took no action or sent a warning letter to the offender (Morrison, 2001). Today, the TSA also does not track air rage incidents, noting that arrests are conducted in various jurisdictions and that it is difficult to track what offenses the individuals were arrested for and what they were ultimately charged with. However, Andrew Thomas (2001, 2003, 2006), author of the books *Air Rage* and *Aviation Insecurity*, cites numerous cases of air rage on his website (see *http://www.airrage.org*), as does Philip Baum, an internationally recognized security expert and the editor of *Aviation Security International* magazine (see *http://www. avsec.com*).

Air rage incidents in the post-9/11 world commonly involve alcohol, drugs, or individuals with diminished mental capacity. Passengers today are more aware that other passengers and flight crew are actively observing their behavior and that air marshals may be onboard. This has not stopped air rage incidents entirely, even against federal officers. In June 2004, a woman was arrested on charges of slapping a federal air marshal after he asked her to turn off her cell phone before departure. Another incident occurred on December 31, 2003, on a Northwest Airlines flight from Pittsburgh, PA, to Minneapolis, MN, when a woman caused enough of a disruption that she was reseated next to an air marshal. She then threatened to kill the air marshal, punched him, attempted to choke him, and after being handcuffed, kicked him. She pleaded guilty to assault on a federal officer and interfering with a flight crew and was sentenced to eight months in prison.

Shortly after the liquid prohibitions went into effect in August 2006, the aviation industry experienced a sharp increase in air rage incidents; within just a few days of each other, up to six aircraft were forced to land because of disruptive and disorderly passengers. On August 18, a Continental Airlines flight was diverted and an actor was arrested when he attempted to take his young son to the lavatory; a flight attendant alleged she was assaulted, an allegation that the actor denied. Also on August 18, an Air France flight was forced to land in San Francisco, CA, after a man assaulted crewmembers for refusing to serve him alcohol. On August 19, a British Airways flight was diverted after a couple got into a violent argument. On August 20, a United Airlines flight was diverted when a 20-year-old man assaulted crewmembers. Although this spate of air rage incidents could have been coincidental, it is noteworthy that they all occurred just shortly after new prohibitions went into effect restricting passengers from carrying liquids onto an aircraft and being forced to check most of their usual carry-on baggage.

Air rage incidents are a threat to the lives of those onboard, a threat to the safety of the flight, and a cost to the airline in time and delay. Preceding 9/11, a pilot would often be called from the cockpit to handle air rage incidents. Now, the need for the flight crew to stay in the cockpit and for the cockpit to be protected from intruders makes this an ineffective and dangerous option.

Many international airlines employ their own security personnel to help protect the aircraft, the crewmembers, and other passengers. The United States relies on flight attendants to handle air rage incidents and perhaps a few federal air marshals who may be available. Federal air marshals and other federal law enforcement personnel may be reluctant to compromise their cover unless the safety of the flight is threatened, as this could be a tactic used by hijackers to expose air marshals. Flight attendants cannot rely on air marshals being present and willing to assist and must be trained to handle air rage incidents. Part of this training includes how to differentiate between a person who is just upset and one who represents a threat to the safety of the flight. Air rage is more common than terrorism and strategies for dealing with an intoxicated passenger are necessarily different than the strategies for dealing with a hijacker. Flight crews and aviation security practitioners must understand these differences.

Protecting crewmembers against air rage is often a combination of several elements including training in verbal deescalation of an incident, training in physical self-defense, enlisting the help of other passengers, communication with the flight deck, and techniques to safely restrain a hostile individual. Many airlines have contracted with private industry to provide verbal and physical self-defense instruction. Hands-on self-defense training for flight crews also serves a protective purpose for terrorist incidents as demonstrated in the intercession by crewmembers and passengers when Richard Reid tried blowing up American Airlines Flight 66 in December 2001.

Flight Crew Protection

The passengers on United Airlines Flight 93 fought back. When they learned the hijackers real intent, they organized a resistance and died fighting for their lives and the lives of possibly thousands on the ground. Although all onboard died, the hijackers were not able to fly the airplane into its primary target—the White House or the U.S. Capitol. Before 9/11, the common strategy for handling a hijacked aircraft called for crewmembers to cooperate with hijackers. The goal was to get the aircraft on the ground where negotiations could begin. The 9/11 hijackers counted on this cooperation to be able to maintain control of the passengers and flight crew. The common strategy has changed from passive compliance to active resistance. Neither passengers nor crewmembers can safely assume that the intent of a hijacker is to do anything other than use the airplane as a weapon. In response to this new paradigm, the U.S. Congress passed legislation authorizing certain airline pilots to carry firearms and be deputized as federal flight deck officers with the authority to use deadly force to defend the cockpit. Additionally, ATSA 2001 called for mandatory self-defense training for pilots and flight attendants.

Airline employees travel for a living, going to different cities around the nation and the world. Even crews flying regularly to familiar cities often take advantage of their flight benefits to take family and friends to vacation destinations. Travel security is a concern for the individual crewmember and for the airline employing that crewmember.

In-flight security and response by crewmembers must be addressed by considering the perceived threat and applying the appropriate response strategy. A passenger who is upset and argumentative should receive a verbal response in an attempt to deescalate the situation. If the passenger becomes physically violent, then crewmembers need to be able to respond to protect themselves and the safety of the other passengers. If the passenger is threatening the lives of the crewmembers or passengers or the passenger is a terrorist threatening the safety of the flight, crewmembers must have the resources and knowledge to defend the aircraft and themselves. In all cases, from verbal response to deadly force to protect the aircraft, crewmembers should be trained in each area and know when to escalate from one stage to the next.

Protecting Airline Personnel While Traveling

"No matter how many times I tell flight crews how dangerous some of the cities they travel to are, and that they should stay near the hotel, they still wander off to see the sights."

—Anonymous airline security coordinator

Pilots and flight attendants generally enjoy travel. Either as part of the job or their flight benefits, flight crews travel to thousands of destinations every year. The very nature of frequent travel exposes flight crews to more risks than most other individuals. Flight crews attract criminals and terrorists, as their uniforms or identification can be exploited to access an airport or an aircraft. Unfortunately, no matter how dangerous a particular city is, there will be flight crewmembers ready to risk their own personal safety for relaxation or recreation. Although it would seem easy to ban flight crews from leaving the safe environs of their hotel, this is not practical. Awareness and self-defense training along with other safe traveler practices are perhaps better solutions.

It is in the interest of both employees and employers to be aware of traveler safety practices. General travel security requires the traveler to have a high level of awareness of location and the nature of his or her surroundings. Airline personnel should be acutely aware of their responsibility to protect their travel documents, uniforms, and equipment. The theft of these items could facilitate criminal or terrorist activity.

Hotel room and street theft constitute nearly 70% of all travel theft. Theft from vehicles and at airports accounts for almost 30% of total thefts (Foster, 2004). Tourists in particular are targets, and there are usually visual indicators that a traveler is a tourist. Travelers should minimize these indicators, such as reading a map in public, openly carrying travel guides, dressing differently from local styles, opening a hotel room door to strangers, and not taking extra precautions when using the fitness or pool facilities.

Once at a hotel, flight crews should change from airline attire and dress to blend in to the local culture. Keep valuables locked in room safes or at the hotel's safety deposit boxes. Federal flight deck officers must be extra cautious when traveling with firearms. It is best to travel with small amounts of cash on hand and use a security wallet. International travelers should always carry their passports. Hotel staff are familiar with the surroundings, the areas to frequent, and areas to avoid. Therefore, travelers should seek directions to restaurants and other sites from hotel staff and get clear directions written down to the destination or from police officers or business owners (Foster, 2004).

Airport crime is particularly serious, as terrorists may be looking to access a traveler's luggage to place an explosive device. Travelers should never leave bags unattended nor accept or carry anything for another person (McAlpin, 2003). If bags have been out of a crewmember's control, such as in a hotel shuttle van trunk area, it is wise to search the luggage after you have reclaimed it. If there is a suspicion that the luggage has been tampered with, immediately notify airport police, and do not touch the bag until it has been cleared by proper authorities.

The U.S. Department of State lists information about travel to foreign countries and travel warnings on its website (see *http://www.state.gov*). The website provides tips for safe international travel, including leaving expensive or flashy jewelry at home, and references the Department of State's consular information sheets on each country. U.S. citizens traveling to foreign countries can register with the nearest U.S. embassy or consulate through the State Department's travel registration website (see *https://travelregistration.state.gov*), which will make a U.S. traveler's presence known if it is necessary to contact him or her in an emergency.

Travelers are encouraged to "stay on Bourbon Street." Bourbon Street is a metaphor for the most popular street in a U.S. city or town. Every major city in the United States has its own version of Bourbon Street. Generally, the worst crime that can happen to a traveler on Bourbon Street in New Orleans (in the United States) is to be the victim of a pick-pocket. The more serious crimes—armed robbery, murder, rape, assault—are more likely to occur off the main street where there are fewer police patrols and fewer witnesses. In Las Vegas, NV, the most popular location is the Las Vegas Strip. In downtown Denver, CO, it is the area known as LoDo, or the 16th Street Mall. Every traveler should find their destination's version of Bourbon Street and stay on it. Hotel concierges can tell travelers where the main street is in a city. Going into parts of the city away from tourist attractions without local escorts invites trouble. When traveling outside a hotel in an unfamiliar area, it is best to travel in groups because groups are less likely to become targets for criminals.

Airline personnel should always carry a cell phone and ensure the battery is charged before going out. If venturing out from a hotel alone or with just one other person, advise the hotel staff of the destination and expected return time. If confronted with a violent situation, the general idea should not be to win the fight but to safely escape the situation and report the incident to law enforcement.

Airlines travel to both safe and dangerous countries, including countries with very dynamic environments where gunfire and explosions are the norm, not the exception. Gunfire and explosions from hand grenades and IEDs are characteristic of terrorist acts. People can increase their chances of surviving an attack by following some basic principles:

- The best response to gunfire is usually to drop to the ground. Shooters look for targets in their line of sight, so staying below this line may improve chances for survival (Aviv, 2003). Furthermore, when crawling away from danger, use a low belly crawl and avoid placing one's spine or head too far above the ground (as in a traditional four-point crawl).
- When grenades and other IEDs explode, the blast travels upward and outward in a conical shape. Dropping to the ground is the recommended protection, unless protective cover is immediately available (Aviv, 2003). Following any blast, if it is safe to move, leave the area as soon as possible as there may be secondary devices ready to explode.
- Avoid eye contact. Women should consider wearing sunglasses as it projects confidence, travel in groups of two or more, and always look like you know where you are going (Aviv, 2003).

Self-Defense for Crewmembers and Passengers

Throughout the history of aviation, there have been attacks on flight crew, flight attendants, and other passengers by passengers. The terrorist attacks on 9/11 brought to light a long ignored subject—aircrew self-defense training. ATSA 2001 (and ICAO Annex 6 addressing the operation of aircraft) outlined eight elements to be included in flight crew security training:

1. Determination of the seriousness of any occurrence
2. Crew communication and coordination
3. Appropriate responses to defend oneself
4. Use of protective devices
5. Psychology of terrorists to cope with hijacker behavior and passenger response
6. Live situational exercises regarding various threat conditions
7. Flight deck procedures or aircraft maneuvers to defend the aircraft
8. Any other subject matter deemed appropriate by the administrator

ICAO Annex 6 includes aircraft search procedures and least-risk bomb locations as an additional measure.

Self-defense for flight crewmembers involves three basic levels. Upset and argumentative passengers may simply require a nonthreatening verbal response. As an individual gets progressively upset and initiates (or indicates) physical aggression, notification to law enforcement officers either on the ground or in flight and the use of restraints may be necessary. Passengers under the influence of drugs or alcohol or of diminished mental capacity may also require more physical control measures depending on the level of perceived threat. Extremely violent passengers or terrorists require the highest level of response, which may include the use of deadly force to protect other passengers, the flight crew, and the aircraft. Understanding which level of response is necessary in a given situation is essential to successfully managing in-flight incidents.

Another consideration is the psychology of confrontation from one human to another. Extensive studies have been conducted expanding the response options from simply "fight or flight" to also include posture, submit, or detach. An understanding of how and what to expect from others and oneself during a high-stress, possibly deadly encounter is essential to successfully resolving minor in-flight incidents or actual terrorist attacks.

Flight attendants and flight crews are in a difficult position when handling a drunk or disorderly passenger because that person is a paying customer and a citizen with rights. Humans make mistakes—they get angry and sometimes do or say stupid things that do not necessarily warrant an extreme response. Crewmembers must make judgment calls as to whether someone is merely upset and can be reasoned with or whether a warning or even physical restraints are in order. In some cases, the safety of the entire flight is in jeopardy and the plane should land immediately. After a 2001 air rage incident on a Lufthansa flight, renowned aviation expert and editor of *Aviation Security International* magazine Philip Baum commented that "crewmembers need passenger profiling capabilities in order to detect and interpret warning signs early, before incidents escalate to the point where we compromise both personal and flight safety" (Baum, 2001, p. 27).

The three levels of response that should be addressed in self-defense training are verbal, physical, and extreme. In verbal self-defense, a person is trained in methods and procedures of talking to upset individuals as a way to calm them and move toward a peaceful resolution. This has the benefit of turning a potentially violent situation into one where no one gets hurt and may even build good customer relations. Someone who stays in control of his or her emotions is in control of an event and can better control the outcome. In their book *Verbal Judo*, George Thompson and Jerry Jenkins addressed managing upset people and deescalating tense situations in a way that maintains respect for all parties. Thompson's key principle is to be sincerely empathetic with the upset party; he noted that there are many ways to calm someone, but the underlying principle is that "the one hoping to do the calming must project empathy" (Jenkins and Thompson, 2004, p. 30).

Jenkins and Thompson stated, "Empathy is the quality of standing in another's shoes and understanding where he's coming from: empathy absorbs tension" (2004, p. 64). Flight crew personnel should respect the upset individual by not embarrassing the individual in front of others, insulting the person, or making him or her feel powerless. By allowing people to say what they want as long as they continue to do what they are asked, crewmembers show respect while ensuring that a situation stays under control. Thompson and Jenkins further discussed the difference between "nice" people, "difficult" people, and "wimps."

Nice people generally do what they are asked. Difficult people do not usually do what they are asked the first time but will usually comply if shown the positive benefit of compliance. Jenkins and Thompson (2004) described "wimps" as difficult people disguised as nice people. They may comply in front of the crewmember and then defy the orders once the crewmember has turned away. This type of behavior must be exposed and dealt with quickly.

The physical level of self-defense deals more with a drunk or unruly passenger. This level involves contact between a crewmember and a passenger (or sometimes enlisting the assistance of other passengers). Physical contact does not warrant deadly force. A variety of holds and restraints can be used to force an individual to comply with directions, move from a seat or location, or defeat an attack. Police officers are well trained in these techniques, but crewmembers and passengers should not attempt them unless they have received formal training. Many community colleges offer programs, often taught by police officers, in basic grappling and hold techniques, and numerous training courses can be found throughout the country via the Internet.

The third level of self-defense is when life or the safety of the aircraft is in jeopardy. This level should only be used when there are no other options. Unfortunately, simple arguments can move quickly through a level 2 physical event into a level 3 life-threatening situation within seconds (Jenkins and Thompson, 2004).

Violent Passenger or Terrorist

Once a situation has escalated to a level requiring a violent response, then the reasonable use of force, lethal force, and passenger restraint must be considered. Passenger restraints present a serious issue considering the legalities involved in false imprisonment and the possibility of aircraft evacuation in an emergency. It is recommended that "Effective passenger restraint skills require restraint tools that are safe and effective, and crewmembers that are trained to apply them correctly" (Baum, 2001, p. 31). Before a passenger can be restrained, that person must be controlled. This often involves an effective hold such as an arm bar or wristlock. These methods are traditionally the domain of the law enforcement officer, but their techniques have been integrated into the training programs for airline personnel.

A "hold" of any sort requires a level of consent on the part of the person being held or overwhelming force applied by the defenders. Even then, if a hold is not properly applied, it can easily turn to a disadvantage for the defender or perhaps result in an application of unnecessary force on the attacker.

One popular form of self-defense in the United States is Krav Maga (KM). Developed by the Israeli Defense Force, KM is a basic form of fighting and self-defense that uses natural instincts to its advantage. KM is useful for women and men of any size. Its techniques are not difficult to master and can also be effective against attackers with knives, sticks, and guns. One basic premise of KM is the use of distractions. Whenever there is a confrontation between individuals, there can be a great deal of confusion. KM teaches some basic verbal defense skills that the defender can use to grab the initiative by temporarily distracting the attacker (terrorist or otherwise).

Strategies used in security-related situations are warranted by circumstances. A terrorist with a gun in hand and screaming at passengers on an airplane may require an immediate physical response, but some training in how to verbally distract an attacker for the purpose of taking over the situation and initiating a counterattack may also be useful. However, a terrorist shooting people in an airport terminal requires an immediate physical response to prevent the further loss of life.

In *Aircrew Security*, Waltrip and Williams (2004) summarized some basic safety practices to apply when dealing with upset passengers showing signs of becoming violent. Crewmembers should maintain distance from the passenger, keep a balanced stance, keep hands above the waist in a nonthreatening manner, choose a retreat path, watch the passenger's hands, not turn away from the passenger if possible, and recognize when the situation is deteriorating.

An airline's use-of-force policy should be clearly articulated and integrated into the aircrew security training program. A common law enforcement escalation-of-force policy progresses from presence to voice, hands, chemical spray, stun gun or Taser, impact weapon, and, finally, lethal force. Considering that chemical spray, Tasers, and impact weapons are prohibited items, an airline use-of-force policy may include the use of presence, voice, hands (or asking for help from others), makeshift weapon if the person is becoming violent, and, finally, force (Waltrip and Williams, 2004). Although this scale is useful for training, it is understood that each incident is different and crewmembers may have to skip to higher levels of response.

Another consideration is the reasonable application of force. Crewmembers and passengers must keep in mind that the vast majority of airline security incidents are not terrorist threats but the behavior of upset or intoxicated passengers. Section 144 of ATSA 2001 states:

> An individual shall not be liable for damages in any action brought in a Federal or State court arising out of the acts of the individual in attempting to thwart an act of criminal violence or piracy on an aircraft if that individual reasonably believed that such an act of criminal violence or piracy was occurring or was about to occur.

As Waltrip and Williams (2004) pointed out, the operative term in the above quote is *reasonable*. A passenger or flight crewmember can use a variety of items to fight back against a hijacking attempt. Everyday travel items such as shoes, pens, briefcases, laptops, key rings, bottled water, unopened soft drink cans, hot coffee, carafes, or even a large wallet or purse can be used as a weapon to defend or fight back. Commercial airplanes also have items onboard that can be used as weapons, such as seat cushions and blankets (which can be used as buffers against knives), fire extinguishers, and first-aid kits (Holt,

2002). Fire extinguishers can also provide a temporary smoke screen. Megaphones and life jackets are other useful items for either hitting or fending off a knife attack.

Pharmaceuticals can also be used to keep a dangerous individual subdued until landing, whether they are provided by passengers with prescriptions or are onboard for emergency use. Commercial-service aircraft have first-aid kits that contain basic items (bandages, etc.) and some pharmaceuticals such as painkillers and antinausea drugs, which could be used in an emergency. When Richard Reid attempted to destroy an American Airlines flight in 2001, three doctors onboard injected him with Valium and Narcan, both sedatives, and an antihistamine in an attempt to subdue him (CNN, 2002).[8]

Passengers and flight crew should observe other passengers for signs of suspicious behaviors such as extreme nervousness, silently communicating, or sending signals to other passengers. Passengers in the cabin should also keep their seat belts on at all times during incidents as the pilot may elect to maneuver the aircraft so as to throw hijackers off balance. Although this is not recommended in most airline flight manuals, it is a strategy that pilots have used in the past (Barrett, 2003).

If a hijacking begins, passengers should wait to see if an air marshal or other law enforcement officer takes action. Passengers jumping up in the middle of an armed response may be misidentified as a hijacker and shot. If it's apparent that a law enforcement officer is either not onboard or not going to immediately respond, anyone taking action should understand that the most confusing part of a hijack is at the very beginning. This can be used to a defender's advantage.

In the book *On Killing*, author David Grossman (1996) described posturing as an effective method of controlling one's opponent. It is likely that the hijackers will be yelling and moving quickly at the beginning of the hijacking as a method of posturing, instilling fear into passengers. Defenders should also yell when attacking hijackers as a way to posture themselves and instill fear into the hijackers.

Federal Flight Deck Officer Program

A response to an individual attempting to gain access to the cockpit is allowing the flight crew to use deadly force. Originally, some airlines purchased Tasers to give to flight crew, but in the spirit of defending the flight deck, the Air Line Pilots Association (ALPA) and other pilot groups lobbied for the right to carry a gun in the cockpit (Sloan, 2002). The Arming Pilots Against Terrorism Act was incorporated into the Homeland Security Act of 2002 and passed by the U.S. Congress, thus creating the Federal Flight Deck Officer Program. FFDOs are passenger and cargo airline pilot volunteers; they are provided a firearm and a week-long training session taught by the TSA. FFDOs receive training in the use of deadly force, the care and handling of the weapon, and some basic self-defense tactics. The jurisdiction of the FFDOs is limited strictly to the cockpit. FFDOs

[8]Reid's attorneys later argued, unsuccessfully, that the drugs prevented Reid from understanding his Miranda rights.

are not authorized to use their firearms in the cabin, nor are they allowed to use them on airport property. Airline and air cargo operators with security programs under Part 1544 are eligible for the FFDO program.[9]

Debate regarding the FFDO Program also included concern over damage to aircraft from onboard gunfire, however, the risk seems to be quite minimal as described here:

> *A Boeing Commercial Airplanes representative on May 2 told a Congressional subcommittee that the risk of losing an airplane from a stray bullet is "very slight." Speaking at a U.S. House of Representatives Aviation Subcommittee hearing on whether flight crews should be allowed to carry guns aboard commercial transports, Ron Hinderberger, Boeing Commercial Airplanes director of Aviation Safety, said the commercial service history of Boeing "contains cases of gunfire onboard in-service airplanes, all of which landed safely." Boeing is not taking a position on the issue of whether flight crews should be armed. In late May, the Bush administration announced a ban on firearms in cockpits; some members of Congress vowed to fight to overturn the decision. (Boeing, 2002)*

Some pilots have criticized the program for the TSA's tight restrictions on who is selected to be an FFDO. Bob Lambert, president of an airline pilot's lobby in 2002, estimated that 40,000 pilots would volunteer for the program if the TSA removed the psychological screening exam requirement (Goo, 2003). However, Duane Woerth, president of the ALPA, agreed with the psychological exam because of different types of emotional responses between flying an aircraft and firing a weapon (Goo, 2003). Today, the program mostly suffers from a lack of funding and support that will allow its ranks to increase and support the existing cadre. The program is cost-justified as pilots are volunteers, draw no salary, and pay out-of-pocket expenses to be in compliance with requalification and recurrent training aspects.

FFDOs must carry credentials while exercising FFDO duties and transporting firearms en route to or from assigned aircraft.

Foreign Aircraft Operations

Foreign air carrier operations introduce unique challenges to the aviation security system. Foreign operators must adhere to U.S. aviation security regulations. For airport operators, the foreign air carrier arrival area represents the U.S. border. Therefore, customs, immigration, and agriculture protection agents become a part of the airport security program, and the airport operator must provide these agencies with the proper facilities to secure the U.S. border. The regulations for foreign aircraft operations mirror those for domestic aircraft operators and are designed to ensure that security procedures applied to inbound

---

[9]Since 1960, any pilot carrying U.S. mail had the authority to carry a firearm, a policy that withstood until 1987 (Lott, 2004).

U.S. flights meet the same standards as domestic operations. In certain cases, where aircraft are arriving from a location where screening or other security measures were not conducted to U.S. standards, passengers and their baggage may have to be rescreened before accessing U.S. soil. Airport operators may have to build additional screening areas to handle reverse screening.

Title 49 CFR Part 1546 Foreign Air Carrier Security

Regulations for foreign air carriers are similar to those for domestic aircraft operators, but there are fewer mandated in Part 1546 Foreign Air Carrier Security regulations.[10] These do not address security coordinators, the carriage of accessible weapons, the carriage of prisoners, the transportation of federal air marshals, the security of aircraft and aircraft facilities such as administrative offices and hangars, exclusive area agreements, conducting CHRCs, contingency plans, or security directives and information circulars.

One major difference in foreign air carrier security versus domestic is that foreign air carriers may not access SSI materials unless specifically authorized by the TSA. Foreign air carriers do not follow, nor are they allowed access to, the AOSSP, which is only for domestic air carriers. Foreign air carriers must adhere to the Model Security Program (MSP), which outlines the relevant required security actions and programs.

Foreign air carriers are allowed to develop their own practices for the carriage of prisoners, requirements for carrying weapons in checked baggage, and protection of aircraft and facilities. Because there is no requirement for foreign air carriers to have a personnel identification system, foreign air carriers operating within the SIDA of a U.S. commercial-service airport must undergo an airport-conducted CHRC and be issued approved airport access/ID media. A known discrepancy with conducting the CHRC on a foreign air carrier employee is that if the foreign employee has committed felonies in another country, but none in the United States, it is unlikely that the CHRC will prevent issuance of an airport access/ID badge. Individuals applying for an access/ID badge do have to supply other identification information such as Social Security numbers, passport identification numbers, or resident alien identification numbers. This data may be checked by the TSA and other U.S. federal law enforcement agencies, which may lead to the discovery of criminal activity outside the United States.

Noteworthy is the exclusion of exclusive area agreements for foreign air carriers. Foreign air carriers cannot take responsibility for security in certain areas of an airfield or sections of an ASP. As with domestic aircraft operator security programs, foreign air carriers operating within the United States must allow the TSA to enter secure areas to make inspections, conduct tests, and copy the aircraft operator to establish compliance with Part 1546.

---

[10]Part 1546 also includes screener qualifications when the foreign air carrier conducts screening and is materially the same as Title 49 CFR Part 1544 Subpart E.

Security programs and security requirements are the same for foreign aircraft operators as for domestic aircraft operators. Foreign air carrier security programs must be approved by the TSA and ensure that the level of security is equal to or greater than that for U.S. aircraft operators serving the same airport. The program must be drafted in English unless otherwise specified by the TSA. The procedure for amending the security program is the same for foreign operators as for domestic operators.

The requirements for screening individuals and property are similar to domestic screening requirements. Title 49 Part 1546.203 grants wide latitude to foreign air carriers in determining the requirements for the transportation of weapons in checked baggage, provided foreign air carriers ensure that checked baggage has been properly screened and refuse to transport any item not subjected to screening. The regulations require a passenger to notify the foreign air carrier that he or she is checking a firearm but allows the foreign air carrier to determine how to prepare the weapon for transport. Part 1546.203 only requires that the weapon be transported in an area of the plane that is inaccessible to passengers. The requirements such as loaded versus unloaded, hard-sided container, locked, and so forth required of domestic air carriers are not required of foreign carriers by the transportation security regulations. Discretion is left to the foreign air carrier.

With respect to air cargo, foreign aircraft operator regulations do not refer to an all-cargo or 12-5 cargo program. However, regulations still require the foreign air carrier to ensure that any cargo placed onboard a commercial aircraft carrying passengers has been subjected to the required screening and that cargo received in the United States has only been accepted from a known shipper. Additionally, Part 1546.213 requires that those with access to cargo undergo a security threat assessment (STA). Foreign air carriers can decide who is a known shipper based on the requirements that each determine the shipper's validity and integrity as provided in the foreign air carrier's security program, separate known shipper cargo from unknown shipper cargo, and ensure that cargo is screened or inspected.

Threat and Threat Response

Foreign air carrier regulations do not have a section related to contingency plans as do U.S. domestic aircraft operators, and foreign carriers are not required to increase their security posture when the DHS color-coded alert system is raised. All foreign air carriers are required to have their security programs approved by the TSA. The operation of the foreign air carrier is at the pleasure of the FAA and the TSA. The TSA can therefore require that foreign air carriers take certain increased security measures as deemed appropriate. Foreign air carriers are required to report to the TSA any threat to an aircraft that is either operating in the United States or is intending to fly to the United States. Upon receipt of a threat, the foreign air carrier must notify the PIC and conduct a security inspection at the earliest opportunity. Any foreign air carrier flight within the United States or heading to the United States that has received a bomb threat is not allowed to take off unless the aircraft has been searched. Any foreign aircraft in flight that receives a threat must land as soon as possible, notify the appropriate authorities, and search the aircraft before continuing the flight.

Additional Considerations of Foreign Air Carrier Operations

Foreign air carriers have additional requirements such as providing passenger manifests to the U.S. government. Foreign air carriers or domestic air carriers flying into, out of, and over the United States must provide passenger manifest lists to the TSA within 15 minutes after the departure of the flight. The TSA inputs the data into the Advance Passenger Information System (APIS). Carriers must also provide a master crew list (MCL) and crew manifest data to the TSA on the Customs and Border Protection (CBP) website. The TSA reviews the data against the no-fly lists and known terrorists watch lists. In some cases, flights inbound to the United States have been diverted and a passenger taken off the plane by law enforcement.

Border Protection

Airport operators must design and operate their federal inspection service (FIS) areas (customs, immigration, Department of Agriculture) to ensure that passengers entering the United States via air are separated from domestic flight operations until FIS officials can conduct the proper inspections of the passengers and their baggage. At the G8 conference in 2005, the United States announced the creation of a Smart Border of the Future, which is based on the following rationale:

> *The border of the future must integrate actions abroad to screen goods and people prior to their arrival in sovereign U.S. territory, and inspections at the border and measures within the United States to ensure compliance with entry and import permits. . . . Agreements with our neighbors, major trading partners, and private industry will allow extensive pre-screening of low-risk traffic, thereby allowing limited assets to focus attention on high-risk traffic. The use of advanced technology to track the movement of cargo and the entry and exit of individuals is essential to the task of managing the movement of hundreds of millions of individuals, conveyances, and vehicles. (White House, 2002)*

To this end, the design of FIS facilities is largely subject to the discretion and desires of the CPB agency. Any airport receiving flights from outside of the United States must have an FIS area:

> *Passenger processing facilities are provided by the airport at no cost to the government and inspection services are normally furnished by the government at no cost to the airport. By law, airports are required, at airport expense, to provide adequate passenger and baggage processing space, counters, hold rooms, office space,*

equipment, utilities, vehicle parking, and other facility-related support required for the FIS agencies to function properly. (TSA, 2006)

Through Section 233(b) of the Immigration and Nationality Act (INA), the aircraft operator is responsible for the costs of the FIS facilities. This is usually through the collection of fees by the airport (as the owner/operator of the facility). The FIS facility represents the border of the United States. Although airport borders are well within the geographic borders of the United States and are often not as clearly delineated as a sea or land border, they are borders nonetheless, and access must conform to U.S. laws and treaties. Design of the FIS facility is guided by the CBP Airport Technical Design Standards (also called the Air Technical Design Standards and available from the CBP division of the DHS). This document contains CBP requirements for the design of new or remodeled airport terminal building facilities to accommodate CBP inspection of aliens and their luggage arriving into the United States at U.S. ports of entry and at preclearance, pre-inspection, and predeparture sites outside the United States (Figure 8.1). It describes the physical characteristics of an FIS area, including passenger and baggage flow, terminal building space use, guidance for processing arriving international passengers and baggage, and administrative areas such as offices, inspection booths, holding cells, security requirements, X-ray systems, access control, and other equipment to support the monitoring, control, and operation of the facility.

The primary mission of the CBP is to prevent terrorists and weapons from entering the United States. Efficient processing of legitimate visitors to the United States is also a priority, but not at the expense of the first priority. The FIS facility design must consider both priorities. Passengers arriving from outside the United States into an FIS facility first go through immigration, a series of counters where immigration personnel check the validity of the passengers' travel documentation, review their reason for entering the United States, and then determine the visitors' expected duration of stay. The design of the FIS facility must prevent passengers from bypassing this process.

After immigration processing is complete, passengers enter a baggage claim area to pick up checked baggage. Customs agents watch for individuals attempting to bring contraband, which could include illegal weapons or drugs, into the United States. Agents decide who to check based on prior intelligence, profiling, and random checks. Department of Agriculture agents are often on hand to ensure that certain foods and other items are not brought into the country. On some occasions, agents of the U.S. Fish and Wildlife Service, the Public Health Service, and other federal agencies may be present and conducting inspections for a variety of reasons, including the illegal trafficking of protected fish, wildlife, and plants, and to prevent the introduction, transmission, and spread of communicable diseases. At larger airports, Immigration and Customs Enforcement (ICE) may be present and conducting ongoing investigations of criminal activity through the airport's borders.

The size of an FIS facility is determined by the number of passengers processed at the peak hour of operation and the number of aircraft arriving during a set period (TSA, 2006).

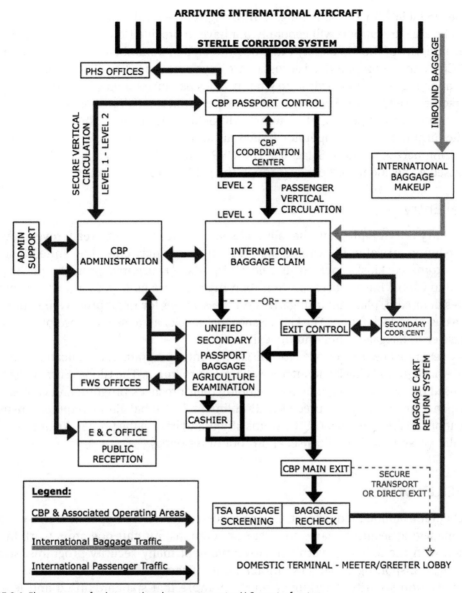

FIGURE 8.1 Flow process for international passengers at a U.S. port of entry.

In some instances, airport operators may have to incorporate additional TSA checkpoints into the overall design of the FIS facility to screen passengers and their carry-on baggage if they are arriving from an airport that does not meet the U.S. standard for screening. Whether or not screening facilities are used depends on where passengers are able to go after leaving the FIS facility. If the facility design is such that passengers enter the sterile area of the airport after clearing the FIS facility, then the airport must incorporate space into the FIS facility for TSA screening. If the design of the facility forces all passengers into the public area, the airport will generally not have to provide for a screening checkpoint.

The integrity of the FIS facility must be protected in a number of ways, including the use of CCTV and a computerized access control system. The CBP will assist in access control by applying strict standards as to who (besides passengers) may be in the FIS facility, often requiring airport workers, such as airline personnel and airport operations, security, and law enforcement personnel, to wear additional identification. The CBP may require that the airport or aircraft operator issued an access/ID badge bear a specific symbol authorizing access to the FIS facility. The CBP usually issues this icon after someone has passed a specific record check through the CBP computer systems.

Global Entry

Global Entry is a CBP program that allows expedited clearance for preapproved, low-risk travelers upon arrival in the United States. Participants may enter the United States by using automated kiosks located at select airports. At airports, program participants proceed to Global Entry kiosks, present their machine-readable passport or U.S. permanent resident card, place their fingertips on the scanner for fingerprint verification, and make a customs declaration. The kiosk issues the traveler a transaction receipt and directs the traveler to baggage claim and the exit.

Travelers must be preapproved for the Global Entry program. All applicants undergo a rigorous background check and interview before enrollment. The Global Entry program is being used by the TSA as a guide in developing their precheck program and Global Entry participants are eligible for precheck. Enrollment in Global Entry requires an online application, an interview at a U.S. Customs office, and having the proper documents, such as a valid passport, driver's license, and proof of residency.

Conclusion

This chapter examined requirements and processes that aircraft operators must address when managing security related to aircraft operations. Aircraft operators work in conjunction with the federal government to implement many security programs, such as screening, transporting armed law enforcement officers, or conducting CHRCs.

The aviation security practitioner must be aware that standards used and levels of performance demonstrated by aircraft operators can vary when managing security processes and systems. In response to these concerns, the U.S. government routinely publishes security guidelines and regulations such as the Aircraft Operators Standard Security Program (AOSSP) and Title 49 CFR Part 1544. In addition to delineating responsibilities for managing a safety program, the AOSSP also provides standards or a *common strategy* for airlines to respond to many forms of security threats. CFR Part 1544 specifies the security requirements an aircraft operator must follow to provide commercial operations in the United States. Supplementing government directives, airline security officers conduct their own inspections of airports to verify that the airports serviced will also meet minimum security standards.

All airlines operating within the United States must have an approved security program describing their facilities, equipment, and related security responsibilities. Fundamental to these programs is the responsibility of the aircraft operator to prevent or deter the carriage of any weapon, explosive, or incendiary device on a person or carry-on baggage before the person or bag boards an aircraft or enters a sterile area. When a U.S. air carrier is operating in another nation or at certain small air carrier stations within the United States, the aircraft operator must still ensure that all applicable measures are taken to properly screen individuals and carry-on baggage. All-cargo aircraft operators who carry passengers must ensure that only authorized passengers and their carry-on baggage are allowed to board and that they do not bring any weapons, explosives, incendiaries, and other destructive devices, items, or substances onboard.

This chapter also introduced different types of approved security programs, such as a full security program, an all-cargo program, the 12-5 program, or a private charter program. All airlines operating under any of these programs must ensure that cargo is properly screened and inspected to prevent or deter the carriage of unauthorized persons, unauthorized explosives, incendiaries, and other destructive substances or items in cargo onboard an aircraft.

Aviation security practitioners must understand the organizational matrix of personnel involved in aircraft operations. Therefore, this chapter examined and compared the roles of various security personnel such as GSCs and LEOs. Of importance are the ASCs serving as the primary contact for their airline to the TSA. ASCs also coordinate the transportation of LEOs and their prisoners. These types of activities are managed under regulatory procedures and can have a significant effect on aircraft operations. How these and other personnel manage ground security in boundaries such as exclusive areas or SIDAs was also described.

Aircraft operators must also consider the security implications related to other threats, such as disgruntled employees, baggage and cargo theft, abuse of employee flying privileges, assault from passengers, and breaches in computer information security. Strategies for managing these and other threats usually include training programs. These programs include topics such as learning the psychology of aggressive behavior (e.g., air rage), self-defense techniques, and processes for handling secret or sensitive information. Not only are airline security personnel concerned with the safety of passengers during flight, but they are also concerned for the safety of their flight crews both onboard and while staying in cities around the globe. Aviation security practitioners should remember that upset or intoxicated passengers cause the majority of airline security incidents.

Conducting aircraft operations by or within other nations creates higher levels of complexity in managing aviation security. For example, customs and immigration agents become a part of the security program used by each airline that is operating internationally. Regulations for foreign aircraft operations within the United States ensure that security procedures applied to inbound U.S. flights meet the same security standards as domestic aircraft operations.

References

ABC ActionNews, 2005. Investigation Finds Above-average Rate of Luggage Theft at TIA. ABC Action News, retrieved Dec. 21, 2006, from http://wfts.com/stories/2005/02/050221tsa.shtml.

ABC News, Nov. 19, 2004. TSA Under Fire for Rising Theft by Baggage Screeners. ABC World News, retrieved Aug. 8, 2008, from http://abcnews.go.com/WNT/story?id=26657.

Associated Press, April 4, 2006. Air Marshals Admit Drug Smuggling. CBS News, retrieved Aug. 8, 2008, from http://www.cbsnews.com/stories/2006/04/04/national/main1469729.shtml.

ATSA, 2001. Aviation and Transportation Security Act of 2001. P.L. 107–71 Sec. 101.

Aviv, J., 2003. The Complete Terrorism Survival Guide. Juris, New York.

Barrett, R., July 28, 2003. Protect the Cockpit at All Costs: Shoot to Kill Replaces Negotiation as Mantra for Pilots Training to Use Handguns. Milwaukee Journal Sentinel 1.

Baum, P., 2001. Independence Day Rage: A Victim's Tale. Aviation Security International, retrieved Aug. 8, 2008, from http://www.avsec.com/interviews/independence-day.htm.

Baum, P., 2011. In Flight Security. In: Baum, P. (Ed.), Aviation Security: Challenges and Solutions, vol. 1 AVSECO, Lantau, Hong Kong, pp. 193–207.

Boeing, June 2002. Boeing Testifies on Guns and Flight Crews. Boeing Frontiers 1, retrieved Aug. 8, 2008, from http://www.boeing.com/news/frontiers/archive/2002/june/i_bitn.html.

CNN, June 12, 2002. "Shoe bomb" Suspect Drugged during Flight. CNN.com, retrieved Aug. 8, 2008, from http://archives.cnn.com/2002/LAW/06/12/shoe.bomber/index.html?related.

Drug Enforcement Administration (DEA), 2003. Statement of Rogelio E. Guevara, Chief of Operations, Drug Enforcement Administration before House Judiciary Committee Subcommittee on Crime, Terrorism, and Homeland Security, May 6, 2003. U.S. Drug Enforcement Administration, retrieved Aug. 8, 2008, from http://www.usdoj.gov/dea/pubs/cngrtest/ct050603.htm.

Forgione, M., 2011. TSA Luggage Theft at LAX: How Common Is This? Los Angeles Times Press, Los Angeles.

Foster, S., 2004. Smart Packing for Today's Traveler. Smart Travel Press, Portland, OR.

General Accounting Office (GAO), 1987. FAA's Implementation of a Performance Standard for Passenger Screening Process (Publication No. T-RCED-88-4), retrieved Aug. 8, 2008, from http://archive.gao.gov/d39t12/134217.pdf.

Goo, S.K., Aug. 27, 2003. Hundreds of Pilots Trained to Carry Guns; but Lobby Says TSA's Program Restrictions Discourage Potential Applicants. The Washington Post A.10.

Grossman, D., 1996. On Killing. Time Warner Book Group, New York.

Holt, R.L., 2002. Stop Sky Jackers. California Financial Publications, Cardiff by the Sea, CA.

Jenkins, J., Thompson, G., 2004. Verbal Judo: The Gentle Art of Persuasion. HarperCollins, New York.

Lott, J.R., Jan. 2, 2004. Marshals Are Good, but Armed Pilots Are Better. The Wall Street Journal Europe, retrieved Aug. 8, 2008, from http://johnrlott.tripod.com/op-eds/ArmedMarshalsWSJE.html.

McAlpin, A., 2003. Pack It Up: Traveling Smart and Safe in Today's World, third rev. ed Flying Cloud, Stow, MA.

Morrison, B., Dec. 5, 2001. FAA Seldom Punished Violence. USA Today, retrieved Aug. 8, 2008, from http://www.usatoday.com/news/sept11/2001/12/05/air-violence.htm#more.

National Drug Intelligence Center (NDIC), Nov. 2002. Marijuana, New York Drug Threat Assessment. National Drug Intelligence Center, retrieved Aug. 8, 2008, from http://www.usdoj.gov/ndic/pubs2/2580/marijuan.htm#Top.

Price, J., 2008. Airport Certified Employee (ACE)—Security Module 3 Aircraft Operator, rev. ed. American Association of Airport Executives, Alexandria, VA.

Sloan, J., Dec. 16, 2002. Cockpit Guns Called Last Line of Defense. Tampa Tribune 1.

Thomas, A., 2001. Air Rage: Crisis in the Skies. Prometheus Books, Amherst, NY.

Thomas, A., 2003. Aviation Insecurity: The New Challenges of Air Travel. Prometheus Books, Amherst, NY.

Thomas, A., 2006. Breaking News. AirRage.org and AviationInsecurity.com, retrieved Aug. 8, 2008, from http://www.airrage.org.

TSA, 2006. Recommended Airport Security Guidelines for Airport Planning, Design and Construction. Transportation Security Administration, U.S. Department of Homeland Security, retrieved Aug. 8, 2008 from http://www.tsa.gov/assets/pdf/airport_security_design_guidelines.pdf.

Waltrip, S., Williams, C., 2004. Aircrew Security: A Practical Guide. Ashgate, Burlington, VT.

White House, Jan. 2002. Securing America's Borders Fact Sheet: Border Security. The White House, retrieved Aug. 8, 2008, from http://www.whitehouse.gov/news/releases/2002/01/20020125.html.

9

General Aviation Security

Objectives

This chapter examines potential security threats to general aviation (GA) airports and flight operations. Strategies for protecting GA airports and aircraft are provided and an overview of changes in aviation security that affected GA operations after 9/11. You will learn about challenges to developing and implementing security regulations for GA airports. We also discuss the security strategies used by the Transportation Security Administration (TSA) and various GA airport and aircraft operators.

Introduction

Since 9/11, the general aviation community has struggled to help the public understand the nature of GA. These efforts have focused on greater public understanding for how GA benefits the U.S. economy and communicating ways to prevent onerous legislation that may cause irreparable damage to the industry. The theft and illegal use of general aviation aircraft predate 9/11. General aviation aircraft have long been used as platforms to smuggle narcotics and weapons, and for human trafficking operations. There have been several incidents that have called GA security into question, including:

- GA aircraft have long been used in illegal drug smuggling operations; ultra-light aircraft, which fly low enough to avoid radar and are very difficult to detect, are a recent addition to the smugglers tool kit.
- In 1993, the FBI indicated that Osama bin Laden assessed the possibility of using an agricultural aircraft to spread a chemical agent on a ground target.
- In 1994, part of Operation Bojinka included a plan to crash a Cessna small aircraft filled with explosives into the headquarters of the Central Intelligence Agency (CIA).
- Also in 1994, an individual crashed a Cessna 150 light training aircraft into the White House.
- In 1998, author James Canton presented research to the FBI on a hypothetical attack on the National Mall using a GA aircraft to spread anthrax.
- The 9/11 hijackers trained at U.S. flight schools on GA aircraft and had used GA aircraft to scout out the New York City airspace in preparation for the attacks.
- In 2002, a teenager crashed a Cessna 172 into the Bank of America building in Tampa, FL.
- According to the Federal Bureau of Investigations (FBI) (Government Accountability Office, 2004), terrorists have assessed the feasibility of using GA aircraft as a possible medium for attacks; in a CIA report to Congress, the agency director

suggested that bin Laden had assessed whether to use GA aircraft to carry out the 9/11 attacks, but instead went with commercial airplanes, suggesting the terrorists exercised a deliberative process of weighing the pros and cons of using general aviation (DHS OIG, 2009, p. 12).

- In 2005, an individual crashed an ultra-light into the German Parliament building in an apparent suicide.
- In 2005, there were two embarrassing moments for the GA industry that included inebriated pilots stealing planes they had access to fly to go joyriding.
- In 2010, an individual flew his own small aircraft into the local offices of the Internal Revenue Service in Austin, TX.
- In 2011, the so-called "Barefoot Bandit" stole several GA aircraft and led U.S. law enforcement officers on a chase across the country. The individual did not have a pilot certificate and taught himself to fly using a home-based computer flight simulator.
- On the tenth anniversary of the 9/11 attacks, the Department of Homeland Security (DHS) issued an alert for the possible use of a GA aircraft in an attack on the United States, which brought the whole issue of GA security to the forefront of the public's mind.
- Adding to all this are continuing stated beliefs on the part of security officials warning that since many of the holes in commercial security have been fixed, terrorists will look to other means of attack such as GA and air cargo.

GA provides vital services to the United States and greatly enhances the U.S. economy. GA accounts for some 77% of all flights in the United States. With more than 200,000 aircraft, 650,000 pilots, and 19,000 airports and landing strips, the GA industry provides jobs and opportunities for thousands of people. One estimate put the value of GA at 1% of the U.S. gross domestic product.

General aviation flights account for more passengers and hours flown than commercial-service operations. Approximately 24% of all GA flights are conducted for business or corporate use (GAO, 2004); nearly two-thirds of all business flights carry passengers in midlevel management positions, sales representatives, and project teams for major corporations. GA accounts for three-quarters of all takeoffs and landings in the United States, contributes about $100 billion to the U.S. economy, and has about 1.3 million jobs (GAO, 2004). These estimates do not include multipliers, such as the number of jobs created by the hotel, rental car, restaurant, and tourist industries as a result of a GA airport in a community.

The creation of GA security policies is important to protect these interests.[1] Many career airline, corporate, and military pilots were first inspired by an airshow or by visiting a GA airport, and then eventually conducting their flight training at a GA airport. The United States has one of the most robust GA industries in the world and members of the aviation community have looked to protect that industry since 9/11.

---

[1]In 2008, the FAA lifted the age 60 rule to address the upcoming shortage of airline pilots. Future airline pilots train in GA aircraft, so any negative impact to the GA's flight community may inadvertently impact flight training, which is essential to filling tomorrow's airline pilot ranks.

GA provides essential transportation services to a community, including:

- Relieving commercial-service airports of GA traffic
- Providing air cargo and U.S. mail services
- Providing essential medical transportation flights
- Providing a location for numerous businesses to operate, employing local citizens, and providing revenue streams into a city or county
- Providing a location for flight training services
- Allowing safe, secure, and essential corporate and business flight operations

The domain and mission of GA is often misunderstood by the public, policy makers, and regulators. General aviation consists of any flight operation that is not commercial service or military. These operations include flight training, personal or recreational flights, business aircraft operations, experimental aircraft, glider operations, sky diving, and other forms of flight operations. The FAA also has yet to define a category of operation for remotely piloted vehicles (RPVs) or unmanned aeronautical vehicles (UAVs), so for the purposes of this chapter, we will consider them to be GA operations as well. There is also debate within the industry as to which acronym is more viable—RPV or UAV. For our purposes, we have selected to use RPV.

This chapter will also address helicopter security. While there are commercial and even scheduled helicopter flight operations, these are normally operated under Title 14 CFR Part 135 Air Taxi or Air Charter Operations and the majority of helicopter operations are flown as GA flights.

Securing General Aviation

Securing GA is challenging considering the magnitude and nature of GA operations. There are approximately 450 commercial-service airports in the Untied States and more than 5,000 GA airports, plus an additional 14,000 private airstrips (also categorized as GA operations). Those 14,000 airports are not regulated by the Federal Aviation Administration (FAA) or the TSA.

In 2009, reporters from a Houston TV station aired a segment highlighting how they were able to access a GA airport without being interdicted and had access to aircraft. The news story resulted in an investigation by the DHS Office of the Inspector General, which concluded that the allegations were not compelling and, while the reporters were able to access the airport, they were unaware of aircraft locking, grounding, and closed-caption television (CCTV) layers of security that would have made it difficult to actually take the aircraft (DHS OIG, 2009, p. 4).

The report stated that GA presents only limited and mostly hypothetical threats to aviation security, but also noted that the potential for a terrorist group to use GA aircraft to conduct an attack remains a possibility that cannot be ignored.

Most GA aircraft are too light to be used as a platform for conventional explosives and heightened vigilance by GA airport and aircraft operators would make it difficult for someone to load the necessary quantity of explosives without detection. A 1,300-pound bomb, like what was used against the World Trade Center in 1993, would likely exceed the payload capacity of most light aircraft. The benefit GA aircraft do have over a truck bomb or ground-based attack is the ability to fly over most physical barriers and access points, so smaller-scale attacks into sensitive, but heavily guarded (from the ground) facilities is a possibility.

The potential exists for light aircraft to be used as a delivery platform for chemical, biological, radiological, and nuclear (CBRN) forms of weapons. General aviation aircraft are capable of slow flight at relatively low altitudes above densely populated areas and large congregations of people on the ground (Elias, 2009, p. 14). Their slow speed and the ease at which doors and windows on nonpressurized airplanes and helicopters can be operated in flight may actually pose a greater threat from CBRN attacks. Agricultural aircraft used for spraying crops with pesticides and fertilizers pose a unique threat as a platform for a biological or chemical attack because they are specifically designed for aerial dispersal and could be exploited by terrorists for this specific purpose (Elias, 2009, p. 14). However, many chemical agents must be released in large concentrations to have a significant impact, and some, such as cyanide, may only be effective if released in an enclosed area. The effectiveness of other weapons, such as mustard gas and nerve agents, may be effective in open-air dispersal but the limited payload capacity of most light GA aircraft could limit that effectiveness (Elias, 2009, p. 14). Even in a crowded stadium a rapid response to an attack by the medical community could limit the effectiveness of the attack, but it would likely still cause fear, panic, possibly thousands of injuries, and hundreds of deaths (Elias, 2009, p. 14).

The threat of a radiological or nuclear release by a light GA aircraft is tempered somewhat by the fact that other means of attack (e.g., a suitcase, a car, etc.) are just as effective as aircraft in delivering such a device to a target site. A report prepared for the Aircraft Owners and Pilots Association (AOPA) by a nuclear safety consultant noted that the threat to a reactor from a small GA aircraft is practially nonexistent (Elias, 2009, p. 14).

There are three primary issues to consider when addressing security of GA airports and aircraft. First, what is the threat to GA airports and aircraft; second, what is the threat to the public or infrastructure from GA aircraft; and third, what security measures should be required or recommended to prevent both of the aforementioned threats from occurring? The GA community also considers the possibility of legislation that could economically imperil GA flight operations as a threat to the viability of industry and the safety of the public.

Key threats include:

- The misuse or theft of a GA aircraft, either as a weapon of mass destruction (WMD), a delivery platform for an improvised explosive device (IED), or a CBRN element, flown into another aircraft in an intentional midair collision.
- Small aircraft (below 12,5000 pounds) may represent a threat from their ability to carry explosive elements, or CBRN, to a target such as an open-air stadium; explosives could be combined with a chemical agent, such as chlorine, multiplying the casualties.

- A large business aircraft could be hijacked and used as a WMD but arguments exist as to the effectiveness of such an attack based on aircraft fuel capacities and kinetic energy of the aircraft, as compared to the larger scheduled-service aircraft fleet.
- The use of a GA aircraft for elicit activities such as drug, weapon, or human smuggling.
- Theft or sabotage of a GA aircraft, theft of aircraft rescue and firefighting equipment or fuel trucks, or sabotage or damage to a GA airport.

In addition to these primary threats, there are secondary threats and uses of GA, such as a helicopter being used to coordinate a ground attack, or as an observation platform to survey a potential target. Helicopters have also been hijacked with the intent to use them for prison escapes, or kidnapping executives on a corporate helicopter.

The biggest challenge in preventing attacks either on or with GA aircraft is that GA operations are vastly different than flight operations at a commercial-service airport. Therefore, GA operations cannot usually be secured with the same strategies and processes as used at commercial-service airports. While there are some similarities between a GA private or charter operation and a scheduled-service operation (i.e., passengers board a flight, bags are carried on board, etc.), security processes that are effective in one setting may be ineffective and even inappropriate in another. One notable difference is that in a scheduled-service operation most of the passengers do not know the pilot personally. In a GA private flight operation, many times the pilot is the employee of the passenger(s). To require passengers on a GA flight to be screened essentially puts the pilot in charge of screening his or her employer before being allowed onto an aircraft that is owned by that individual. However, in the case of an aircraft charter it would be more reasonable to accept that the passengers should be screened, as the flight operation does bear significant similarities to a scheduled-service operation. These are just a few examples of where a "blanket approach" to aviation security practices would most likely not work in GA.

Operational policies and management philosophies between commercial-service airports and GA airports frequently vary. Therefore, commercial airport security measures are typically not effective when applied in GA settings. Many GA tenants do not see the need for the complexity or expense associated with commercial airport security systems. Additionally, a commercial airport is designed to move passengers from land-based transportation, to air-based transportation, and back to land-based. A GA airport is more akin to a service center than as a hub for transferring passengers. Privately owned fixed-base operators (FBOs) provide fuel and maintenance services for aircraft and are many times the only "terminal" building on the airfield. Many GA airports have small to very little staff and the vast majority of GA airport managers do not have experience or training in airport security, since none are required to undergo the airport security coordinator (ASC) training that is required at a commercial-service airport.

Barriers to improving security measures within the GA system are amplified in that the public has traditionally assumed that risks or threats to GA operations can be entirely eliminated through legislation and regulations. Legislative or regulatory actions alone are not able to completely ensure security at any airport, regardless of the category of operation.

For example, a GA owner flying his or her personal aircraft into a building is an action outside of the traditional airport's security system to detect or mitigate.

Whether threats to or associated with GA constitute viable concerns to national security remains unclear. Hundreds of thousands of GA aircraft operations are conducted every year, the vast majority of which pose little or no threat.

Although commercial-service airports operate under CFR Title 49 Part 1542, GA airports do not have a counterpart regulation. GA aircraft operations are regulated under Title 49 CFR Part 1550 for those aircraft operators conducting charter operations with aircraft in excess of 12,500 pounds maximum gross takeoff weight (MGTW) or aircraft operators whose passengers deplane into a sterile area. There are also GA operations at commercial-service airports. About 300 out of the 5,000 GA airports in the United States are classified as GA but periodically host some commercial-service operations, often for a few days a week or during certain seasons. Where GA operations occur on regulated commercial-service airports, the tenants are required to adhere to both Part 1542 and the airport security program (ASP). In 2008, the TSA began working on the Large Aircraft Security Program (LASP), which would have regulated GA airports similarly to small commercial-service airports. However, negative industry response was so overwhelming (nearly 10,000 comments to the notice of proposed rulemaking, or NPRM) that the TSA withdrew their initial proposal and began writing a new LASP, this time with industry input as part of the process. The new LASP places more focus on the aircraft operator as the responsible party for aircraft security, and less on the airport operator. This change in policy reflects the operational realities and differences between commercial-service and GA airports.

GA airport operators should pay attention to rulemaking that impacts their GA operators, not just from a security perspective but also from an economic perspective. When the private charter rule was implemented, some GA airports saw their large aircraft operations shift to nearby commercial-service airports where screening equipment and personnel were located. This migration reduced fuel revenue at those GA airports, normally a major source of a GA airport's revenue stream.

How 9/11 Changed General Aviation

The Aviation and Transportation Security Act of 2001 (ATSA 2001) and several other security measures aimed at restricting GA operations were implemented after 9/11. These regulations cover a variety of GA security concerns including security regulations addressing public and private charter flights, airspace restrictions, and new rules for flight training operations (Title 49 CFR Part 1552).

Pilots

A significant change to aviation (GA and commercial) affecting pilots was the change in document format to the long-standing pilot certificate. Before 9/11, pilot certificates (commonly and erroneously called "licenses") were simple in design and fairly easy to duplicate. Initially, the FAA discussed switching to a pilot certificate that included the pilot's photograph, but the AOPA objected on the grounds that there was a simpler and less costly alternative—to require all pilots to have both the pilot certificate and a government-issued photo ID while flying. Title 14 CFR Part 61.3(a)(1) already required the pilot in command (PIC) or flight crewmember to have a valid flight certificate in their possession when operating an aircraft. Government-issued photo ID must now be in the pilot's possession (Title 14 CFR Part 61.3(a)(2)(i–vi)). The FAA also issued redesigned pilot certificates, which are difficult to forge (Figure 9.1). When a pilot applies for a medical

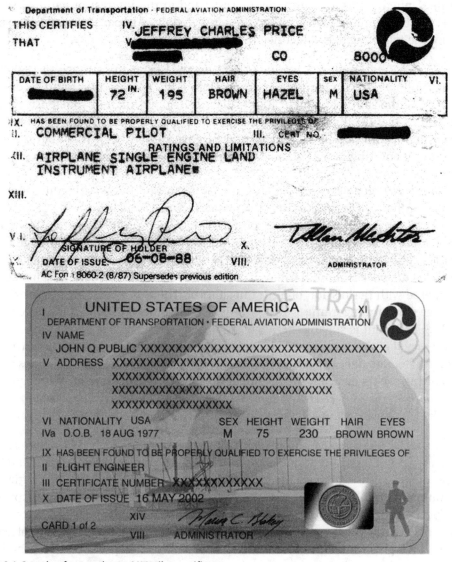

FIGURE 9.1 Sample of pre and post 9/11 pilot certificate.

certificate or any new flight rating or certificate, the TSA checks the applicant against the no-fly and selectee lists, and possibly other terrorists watch lists.

Airspace

Since 9/11, the U.S. government has maintained that an efficient way to protect lives and facilities where large numbers of people gather is to restrict flight in the airspace around those venues. In addition to these temporary flight restrictions (TFRs), several areas within the United States are permanently restricted in the interest of national security. The 9/11 attacks significantly expanded these restrictions. Airspace restrictions primarily fall under the following five categories:

- The *air defense identification zone* (ADIZ) extends from the surface to 18,000 feet. A contiguous ADIZ surrounds the entire United States. For an aircraft to enter an ADIZ, either an (instrument flight rules (IFR) flight plan or a defense visual flight rules (DVFR) flight plan is required. ADIZs are established over other areas such as military bases and centers of government (e.g., Washington, DC). VFR traffic entering an ADIZ must have permission from an air traffic control center.
- A *flight restriction zone* (FRZ) extends in a circle around the Washington Monument. Entry is authorized only to those who have received a waiver from the TSA.
- *Prohibited airspace* is continuously off limits. P-56, the prohibited area above the U.S. White House, extends from the ground to 18,000 feet. P-40 is the designation for the prohibited airspace above Camp David.
- *Restricted airspace* is more flexible than prohibited airspace. Restricted airspace can be entered with permission of air traffic control. There is a large area of restricted airspace surrounding Camp David, MD, which becomes prohibited airspace when the U.S. president is staying there.
- *Temporary flight restrictions* (TFRs) are areas temporarily designated as restricted or prohibited airspace. These are frequently established for special activities such as sporting events or when the president flies into a city outside of Washington, DC. There are TFRs over Disneyland in California and Walt Disney World in Florida, U.S. power plants, various air shows, and other activities.

Other special-use airspace also exists and must be observed by GA pilots. National security areas, military operations areas (MOAs), and military training routes (MTRs) are areas of special-use airspace, which may be generally accessed by VFR traffic.

Operations in the Washington, DC, ADIZ (and other areas of restricted airspace) fall under Special Federal Aviation Regulations (SFARs). The establishment of TFRs and the expanded ADIZ over Washington, DC, resulted in numerous violations as general aviation pilots on VFR flights wandered into restricted or prohibited airspace. The AOPA and other GA groups have been strong opponents of unnecessary TFRs, although the AOPA remains a strong supporter of sound GA security measures.

Temporary Flight Restrictions

Airspace restrictions are one strategy that the government uses to protect facilities and congested populations during special events. The issuance of TFRs has become a confusing issue for the GA community and has resulted in thousands of unintentional violations. The FAA has a website where pilots can find information on TFRs, but it is still difficult for pilots to stay current with new and changing TFRs (Figure 9.2).

As of December 2006, there were 6,658 TFR violations since 9/11. More than 1,600 were within TFRs over locations where the U.S. president was traveling, more than 3,000 were related to Washington, DC, security breaches, and 2,600 were related to pilots inadvertently flying into the Washington, DC, ADIZ (Trautvetter, 2006). In many cases, the aircraft in violation had to be intercepted by a U.S. military or Customs and Border Protection (CBP) aircraft. To date, no infraction has resulted in a civilian aircraft being shot down. Part of the reason there are so many airspace violations is that pilots often misinterpret or do not understand when a TFR has been issued, or the flight requirements associated with the TFR. Another problem contributing to airspace violation is that prior to 9/11, pilots were used to restrictions being designated on navigational charts—TFRs were extremely rare. Pilots must now diligently read each applicable NOTAM (notice to airmen) and check additional information online to ensure that he or she has all of the information essential for the flight, including TFRs, which are temporal in nature. The FAA recently started a Washington, DC, ADIZ and TFR training information in an attempt to reduce the number of violations (see *http://www.faa.gov/pilots/safety/notams_tfr/media/tfrweb.pdf*).

One notable airspace violation occurred on May 11, 2005, when two pilots in a Cessna 172 strayed into the ADIZ over Washington, DC. The aircraft flew within three miles of the U.S. White House, prompting officials to evacuate several government buildings. For eight minutes the alert level at the White House was raised to red. A Blackhawk helicopter and two F-16 s escorted the aircraft to landing, and the pilots were taken into custody (CNN, 2005).

General Aviation Airport Security Program

Since 9/11, the TSA and the U.S. Congress have focused on creating security regulations for commercial-service airports and the airline industry. These regulations have had minimal impact on GA operations. In cases where Congress and the TSA have addressed GA security issues, the focus has been primarily on GA aircraft operators and not GA airport operators. The TSA published GA airport security guidelines in 2004 (Figure 9.3). However, in 2008, the TSA started implementing more security outreach programs to GA airports. Without the impetus of an attack directed toward the GA community, it is difficult to predict whether GA airports will eventually be regulated. However, there are numerous GA security-related best practices that can be implemented. GA airport operators should consider implementing security measures to prevent GA from being used in a terrorist attack, which would subsequently create a need for new transportation security regulations specific to GA.

With a lack of security regulations covering GA airport operations, GA airport sponsors (owners and operators, such as city and county governments) routinely question why

NOTAM Text

NOTAM : 8/1522

FDC 8/1522 ZDV PART 1 OF 5 CO.. FLIGHT RESTRICTIONS. DENVER, CO, AUGUST 25-28, 2008 LOCAL. THIS NOTAM REPLACES NOTAM 8/1191 DUE TO ADDITION OF PROCEDURES FOR CENTENNIAL AIRPORT, CO (KAPA). PURSUANT TO 49 USC 40103(B), THE FEDERAL AVIATION ADMINISTRATION (FAA) CLASSIFIES THE AIRSPACE DEFINED IN THIS NOTAM AS 'NATIONAL DEFENSE AIRSPACE'. ANY PERSON WHO KNOWINGLY OR WILLFULLY VIOLATES THE RULES CONCERNING OPERATIONS IN THIS AIRSPACE MAY BE SUBJECT TO CERTAIN CRIMINAL PENALTIES UNDER 49 USC 46307. PILOTS WHO DO NOT ADHERE TO THE FOLLOWING PROCEDURES MAY BE INTERCEPTED, DETAINED AND INTERVIEWED BY LAW ENFORCEMENT/USSS/SECURITY PERSONNEL. PURSUANT TO TITLE 14 CFR SECTION 99.7, SPECIAL SECURITY INSTRUCTIONS, ALL AIRCRAFT FLIGHT OPERATIONS ARE PROHIBITED WITHIN A 10 NMR OF 394456N/1050024W OR THE BJC137011.6 UP TO BUT NOT INCLUDING 18000 FT MSL. THE AIRSPACE EAST AND SOUTH OF A LINE FROM 393705N/1045230W TO 394107N/1045238W TO 394105N/1044830W UP TO AND INCLUDING 8000 FT MSL IS EXCLUDED FOR IFR ARRIVALS AND DEPARTURES ONLY TO/FROM CENTENNIAL AIRPORT, CO (KAPA). FFECTIVE 0808252000 UTC (1400 LOCAL 08/25/08) UNTIL 0808260500 UTC (2300 LOCAL 08/25/08), 0808262000 UTC (1400 LOCAL 08/26/08) UNTIL 0808270500 UTC (2300 LOCAL 08/26/08), 0808272000 UTC (1400 LOCAL 08/27/08) UNTIL 0808280500 UTC END PART 1 OF 5 FDC 8/1522 ZDV PART 2 OF 5 CO.. FLIGHT RESTRICTIONS. DENVER, CO, (2300 LOCAL 08/27/08), AND 0808282200 UTC (1600 LOCAL 08/28/08) UNTIL 0808290500 UTC (2300 LOCAL 08/28/08). WITHIN A 30 NMR OF 394456N/1050024W OR THE BJC137011.6 UP TO BUT NOT INCLUDING 18000 FT MSL EFFECTIVE 0808252000 UTC (1400 LOCAL 08/25/08) UNTIL 0808260500 UTC (2300 LOCAL 08/25/08), 0808262000 UTC (1400 LOCAL 08/26/08) UNTIL 0808270500 UTC (2300 LOCAL 08/26/08), 0808272000 UTC (1400 LOCAL 08/27/08) UNTIL 0808280500 UTC (2300 LOCAL 08/27/08), AND 0808282200 UTC (1600 LOCAL 08/28/08) UNTIL 0808290500 UTC (2300 LOCAL 08/28/08). EXCEPT AS SPECIFIED BELOW AND/OR UNLESS AUTHORIZED BY THE AIR TRAFFIC SECURITY COORDINATOR VIA THE DOMESTIC EVENTS NETWORK (DEN): A. ALL AIRCRAFT OPERATIONS WITHIN THE 10 NMR AREA(S) LISTED ABOVE, KNOWN AS THE INNER CORE(S), ARE PROHIBITED EXCEPT FOR: APPROVED LAW ENFORCEMENT, MILITARY AIRCRAFT DIRECTLY SUPPORTING THE UNITED STATES SECRET SERVICE (USSS) AND THE OFFICE OF THE PRESIDENT OF THE UNITED STATES, APPROVED AIR AMBULANCE FLIGHTS, AND REGULARLY SCHEDULED END PART 2 OF 5 FDC 8/1522 ZDV PART 3 OF 5 CO.. FLIGHT RESTRICTIONS. DENVER, CO, COMMERCIAL PASSENGER AND ALL-CARGO CARRIERS OPERATING UNDER ONE OF THE FOLLOWING TSA-APPROVED STANDARD SECURITY PROGRAMS/PROCEDURES: AIRCRAFT OPERATOR STANDARD SECURITY PROGRAM (AOSSP), FULL ALL-CARGO AIRCRAFT OPERATOR STANDARD SECURITY PROGRAM (FACAOSSP), MODEL SECURITY PROGRAM (MSP), TWELVE FIVE STANDARD SECURITY PROGRAM (TFSSP) ALL CARGO ONLY, OR ALL-CARGO INTERNATIONAL SECURITY PROCEDURE (ACISP) AND ARE ARRIVING INTO AND/OR DEPARTING FROM 14 CFR PART 139 AIRPORTS. ADDITIONALLY ALL WAIVERS ARE TEMPORARILY SUSPENDED DURING THE TIME OF THIS NOTAM EXCEPT FOR ELO, GOV AND SPO WAIVERS. B. FOR OPERATIONS WITHIN THE AIRSPACE BETWEEN THE 10 NMR AND 30 NMR AREA(S) LISTED ABOVE, KNOWN AS THE OUTER RING(S): ALL AIRCRAFT OPERATING WITHIN THE OUTER RING(S) LISTED ABOVEARE LIMITED TO AIRCRAFT ARRIVING OR DEPARTING LOCAL AIRFIELDS, AND WORKLOAD PERMITTING, ATC MAY AUTHORIZE TRANSIT OPERATIONS. AIRCRAFT MAY NOT LOITER. ALL AIRCRAFT MUST BE ON AN ACTIVE IFR OR VFR FLIGHT PLAN WITH A DISCRETE CODE ASSIGNED BY AN AIR TRAFFIC CONTROL (ATC) FACILITY. AIRCRAFT MUST BE SQUAWKING THE DISCRETE CODE PRIOR TO DEPARTURE AND AT ALL TIMES WHILE IN THE TFR AND MUST REMAIN IN TWO-WAY RADIO COMMUNICATIONS WITH ATC. C. FOR OPERATIONS WITHIN THIS TFR, ALL USSS VETTED AIRCRAFT OPERATORS END PART 3 OF 5 FDC 8/1522 ZDV PART 4 OF 5 CO.. FLIGHT RESTRICTIONS. DENVER, CO, BASED IN THE AREA AND ALL EMERGENCY/LIFE SAVING FLIGHT (AIR AMBULANCE/LAW ENFORCEMENT/FIREFIGHTING) OPERATIONS MUST COORDINATE WITH ATC PRIOR TO THEIR DEPARTURE AT 303-626-6580 TO AVOID POTENTIAL DELAYS. D. THE FOLLOWING OPERATIONS ARE NOT AUTHORIZED WITHIN THIS TFR:

FIGURE 9.2 Example of Notice to Airman about a Temporary Flight Restriction over Denver, Colorado.

Security Guidelines for
General Aviation Airports

Information Publication A-001
May 2004

Version 1.0

FIGURE 9.3 TSA threat advisory issued by the DHS to GA airport and aircraft operators was one of the first attempts to secure general aviation (Transportation Security Administration, 2004).

security measures and related funding should be of concern. GA sponsors also note that there has been minimal federal funding for GA airport security measures and that FAA AIP grants receive safety and other airfield projects first, and security second, third, or fourth.[2]

[2]In 2008, the DHS began assessing possible federal funding programs for GA security measures.

Since 9/11, funding for GA airport security has come primarily from the airport and local and state government agencies, or the occasional TSA pilot program.

The DHS has identified three security threats related to the general aviation community: using a GA aircraft as a WMD, using GA aircraft to transport dangerous individuals into the country, and using GA aircraft to transport dangerous materials into the country. There is some evidence to support these concerns as, again, the GA industry was heavily used throughout the 1970s, 1980s, and 1990s to smuggle drugs into the United States. Additionally, it was within the GA flight training industry where the 9/11 hijackers trained. GA airports house fuel trucks and fire trucks, any of which can be stolen and used for other purposes, such as a bomb or to gain or force access into a secure area. Liquefied natural gas tankers, nuclear facilities, and similar targets that have naturally occurring standoff distances from ground attacks, remain vulnerable to aerial attacks from both GA and commercial-service aircraft.

GA airports may also be targets for domestic terrorism, acts of violence from radical environmental or wildlife activists groups, criminal activities, or even workplace violence.[3]

Certain structures on airports may be considered high-value targets, particularly air traffic control facilities and airway navigation equipment. A small air traffic control tower at a remote GA airport may not seem a likely target for an international terrorist organization because of its limited impact on the national aviation system, however, its destruction could be a significant achievement for a domestic terrorist organization determined to make a statement about the impact of aviation on the environment or to inflict harm to a society and its government.

GA airports are often sources of concern and discussion by environmental organizations. Airport managers routinely find themselves involved in community noise issues, wildlife management, development projects, wetland protection, and management of storm water runoff. These issues are often interrelated with other environmental concerns such as deicing chemicals or pesticides. These and many other issues are environmentally important, making airports and associated facilities potential targets for ecoterrorist groups, vandalism, and civil disturbances.

A study conducted by the Government Accountability Office (GAO) in 2004 pointed out the security risks posed by general aviation aircraft:

In 2004, the Secretary of the Department of Homeland Security acknowledged that the department, along with the Central Intelligence Agency (CIA), FBI, and other agencies, lacked precise knowledge about the time, place, and methods of potential terrorist attacks related to general aviation. Additionally, industry and TSA officials stated that the small size, lack of fuel capacity, and minimal destructive power of most general aviation aircraft make them unattractive to terrorists and, thereby, reduce the possibility of threat associated with their misuse. Historical intelligence

---

[3]For example, in 2006 at Grand Junction, CO, a former air traffic controller constructed pipe bombs and detonated them outside the homes of his former coworkers and an FAA employee.

indicates that terrorists have expressed interest in using general aviation aircraft to conduct attacks. The following are examples of intelligence information indicating terrorist interest in general aviation: (a) CIA reported that terrorists associated with the September 11 attacks expressed interest in the use of crop-dusting aircraft (a type of general aviation aircraft) for large area dissemination of biological warfare agents such as anthrax; (b) CIA reported that one of the masterminds of the September 11 attacks originally proposed using small aircraft filled with explosives to carry out the attacks; (c) in May 2003, the Department of Homeland Security issued a security advisory indicating that al Qaeda was in the late stages of planning an attack, using general aviation aircraft, on the U.S. Consulate in Karachi, Pakistan, and had also planned to use general aviation aircraft to attack warships in the Persian Gulf. (GAO, 2004, p. 16)

The GAO also noted that many of the approximately 19,000[4] GA airports have unique characteristics that might make them viable for terrorist activity. With such a large number of GA airports, it would be difficult to conduct individual onsite vulnerability assessments.

There is an economic benefit to securing a GA airport. A secure airport can attract corporate operators who are looking to protect their multimillion-dollar investment in aircraft transportation. Without proper security measures, corporate operators often will land at a GA airport to drop off or pick up passengers, then ferry the aircraft to another nearby airport that does have adequate security (e.g., perimeter fencing, security patrols, and airfield access control measures). In these cases, the unsecured GA airport often loses the fuel sales and service fees.

Airfield security measures usually have an added safety benefit by restricting access to those who have business on the airport. Access control measures have been successfully used to reduce runway incursions for several years, and fencing helps keep wildlife and the public from inadvertently coming onto the airfield.

Some airport tenants oppose security measures at their GA airport. Airport operators can argue that security measures not only protect the investment a tenant has made in facilities and aircraft but also the investment in their aviation career or hobby.

There are few in the GA airport community who want to administer new TSA regulations. With every new GA security incident, lawmakers again examine GA as a potential concern for national security. Some legislators would like to see greater TSA regulations on GA operations, and some would even like to eliminate GA in the United States.[5]

Without regulatory guidance, the creation of airport security programs for GA airports can be difficult. Most GA airport operators attempt to follow the format and guidelines for

---

[4]This number includes an estimated 14,000 private airstrips subject to no regulatory authority.

[5]Even nonsecurity incidents involving GA aircraft can create a political firestorm. In 2006, a small airplane, piloted by New York Yankee's pitcher Cory Lidle and his flight instructor, crashed into an apartment building. The crash created a public relations problem in regard to the safety and security of GA aircraft. Industry organizations such as the AOPA worked hard to convince lawmakers that despite the crash, placing further restrictions on GA was not necessary.

the commercial-service airport security programs. There are thousands of GA airports, which have an endless variety of airport operating characteristics, and this makes the creation of a "model" GA airport security program problematic. Most GA airport operators attempting to construct a security program use the Airport Characteristics Measurement Tool provided within the TSA's IP-001 to determine baseline security measures.[6]

Robinson Aviation, Inc. (RVA), a Virginia-based company that conducted GA security assessments throughout Virginia and other states, developed an ASP format for GA. Its guidelines are summarized here[7]:

- RVA's GA Airport Security Program Contents Notification that the document is protected as confidential information.
- Distribution of the security plan and record of amendments.
- Airport security coordinator contact information and security responsibilities.
- General airport information: description of the airport, organizational structure, airport activities and characteristics, emergency contact telephone numbers, airport administration, and airport address.
- Definitions and terms.
- Description of the aircraft movement area (AMA). Some GA operators may elect to use the AMA in lieu of using the Part 1542 terminology of air operations area, secured area, and so on to delineate the location that must be protected (i.e., the airfield).
- Description of the airport security procedures, including the designation of areas if so desired, access control procedures, airport security processes, and perimeter barriers/fencing.
- A copy of the airport emergency plan grid map. Normally, this map is used by off- and on-airport fire response personnel to locate crash sites. It is also helpful for law enforcement officers (LEOs) when they need to respond to a particular part of the airfield and should be supplied to those local LEOs who normally patrol or respond to the airport.
- Identification of airport personnel. This section describes the personnel identification and vehicle identification system. At airports with formalized badging, this section describes the background checks that individuals are required to undergo to receive a badge, and any security training they must complete, along with recurrent certification procedures. Although many GA airports do not have a formalized badging process (some do), other forms of ID can be used to properly identify who is on the airfield. Some examples are government-issued photo IDs and FAA-issued pilot certificates, law enforcement or emergency worker identification, FAA or TSA inspector credentials, military identification, or a driver's license. Vehicle identification requirements are also described in this section, along with the requirements to obtain appropriate vehicle identification.

---

[6]The American Association of Airport Executives has also created a series of GA airport security training films available through Digicast video programs (TSA, 2004).

[7]With permission of Robinson Aviation, Inc. (RVA); see *www.rvainc.net*.

- Law enforcement support. This section includes any memorandums of understanding or letters of agreement that exist between the airport and any law enforcement agency, plus descriptions of the roles and responsibilities of federal, state, and local law enforcement agencies with respect to the airport. This section includes communication protocols with law enforcement, records of LEO activity on airport property, and any training programs for LEOs for airport security response procedures.
- Airport security committee. This section describes the makeup, mission, and meeting schedule of the airport security committee. The committee should include the airport operator, local law enforcement and any contracted airport security personnel, representatives from the airport tenants including fixed-base operator management, corporate operators, small aircraft operators, flight schools, aircraft maintenance operators and similar personnel, aircraft rescue and firefighting personnel, and local community representatives. To keep the size of the committee manageable, airport tenants with similar business structures may wish to appoint a designee to represent their interests collectively. For example, at airports where there are dozens of based corporate flight operators and perhaps hundreds of small aircraft operators, they will usually form their own associations on the airport. These associations should be represented on the security committee.
- Security training programs for airport tenants.
- Local TSA and FBI contact information for individuals who can work with the airport operator on security-related issues.
- Information technology security procedures to ensure computer security is maintained, including management and operational controls.
- Most GA airports have an airport emergency plan (AEP) that describes procedures to be used in the event of a number of emergencies, including aircraft crashes, power outages, hazardous material incidents, and bomb threats. When an ASP is developed for a GA airport, the security sections—hijacking, bomb threat, IED detection and mitigation, and so on—should be extracted from the AEP and placed in the security plan. This will prevent the inadvertent release of sensitive security information (SSI) to the general public as the AEP is usually a public document and susceptible to open records requests.

Should GA airports be regulated, it is likely that the regulations will contain elements of, if not completely mirror, Part 1542 Partial Security Programs. In developing an ASP for a GA airport, the operator should take this into consideration.

GA Airport Security Best Practices

Creating security procedures at GA airports requires operators to consider numerous logistical concerns. GA airport operations vary widely in size and nature. Some may be single dirt strips with a couple of flight training aircraft conducting three or four operations a month. Others may be large, busy corporate GA airports with numerous runways located near large cities, accommodating 20,000 operations or more a month, mostly with large business jet aircraft. Many smaller GA airport operators believe they pose very little security risk, whereas larger GA airports provide access to aircraft carrying tens of thousands of gallons of fuel, making them potential weapons.

Staffing at GA airports also varies. Smaller airports tend to have one person who acts as the airport manager and fulfills all roles related to keeping the airport operational. At some of these small facilities, the airport manager may not be a full-time employee. In some locations, the manager is a staff member from the city or county public works department who has been assigned the duty of managing the airport. Medium-sized GA airports usually have staff ranging from 2 to 10 personnel, whereas the largest GA airports can employ 20 or more, allocated into divisions such as operations, maintenance, planning, and administration. Large GA airports have organizational structures that can usually focus more effort on managing security.

Funding for security at GA airports is also problematic. The cost of providing security can be prohibitive to a GA airport, as the GAO (2004) noted in the following report:

> General aviation airports have received some federal funding for implementing security upgrades since September 11, but have funded most security enhancements on their own. General aviation stakeholders we contacted expressed concern that they may not be able to pay for any future security requirements that TSA may establish. In addition, TSA and FAA are unlikely to be able to allocate significant levels of funding for general aviation security enhancements, given competing priorities of commercial aviation and other modes of transportation. (p. 39)

Several airport directors interviewed for the GAO (2004) report noted the use of AIP funds normally reserved for airport safety capital improvements, to pay for security upgrades:

> Because general aviation airports are generally not subject to any federal regulations for security, in order to meet eligibility requirements for their grants, general aviation airport projects are generally limited to those related to safety but have security benefits, such as lighting and fencing, as well as the acquisition and use of cameras, additional lighting, and motion. (p. 40)

Funding may be available from the DHS through the State Homeland Security Grant Program. The State Homeland Security Grant Program provides funding to states to purchase equipment to protect critical infrastructure. This includes funding safety measures applied to GA airports. For example, Wisconsin "plans to use at least $1.5 million of its $41 million Homeland Security Grant in 2004 to enhance security at general aviation airports located along the Great Lakes" (GAO, 2004, p. 41).

Preventing aircraft theft is primarily the aircraft owner's responsibility. Simple precautions such as locking the aircraft when not in use, securing the flight controls with throttle or control locks, putting the aircraft in a hangar, and placing antitamper tape over the aircraft doors can significantly reduce the risk of theft. These security practices should be encouraged in airport literature, signage, and with any airport and FBO personnel.

State of GA Security

In 2007, the Airport Cooperative Research Program (ACRP)/Transportation Research Board (TRB) completed a study of GA safety and security practices. The report discovered several interesting perspectives related to GA security:

- Federal regulation of general aviation airports is very limited and often left to the states, more than half of which have licensing and inspection requirements for GA airports (ACRP, 2007, pp. 6–7). These requirements are most often related to safety and follow the Part 139 guidelines, but they are not necessarily related to security.
- GA airports frequently enter into agreements with local emergency responders and other federal and state law enforcement agencies to assist with security (ACRP, 2007, p. 14).
- Although both the FAA and the TSA have promulgated legislation to address general aviation airports, few actual regulations have been generated (ACRP, 2007, p. 15).
- The majority of GA airports (80%) that responded to the TRB's survey reported that they do have a security plan in place. However, many do not have an active airport security committee (ACRP, 2007, p. 15).
- With respect to perimeter fencing, most of the airports were greater than 40% fenced, and fencing was reported to be one of the primary upgrades to security. The report also noted that most fencing was in place before 9/11 and was intended to deter wildlife from the airfield (ACRP, 2007, p. 16).
- Most airports participate in the Airport Watch program of the AOPA (discussed later in the chapter), and one airport, Centennial Airport in Denver, CO, takes it a step further and offers a reward program for providing information that leads to an arrest or investigation.
- Several states, including Virginia and Colorado, had taken additional measures related to GA airport security. Virginia now has a voluntary security certification program and airports may receive education as well as technical and resources assistance to secure their facilities. Colorado's Office of Preparedness and Security developed the Terrorism Protective Measures Resource Guide, and one airport operator called in the National Guard's Full-Spectrum Integrated Vulnerability Assessment team to assess the airport's vulnerabilities.
- Several industry organizations including the National Business Aviation Association and Airport's Council International developed security guidance for their members.

Attempts to Secure General Aviation

In 2003, the TSA commissioned an aviation security advisory committee comprised of numerous industries such as the American Association of Airport Executives, the National Business Aviation Association, and the National Air Transport Association. The committee developed a comprehensive list of recommendations, which were forwarded to the TSA for consideration resulting in Information Publication (IP) A-001 Security Guidelines for General Aviation Airports (TSA, 2004). IP A-001 is publicly available through the TSA. It is not mandatory for GA airports.

IP A-001 focused on security enhancements in six areas: personnel, aircraft, airports/facilities, surveillance, security procedures and communications, and specialty operations (e.g., agricultural flights). Among the recommendations are the following:

1. Personnel:
   a. The PIC should verify the identities of his or her passengers.
   b. Restrict access to aircraft by student pilots.
   c. Fixed-base operators should implement sign-in/sign-out procedures for transient pilots.
2. Aircraft:
   a. Aircraft should be secured using existing mechanisms such as door locks, keyed ignitions, aircraft stored in hangars, and the use of propeller or throttle locks or security tape.
3. Airports/facilities:
   a. Implement reasonable vehicle access controls to facilities and ramps, including signage, fencing, gates, or other positive control techniques.
   b. Install outdoor lighting covering aircraft parking and hangar areas, fuel farm and fuel truck parking areas, and airport access points. Fuel trucks and fire trucks should require higher levels of protection.
   c. Secure hangar doors when unattended.
   d. Post signs warning against trespassing and unauthorized use of aircraft, and post phone numbers to airport operations and the nearest law enforcement agency (Figure 9.4).

FIGURE 9.4 Signs are posted at airports listing the GA security hotline.

4. Surveillance:
   a. Implement an airport community watch program such as the AOPA's Airport Watch.
   b. Familiarize local law enforcement officers with the airport facilities. This should include advising law enforcement of who is authorized to be on the property, how to drive on the airfield without endangering aircraft, and the normal operations for that airfield.
5. Security procedures and communications:
   a. Create a security plan for the airport including an emergency locator map (often used by the fire department to quickly find facilities at an airport), and create procedures for handling bomb threats and suspected stolen or hijacked aircraft.
   b. Develop a threat communications system including 24-hour phone numbers for the airport director, local police or sheriff's department, fire department, FBI, TSA, and other organizations.
6. Specialty operations:
   a. This section addresses agricultural aircraft operations and the importance of aircraft owners to control access to their aircraft and chemicals.

The TSA also included an airport self-assessment that airport operators could use to determine security measures they should be implementing. The TSA has attempted to put the assessment tool into a computer program, but that effort has not been successful because of the variety of operational characteristics in GA airports. Frustrated with what was perceived as a lack of effort in GA security, Congress included in the 9/11 Act language requiring the TSA to assess the threats and vulnerabilities of GA airports by August 2008. The TSA missed the deadline. However, the TSA is taking an additional step with the assessment tool by giving airport operators "credit" for security actions they have previously taken. The assessment tool rated airports based on their vulnerability, but it did not have an offset score if the airport operator had taken prior security mitigation measures.

State Actions

Several states have started to adopt their own general aviation airport regulations. Many of these programs have been in response to specific aviation security incidents such as the theft of an aircraft from the Danbury Airport in Connecticut in 2005. A 20-year-old intoxicated man hijacked the aircraft and then flew himself and three friends from the GA airport to Westchester County Airport in New York. Because of this incident, a WABC–New York Channel 7 investigation discovered lax security at several GA airports. WABC–New York reported observing open gates along with fencing in various states of disrepair at these GA airports. These conditions should not be altogether surprising, as there is presently not a mandate for GA airports to have security measures, nor is there funding to establish security measures at GA airports.[8]

---

[8]Danbury's airport management had implemented various security measures before the theft.

With high fuel capacities, many GA business jets could serve as effective weapons if used in a manner similar to the airliners on 9/11. These incidents have prompted states to take various security-related actions at GA airports. For example, Colorado conducted a study of its 64 GA airports to determine what TSA security measures should be applied. In 2004, New York passed the Anti-Terrorism Preparedness Act, which required all GA airports in the state to register with New York's Department of Transportation and create a security plan. New York passed a bond measure that provides $30 million in capital improvements for security measures at GA airports. In Connecticut, the AOPA worked with the Governor's Aviation Task Force to promote GA security. In Virginia, the state requires GA airports to conduct vulnerability assessments by different agencies, such as law enforcement and security consultants. At Centennial Airport in Colorado, airport director Robert Olislagers raised fuel fees with the consent of his FBOs to pay for security enhancements and worked with the TSA on testing new GA-related security programs.

Historically, it has been illegal for states to impose regulations on federally funded airports. Although this has occurred in some states, such as California, the FAA is very careful that state restrictions or regulations on airports do not interfere with the national air traffic control system or intrastate commerce. New York attempted to require pilots in their state to undergo criminal history record checks (CHRCs). The AOPA sued New York, contending that any attempts to regulate security would be a violation of the supremacy clause of the U.S. Constitution, which invalidates state laws that conflict with federal laws. The AOPA contends that the federal laws on pilot background checks were adequate. In 2002, several states attempted to pass laws placing severe restrictions on GA aircraft operators but met strong opposition by the FAA.

AOPA's Airport Watch

One of the more popular GA security programs is the AOPA/TSA partnership Airport Watch. This program enlists the support of 550,000 GA pilots to watch for and report suspicious activities that might have security implications. The AOPA has distributed Airport Watch materials to 5,400 public-use GA airports, pilot groups, and individual pilots. To build on the success of these local efforts, the program includes special materials such as a video to train pilots to be alert for suspicious people or activities at the airport. Airport Watch encourages GA pilots and airport operators to call 911 or the GA security hotline (1-866-GA-SECURE) when they see suspicious activity at their airport (see Figure 9.4).

Airport Watch recommends that people watch for the following as indications of security-related concerns:

- Pilots who appear to be under the control of another party.
- Anyone trying to access an aircraft through force—without keys, or using a crowbar or screwdriver.
- Anyone who seems unfamiliar with aviation procedures trying to rent an airplane.
- Anyone who misuses aviation terminology or jargon.

- People or groups who seem determined to keep isolated, including stakeholders to the airport who strive to avoid contact with you or other airport tenants.
- Anyone who appears to be just loitering, with no specific reason.
- Any out-of-the-ordinary videotaping of aircraft or hangars.
- Aircraft with unusual or obviously unauthorized modifications.
- Dangerous cargo or loads—explosives, chemicals, openly displayed weapons—being loaded into an airplane.

In 2011, the TSA established See Something, Say Something, which is a nationwide effort adopted originally from the Port Authority of New York and New Jersey. The program uses videos and signs to promote the reporting of suspicious activity and personnel to the TSA and the FBI.

Fixed-Base Operator Security

FBOs serve as the "truck stops" for GA aircraft. FBOs provide fueling, aircraft handling, and hangar storage services. Many FBOs also provide aircraft charter services. FBOs and business aircraft operators share the responsibility and work together to secure the business aviation industry.

FBOs are predominantly represented by the National Air Transport Association. Signature Aviation Flight Support was one of the first FBOs to develop its own security procedures, which include the following recommendations:

- Flight crews must provide photo identification to gain access to the Signature ramp.
- Flight crews must identify all passengers before they are allowed access to Signature's ramp and escort passengers to and from their aircraft.
- FBO personnel must maintain line-of-sight surveillance of passengers on the ramp who are not escorted by the pilot.
- Passengers are asked to walk directly, and as a group, to their aircraft and not delay entering their aircraft.
- Transient flight crews are issued special codes upon their arrival to guarantee that the same flight crews return to their respective aircraft.
- The use of personal cars or limousines on Signature's ramp is not permitted.

Secure Fixed-Base Operator Program

TSA's Secure Fixed-Base Operator Program is a proof of concept, public–private sector partnership program that will allow FBOs to check passenger and crew identification against a manifest or eAPIS (electronic Advance Passenger Information System) filings for positive identification of passengers and crew onboard general aviation aircraft. eAPIS checks are becoming an increasing part of GA security as DHS focuses more on deterring or catching individuals with ill-intent coming into the United States.

Helicopter Security

In 2009, U.S. and Canadian officials intercepted a large drug smuggling operation between the two countries that resulted in eight arrests, and the seizure of cocaine, marijuana, and over 200,000 ecstasy tablets, plus two Bell Jet Rangers helicopters. Known as Operation Blade Runner, the Drug Enforcement Agency (DEA) followed the flow of narcotics from British Columbia to the state of Washington, to a remote forest heliport (Price, 2009, p. 23). While helicopters have not historically been the target of bombs or hijackings at the same rate as aircraft, the use of helicopters in criminal and terrorist activities must still be considered.

Helicopters are able to fly at slow speeds through congested city areas making them excellent platforms to fly in armed personnel to conduct a terrorist active shooter incident inside a stadium, airport, or any open-air assembly, bypassing any ground defenses. Helicopters have mostly been used for prison escapes, and as observation platforms for follow-up criminal activity. Business executives are vulnerable as well with attempted helicopter hijackings to kidnap valuable executives. In 2004, Pakistan provided the U.S. government with information that revealed a plot to attack New York City and other targets in the country using helicopters.

As with other forms of aviation, security measures should not eliminate the inherent benefit and use of this form of transportation or use. Helicopters are profitably operated as tourist platforms and for all sorts of other uses such as oil rig transport, medical and law enforcement, power line maintenance, and so on. Good helicopter security should include suspicious awareness training for helicopter pilots and helicopter operations personnel. Pilots should be aware of questions that could be asked by passengers that would generate suspicion.

Security programs should be part of any helicopter operation with procedures for pilots under duress (code words, panic alarms), being hijacked, or feeling that there is a threat on their helicopter. Operators may wish to install CCTV into their helicopters so that the flight can be monitored remotely, and helicopter operators should conduct background checks on their pilots prior to hiring. For corporate helicopters carrying VIPs, additional threats may need to be addressed such as surface-to-air missile defenses and executive security protection practices. In certain cases, commercial helicopter operations feed directly into aircraft flight operations, so any charter or scheduled helicopter flight that transfers baggage and passengers into a commercial airliner under a security program should ensure that all personnel and items have undergone the appropriate screening.

Training for GA Airport Security and Law Enforcement Personnel

In 2006, the Waukesha County Technical College received a federal grant to study and develop training programs for general aviation stakeholders. They implemented the training in 2007 and are continuing the program, both in face-to-face sessions and online. The training is aimed at GA airport operators who may not have a significant security background, newcomers to the GA security industry, aviation businesses, pilots and aircraft owners, and law enforcement and emergency workers with GA airports in their jurisdiction. GA security-related training programs include topics such as:

- How to recognize GA aircraft and facilities that could be used for illegal purposes
- How to apply crime prevention through environmental design concepts to GA airports
- Establishing an Airport Watch program
- Establishing an aircraft key control system
- Antitheft devices for GA aircraft
- Security signage and marking plans
- How to orient local LEOs to the airport environment and aircraft operations
- Creation of an airport security committee
- Creation of an emergency notification system
- National Incident Management System fundamentals
- How to create a business continuity plan
- Developing instruction detection, integrated security, and CCTV systems
- Troubleshooting airport security plans

Most GA airport operators do not have aviation security experience or may only have commercial airport security experience. Therefore, training in GA security should be required for anyone without a GA security background.

Alien Flight Student Program

Title 49 CFR Part 1552 addresses the security of flight schools and awareness training for flight school employees. Considering that the 9/11 hijackers were trained at U.S. flight schools, it made sense to begin applying security regulations for flight schools. Shortly after 9/11, the FAA sent security guidance to all flight schools,[9] much of which was later formalized into the rulemaking under this section. Originally known as the Alien Flight Training Rule (now known as the Flight Training Candidate Checks Program), this section requires every flight student to prove his or her U.S. citizenship before beginning flight training in an aircraft weighing 12,500 pounds or less. Foreign flight students are required to complete a CHRC and submit their fingerprints to the TSA. This rule required flight schools and flight instructors to provide security awareness training to each ground and flight instructor and any employee who has direct contact with a flight school student (regardless of citizenship or nationality) and to issue and maintain records of this training (e-CFR, 2008). The regulation not only applies to the thousands of flight schools across the United States but also to every individual certified flight instructor, every airline

[9]The FAA sent the notice to FBOs, which the FAA defined as any aviation business, whether a flight school, aviation fuel provider, charter service, or aviation maintenance operator.

flight training academy that uses flight simulators, and any aviation college or university conducting flight training.

By regulatory definition, flight training consists of aircraft simulators (a flight training device as defined under Title 14 CFR Part 61.1), flight schools (any flight training facility operating under Title 14 CFR Parts 61, 121, 135, 141, or 142), flight training (instruction received in an aircraft or simulator not including recurrent or ground training), ground training (classroom or computer-based instruction not including instruction in a simulator), or recurrent training (periodic training required under Title 14 CFR Parts 61, 121, 125, 135, or Subpart K of Part 91). Collegiate aviation flight training programs are also covered under this regulation.

Section 1552.3 applies specifically to instruction given in an aircraft with a MGTW of more than 12,500 pounds. The regulation requires flight schools to notify the TSA that a candidate has requested such flight training. A form must be submitted to the TSA that includes valid personal identification information, proof of citizenship, the type of training being applied for, fingerprints (unless already on file with the TSA), and a photograph of the candidate. Before flight training can begin, the TSA must inform the flight school that the candidate does not pose a threat to aviation or national security, or more than 30 days have elapsed since the TSA received the information. In either event, the flight school can begin the candidate's flight training within 180 days. A flight school may request expedited processing if the previous conditions are met, the candidate is eligible for expedited processing, and information establishing eligibility is presented.[10]

For candidates wishing to receive flight instruction on an aircraft that is 12,500 pounds or less, the flight school or flight instructor must verify the student's U.S. citizenship; if the candidate is not a U.S. citizen, the flight school or flight instructor must require the individual to register with the TSA before beginning flight training. If the student is attending an aviation university or college using aircraft or simulators to conduct flight training, the same information is submitted to that entity. The flight school must immediately terminate or cancel a candidate's flight training if the TSA notifies the flight school at any time that the candidate poses a threat to aviation or national security.

Flight School Security Awareness Training

Good security of flight school aircraft or any aircraft at a GA airport should include practices such as locking aircraft; using throttle, prop, and tie-down locks; securing hangar doors; not leaving ignition keys in the aircraft, even momentarily; reporting suspicious

---

[10]A candidate is eligible for expedited processing if he or she already holds an airman's certificate from a foreign country that is recognized by the FAA or a military agency of the United States, or the candidate is employed by a foreign air carrier that operates under Title 14 CFR Part 129 and has a security program approved under Title 49 CFR Part 1546. The individual must also already have successfully completed a CHRC in accordance with Title 49 CFR Part 1544.230, or is part of a class of individuals who the TSA has determined poses a minimal threat to aviation or national security because of the flight training already possessed by that class of individuals.

activity to the authorities immediately; and vigilance on the part of the flight instructor over his or her new students until trust has been established.

The TSA has also created a flight school awareness training program. Title 49 CFR Parts 1552.21 and 1552.23 are specific to flight schools operating under Title 49 U.S.C. Subtitle VII. It introduces the flight school employee, defined as a flight instructor, chief flight instructor, director of training, or an independent contractor, who has a contract with a flight school to provide flight instruction. A flight school employee also includes a ground instructor at a college or university that provides flight instruction. Part 1552.23 requires that each flight school employee must complete the TSA's Flight School Security Awareness Training Program (see *http://download.tsa.dhs.gov/fssa/training*). Flight schools must maintain records of the security awareness training completed by their personnel. The training includes situational scenarios, suspicious behavior of personnel, signs that an aircraft has been altered for illegal uses, and other information useful for identifying flight candidates who may be pursuing illegal activities. Flight schools may elect to use their own training program, which must include everything in the TSA's online training.

Shortly after 9/11, the FAA issued guidance for flight school security. The TSA followed up with its own handout in 2006 (Figure 9.5).

Maryland-3

Title 49 CFR Part 1562 addresses three GA airports located close to the center of Washington, DC: College Park Airport, Washington Executive/Hyde Field, and Potomac Airfield (Figure 9.6). They are known as the Maryland-3 and are presently the only GA airports with specific security regulations.

Each of the Maryland-3 airports must have an airport security coordinator who, similar to an ASC at a commercial-service airport, administers the TSA's security requirements. These include maintaining a copy of the airport security procedures and a copy of the FAA procedures applicable to GA operations at the Maryland-3 airports. The TSA has the authority to inspect the airport and related security program documentation. Security coordinators at these airports must provide their names and other identifying information including fingerprints and social security numbers to the TSA. They must also have successfully completed a security threat assessment (STA) and not have been found guilty (or not guilty by reason of insanity) of any of the disqualifying offenses as outlined in Title 49 CFR Part 1542.209.

A Maryland-3 ASC must implement a set of security procedures. These include keeping a record of those authorized to use the airport, monitoring the security of aircraft during operational and nonoperational hours, alerting aircraft owners/operators and the TSA of unsecured aircraft, implementing and maintaining security awareness procedures at the airport, and limiting approval for pilots who violate the Washington, DC, metropolitan area flight-restricted zone and are forced to land at the airport. They also contain any additional procedures required by the TSA to provide for the security of aircraft operations into or out of the airport.

U.S. Department of Homeland Security
601 South 12ᵗʰ Street
Arlington, VA 22202

**Transportation
Security
Administration**

Advisory – Security Information for Flight Schools and Flight Training Providers

September 1, 2006

The Department of Homeland Security and the Transportation Security Administration continue to monitor reports on potential terrorist threats in the United States. Based on a recent interagency review of available information, we remain concerned about Al-Qaeda's continued efforts to plan multiple attacks against the United States. These attacks may involve aviation.

Recently, an aviation related incident was reported surrounding suspicious activities at flight schools. At this point there is no indication that this occurrence is terrorist related, however, we request airport managers, flight schools, flight training providers, and aircraft operators remain vigilant for suspicious behavior and activities.

TSA reminds general aviation aircraft and airport owners and operators to review the security measures contained in the TSA Information Publication, Security Guidelines for General Aviation Airports (available at http://www.tsa.gov/public/interapp/editorial/editorial_1113.xml), and the Aircraft Owners and Pilots Association's Airport Watch Program materials (available at www.aopa.org/airportwatch).   In addition, general aviation aircraft and airport owners and operators are encouraged to consider the following:

- Secure unattended aircraft to prevent unauthorized use.
- Verify the identification of crew and passengers prior to departure.
- Verify that baggage and cargo are known to the persons on board.
- Where identification systems are in place, encourage employees to wear proper identification and challenge persons not wearing proper identification.
- Direct increased vigilance to unknown pilots and/or clients for aircraft rental or charters – as well as unknown service/delivery personnel.
- Be alert/aware of and report persons masquerading as pilots, security personnel, emergency medical technicians, or other personnel using uniforms and/or vehicles as methods to gain access to aviation facilities or aircraft.
- Be alert/aware of and report aircraft with unusual or unauthorized modifications.
- Be alert/aware of and report persons loitering in the vicinity of aircraft or air operations areas – as well as persons loading unusual or unauthorized payload onto aircraft.
- Be alert/aware of and report persons who appear to be under stress or the control of other persons.
- Be alert/aware of and report persons whose identification appears altered or inconsistent.

The theft of any General Aviation aircraft should be **immediately** reported to the appropriate authorities and the TSA General Aviation Hotline at 866-GASECUR (866-427-3287). In addition, persons should report any suspicious activity **immediately** to local law enforcement and the TSA General Aviation Hotline.

FIGURE 9.5 TSA advisory to flight schools.

Each pilot flying into and out of the Maryland-3 must provide the TSA with his or her name, social security number, date of birth, current address and phone number, current airman or student pilot certificate, medical certificate, a government-issued photo ID, and fingerprints. Pilots must provide a list of the make, model, and registration number of each aircraft that he or she intends to operate at the airport. Each pilot must successfully

FIGURE 9.6 Example of Temporary Flight Restriction over Denver, Colorado.

complete an STA and receive a briefing acceptable to the TSA and the FAA of the operating procedures for the airport. Pilots may not have been convicted or found not guilty by reason of insanity, in any jurisdiction, during the 10 years preceding application for authorization, of any of the disqualifying offenses listed in Title 49 CFR Part 1542.209. A pilot must not have a record on file with the FAA of a violation of a prohibited area, a flight restriction, any special security instructions issued under Title 14 CFR Part 99.7, a designated restricted area, any emergency air traffic rules issued under Title 14 CFR Part 91.139, or a temporary flight restriction. If a pilot or ASC violates any of these or commits a disqualifying offense, he or she must notify the TSA within 24 hours of conviction. Pilots flying into the Maryland-3 must secure their aircraft after flight, file a flight plan (IFR or VFR), and obtain an air traffic control clearance and transponder code.

Aircraft Security under General Operating and Flight Rules

Title 49 CFR Part 1550 applies to general aviation aircraft operations. It provides for the TSA to inspect any aircraft operations into sterile areas of commercial-service airports. It includes requirements of the 12-5 security program for aircraft engaged in scheduled service, charter, or cargo operations that are more than 12,500 pounds and not under any other security program. Although the regulations in this section are not extensive, they do provide a foundation for future regulation of these operations. As with other aircraft operator regulations, any aircraft operator must allow TSA personnel to inspect and copy records related to activities under this regulation.

For aircraft operations where a sterile area is involved, either if a passenger or crewmember enplanes from or deplanes into one, if the operation is not already covered under a scheduled passenger service security program (full or partial) or public or private charter operations under Title 49 Part 1544 or 1546, screening must be done. Operations covered by this section include private aircraft operations and operations under the 12-5 security program when enplaning from or deplaning into a sterile area.

12-5 Standard Security Program

Aircraft of 12,500 pounds or more not operating into or out of a sterile area and not covered under a program under Title 49 CFR Part 1544 or 1546 (Domestic or Foreign Air Carrier Operations) must adhere to the 12-5 rule. Any aircraft operator conducting commercial aircraft operations (scheduled service, private or public charter, or cargo) where the operator owns or has managing control of the aircraft must search the plane before departure and screen passengers and carry-on items. The TSA must first notify the aircraft operator that such security measures are necessary through NOTAM, a letter, or other communication. Aircraft operators can apply for waivers from the TSA for these procedures. Presently almost 1,000 aircraft operators must adhere to the 12-5 rule.

Agricultural Aircraft Security

The FBI indicated that in 1993 Osama bin Laden assessed the possibility of using an agricultural aircraft as a method to disperse chemical or biological weapons. In 2001, before 9/11, Canton (2006) in *The Extreme Future* discussed a model he had completed on an anthrax attack on the National Mall in Washington, DC, using an aircraft to disperse the anthrax agent. The results showed that nearly 100,000 people could be at risk, and the resultant shut down of the facilities in the affected area, including hospitals, would be devastating.

After 9/11, the FBI visited the thousands of agricultural pilots across the United States to assess the potential risk from this industry. In an effort to prevent the theft of their planes for potential use in a terrorist attack and to ward off stifling federal regulations, the National Agricultural Aviation Association developed several security recommendations and participated on the TSA advisory committee. Although the threat from a stolen agricultural aircraft exists, the aviation agricultural community is small and well known to the local farmers and chemical distributors.

Security recommendations for agricultural operators include securing aircraft and chemical agents when not in use, enhancing the lighting around aircraft and chemical storage areas, parking farm equipment such as loaders and combines in areas that prevent taxiing of aircraft when not in use, and establishing contact with local law enforcement agencies. Immediately reporting stolen aircraft or chemical agents is fundamental to agricultural aviation security. Agricultural aircraft could be used to spread chemical, biological, or radiological weapons, and many of the ingredients can be found within the same community.

Corporate Aviation Security

There is no hard-and-fast definition of a corporate aircraft. A corporate aircraft could be a Cessna 172 that a business uses to fly for servicing regional accounts. A corporate aircraft could also be a Gulfstream V, a large business jet, owned by a large corporation and that flies to destinations around the world. Although the security measures discussed in this section are directed primarily at the larger business aircraft, they could also apply to smaller aircraft operations.

Overall, the business aviation community may not have the security measures of commercial aviation operations, such as passenger and baggage screening, but it is a mistake to think that no security measures are present. A corporate aircraft is a multimillion-dollar investment for a business, and businesses owners want to ensure that investment is protected. Long before 9/11, corporations already took many preventative measures to ensure the safety and the security of their aircraft and passengers. It is not uncommon for a business aircraft to fly into a small airport that is more convenient for the passengers but lacks the hangar space to store the aircraft, then shuttle the aircraft to another larger airport to hangar the aircraft until it is needed again, hours or days later.

Another factor with corporate aircraft security is that corporations often have business dealings throughout the world. Corporate aviation security considerations must include differences in culture, customs laws, currency, immigration, and the threat to the aircraft and its occupants in various parts of the world.

In 1992, Harry Pizer and Stephan Sloan published *Corporate Aviation Security.* Although the text is dated and the perceived threats to corporate aviation have not occurred, their book provided a good foundation for corporate flight departments. Owners of small aircraft used for business purposes can also benefit from their lessons. Their key points are summarized here, together with security best practices from the National Business Aircraft Association (NBAA) and other industry practices.

Before 9/11, the main security concerns for a corporation with respect to its aircraft were protecting the asset from theft and vandalism and protecting personnel from kidnapping, particularly in certain less stable parts of the world. Corporations also worried about their aircraft being used, either knowingly or unknowingly, to transport illegal goods. After 9/11, the security concerns increased to include bombings and hijackings with the intent of using the aircraft as a weapon.

Corporations are also motivated to provide for security of their flight operations as a litigious society has made certain that were anything to happen to the safety and security of those carried onboard a corporate plane, the corporation would have to demonstrate in court that diligent and reasonable security measures were taken (Pizer and Sloan, 1992).

Although most corporations that own aircraft have just one, a few have dozens, and even fewer have more than a hundred. Some operators, such as NetJets®, a fractional operator that provides corporate aircraft for companies, manage several small fleets of aircraft of various sizes and capabilities.

Regardless of the size of the company, every corporate aircraft operator should have a designated security director and a standard aircraft operations security program administered by the coordinator. Depending on the size of the fleet, the security director may have a small staff to handle administrative and training issues and may have a staff of certified protection professionals who are armed and trained in the executive protection techniques. A few aircraft operators are large enough and have enough cause to go into Reagan National Airport that they employ their own armed security officers (ASOs) who are certified law enforcement officers. ASOs are required to accompany GA aircraft flights into Reagan National Airport. A security committee should be established. It should be chaired by the security director and include representatives from the corporation's pilots, flight attendants and mechanics, dispatch and flight planning personnel, and representatives from management. Other members may be retired or active federal, state, and local law enforcement officers and security consultants.

The corporate security program should address airport security measures, predeparture procedures, operational security (OpSec), in-flight security, arrival and destination security, intelligence, security screening (where appropriate), measures to keep prohibited items off company aircraft, executive protection measures, contingency plans, and security training for flight crew.

Airport security measures should be addressed with the respect of protections the airport offers to its tenants. As a minimum, the airport should meet the TSA's GA airport security guidelines and have implemented the AOPA's Airport Watch program. Perimeter fencing, appropriate hangars for aircraft, adequate ramp lighting, CCTV monitoring of the flight line, 24/7 security, law enforcement, FBOs, and airport operations personnel should also be on hand. A database should be established that profiles the airports the corporation typically flies into. This should include security threats and mitigation measures taken by the airport operator and a liaison maintained with airport and FBO management personnel at these facilities, particularly at foreign airports. When a new airport is used, the security director should create a new profile and provide a predeparture briefing to the flight crew and passengers.

The following list is provided by the NBAA (1995–2008) and should serve as minimum security requirements at the corporate operator's home base:

- Ensure home facility perimeter security with effective fencing, lighting, security patrols, gates, and limited access areas
- Ensure street-side gates and doors are closed and locked at all times
- Require positive access control for all external gates and doors
- Close and lock hangar doors when the area is unattended
- Secure all key storage areas (food and liquor, parts and tools, etc.)
- Have an access control management system for keys and passes
- Confirm the identity and authority of each passenger, vendor, and visitor prior to allowing access to facilities and aircraft
- Escort all visitors on the ramp and in the hangar area
- Use a government-issued photo ID to verify the identity of any visitor or vendor
- Post emergency numbers prominently around the facility
- Ensure easy access to phones or "panic buttons" in various facility locations (break room, hangar bay, etc.)
- Confirm security of destination facilities
- Be aware of your surroundings and do not be complacent—challenge strangers

Operational security, or OpSec, means keeping private matters private. Only those with a "need to know" should be provided with information relating to operational security. OpSec includes the details of any corporate travel, security of computer information systems including laptops and personal digital assistants that may include corporate proprietary information, and the protection of any written materials that may be kept onboard during the flight. Dissemination of information such as the departure and arrival times, who will be onboard, what will be carried on the flight, the flight's destination, and the route of flight should only be divulged to those with a need to know (i.e., those who need to know certain information to perform their duties). OpSec not only protects the aircraft operator and its occupants, but may also protect the company from divulging information to potential competitors (Pizer and Sloan, 1992).

Predeparture security procedures are completed whether the aircraft is at its home base or at a transient location. These include observing all maintenance on the aircraft during the time it is in a foreign airport and observing catering, fueling, and other activities to ensure that no dangerous or illegal items are placed onboard. Any stores, such as food, beverages, cleaning materials, etc., should be inspected before being brought onboard. People not associated with the aircraft should be kept away from it (Pizer and Sloan, 1992).

If the aircraft has been unattended for any time, particularly if left outside, a complete preflight inspection should include checking the aircraft for unauthorized items. In some cases, a corporation may have its own security personnel who are responsible for watching the plane when it is not in service. Some operators have also installed video cameras on the exterior of their aircraft to watch for suspicious activity while on the ramp.

Predeparture procedures include checking the identity of all personnel onboard the flight, and, where appropriate, screening passengers and bags. Passengers should be either on a list of those authorized to fly on the corporation's plane or be vouched for by a passenger on the authorization list.

Predeparture procedures include collecting as much information as possible about the destination airport(s) and disseminating it to the passengers and crew via an intelligence briefing. The briefing is prepared by the security director and should include the following:

- Potential threats including street crime statistics; areas of town to be avoided; known recent hijackings, kidnappings, or extortion attempts of executives traveling to that region; situations of general unrest; rioting; and the political and social climate.
- Airport security measures at the destination airport.
- Information on FBOs including any credentialing or background check requirements on FBO personnel, including caterers, fuelers, and mechanics.
- Emergency contact information for the U.S. embassy, Department of State, local law enforcement agencies, and U.S. military bases that may be nearby.
- Information on ground transportation to and from local hotels, meeting locations, and so on.
- Companies that carry well-known celebrities or political personalities must also be aware of the presence of unwanted news media and paparazzi.

When traveling to foreign airports, many corporations elect to use a ground-handling agent who can assist with customs and immigrations paperwork and other laws and regulations for traveling to a specific country. A basic background check should be done on any company or individual who acts as a handling agent for the corporation.

Intelligence

Corporations, particularly large corporations, should generate their own intelligence briefings and reports that are relevant to the operations of the company. This information can be used in flight planning and in developing security awareness processes. Jane's Information Group is a common source of this type of intelligence information and can keep corporations notified about worldwide news, country risk assessments, and other public safety and security information. Companies such as Jeppesen also offer extensive international trip planning services. Intelligence information should include a general overview of the destination and information relevant to the specific location of the flight crew and passengers.

In-flight Security

The events of 9/11 put a new face on the nature of a hijacking. Previous guidance to comply with hijackers' requests and work to land the plane and negotiate can no longer be trusted.

Flight crewmembers should assume the worst-case scenario if their aircraft is hijacked and take whatever action is necessary to defend the flight deck from intrusion and takeover. Although most corporate pilots know their passengers personally, and the passengers may be the employers of the flight crew, there are instances where a hijacking could occur. A hijack could be attempted by someone who has duped a corporate officer into being allowed onto the plane or by a disgruntled company employee who is on the authorized passenger list. With the increase of fractional operators, the pilots may not know whom they are carrying as much as a corporate flight crew who is flying the boss's plane. An aircraft could also be taken over while it is still on the ground as the flight crew makes its departure preparations.

Flight Crew Training

Flight crew personnel should receive self-defense training and conduct exercises in how to handle a hijacking attempt. After 9/11, some corporate flight crewmembers were carrying their own personal firearms for protection. Local regulations will govern this procedure, as the federal regulations do not address the carriage of firearms in an aircraft under a Part 91 operation.

Other Security Practices

Other security practices for corporate operators include the establishment of contingency plans for unattended items, bomb threats, hijackings, and theft attempts. Corporations should conduct regular security awareness training for flight crews and passengers and exercises and simulations for certain aircraft incidents.

NBAA TSA Access Certificate Program

A major benefit to corporate flying is the ability to access more than 5,000 U.S. public-use GA airports instead of being restricted to 450 commercial-service airports. After 9/11, the U.S. government put tight restrictions on access to U.S. airports by GA aircraft coming into the United States from foreign locations.

To alleviate the problem, the NBAA pioneered the TSA Access Certificate (TSAAC) program. TSAAC allows operators under Title 14 CFR Part 91 (Private Operator/Owner) to fly internationally without having to pass through a "portal" country as outlined in FDC 2/5319. Presently, the portal countries include Canada, Mexico, the Bahamas, England, Scotland, Wales, and Northern Ireland. Under the current pilot program, aircraft operators with a TSAAC certificate may fly out of three airports: Teterboro, NJ, Morristown, NJ, and White Plains, NY.

According to the NBAA website, on August 23, 2006, the FAA issued NOTAM 6/7435, which allows operators of aircraft under 100,309 MGTW to fly internationally without having to pass through a portal country (National Business Aviation Association, 2006). As a result, operators meeting the requirements of this NOTAM no longer need to request waivers or obtain a TSAAC to avoid passing through a portal country.

DCA Access Standard Security Program

Section 1562.21 applies to aircraft operations into or out of Ronald Reagan Washington National Airport (DCA). It applies to FBOs located at DCA or gateway airports, the aircraft operator security coordinator and crewmembers, passengers, and armed security officers on the general aviation aircraft. This section does not apply to medical evacuation flights, commercial-service or military flights, or all-cargo operations.

To operate into and out of DCA, aircraft operators must designate a security coordinator who must have passed a CHRC/STA and implement the DCA Access Standard Security Program (DASSP). Flight crewmembers are required to pass a CHRC/STA and not have any airspace restriction violations on record with the FAA. The aircraft operator must provide passenger manifest information to the TSA at least 24 hours before a flight and a STA will be conducted on each. The aircraft operator may not carry a passenger into DCA who has not passed an STA.

For any flight, an aircraft must fly directly from a designated gateway airport to DCA (no interim stops), and the operator must ensure that passengers and carry-on bags have been screened. Checked baggage and cargo must be screened and be inaccessible to the passengers throughout the flight. Aircraft operating out of DCA must be equipped with a cockpit door and ensure it is closed and locked at all times. Pilots must notify the National Capital Region Coordination Center before departure from DCA or a gateway airport into DCA. The TSA has the authority to inspect the aircraft and the security programs.

Each flight must carry a federal air marshal or an ASO. The ASO program is administrated by the Federal Air Marshal Program. Each ASO must be qualified to carry a firearm and complete an STA and a CHRC. ASOs include active law enforcement officers employed by a governmental agency and retired law enforcement officers who are authorized to carry out the duties of an LEO. Other ASOs must meet the requirements determined by the Federal Air Marshal Program.

In addition to basic law enforcement training, each ASO must complete TSA-specific training related to operations into and out of DCA. Each has the authority to use force, including deadly force. ASOs may not drink alcohol or use intoxicating or hallucinatory drugs during the flight and within eight hours before the flight. ASOs must identify themselves to the flight crew before the flight.

Each FBO operating out of or into DCA (i.e., a gateway airport) must implement an FBO security program, which includes designating a security coordinator, provisions to support the screening of persons and property, and the aircraft search for aircraft heading into or out of DCA.

Conclusion

This chapter analyzed various strategies for protecting general aviation airports and aircraft flight operations. Strategies for securing GA used by government and stakeholders to GA were presented. A key challenge to developing legislation, policies, and processes for securing GA airports is that stakeholders have different perceptions regarding the nature of GA. GA operations can vary greatly in the types of aircraft used and purposes for GA flight operations.

Congress and the TSA have focused GA security efforts on GA aircraft operators rather than GA airport operators. GA aircraft operations are regulated under Title 49 CFR Part 1550 for those aircraft operators conducting charter operations with aircraft in excess of 12,500 pounds MGTW or aircraft operators whose passengers deplane into a sterile area. In 2008, the TSA began working on the Large Aircraft Security Program, which may result in security regulations for GA aircraft private operations conducted under Title 14 CFR Part 91.

The DHS has expressed concern that a GA aircraft could be used as a WMD or to transport dangerous individuals or materials into the United States. However, the TSA has stated that the small size, lack of fuel capacity, and minimal destructive power of most GA aircraft make them unattractive as a WMD. Frustrated with what was perceived as a lack of effort in GA security, Congress included in the 9/11 Act language requiring the TSA to assess the threat and vulnerabilities of GA airports by August 2008.

Government agencies recommend that GA airport operators use security strategies such as perimeter fencing, security patrols, and airfield access control systems. GA airport operators should also create a security plan that includes an emergency locator map and procedures for handling bomb threats and suspected stolen or hijacked aircraft. Some U.S. states require GA airports to conduct vulnerability assessments by different security agencies.

The TSA and the FAA are challenged to allocate significant levels of funding for general aviation security enhancements because of security demands of commercial aviation and other modes of transportation. The TSA does offer a Secure Fixed-Base Operator Program allowing FBOs to check passenger and crew identification against manifest or eAPIS filings for positive identification of passengers and crew onboard GA aircraft.

Without regulatory guidance, the creation of airport security programs for GA airports will continue to be challenging. Although policies and legislation are debated regarding GA security issues, most GA airport operators will continue to use commercial-service airport security programs as models for enhancing security at GA airports.

References

ATSA, 2001. Aviation and Transportation Security Act of 2001, P.L. 107–71 Sec. 132.

Airport Cooperative Research Program (ACRP), 2007. ACRP, Airport Cooperative Research Program, Synthesis 3: General Aviation Safety and Security Practices. Transportation Research Board, Washington, DC, retrieved Aug. 10, 2008, from http://onlinepubs.trb.org/onlinepubs/acrp/acrp_syn_003.pdf.

Canton, J., 2006. The Extreme Future. Penguin Group, New York.

CNN, 2005. Intruding pilots released without charges, May 12, 2005, retrieved Aug 13, 2008, http://www.cnn.com/2005/US/05/11/evacuation/index.html.

DHS OIG, 2009. TSA's Role in General Aviation Security. DHS, Office of the Inspector General, Washington, DC.

e-CFR, 2008. Code of Federal Regulations Title 49 CFR 1552.23 – Security Awareness Training Programs.

Elias, B., 2009. Security General Aviation. Congressional Research Service, Washington DC: General Aviation Security: Aircraft, Hangars, Fixed-Base Operators, Flight Schools and Airports, Dr. Daniel J. Benny, PhD.

Government Accountability Office (GAO), 2004. General Aviation Security: Increased Federal Oversight Is Needed, but Continued Partnership with the Private Sector Is Critical to Long-term Success (Publication No. GAO-05-144), retrieved Aug. 10, 2008, from www.gao.gov/new.items/d05144.pdf.

National Business Aviation Association (NBAA), 1995–2008. Security: Best Practices for Business Aviation Security. NBAA, retrieved Aug. 13, 2008, from web.nbaa.org/public/ops/security/bestpractices.

NBAA, Aug. 23, 2006. Airspace/Air Traffic TFRs Airspace Alerts and NOTAMs, NOTAM FDC 6/7435—U.S. Entry and Overflight Requirements. NBAA, retrieved Aug. 13, 2008, from web.nbaa.org/public/ops/air space/restrictions/2006/200608237435.php.

Pizer, H., Sloan, S., 1992. Corporate Aviation Security: The Next Frontier in Aerospace Operations. University of Oklahoma Press, Norman.

Price, J., June 1, 2009. Helicopters and Heliports: Heli-security. Aviation Security International 15 (3).

Trautvetter, C., Dec. 6, 2006. Stunning Number of TFR Violations Since 9/11. AVweb, retrieved Aug. 13, 2008, from www.avweb.com/avwebflash/briefs/TFR_Violations_193937-1.html.

Transportation Security Administration (TSA), 2004. Security Guidelines for General Aviation Airports. Transportation Security Administration, Department of Homeland Security, retrieved Aug. 10, 2008, from www.tsa.gov/assets/pdf/security_guidelines_for_general_aviation_airports.pdf.

10

Air Cargo Security

Objectives

This chapter examines issues related to aircraft operator security along with risks and processes associated with air cargo. An overview and assessment of the vulnerabilities of aviation systems in relation to air cargo is also discussed. Policies, methods, and regulations are examined for managing security within the air cargo supply train. The 9/11 bill, along with Title 49 CFR Part 1548 Indirect Air Carrier Security legislation, are examined in relation to air cargo security. We also look at a synopsis of where air cargo security is today and examine concerns regarding future legislation and methods for managing air cargo security.

Introduction

The business and logistics for supporting air cargo is a highly complex system of global infrastructure that is subject to risk from crime and terrorism. The U.S. Government Accountability Office (GAO) describes the nature of *air cargo* as follows:

> *Air cargo includes freight and express packages that range in size from small to very large, and in type from car engines, electronic equipment, machine parts, apparel, medical supplies, human remains, to fresh-cut flowers, fresh seafood, fresh produce, tropical fish, and other perishable goods. Cargo can be shipped in various forms, including in unit-loading devices, wooden crates, assembled pallets, or individually wrapped/boxed pieces, known as break bulk cargo. (GAO, 2005, October, p. 9)[1]*

In 2010, an attempt was made to bomb commercial airliners using explosive devices hidden in computer print cartridges shipped as air cargo. This effort in terrorism, commonly referred to as the Yemen air cargo plot, brought to public attention that air cargo is vulnerable as a target for terrorism. The Yemen air cargo plot was the first known terrorist activity using scheduled air cargo service as a mode for implementing an attack.

Most shipping customers assume that express or overnight delivery always utilizes air cargo service. However, only a small percent of packages travel by air and an even smaller amount is placed on a passenger-carrying plane as cargo. These small percentages still represent over 10.5 million tons of cargo shipped by air every year within the United States.

---

[1]Break bulk cargo includes packaged loose cargo such as cartons or boxes. Bulk cargo includes unpackaged loose cargo such as grain and is rarely shipped via air. Break bulk cargo comprises nearly 60% of all air cargo shipped on commercial airliners.

Of that capacity, over 8 million tons are shipped as cargo on international flights to and from the United States, along with over half a million tons of mail (Elias, 2010).

All-cargo air carriers focus on the transportation of cargo and usually do not carry paying passengers.[2] However, passengers on all-cargo aircraft are known as *supernumeraries*, and are subject to background vetting as prescribed by the Transportation Security Administration (TSA). The U.S. GAO estimated that at least 22% of a U.S. passenger airliner's hold is, on average, cargo, with the remaining cargo transported by all-cargo aircraft. The vast majority of air cargo, more than 9 million tons, is transported each year (FY 2000 average) by all-cargo operators such as FedEx, UPS, and DHL. The remaining 2 million-plus tons are carried by passenger carrier aircraft, just beneath the passenger cabin floor (GAO, 2002). Passenger planes often specialize in carrying "just-in-time" cargo, which consists of perishable items such as seafood and flowers or high-value fragile items like computers, jewelry, and artwork. Manufacturers also rely on just-in-time air cargo shipments to reduce inventory costs and stay competitive in global markets (Airforwarders, 2006). The individual size and weight of items carried as air cargo are usually smaller and lighter as compared to cargo carried by rail, vessel, or truck. Air cargo is usually needed at the destination more quickly than alternative forms of transportation.

There are three primary risks associated with terrorism that apply to the air cargo industry:

1. Hijacking an all-cargo aircraft and using it as a weapon of mass destruction.
2. Introducing an explosive to a passenger-carrying aircraft via the air cargo supply chain.
3. The illicit shipment of weapons, explosives, or chemical, biological, radiological, and nuclear (CBRN) agents via air cargo.

The industry has already experienced examples of the first two types of threats with the Yemen air cargo bombs in 2010 and in 1994 when a former FedEx employee attempted to hijack a FedEx flight for the purpose of crashing into the ground. Other security risks include theft and smuggling and shipment of undeclared or undetected hazardous materials aboard an aircraft (Elias, 2010). Since 9/11, two Congressional acts have attempted to address the security of air cargo: the Implementing the Recommendations of the 9/11 Commission Act (2007) and the National Intelligence Reform Act (2004).

Terrorism and Crime in Air Cargo

In 1979, Ted Kaczynski (a.k.a. the Unabomber) placed a bomb in the hold of an American Airlines passenger aircraft in a parcel shipped via U.S. mail. This was the first identifiable incident of a bomb placed as cargo aboard an aircraft versus checked baggage. The bomb failed to detonate and began smoking, which alerted the flight crew. The flight crew

---

[2]Passengers are occasionally carried on all-cargo airlines to accompany cargo such as human organs, animals, or perishable high-value items that need constant monitoring and attention.

immediately landed the aircraft without further incident. Although his identity was not known at the time, Kaczynski was also responsible for mailed bomb attacks to university professors. The Federal Bureau of Investigation (FBI) suspected that the attempt on American Airlines was implemented by the same suspect that had attacked the university professors. Therefore, the FBI nicknamed the suspect the Unabomber—based on the acronym UNABOMB, which stood for "university and airline bomber."

After 9/11, the Aviation and Transportation Security Act of 2001 (ATSA 2001) required the federal government to provide for "the screening of all passengers and property, including U.S. mail, cargo, carry-on and checked baggage, and other articles, that will be carried aboard a passenger aircraft operated by an air carrier or foreign air carrier in air transportation or intrastate air transportation" (ATSA, 2001). Foreign air carriers operating within the United States must abide to All-Cargo International Security Procedures (ACISP).

ATSA 2001 also specified that "a system must be in operation to screen, inspect, or otherwise ensure the security of all cargo that is to be transported in all-cargo aircraft in air transportation and intrastate air transportation as soon as practicable after the date of enactment of the Aviation and Transportation Security Act" (ATSA, 2001). However, this language was not interpreted to require physical screening or inspection of cargo shipments; rather, the TSA relied on the use of the Known Shipper program to prevent the introduction of unknown sources aboard passenger aircraft (Elias, 2010).

Air cargo security continued to be a national security concern when, in 2003, the Department of Homeland Security (DHS) issued warnings that al-Qaeda might be plotting to fly cargo planes from overseas into U.S. targets, such as nuclear power plants, bridges, and dams. Also in 2003, Charles McKinley hid in a shipping crate and mailed himself by air on a cargo flight from Newark, NJ, to Dallas, TX, to visit his parents (CNN, 2003). He miraculously survived the trip, was arrested, charged, and served a year in prison. In February 2004, three individuals shipped themselves in a shrink-wrapped pallet on an all-cargo flight from Santo Domingo, Dominican Republic, to Miami, FL. They were caught in a warehouse at Miami International Airport. In August 2004, a woman shipped herself in a cargo crate from Nassau, Bahamas, to Miami, FL. Both incidents again spotlighted the apparent lack of security in the air cargo industry (GAO, 2005).

Critics of the lack of air cargo security underscored that the preceding incidents demonstrated how easily a weapon, bomb, or even a terrorist could be transported by air cargo. Prior to 2010, many aviation security experts had predicted nonpassenger-related air cargo would become a target for terrorists because of the traditionally fewer screening requirements for air cargo, combined with the improvements made in passenger and baggage screening. As stated in a 2003 report to Congress:

> *While using cargo as a means to place explosive or incendiary devices aboard aircraft has historically been rare, heightened screening of passengers, baggage, and aircraft may make cargo a more attractive means for terrorists to place these devices aboard aircraft in the future. (Elias, 2008, p. 8)*

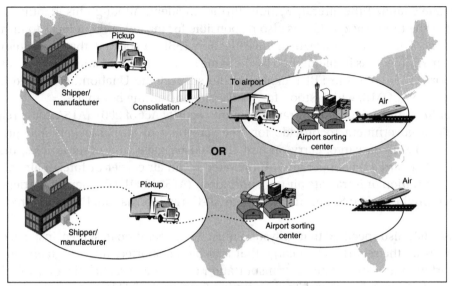

Source: GAO analysis of TSA information.

FIGURE 10.1 Flow of cargo from shipper to air carrier. The top path represents shipping via an indirect air carrier (i.e., a freight forwarder, or airforwarder). The bottom path represents a direct shipping operation. *(Source: Government Accountability Office.)*

Some have speculated that the somewhat unpredictable nature of logistics in the air cargo supply chain may in some cases reduce the probability of success to anyone attempting to place a bomb onboard a commercial aircraft via air cargo (Figure 10.1). However, the probability of continued use of air cargo as a mode for terrorism is significant since even a failed attack, such as the Yemen air cargo plot, can have a significant economic impact on society.

Placement of an improvised explosive device (IED) onboard an all-cargo aircraft for the purpose of destroying the flight and thus harming the economy was originally part of the 9/11 attacks. That strategy was rejected by Osama bin Laden as he felt it was an unnecessary complication to the plan. The fact that al-Qaeda considered bombing all-cargo carriers demonstrates that terrorists understand the significant impact air cargo has on the global economy. In the following passage, the TSA agrees that air cargo will be a threat to commercial aviation:

> *It has been reported that TSA considers the likelihood of a terrorist bombing of a passenger airplane to be between 35% and 65% based on year-old intelligence reports, and TSA believes that cargo is either likely to become, or already is, the primary aviation target for terrorists in the short term. (Elias, 2008, p. 5)*

According to *Inspire*, an al-Qaeda propaganda publication, the cost to al-Qaeda of the Yemen air cargo plot was about $4,200, whereas the United States has spent billions in attempting to secure the air cargo supply chain. The predictions of security experts were accurate when two explosives were introduced to the aviation supply chain out of Yemen.

In 2006, finalized rulemaking of the Air Cargo Security Act established the Full All-Cargo Aircraft Operator Standard Security Program (FACAOSSP) and the Indirect Air Carrier Standard Security Program (IACSSP). The act specifically required security threat assessments (STAs) for individuals handling air cargo, and criminal history record checks (CHRCs) and STAs for individuals working in cargo operations areas with access to aircraft. The rulemaking also required airports to make their cargo areas secure identification areas (SIDAs) and aircraft operators the final responsible party for what is carried on their aircraft. Remissions for this act were that it did not require CHRCs for aviation workers with unescorted access to all-cargo aircraft, ramps, and cargo, but instead gave priority to the use of STAs as a means to mitigate CHRCs and other risks. Also, the act did not require training of crewmembers in the all-cargo common strategy plan as is required in passenger airline domain. Additionally, no requirements for reinforced flight deck doors were specified (as in passenger airline operations) and, perhaps most important, no requirement for inspection of cargo loaded on all-cargo aircraft was stipulated.

Unfortunately, major cargo and baggage theft rings have been uncovered at airports throughout the United States and the smuggling of contraband, counterfeit, and pirated goods has also been a problem for air cargo security. A large portion of air cargo crime is committed either by cargo workers or with the assistance of cargo workers, so increased security of cargo operations areas and improved background checks of cargo workers may help reduce crime and terrorism associated with air cargo (Elias, 2010). The USA PATRIOT Act required the U.S. Department of Justice (DOJ) to establish a separate category for cargo theft in the Uniform Crime Reporting System and also refines relevant statues and increases criminal penalties for cargo theft and stowaways (Elias, 2010).

In 2007, the Implementing the 9/11 Commission Recommendations Act (known as the 9/11 Act) was signed into law. This required the DHS to establish 100% screening of cargo transported on passenger aircraft and further clarified that screening in this context means the physical examination or use of other nonintrusive methods (e.g., X-ray, explosive detection system [EDS], explosive trace detection [ETD], K-9, etc.) (Elias, 2010). The legislation mandated an interim milestone to screen 50% of all cargo by 2009. The goal is eventually to screen 100% of all cargo carried on passenger aircraft both domestically and internationally. The Certified Cargo Screening Program (CCSP) and several other programs have been created by the TSA to meet these objectives. One focus has been to look at screening cargo at various points throughout the supply chain, rather than just at the airport.

Historically, terrorists select specific targets for specific reasons, thereby making the element of time and scheduling important to the success of the attack. When using an indirect carrier (IAC), the shipping party most likely will not know what schedule or flight the cargo will be shipped on, which makes targeting a specific flight or targeting any passenger aircraft difficult. IEDs that use timing devices may also not be as effective, as there is no guarantee when the cargo will move from the air carrier sorting facility to the aircraft, or whether it is on a passenger aircraft or an all-cargo flight. IEDs detonated with a barometric trigger[3] provide some assurance that the bomb will detonate in flight—again, there is no guarantee what flight the bomb will be on. Conceivably, terrorists could place bombs in air cargo with barometric trigger switches and not care what flights are attacked, which, as we have learned, may be a viable strategy for certain outcomes, such as social fear and general economic impact or increased spending for new security measures by governments.

The cargo areas of a commercial airliner are hardened for structural and safety reasons. Additionally, aircraft cargo areas have relatively large objects of mass that can buffer an explosion. Therefore, a bomb of sufficient force is needed to cause serious damage or destruction. This usually requires a device that is more easily detected by screening technology. Bombs created for detonation in the passenger cabin do not have to be as big, as the bomber can ensure the device is placed right next to the skin or window of an aircraft.

Prior to 9/11, methods and policies related to air cargo security did exist. While most air cargo was not inspected using methods similar to those used for checked-baggage screening, the TSA had taken the following steps (Figure 10.2):

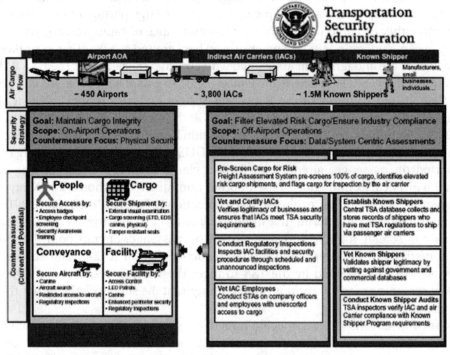

FIGURE 10.2 The TSA's perspective of the layers involved in air cargo security.

[3]An instrument that detonates a bomb once the aircraft achieves a specified barometric altitude.

1. Issuing security directives and information circulars to air cargo entities.
2. Required air cargo operators to have security programs.
3. Adopted an internal air cargo strategic plan to provide guidance for the future of air cargo security.
4. Initiated a new cargo profiling system to identify high-risk freight.
5. Convened the Aviation Security Advisory Committee (ASAC) Working Group and a Freight Assessment System (FAS) Working Group to develop ways to identify suspect air cargo.
6. Increased funding for air cargo security research and the development of advanced inspection technology.

Since 9/11, the air cargo industry has taken proactive measures to further secure air cargo. These initiatives include the following:

1. Participated in the TSA's ASAC Working Group offering its facilities as test-beds for new inspection technologies and security programs.
2. Advocated the use of canines for inspections.
3. Maintained barriers to protect the flight decks of all-cargo aircraft.
4. Implemented many of the ASAC Working Group recommendations without waiting for TSA mandates.

The Nature of Air Cargo

To better understand challenges to managing security as related to air cargo, it is necessary to understand the complex nature of the air cargo supply chain. Commercial-service airlines carry passengers along with some amount of baggage and cargo in the hold of the aircraft, which is provided by either a direct shipper or a freight forwarder. Some companies, such as pharmaceutical firms, fresh food, and fresh flower/plant companies that rely on short delivery cycles contract directly with the passenger airlines and are known as direct shippers. A direct shipper is not in the business of shipping goods, but must ship them as part of their business model. By contrast, freight forwarders are businesses that act as transportation agents to companies desiring to ship cargo by air. Freight forwarders make their money by shipping goods for other companies. Freight forwarders manage the logistics of picking up, shipping, and dropping off packages on behalf of their customer.

An indirect air carrier is a freight forwarder that solicits or receives freight from other companies and consolidates the cargo into larger shipping units for air transport. IACs usually provide pickup and delivery of the freight from the shipper to the recipient (Airforwarders Association, 2005) and frequently use vehicles, vessels, and rail to move some freight. Freight forwarders must have an IACSSP and are part of the Indirect Air Carrier Management System (IACMS), which the TSA uses to approve and validate new and existing IACs. Freight forwarders are also known as *airforwarders* and are characterized by the Airforwarders Association as follows:

Airforwarders are best understood as "travel agents for cargo" as they schedule the movement of cargo using other company's planes. Airforwarders can be very large businesses responsible for moving big shipments for companies like IBM or the Department of Defense, or small, family-owned companies that move cargo through their region. Airforwarders, who can also be referred to as freight forwarders or indirect air carriers (IACs), were traditionally responsible for organizing the travel of cargo from Point A to Point B. Today, however, airforwarders are tasked with the planning, oversight and responsibility for transporting companies' goods and products—anything from flowers and seafood to pallets of humanitarian supplies—from one end of the supply chain to the other. The parameters of this job include steps such as pick up of goods, customs clearance, transportation, warehousing, regulatory compliance and delivery. (Airforwarders, 2006, p. 1)

Companies who utilize airforwarders typically have cargo requiring transportation directly to a customer, or in less time than a traditional overnight all-cargo operator can provide (e.g., fresh seafood or perishable medical supplies). Airforwarders must ensure cargo is screened and secure throughout the supply chain. Airforwarders accept cargo at their facilities or pick it up from the customer and deliver it to the air carrier facility. The aircraft operator (passenger air carrier or all-cargo) then takes receipt of the cargo, loads it on an aircraft, and flies it to the desired location. The cargo is then picked up by an agent of the airforwarder and transported to the final destination (Airforwarders, 2006).

The screening and inspection of cargo is a serious issue with significant financial ramifications should the system be disrupted by attack or overly burdensome regulatory processes, as described in the following:

The air cargo system consists of a large, complex distribution network linking manufacturers and shippers to freight forwarders to airport sorting and cargo handling facilities where shipments are loaded and unloaded from aircraft. Business and consumer demand for fast, efficient shipment of goods has fueled the rapid growth of the air cargo industry over the past 25 years. In fiscal year 2001, about 13.9 billion revenue ton miles (RTMs) of cargo were shipped by air within the United States, and another 14.5 billion RTMs of cargo were shipped by air on international flights to and from the United States. It is estimated that air cargo shipments will increase by 49% domestically and by 86% internationally over the next ten years. In 1999, air cargo comprised about 0.4% of all freight movement in the United States. While this percentage may seem small, it is much greater than the 0.07% of freight that traveled by air in 1965, indicating that not only is the volume of air cargo increasing significantly, but so is the percent of total freight movements that travel by air. Also, cargo shipments by air comprise a significant percent of the total value of cargo shipments. In fact, in 2000, air cargo accounted for 29.7% of international trade by value, surpassed only by maritime shipping which accounted for 37% of the import/export value of cargo. (Elias, 2008, p. 1)

To address adequately concerns related to air cargo security, all elements of the air cargo supply chain must be considered. From the time an item is sealed and sent out, the integrity of the item must be maintained while it is moved to the airport (or possibly an interim warehouse facility if the item is being moved via a freight forwarder). The security of the cargo must be preserved as it is moved from the truck to the air cargo sorting facility and, finally, to the aircraft.

How air cargo should be screened has been a topic of debate internationally for several years and domestically since 9/11. Some have suggested that an expedient solution is to screen all cargo placed on commercial aircraft just like baggage is screened. However, this solution is not practicable. To do so would cost billions of dollars and bring air cargo operations to a halt, as noted in a 2003 Congressional Research Service report on air cargo security:

> *Given the sheer volume of cargo that must be expediently processed and loaded on aircraft, experts generally agree that full screening of all air cargo, as is now required of checked passenger baggage, is simply not feasible with available screening technologies and procedures. In fact, it has been reported that TSA computer models estimated that if full physical screening is implemented, only 4% of the daily volume of freight at airports could be processed due to the time that would be required to breakdown shipments, inspect them, and reassemble them for transport. Currently, less than 5% of cargo placed on passenger airplanes is screened. TSA staff have recently made recommendations to increase these inspections to 5%, acquire at least 200 additional explosive trace detectors system wide to support this effort, and conduct focused audits at freight forwarder and air cargo operations facilities. (Elias, 2008, p. 2)*

In 2005, the GAO assessed the TSA's progress in developing and implementing proposed security measures and developing a system to target and inspect higher-risk cargo. Both the TSA and U.S. Customs and Border Protection (CPB) agencies were involved. In 2006, the TSA was in the process of developing a freight assessment system, which is a component of its air cargo database that is used to identify high-risk cargo (GAO, 2005). In 2004, the TSA initiated testing on using EDSs (the same systems used to inspect checked baggage) to inspect break bulk cargo at Ted Stevens Anchorage International Airport in Alaska, Atlanta Hartsfield-Jackson International Airport in Georgia, Chicago O'Hare International Airport in Illinois, Dallas/Fort Worth International Airport in Texas, Los Angeles International Airport in California, and Miami International Airport in Florida. The testing was expanded to include the inspection of mail and some all-cargo operations. Initial testing demonstrated that the inspection by EDS of break bulk cargo and mail is feasible, but there were certain technological limitations such as false alarm rates that had to be resolved using ETD equipment.

In 2005, the TSA's air cargo strategic plan included the following four major components:

1. Enhancing the Known Shipper program.
2. Establishing a cargo prescreening system that identifies and inspects high-risk cargo.
3. Instituting major air cargo research and development programs.
4. Partnering with airlines and others to implement additional measures such as enhanced background checks on persons with access to cargo and new procedures for securing aircraft between flights.

The TSA committed $85 million in research and development to enhancing air cargo safety. The TSA established the following allocations:

- $26 million to evaluating EDSs and operating test programs at several airports.
- $21.5 million for research and development to identify how current technologies can be applied to air cargo.
- $7.5 million for research and development to identity existing technology that can be used to build automated inspection systems for U.S. mail.
- $30 million to develop new technologies for inspecting cargo for explosives and radiation, chemical, and biological agents.

Part of the TSA's research and development program studies the feasibility of alternate inspection systems. Some systems feature additional trace element detection systems that can detect explosive substances by sampling air molecules within break bulk cargo packages and shipping containers. Other systems use quadruple resonance imaging, or high-energy computed tomography, which can generate three-dimensional images of oversized boxes and palletized cargo and containers. Backscatter X-ray, Terahertz spectrometer-based trace detection that detects residues based on spectrum analysis, and nuclear detection systems that measure radioactive elements and ratios within a container are also in development.

The benefit of shipping cargo by air is the ability to move an item much faster than by land or sea transportation. Therefore, the ability to facilitate ground vehicles that have to deliver and pick up packages through the airport cargo facilities in a short time is essential. Although it may only take a minute or two to run packages through an EDS system or to run trucks through a backscatter X-ray system, each minute adds additional time to loading and unloading and thus additional delays. Over the course of an hour, these minor delays will add up to a significant overall delay. This is the reason the air cargo industry and the government strive to identify high-risk cargo for inspection rather than inspecting all cargo.

Known Shipper Program

Historically, the primary air cargo security program has been the Known Shipper program, a by-product of the 1996 Gore Commission report and a mainstay air cargo security program (called registered agent) internationally. A *known shipper* was originally based on the airline knowing the shipper through the course of business and being approved by a particular airline to ship cargo on that airline. Before 9/11, each airline had its own protocols for known shippers. In most cases, a representative of the airline would physically visit and inspect the shipper's facility, which could be anything from a large warehouse in the industrial district to the basement of someone's house. Once approved by the airline, the shipper could then ship on that airline. However, the shipper had to repeat the same protocol with each airline used to transport air cargo.

After the air cargo security changes in 2006, the TSA began developing the Known Shipper Data Management System (KSMS). Through this system, the TSA took over the vetting process including facility inspections and compliance with security programs. Known shippers are kept in a centralized database whereby the TSA can conduct further background and intelligence database checks. The shipping industry no longer needs to be approved to do business on each air carrier, as once a shipper is approved by the TSA and added to the KSMS, any airline will be able to accept that shipper's goods (P. Hamilton, assistant general manager, Air Cargo Programs, TSA, personal communication, 2006). Since 2008, the TSA has processed more than 1.4 million entities through the KSMS.

Government Accountability Office on Air Cargo

The GAO issues reports on air cargo security and the ability of the TSA to meet established U.S. government mandates. A GAO (2002) report noted that in a TSA investigation, inspectors found numerous security violations by freight forwarders and air carriers, both of which are required under federal relations to have security programs. The second area of vulnerability included tampering with cargo during transport by vehicle to the airport or at the cargo-handling facilities. The report stated that cargo theft has been a major problem with the industry, accounting for more than $10 billion in losses from all forms of transportation annually. For example, in one notable incident at Brussels, Belgium, airport in 2001, thieves stole $160 million in diamonds from a Lufthansa aircraft (Price, 2003). The losses generated by theft within the air cargo community also demand that aircraft operators pay significant attention to protecting the air cargo supply chain.

The GAO (2002, p. 3) recommended a three-phase strategy called a *risk management approach* to air cargo security. Phase 1 was to conduct a threat assessment identifying potential threats to air cargo, phase 2 was a vulnerability assessment to identify weaknesses that may be exploited, and phase 3 was a criticality assessment that prioritizes security assets and functions and identifies what systems, methods, or procedures are vital to securing air cargo. From these efforts, the GAO made a series of recommendations to enhance cargo security, which included the following:

1. Use radio frequency electronic "seals" to secure cargo—this technology transmits an alarm if the container has been compromised.
2. Develop or use a variety of systems capable of detecting weapons of mass destruction, including X-rays, radiation, trace and vapor detection systems, and greater use of canines.

3. Use pressure chambers to detect and detonate on the ground any explosive device with a barometric trigger.
4. Develop blast-hardened containers to store cargo within the aircraft.
5. Use access control biometric technology to ensure only authorized persons are able to handle cargo.
6. Use GPS tracking systems to track cargo throughout the transport.
7. Use closed-circuit television (CCTV) to monitor the loading of cargo into the aircraft.

The GAO also recommended better operational practices to improve air cargo security, including cargo profiling, identification cards, background checks for personnel handling cargo, and employing qualified security officers at cargo facilities. Many of these recommendations would be reflected in the TSA's ASAC, which in 2003 focused on air cargo security issues.

In 2005, the GAO (2005) published another report in response to U.S. Congressional inquiries about the progress and status of air cargo security. The report focused on the GAO's investigations of three areas:

1. To what extent has the TSA implemented a risk-based management approach to air cargo security?
2. What actions has the TSA taken to ensure the security of air cargo and what may limit their effectiveness?
3. What are the TSA's plans for enhancing air cargo security, and what financial, operational, and other challenges do the TSA and industry stakeholders face in implementing these plans?

The GAO report determined that although the TSA had completed a risk-based strategic plan, it had not completed risk assessments for air cargo security:

> TSA established an automated Performance and Results Information System (PARIS) to compile the results of cargo inspections and the actions taken when violations are identified. Our [GAO] analysis of PARIS inspection records shows that between January 1, 2003, and January 31, 2005, TSA conducted 36,635 cargo inspections of air carriers and indirect air carriers and found 4,343 violations. (GAO, 2005, p. 38)

The GAO did note that the largest number of violations cited was for failing to adhere to the indirect air carrier security program. The second largest number of violations issued was for not properly documenting air cargo procedures. The GAO also noted that the third largest violation fell into the "other" category (Figure 10.3). This category is used for issues not easily categorized in the violation fields of the agency's PARIS database. The study did show that failing to properly inspect cargo had the lowest number of citations.

The TSA is currently developing a system to compare information on air cargo shipments and the PARIS databases against certain targeting criteria to assign a risk level (GAO, 2005). Cargo identified as posing a higher risk will then be subjected to higher levels of scrutiny and inspection or possibly not carried by aircraft at all.

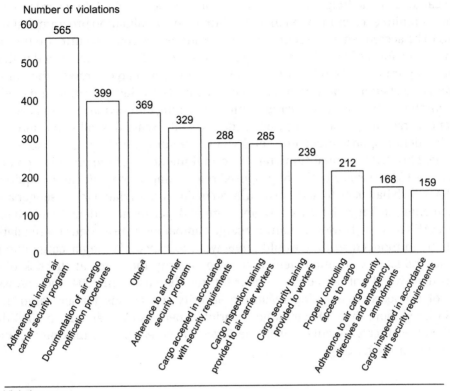

Number of violations

Violation area

Source: TSA.

[a]According to TSA officials, "other" includes information on violations not specifically identified in the agency's PARIS database.

FIGURE 10.3 Top ten areas in which violations were found during air cargo inspections for the period January 1, 2003, to January 31, 2005. *(Source: Government Accountability Office.)*

Aviation Security Advisory Committee: Air Cargo

In 2003, the ASAC Working Group, which included organizations such as the Airline Pilot Association (ALPA), the Cargo Airline Association, the Airforwarders Association, the American Association of Airport Executives (AAAE), the FBI, the National Air Transport Association, the Air Transport Association, the National Customs Brokers and Forwarders Association, and the U.S. Postal Service, studied concerns related to air cargo security. The ASAC Working Group recommended security enhancements to address various risks associated with air cargo. The result was 40 separate recommendations that would eventually form the basis of the TSA's air cargo security rulemaking.

The committee focused on three broad-based security areas: shipper acceptance procedures, IAC security, and security of the all-cargo aircraft. In regard to shipper acceptance, the committee recommended that the Known Shipper program be strengthened, that government databases be coordinated to provide more and better information on threats, that research and development focus on new or enhanced technologies to better screen cargo, and that freight from unknown shippers be screened. In regard to IAC security, the committee recommended that IACs implement a validation program that requires the IAC to be accredited, that security measures address en route security (to the airport), and that some form of security screening for employees of IACs be instituted with necessary security training. As related to the protection of all-cargo aircraft, the committee recommended better airport perimeter access controls, the identification and credentialing of employees with access to cargo ramps, screening of individuals with access to aircraft and cargo ramps, securing unattended aircraft, searching all-cargo aircraft, and better incident response procedures and security training for all-cargo personnel.

At the 2004 AAAE Annual Conference, several members of the air cargo community pressed the TSA to issue a notice of proposed rulemaking for new air cargo requirements. The TSA, which had been focused primarily on major issues related to passenger and baggage screening and organizing its new government department, had been slow to respond to the AAAE's request. However, many air cargo companies were anxious for regulations to at least be introduced so they would know what to expect. Some air cargo companies expressed reluctance to implement their own security measures at the risk of taking actions that would not be required under future rulemaking. In 2004, the TSA issued a notice of proposed rulemaking regarding security regulations for air cargo. In 2005 the TSA hired more than 300 air cargo security inspectors (ACSIs), and in 2006 the TSA published the air cargo regulations in the *Federal Register*, thereby making them official regulations for the air cargo industry.

Summary of TSA Rulemaking

The overall changes with the 2006 TSA regulations have accomplished the following:

1. Called for the requirement of security programs for all-cargo operators using aircraft more than 100,309.3 pounds in mean gross takeoff weight, known as the FACAOSSP.
2. Strengthened security programs for cargo operators using aircraft of more than 12,500 pounds.
3. Extended screening requirements to all-cargo operators.
4. Required airports to extend their SIDAs into the cargo operations areas.
5. Redefined the indirect air carrier definition to include all-cargo carriers.
6. Strengthened foreign air carrier cargo requirements to equal U.S. domestic aircraft operator requirements (specific to ACISP requirements).
7. Strengthened the Known Shipper program, including the TSA takeover of the Known Shipper program vetting process. The TSA also conducts random screening of a percentage of known shipper cargo agents.
8. Implemented methods to identify and screen high-risk cargo.
9. Required STAs on individuals who have unescorted access to air cargo.
10. Screened passengers on cargo flights.

11. Established all-cargo and IACSSP, requiring all IACs to have a TSA-approved security plan in place before operating as an IAC and shipping cargo via air.
12. Accepted cargo only from an entity with a similar security program (in terms of accepting cargo from an IAC or from another aircraft operator).

The TSA also implemented several other air cargo initiatives, including the development of Air Cargo Watch, a program whereby individuals within the air cargo supply chain can identify and report suspicious activity or cargo. Other programs include the increased use of canine explosive detection teams and surge programs using ACSIs. In the air cargo surge program, ACSIs are sent to an airport that is not their home airport to conduct surprise inspections. These inspections provide a second-look approach, whereby inspectors have an opportunity to discover security issues that particular airport security personnel may have missed.

The TSA recognizes that the majority of air carriers and all-cargo operators are not subject to greater risk from crime or terrorism since these operators are usually better able to mitigate security threats and manage security compliance issues than smaller indirect carriers. The agency does worry about the smaller indirect air carriers not having sound security programs. This worry stems not from an unwillingness to comply on behalf of the small IACs, but that with fewer staff, resources, and related business pressures, the small IAC may not be in compliance or have a good understanding of the risks involved with air cargo (P. Hamilton, assistant general manager, Air Cargo Programs, TSA, personal communication, 2006).

Since the implementation of TSA cargo regulations, the inspection process has generally consisted of air carrier personnel conducting visual and manual inspections of air cargo. Visual inspections consist of looking for signs of unauthorized tampering. Examples of these inspections include looking for scratch marks on screws, tampering with packaging tape, unusual odors, and other abnormal signs related to packaging. If X-ray equipment is not available or the item is too large or too sensitive to be cleared by an X-ray device, a manual inspection can be performed provided it is conducted with the shipper present. Air cargo personnel are required to be trained in proper and safe search procedures and have information available to notify local law enforcement if a suspicious item is discovered.

In 2006, the U.S. Congress was still focused on the issue of air cargo security and passed legislation (the 9/11 Act) calling for the 100% screening of all air cargo within three years. Although the intent of the legislation was for all cargo to be screened using the same measures and technologies as that used to screen carry-on and checked baggage, the TSA stated that screening (i.e., physical inspection) could be waived if the cargo came from a certified facility—that is, a facility that meets certain security standards (Savage, 2007). All air cargo service providers have not yet sustained a 100% screening capability, but are making significant progress toward that goal by making better use of new technologies provided for in the TSA's Air Cargo Advanced Screening System (ACAS) program.

Airlines for America (formerly the Air Transport Association) and several cargo operators have argued that putting in facilities and equipment to screen all cargo would effectively shut down the air cargo industry. In contrast, airline pilot and flight attendant unions have supported the 100% cargo screening measure. While attempting to leave very little to regulatory interpretation, the TSA 2006 legislation does not require screening facilities to be placed in the airport, which leaves some interpretation that cargo screening could be conducted at other areas, such as the shipper's warehouse or other facility.

Unknown Shippers and U.S. Mail

Shortly after 9/11, all cargo shipments from unknown shippers were temporarily suspended and subsequently reinstated for a period of time. As of this writing, unknown shippers are only allowed to ship on all-cargo carriers, provided the cargo is not interlined to a passenger carrier. Only known shipper cargo can be interlined from all-cargo airlines to a passenger carrier.

Items that weigh more than 16 ounces are required to be physically inspected before being placed into the U.S. mail system. Postal service personnel look for irregularities in packaging or shipping processes. Items exempt from screening include human organs, human blood, emergency lifesaving drugs, human remains, U.S. mail, diplomatic pouches, and air carrier company mail.

Other forms of cargo include unaccompanied baggage (misplaced or lost) carried by the airlines for return to its owner and certain categories of cargo that are taken on board by a passenger, such as human remains (cargo frequently referred to as "accompanied commercial courier consignment"). The transportation of mail is a unique security challenge since the Fourth Amendment to the U.S. Constitution has established mail as private material protected from unlawful search and seizure (Elias, 2010). The U.S. Post Office does have a screening process that relies on postal clerks who are trained to examine individuals shipping packages that weigh more than one pound. However after 9/11, mail weighing more than one pound was prohibited from being carried aboard passenger aircraft. Items weighing less than one pound are not subject to scrutiny and can be deposited in any mailbox, which easily circumvents the screening or questioning of the sender (Elias, 2010). However, only about 5% to 7% of all domestic mail shipment is transported by passenger or all-cargo aircraft. FedEx is currently the largest carrier of U.S. mail and its operations account for half of the total volume of U.S. mail shipments every year (Elias, 2010).

In 1997, the Gore Commission recommended the U.S. Postal Service obtain authorization to screen shipments weighing more than one pound using EDSs. However, this screening has not yet been implemented, nor has any current legislation enabled this screening criterion. While the TSA has not announced any plans to expand the use of canine teams for the specific purpose of screening mail, it is presently the only means approved by the TSA for screening mail weighing more than one pound that is put on passenger aircraft (Elias, 2010).

Title 49 CFR Part 1548 Indirect Air Carrier Security

Entities engaging in commercial carriage of cargo on passenger air carriers must adhere to Title 49 CFR Part 1548 Indirect Air Carrier Security. As with other sections of the regulations, the TSA maintains the authority to inspect the facilities and records of any entity regulated under Title 49 CFR Part 1548. TSA-compliance inspectors must be allowed access to security areas including the indirect air carrier's facilities on an airfield. IACs must also adopt the IACSSP that provides for the security of the aircraft against acts of violence (explosives, hijacking, etc.) from the time the cargo is accepted by the IAC until it is transferred to an employee other than the IAC (e.g., an airline representative). IACSSP also includes procedures for storing cargo by the IAC. The IACSSP must also provide for security training for personnel employed by the IAC. The IACSSP is a sensitive security information (SSI) document, and access and dissemination must be strictly controlled.

IACSSPs must include measures to prevent the introduction of explosives or incendiaries and refuse to transport any item that has not been subjected to such measures. IACs must ensure that any employee who comes in contact with cargo designated for shipping on an air carrier with a FACAOSSP or a foreign air carrier (in addition to ACISP requirements) must complete an STA or a CHRC before handling such cargo. A cargo employee who completes a CHRC does not have to complete a separate STA as the STA process is integral to the CHRC process. Additionally, anyone involved in handling cargo that will be shipped onboard a commercial air carrier must complete training in the IAC's IACSSP, including applicable airport security requirements, security directives, and information circulars. IACs must appoint an indirect air carrier security coordinator (IACSC) at the corporate level who is responsible for coordination and communication with the TSA. Individuals such as the owner or partner of an IAC who controls more than 25% of the IAC and any corporate officer or director must also complete the STA process. IACs must ensure that cargo accepted from an unknown shipper is kept separate from known shipper cargo.

Certified Cargo Screening Program

The CCSP provides critical supply chain security and extends screening to manufacturing facilities, consolidation points, distribution centers, and independent cargo screening facilities (ICSFs). ICSFs provide a cost-effective avenue for small- and medium-sized freight forwarders to meet regulatory screening requirements (TSA, 2012). The CCSP is supported by the air freight and air carrier industries, leverages best practices from global supply chain programs, and enables businesses to choose the best and most effective model for managing cargo security.

The CCSP allows businesses to:

- Screen cargo where it is packaged
- Maintain in-house packaging integrity
- Avoid screening log jams at the airport
- Build bulk configurations to minimize cost

Certified CCSP facilities must successfully apply, participate, and adhere to strict security standards, including physical access controls, personnel security, and screening of prospective employees and contractors to TSA standards. A secure chain of custody must also be established from the screening facility to the side of the aircraft (TSA, 2012).

Every individual element related to the shipment of cargo carried on passenger aircraft requires screening prior to being transported on any airline passenger aircraft. This could mean that skids and pallets would have to be taken apart, screened, and reconfigured. This process can potentially increase the chance of damaging the cargo and its contents (TSA, 2012).

The TSA developed the CCSP to help industry reach the 100% screening mandate. The program enables freight forwarders and shippers to prescreen cargo prior to arrival at the airport. Most CCSP shipper participants have been able to quickly incorporate physical screening into their shipping process at a small cost relative to their overall operating expenses (TSA, 2012).

Under the CCSP, the TSA certifies cargo screening facilities located throughout the United States to screen cargo prior to providing it to airlines for shipment on passenger flights. Participation in the program is voluntary and designed to enable vetted, validated, and certified supply chain facilities to comply with the 100% screening requirement (TSA, 2012).

CCSP participants must have a process to screen prospective employees and contractors to TSA standards. They must conduct reviews of employees with access to cargo, conduct STAs on all personnel who are part of the CCSP, and provide specialized training for individuals to conduct the screening and handling, or who have access to the screening area. CCSP participants must have procedures in place to prevent unauthorized access to the cargo facility where cargo is screened and stored, including with physical barriers in place (TSA, 2012). Shippers that do not want to become a certified cargo screener but wish to avoid delays or damage to their shipments due to the screening process can use a freight forwarder's IAC services or an independent cargo screening facility (ICSF). An ICSF is authorized by the TSA only to screen cargo that will be transported by a passenger air carrier (TSA, 2012). Most ICSFs are not freight forwarders. The TSA website maintains lists of equipment that is certified for use in ICSF cargo screening and also a list of ICSF facilities.

Certified cargo screening facilities (CCSFs) are entities that are in the business of screening cargo (versus shipping cargo). CCSFs must carry out a TSA-approved security program and adhere to strict chain of custody requirements. Cargo must be secured from the time it is screened until it is placed on passenger aircraft for shipment (TSA, 2012).

Current Air Cargo Security Measures and Issues

Since 9/11, the TSA has implemented numerous measures to secure two primary forms of air cargo: domestic and inbound. Domestic cargo is transported from locations within the United States. Inbound cargo is brought into the United States from an international origin. As of 2012, the TSA estimated that over 90% of domestic cargo transported on passenger aircraft was being screened. However, inbound cargo remains a challenge with significantly lower percentages of cargo confirmed as screened or secure prior to arriving at U.S. security checkpoints.

In 2007, the passage of the 9/11 bill) added more requirements to air cargo security. The TSA worked with the U.S. Congress to significantly strengthen security of air cargo through the 9/11 bill. In the 9/11 bill, the TSA is mandated to screen 50% of air cargo on passenger-carrying aircraft within 18 months and 100% within three years. The TSA uses a multi-layered, high-tech, industry-cooperative approach, utilizing surprise cargo security inspections called *strikes*, covert testing processes designed to evaluate the reliability and validity of security screening measures, security directives, and 100% screening at 250 smaller airports. The bill also required the TSA to eliminate all exemptions to screening of air cargo and increase the amount of cargo that is subject to mandatory screening. With the new regulation, it is anticipated that the TSA will conduct approximately 100,000 more background checks, specifically on air cargo employees who screen cargo or have knowledge of how it is going to be transported or who actually transport the cargo. The rule also requires more robust checks and more visibility on the shipping companies and their employees.

The Air Cargo Division of the TSA currently regulates over 730 charter aircraft operators and 41 all-cargo aircraft operators. TSA principal security inspectors (PSIs) assist air carriers in ensuring system-wide compliance with applicable regulations. The PSI acts as the primary point of contact between TSA and the assigned air carrier on all matters involving aviation security (TSA, 2012).

More than 460 TSA canine teams are now assigned to 25% of each workday in an air cargo–related environment. All-cargo airlines are now subject to random inspection through explosives detection equipment and K-9 detection procedures. Any cargo classified as high risk is also screened using procedures such as manual inspections or K-9 review. Airport operators are now directly involved in protecting their portion of the cargo supply chain, specifically the airfield where cargo is delivered, sorted, and then transported to the aircraft. This effort requires dual responsibility from the airport operator and aircraft operator. Perimeter fencing, access control, and secure storage of freight help to ensure that cargo is not tampered. The airport operator must now also ensure that air cargo areas are designated and managed as SIDA and protected.

The TSA has hired and deployed hundreds of ACSIs who review IAC security programs and conduct routine and unplanned inspections at shipping facilities. In addition, the TSA has instituted a program called Air Cargo Watch, which is a security awareness program designed to give cargo employees and private citizens the ability to detect, deter, and report potential or actual security threats. Air Cargo Watch materials include a presentation, posters, and a training guide. The posters are used to maximize security awareness within the entire air cargo supply chain. Entities displaying the posters should also add phone numbers for local police, sheriff, fire, and rescue services, in which space is provided for on each issued poster. TSA agents now routinely conduct air cargo vulnerability assessments across the United States. These assessments are designed to identify critical air cargo supply chain nodes and identify and evaluate the risks of each node to the air cargo supply chain should a threat be implemented.

Stowaways have been a periodic annoyance to the aviation industry, but generally non-threatening to the safety or the security of the flight. Stowaways hiding aboard commercial aircraft in wheel-wells (the most popular locations for stowaways) are placing themselves in extreme danger as they can be exposed to extreme temperature changes (e.g., $-40°F$) and the significant lack of atmospheric pressure at normal operating altitudes. Many stowaways do not live through their attempt to hide in various areas of the aircraft. Bodies of most stowaways are usually discovered in wheel-well locations, or in some cases, fell out of the plane to their deaths when the landing gear deployed.

Risks caused by stowaways are more of an indication of failures in security than an actual safety hazard. If the security of an airport or aircraft is such that it allows access by a stowaway, then this is also a sign that the security system may be inadequate to prevent an individual from accessing the plane to plant a bomb or commit some other form of terrorism.

Preventing stowaways is a two-fold responsibility: (1) the airport operator has the responsibility to control access to the airfield through the access control system and perimeter protections, and (2) the aircraft operator has the responsibility to ensure that no person or item is placed on board an aircraft without properly being screened.

Since 9/11, there have been instances of individuals taking cover in the cargo hold of aircraft; cargo holds are pressurized and, compared to the wheel-well, much warmer. If a person can access the cargo area, there is a greater threat as they can introduce a bomb into the cargo supply chain, which may result in the device making it onto a passenger flight.

Perimeter transgressions related to the potential of access by a stowaway have recently been experienced. For example, in 2012, an individual ran out of fuel while jet-skiing off the coastline of JFK airport, subsequently swam ashore, climbed the airport's perimeter fence, and walked across the airfield to a concourse before he was stopped by an airline employee; the perimeter intrusion detection system apparently did not alert airport staff to the intruder.

Stowaways are a concern for the TSA, to the extent that the agency is evaluating technology that can search for stowaways in cargo containers. Any such technology would have to be able to see through sealed containers made of wood, plastic, cardboard, fiberglass, or metal, and simultaneously not deter the flow efficiency of cargo.

Conclusion

Securing air cargo and related services is a prime concern for aviation security practitioners. Logistical processes for managing the air cargo supply train are complex and ephemeral. Air cargo security practitioners must embrace these challenges along with managing varying legislation for all-cargo airlines and passenger airlines that also carry cargo. The U.S. Congress and the TSA have worked together to create legislation designed to help prevent cargo aircraft from being used as weapons of mass destruction.

With hijackings becoming more difficult to implement, many stakeholders to the industry believe that air cargo may become a preferred avenue for attack. Therefore, new legislation such as ATSA 2001 and the 9/11 bill were created in response to this concern.

Future concerns related to air cargo security include the potential requirement that all cargo must be inspected using EDS or similar processes. Some have argued that screening all cargo would effectively shut down the air cargo industry, whereas airline pilot and flight attendant unions have supported 100% cargo screening. Furthermore, the TSA is concerned over future risks posed by smaller IACs that may not have valid security programs. Recent legislation is enabling the TSA to mitigate future risks in air cargo. The TSA is developing a multilayered, high-technology, and industry-cooperative approach to managing security risks associated with air cargo.

References

Airforwarders, 2006. Airforwarders Association Industry Facts. Airforwarders Association, retrieved Aug. 6, 2008, from www.airforwarders.org/industry_facts.php.

Aviation and Transportation Security Act (ATSA), 2001. Aviation and Transportation Security Act (an act to improve aviation security, and for other purposes). Senate and House, 107th Congress, retrieved Aug. 6, 2008, from http://frwebgate.access.gpo.gov/cgi-bin/getdoc.cgi?dbname=107_cong_public_laws& docid=f:publ071.107.pdf.

CNN, 2003. Man Shipped from N.Y. to Texas in Crate. CNN.com, retrieved Aug. 6, 2008, from www.cnn.com/2003/US/Southwest/09/09/plane.stowaway/index.html.

Elias, B., 2008. Aviation Security: Background and Policy Options for Screening and Securing Air Cargo, Feb 25, 2008. CRS Report for Congress.

Elias, B., 2010. Airport and Aviation Security: US Policy and Strategy in the Age of Global Terrorism. CRC Press, Boca Raton, FL.

Government Accountability Office (GAO), 2002. Aviation Security: Vulnerabilities and Potential Improvements for the Air Cargo System. GAO-03-344.

GAO, 2005. Aviation Security: Federal Action Needed to Strengthen Domestic Air Cargo (Publication No. GAO-06-76), retrieved Aug. 6, 2008, from www.gao.gov/new.items/d0676.pdf.

Price, J., 2003. Airport Certified Employee (ACE)—Security Module 3 Aircraft Operator. American Association of Airport Executives, Alexandria, VA.

Savage, C., Aug. 24, 2007. No Checks for Bombs in Certified Air Cargo. The Boston Globe retrieved Aug. 6, 2008, from www.boston.com/news/nation/washington/articles/2007/08/24/no_checks_for_bombs_in_certified_air_cargo.

TSA, July 27, 2012. Certified Cargo Screening Program. retrieved July 27, 2012, from http://www.tsa.gov/what_we_do/layers/aircargo/certified_screening.shtm.

11
The Threat Matrix

Objectives

In this chapter, we describe the evolution of terrorist activities and various crimes perpetrated within the aviation system. Aviation security practitioners must be able to detect and deter future unconventional terrorist and criminal acts. The security industry is responsible for designing resiliency into aviation security systems so responses are rapid and effective. Therefore, here we discuss existing and emerging threats to the aviation system and methods for mitigating, preparing for, responding to, and recovering from terrorist and extreme criminal attacks. This chapter examines practical strategies and tactics for accomplishing these requirements against current and future threats against aviation security.

Introduction

The United States has expended significant monetary resources since 9/11 to "fix" the aviation security system; however, as a nation the United States has much more work ahead in developing security systems that can mitigate future threats. Although part of our responsibility is to detect and deter terrorist and criminal acts, it is also part of our responsibility to build resiliency into the aviation security system such that responses are rapid and effective when a terrorist or criminal act occurs.

Terrorism is a way to inflict harm on a country or entity with little risk to the terrorist organization's infrastructure. In addition to direct damage to property and loss of life, a terrorist act produces publicity for a cause, economic damage, and changes to a way of life. Whenever a citizen sees heightened security, he or she is reminded of the attack that caused the new procedures and may continue to fear another attack. Thus, with one attack, numerous objectives have been achieved for the terrorist organization.

Threat Matrix

Modern terrorism uses the Internet and other highly advanced technologies along with unconventional forms of implementing an attack. Therefore, threats can no longer be eliminated through military power. Defeating current and future threats to our global aviation system requires strategies that combine military force where applicable and the construction of a resilient infrastructure. *Airports and airlines must build sustainable resiliency into their security programs.*

In 2006, the United Kingdom's Security Service publicly released figures indicating there were an estimated 200 terrorist cells operating within the United Kingdom, with approximately 1,600 operatives and more than 30 terrorist plots,[1] including plots involving mass casualty suicide attacks in the United Kingdom (Associated Press, 2006). In the United States, in 2006, seven individuals were arrested for conspiring to bomb the Sears Tower in Chicago, IL, and attack the FBI building in North Miami Beach, FL, and other government buildings. At the time of the arrests, at a press conference, U.S. Attorney General Alberto Gonzales noted:

Recent events around the world have demonstrated the challenges posed by home-grown terrorists who live in the area that they intend to attack. The terrorists and suspected terrorists in Madrid and London and Toronto were not sleeper operatives sent on suicide missions. They were students, businesspeople, and members of the community. They were persons who, for whatever reason, came to view their home country as the enemy. And it's a problem that we face here in the United States as well. (Gonzales, 2006, p. 2)

The most dangerous threat to aviation is perhaps the belief that a threat is unstoppable. Throughout the United States, one consistently hears two messages that demonstrate far too many people within the aviation system have this fatalistic view. The first statement is in regard to general aviation (GA) and the use of a GA aircraft to facilitate a terrorist attack. The statement goes something like "Isn't it easier to rent a truck and fill it with explosives than it is to rent or steal an aircraft?" The second statement relates to suicide bombers. The statement goes something like "You cannot stop someone who is totally committed to attacking you and killing themselves in the process."

Although commonly believed, neither of these statements helps prevent the next attack on aviation. The first statement does not negate our responsibility to prevent an aircraft from being used in a terrorist attack, whether it is a commercial-service or GA aircraft. In the second case, whereas some terrorists will die for their cause, they will not "donate their lives cheaply." Many suicide/homicide bomber attacks require significant planning and numerous personnel to be successful, which increases the opportunity for stopping an attack before it starts. Author Stephen Flynn (2007) offered a perspective that the aviation security practitioners should adopt to combat terrorism:

Terrorism will increasingly be like the flu: the only thing we can safely predict is that each season there will be new strains. Thus, it is only prudent to bolster the odds that our immune system can successfully fend off the known strains. This requires our identifying the most likely and attractive targets and striving to protect them from potential attacks. At the same time, we must also assume that not every attack can be prevented. This makes it critical that we be able to respond effectively when disaster strikes and have the means to rapidly restore what is damaged. (p. 18)

---

[1]MI-5 systematically examines each threat, placing threats on a sliding scale and acting as the threat reaches a specific point.

Terrorists intend to conduct attacks that significantly harm their target, primarily economically. Their objective is not just to cause destruction or kill people but to effect massive change—they want to do something that will stay in the news for months if not years.

Another threat to the United States that could stem from a terrorist strike, or repeated strikes, comes from our actions and responses as a society to attacks. Admiral William J. Crowe, Jr., former head of the U.S. Joint Chiefs of Staff from 1985 to 1989, stated, "The real danger lies not with what the terrorists can do to us, but what we can do to ourselves when we are spooked" (Flynn, 2007, pp. 92–93). This quote reminds lawmakers and regulators that the real threat from terrorism comes not in the destruction of property or the death of our citizens, but in our own actions regarding the economic impact we allow to result from these attacks and the civil liberties we compromise in the process. If there is a question as to the real intent of today's terrorists, consider that in a postarrest interview, notorious "shoe bomber" Richard Reid stated that the intended purpose of al-Qaeda is to disrupt and destroy the American way of life by striking at and crippling our economy. Policymaking must be tempered with patience in development and understanding of the resources required to combat a threat.

Modern threats from terrorism fall into three areas: existing threats, emerging conventional threats, and emerging asymmetrical threats. Existing threats are those that have frequently been used. Emerging conventional threats include new yet probable forms of attack, such as the use of surface-to-air missiles. Emerging asymmetrical threats include weapons of mass destruction (WMD) that may cause indiscriminant destruction.

The U.S. Patriot Act defined international and domestic terrorism. *International terrorism* means activities that (1) involve violent acts or acts dangerous to human life that are a violation of the criminal laws of the United States or of any state, or that would be a criminal violation if committed within the jurisdiction of the United States or of any state; (2) appear to be intended to (a) intimidate or coerce a civilian population, (b) influence the policy of a government by intimidation or coercion, or (c) affect the conduct of a government by mass destruction, assassination, or kidnapping; and (3) occur primarily outside the territorial jurisdiction of the United States, or transcend national boundaries in terms of the means by which they are accomplished, the persons they appear intended to intimidate or coerce, or the locale in which their perpetrators operate or seek asylum.

Domestic terrorism means activities that (1) involve acts dangerous to human life that are a violation of the criminal laws of the United States or of any state; (2) appear to be intended to (a) intimidate or coerce a civilian population, (b) influence the policy of a government by intimidation or coercion, or (c) affect the conduct of a government by mass destruction, assassination, or kidnapping; and (3) occur primarily within the territorial jurisdiction of the United States.

The combination of the insider threat, which has been prevalent throughout the history of aviation security, and terrorist threats facing the world today makes it relevant to discuss how terrorist groups operate, as this education may provide an opportunity for airport security practitioners to identify the elements and deter a possible attack.

In *Terrorism and Organized Crime,* Michael R. Ronczkowski discussed methods of dealing with terrorism that are applicable to airport security (Ronczkowski, 2012). Terrorism assumes many guises—political, religious, social, cyber—and can occur in many forms, from bombings to hijackings, to active shooters, to a combination of forms (an active shooter attack combined with a suicide bomber) or other elements such as chemical agents. Often, terrorists are using the protection provided by the laws of the country they are trying to attack as a shield for their activities. This can make dealing with terrorism a frustrating process, one that requires counterterrorism personnel to cooperate throughout the various agencies, react to the last attack to prevent future attacks, and think about future types of attacks and how to detect and interdict the attack, in a lawful manner, before it is launched.

Terrorist operations run the gamut from the unprofessional, unsophisticated, and badly executed, through well-planned and executed attacks conducted by individuals with training and performance that equals that of the world's special operations communities (Sloan and Bunker, 2011). Fortunately, not every attack resembles the planning, financing, and coordination that was prevalent in the 9/11 attacks. Most attacks are amateurish in nature, as seen in the shoe bomber and underwear bomber attacks, occurring less and less at the middle and higher levels of sophistication and professionalism (Sloan and Bunker, 2011).

Professional terrorist groups often reflect the operational theories and practices articulated by former Vice Admiral General William McRaven, and in the OODA loop model developed by John Boyd. The current nature of terrorism—that is, small groups using unconventional warfare methods—reflects many of the same operational characteristics as the nation's special operations groups.

McRaven, seen as one of the world's premier experts on low-intensity conflicts, published one of the most comprehensive studies of special operations warfare, *Spec Ops: Case Studies in Special Operations Warfare: Theory and Practice* (McRaven, 1996). McRaven posits that for a small force to be successful they need to acquire what is known as *relative superiority,* which he describes as a condition that exists when an attacking force, generally smaller, gains a decisive advantage over a larger or well-defended enemy. Relative superiority exists when it is achieved at the pivotal moment in the engagement and, once achieved, sustained to guarantee victory. Once lost, relative superiority is difficult to regain (McRaven, 1996). As an example, the hijackers on 9/11 attained relative superiority for the first three hijackings and then lost it on United Flight 93, which crashed in a Pennsylvania field as passengers attempted to retake the cockpit (Sloan and Bunker, 2011).

McRaven explains how six principles—simplicity, security, repetition, surprise, speed, and purpose—affect relative superiority. Each of these principles relates to the perspective of aviation security practitioners in designing methods to deter or effectively respond to a terrorist or criminal incident.

Simplicity relies on three elements: focusing on a limited number of objectives, having good intelligence, and using innovation (McRaven, 1996). On 9/11, the plan was quite simple: hijack aircraft using knives, put individuals who are capable of flying the aircraft into the cockpit, and aim at a building. Too much complexity can kill an operation, while too little planning may have the attackers show up at a closed venue or one filled with law enforcement and security officers (Sloan and Bunker, 2011).

Security relates to operational security (OpSec) while the attack is being planned and pieces put into place (e.g., training, acquisition of materials, funding). Complex plans involving multiple actors require higher levels of security, which hinders preparation and increases the chances the group will be discovered by law enforcement. An important component of this phase is that in almost all cases, surveillance on a potential target will be done physically, and supplemented with virtual tools such as Google Earth or vacation or facility websites. This puts personnel in a position to be spotted by security and law enforcement personnel, which could not only compromise the operation, but the entire group (Sloan and Bunker, 2011). More complex plots and established terrorist groups will also have a well-developed infrastructure with safe houses, secret armories, bomb factories, motor pools, and methods to finance their operations—some legal, such as charity group fronting the terrorist organization, and some illegal, such as theft, drug or human smuggling, extortion, or street taxation (Sloan and Bunker, 2011).

Repetition is indispensable in achieving success and relates to training to perform the operation. Certain elements, such as tactical skills necessary to execute the fundamental elements of the operation, should be second-nature by the time the attack begins. However, many domestic and international terrorist groups have been caught or eliminated during this phase as it often requires shooting ranges, use of explosives, acquisition of automatic weapons and ammunition, and other higher-profile actions. As an example, less than a month prior to the July 20, 2012, Colorado theater shootings, the alleged shooter was turned down for membership to a local gun club by the owner when the owner heard a "creepy, weird" message on the alleged shooter's voicemail (Winter, 2012).

Surprise is not the same as relative superiority, warns McRaven. Surprise provides for momentary advantage and, while usually necessary for success, alone is not sufficient for success (McRaven, 1996).

Speed, which also relies on proper security and constant repetition (i.e., training), relates to the ability to take action quickly. Speed also relates to John Boyd's OODA (observe, orient, decide, act) loop model. The OODA loop model posits that when two opponents engage each other, the combatant with the faster engagement (or reaction), with all other elements being equal, will win (Sloan and Bunker, 2011). The model was developed by the U.S. Air Force but has been applied to other areas of warfare, and implies that weaker forces can, by speeding up the engagement cycle, defeat superior opponents. The German blitzkrieg launched against France in 1940 is a good example of the OODA loop model.

Purpose means all personnel are focused on a single goal, which reduces extraneous objectives and isolates and limits the intelligence required, which, combined, makes operational security that much tighter.

Each element is interrelated and based on these premises; for an attack to be successful it would have to focus on a limited number of objectives, which also helps maintain operational security throughout the planning cycle. If the operation is too complex it is difficult to conceal and rehearse, making it nearly impossible to execute with surprise, speed, and purpose.

Fortunately, very few terrorist groups will carry out operations at these levels, but when they have, the results have been deadly, noting the 9/11 attacks, the Mumbai shootings, and the Beslan school hostage crisis. Unfortunately, much lower-level attacks can be executed against aviation, such as the Yemen air cargo plot, which likely involved only a few individuals, but there is a difference between the hundreds and, in some cases, thousands who have been killed in high-level attacks versus lower-level lone-wolf or small group attacks.

Once the attack is in motion, there is still not a guarantee of success. Even the Aum Shinrikyo group, which features scientists, engineers, and deep financial resources, made numerous mistakes with its attempts at developing a WMD due to a lack of small-unit operational experiences. While building a bomb is a skill set sought out by terrorist and extremist groups, former military or police officers with firearm and small-unit tactics training are highly valued within. Watch for individuals with prior military or law enforcement experience who may have been forced out of their profession for any reason, and that left them discontented with the organization, the government, or any other reason they would use to justify joining a terrorist or extremist organization.

Adding to the frustration of counterterrorism operations is the difficulty in measuring deterrence. While successful attacks are broadcast throughout the world and some notable near-attacks are given some coverage, such as the liquid bomb plot in 2006, which is more memorable for the Transportation Security Administration's (TSA) 3-1-1 rule rather than the thousands of lives saved by fast-acting MI-5 personnel, most arrests prior to the attack are either not covered or soon forgotten. It is also nearly impossible to measure the number of times a criminal or terrorist decided not to attack aviation (or any other target) due to the deterrent measures put into place.

The TSA has been roundly criticized for not catching any terrorists through their screening and behavior detection processes, but there are two issues here. The TSA is not fundamentally a law enforcement agency or else it would likely reside in the Department of Justice (DOJ). The TSA, like the U.S. Coast Guard, Federal Emergency Management Agency (FEMA), the U.S. Secret Service, and other Department of Homeland Security (DHS) agencies, resides in a sort of netherworld that is focused on protecting the United States from natural and human-made attacks—in most cases, these agencies are the last point of failure, the last line of defense, before a plot becomes a successful attack.

The Federal Bureau of Investigation (FBI) and other law enforcement agencies have caught numerous terrorists and criminals since 9/11; Garrett M. Graff (2011) has cited many of their successes in *The Threat Matrix: The FBI at War in the Age of Global Terror*. The Department of Defense (DOD), the nation's military forces, the Central Intelligence Agency (CIA), and related agencies have eliminated an untold number of terrorists, including most notably Osama bin Laden, potential terrorists, and terrorist support networks since 9/11. The terrorists not getting to the site of the attack (i.e., the airport or aircraft) because they have been arrested or eliminated before even making it to U.S. shores is a success for the entire system. Secondly, if the screening functions did not exist, then it has been shown repeatedly throughout the history of aviation security that criminals and terrorists will continue to exploit that vulnerability until protections are put into place.

Ronczkowski posits that just like the effective SARA model used in law enforcement (scan, analyze, respond, assess), terrorist activities can be organized into a similar model. Terrorists have a planning stage (target selection and analyzing and assessing the mission), a research (reconnaissance) and planning stage, and an execution stage (the attack), and that the use of all the stages are inseparable and integrated (Ronczkowski, 2012). Thus, the four stages within the terrorism model are: *s*elect target, *a*nalyze and assess, *r*esearch, and *a*ttack (SARA).

In the absence of the integration the mission will fail. Terrorists do not randomly and spontaneously pick their targets—they are selective to maximize the tragedy. This provides law enforcement and security personnel the advantage of being able to identify and deter a plot in advance of the attack.

"Training is the foundation on which law enforcement practices are built," says Ronczkowski (2012), but he notes that most terrorism-related training is focused on response to an incident, rather than on what to look for in advance and the types of questions to be asked. In October 2007, the National Strategy for Information Sharing stated, "Whether a plan for a terrorist attack is homegrown or originates overseas, important knowledge that may forewarn of a future attack may be derived from information gathered by State, local, and tribal government personnel in the course of routine law enforcement and other activities."

To promote the training of state and local law enforcement officers in identifying pre-incident indicators to terrorist or criminal activity, there are a couple of programs: Nationwide SAR Initiative and the State and Local Antiterrorism Training (SLATT) program funded by the DOJ.

SLATT provides training and resources to law enforcement personnel on the threats presented by terrorists and violent criminal extremists. SLATT provides antiterrorism detection, investigation, and interdiction training. The program has been in place since 1996, but received little attention prior to 9/11.

The National Security Institute (NSI) training is directed at state, local, and tribal law enforcement and public safety professionals in identifying, reporting, evaluating, and sharing pre-incident terrorism indicators to prevent acts of terrorism. The NSI's programs include documented and verified behaviors and indicators that, when viewed in the totality of circumstances, may indicate terrorism-related criminal activity.

Ronczkowski makes an important point with regards to terrorism when he says that the pieces of the puzzle can be scattered across multiple jurisdictions and databases, which makes interagency cooperation and mutual aid agreements critical (Ronczkowski, 2012). While state and local law enforcement agencies, along with fire departments and emergency medical service (EMS) units, routinely use Memorandum of Understanding (MOUs) for responding to local incidents, their use needs to transcend to spell out guidelines for the gathering of intelligence and investigations into terrorist activities.

While there is some debate how it was developed, the SAR program (suspicious activity reporting), which is promoted through the SLATT and NSI programs, has become an essential resource in the counterterror and counter-criminal efforts. Today, it is the backbone of the Intelligence Reform and Terrorism Prevention Act of 2004, which required the creation of the Information Systems Council (ISC). The ISC was charged with overseeing the development of an interoperable terrorism information sharing system environment.

Tasking local law enforcement to not only perform their traditional policing roles but also to now take on the prevention of terrorist attacks represents a significant shift in their roles (Ronczkowski, 2012). SAR is based on the behaviors and activities that have been historically linked to preoperational planning of and preparation for terrorist attacks, which include acquiring illicit explosive materials, abandoning suspicious packages or vehicles to determine the response time of security or police, taking measurements or photographs of areas not normally of interest to the public, testing security measures, and others. SAR creates a national standard for terrorism-related modus operandi codes (Ronczkowski, 2012). By creating and assigning numbers or codes to terrorism-related behaviors, terrorist activities can be tracked by date, time, and location—similar to the New York Police Department's computerized statistics performance-based management approach (COMPSTAT).

The SAR program allows police to paint their own picture of what is happening in their communities rather than relying on their federal partners. Fusion centers can also benefit from the SAR program as the centers themselves rely on the collection and dissemination of intelligence from a variety of sources. The SAR program does not have a defined starting or ending date, which is intentional and recognizes that the threat and the process is continuous and ever evolving. The goal of the ISC is that by 2014, every federal, state, local, and tribal entity participates in a standardized integrated approach to gather, document, process, analyze, and share terrorism-related suspicious activity information (Ronczkowski, 2012).

Domestic and international terrorist group activities include (Ronczkowski, 2012):

1. *Recruitment:* membership of a group, attendance at rallies and meetings, exposure to Internet recruiting or informational sites, personal recruitment, and accessing extremist literature
2. *Preliminary organization and planning:* identify and clarify roles, exposure to terrorist training materials or actual training, discussion of potential targets, drawings, and assignments
3. *Preparatory conduct:* theft and weapon acquisition, counterfeiting, procuring identification, bomb-related activities, and weapon modifying
4. *Terrorist act:* bombings assassinations, hostage taking, hoaxes, threats, and hijackings

This list addresses strategic phases of operations, but even at lower levels, the tactical phases of an operation must include three phases: planning, preparation, and execution.

Smaller operations require shorter phases, but must still include each element. Even the lone-wolf operator, usually the hardest to interdict, must begin with some type of target selection (planning), acquisition of weapons or explosives (preparation), and the attack (execution). Larger operations have a better chance of being detected due to the larger logistical footprint (Sloan and Bunker, 2011). Another common characteristic is that while logic may dictate that mission intent and target drive the planning cycle, the reality is that many times a group will get good at a particular type of attack, like hijackers were for the Popular Front for the Liberation of Palestine (PFLP) in the 1970s and 1980s, and it becomes their calling card.

The key indicators of terrorist activity have been defined as (1) surveillance, (2) elicitation, (3) testing security, (4) acquiring supplies, (5) suspicious persons, (6) trial runs, and (7) deploying assets. In addition to what has already been listed, these activities should also be regarded as suspicious (with respect to context):

1. Counter-surveillance and testing of security procedures
2. Elicitation of information from security and police personnel
3. Attempts to enter secure facilities, or attempts to smuggle contraband onto the premises
4. Stockpiles of currency (cash), weapons, ammunition, or in possession of multiple forms of identification, passports, and driver's licenses
5. Espouses extremists views in the workplace, in social media, or in personal communications
6. Attempts to acquire or in possession of blueprints or layout plans of sensitive or governmental infrastructure
7. Commits hoaxes or makes statements to determine the response

Crimes in the air cargo area should be taken very seriously. Most cargo crimes are documented as thefts, burglaries, or robberies, but according to the South Florida Cargo Theft Task Force, approximately $25 billion per year is lost in cargo theft. Many valuable materials including computers, pharmaceutical materials, electronic equipment, alcohol, cigarettes, and designer clothing are routinely targeted by both terrorist organizations (for funding) and organized criminal enterprises (Ronczkowski, 2012). Another consideration for cargo crime is *leakage*, defined as the removal of property or insertion of articles, including explosives, without tampering or removing the container's seal. The most common methods are by removing bolts and panels or by drilling out rivets.

Airport police, badging office personnel, TSA screeners (particularly those involved in the document check), airport operations personnel, airline personnel, and airport security managers are all in a unique position to see a variety of activities, including drug smuggling, human trafficking, and the use of theft as a means to finance a terrorist or criminal operation. Plus, with the volume of data and the background checks conducted on aviation personnel, there is the opportunity to identify a bad actor that may be plotting to commit another type of terrorist attack, such as the Denver International Airport shuttle bus driver who intended on attacking the subway systems in New York City.

Existing Threats: Aircraft Bombings, Aircraft Hijackings, and Airport Attacks

The traditional methods of attack on aviation continue to be bombings and hijackings. Although hijackings have numbered more than 700, with the exception of 9/11, approximately 150 airline bombings have resulted in more fatalities. The concept of using an aircraft as a WMD, although not new, was certainly not within the scope of thinking by the intelligence communities throughout the world before 9/11.

Aircraft Bombing: Passenger, Baggage, and Cargo

In 2006, the TSA stated that bombings were a greater threat to aviation security than hijackings. Bombs continue to be one of the most popular weapons of terror and remain an active current threat. The question is, what are the other known methods of introducing a bomb to an aircraft? It is in the answer to this question that aviation security practitioners can develop countermeasures. The first place to assess is current screening systems. It is already known that existing passenger screening technologies do not function well in detecting explosives. However, they are adequate for detecting many prohibited items including guns and knives. The TSA has implemented an operational solution in the form of restricting the amount of liquid that can be taken onboard and conducting random use of explosive trace detection (ETD) technologies. These processes are inconvenient to travelers and not as effective as they need to be to prevent a bombing. The security screening checkpoint must be outfitted with technology that will detect explosives and metal on a passenger and that provides a better look inside carry-on baggage—technology that can distinguish between dangerous liquids and other substances from nonthreatening items. Screening technology also should focus on detecting the elements of an explosive device so as to prevent several collaborators from carrying on a component of a device and then assembling it on the aircraft. However, many explosives can be constructed using materials and items that are not prohibited in an aircraft, so technology is not the only security practice that should be employed. Passenger profiling, beginning with the identification and verification of the passenger when she or he purchases a ticket, to active profiling conducted at the airport, should also be implemented. Checked-baggage screening technologies are effective at identifying many prohibited materials, and these systems should continue to be implemented, inline to the automated baggage systems, wherever possible.

After reviewing and improving the existing screening procedures, focus must shift to the other methods a bomb can be introduced to an aircraft, as terrorists and criminals will shift to the path of least resistance, which right now is represented primarily by cargo, mail, and the placement of a device by an aviation employee. Given the vast amount of cargo that is carried daily, cargo should be separated into high- and low-risk categories, with at least 20% of the low-risk cargo being screened at random and 100% of the high-risk cargo screened. Over time, 100% of all cargo and mail may need to be screened if it is shown that there is an attempt, plot, or actual bombing of an aircraft through the air cargo path. As of this writing, there is a legislative effort to require the 100% screening of all cargo (see Chapter 10).

In regard to the potential for an aviation employee to introduce such a device, there are programs that require random screening of airport workers by TSA personnel. These programs screen items that vendors and caterers bring onto the airfield and aircraft. These programs should continue to be implemented. There is a legislative initiative to require 100% screening of all aviation employees. Although there are several significant logistical and financial ramifications to screening 100% of all personnel, less than 100% screening does present a significant gap in the aviation security system. With several instances of airport workers in the United States and historical evidence that aviation personnel have previously been involved in criminal or terrorist acts on aviation, this program should be implemented.

Many individuals at airports require frequent and repetitive access through the screening checkpoints and in and out of the security areas. Airport police, fire, emergency medical personnel, airport operations, and certain maintenance and security personnel should be excluded from the screening. In lieu of screening, these individuals should undergo thorough and periodic background checks that are required to receive an access/ID badge. This could include an extensive reference check, a polygraph, or other background check similar to those conducted when issuing a military clearance. Just as police officers are thoroughly vetted and entrusted to wear a firearm and make arrests, the operational realities of an airport demand that some level of trust is provided to certain aviation workers to keep the system operating smoothly. Although there are incidents where police officers and others with security clearances and special access have been guilty of using their privilege to commit crime, trained security professionals can employ numerous practices to ensure that those who are trusted remain trusted.

Bomb Threats

A bomb threat disrupts an operation or business, airline or otherwise, for a time, reducing productivity. Bomb threats to an airline, an industry that relies heavily on on-time performance, can be a major disruption as flights are delayed or canceled across the national airspace system, personnel and resources are diverted for hours to handle the crisis, and other airport operations are interrupted. It does not take long for someone to figure out that if a bomb threat stops the operation of an airline every time one is called in, then all one has to do is keep calling in bomb threats. The airline will continue to lose millions of dollars a day responding to threats and, perhaps, the airline will shut down. Therefore, the seriousness of bomb threats must be judged based on a variety of factors. Both the FBI and the Bureau of Alcohol, Tobacco, and Firearms (ATF) publish a card that can be used at airline call centers and other areas of the business that are most likely to receive a bomb threat; it helps the airline and other agencies evaluate the threat (Figure 11.1).

ATF Bomb Threat Checklist

Exact time of call:

Exact words of caller:

Questions to Ask:
1. When is bomb going to explode?
2. Where is the bomb?
3. What does it look like?
4. What kind of bomb is it?
5. What will cause it to explode?
6. Did you place the bomb?
7. Why?
8. Where are you calling from?
9. What is your address?
10. What is your name?

Caller's Voice (circle)

Calm	Rapid
Slow	Stressed
Crying	Nasal
Slurred	Lisp
Stutter	Excited
Deep	Disguised
Loud	Sincere
Broken	Squeaky
Giggling	Normal
Accent	Male
Angry	Female

If voice is familiar, whom did it sound like?

Were there any background noises?

Remarks:

Person receiving call:

Telephone number at which call received:

Date:

Report call immediately to Campus Office of Safety & Security. Off-Campus Site Emergency Coordinator, or call 911.

FIGURE 11.1 ATF bomb threat checklist.

Presently, handling a bomb threat from the air carrier perspective depends on where the call comes in. A call to an airport authority will result in the airport responding by notifying air carriers and the TSA. A call to the airline may not result in notification to the airport or other agencies. If the threat is clearly a hoax or so unspecific as to not warrant further action, determined through a vetting process conducted by the airline security personnel, the airline may elect to not notify other agencies.

When an aircraft operator receives a bomb threat, the in-flight security coordinator of the flight in question must be notified, and any applicable threat measures that are part of the aircraft operator's security program must be implemented. The aircraft operator must also notify the airport operator at its intended point of landing when an aircraft has received a threat. The aircraft operator must attempt to determine whether any explosive or incendiary device is present by conducting a security inspection on the ground before the aircraft's next flight, or if already in flight, as soon as possible after landing. If the aircraft is on the ground, all passengers must be evacuated and the aircraft thoroughly searched. This search is usually carried out by K-9 explosive detection teams since they are generally able to conduct a quick search because of their flexibility and ability to navigate in and around the aircraft cabin and cargo holds.

If an aircraft in flight receives a bomb threat or notification from the ground that a bomb may be onboard, the flight crew must be immediately notified so that in-flight security precautions may be taken. This sometimes involves the movement of a suspicious package or item to an area of the aircraft known as the least-risk bomb location (LRBL). This is an area within the cabin of a commercial aircraft where the aircraft manufacturer has determined that an explosion will result in the least damage to the aircraft. Obviously, some explosive forces can be large enough to destroy the aircraft regardless of where the charge is placed, but use of the LRBL may be effective in most cases. Crewmembers may also have ballistic-proof blankets available, or they may place other bags around the suspect item to mitigate any blast. The strategy for mitigating any bomb threat or suspected device onboard an aircraft will be guided by the aircraft operator security program and the overall totality of the circumstances.

If a threat is received and related to the facilities used by an aircraft operator, the airport operator must be immediately notified, along with the other domestic and foreign aircraft operators at that specific airport. The aircraft operator must conduct a security inspection before continuing to use the threatened facilities or areas. As a practical matter, when a threat to an aircraft operator's facility is received, and that facility is located on airport grounds, the airport operator often takes operational control of the incident including evacuation measures and security inspections by explosive ordinance personnel including K-9 teams, local law enforcement, and the FBI.

The regulations require that aircraft operators notify the TSA of any bomb threat against a flight or an airline facility.[2] However, the aircraft operator can exercise some discretion when determining whether a particular threat is credible enough to stop air carrier operations and contact the TSA. The families of the Victims of Pan Am Flight 103 brought this discretionary policy into question. Before that bombing, Pan Am had received threats that an explosive device was going to be placed on an aircraft departing Frankfurt, Germany. However, the airline did not notify the passengers of Flight 103 that the threat had been received. Even in the postbombing investigation, authorities determined that the threat was without merit. At the time, U.S. President Ronald Reagan defended the practice of keeping bomb threats confidential by noting that to advertise every bomb threat would bring air service throughout the world to a halt.

Aircraft Hijacking

Even though numerous procedures have been implemented to prevent an aircraft from being hijacked and, if hijacked, to prevent it from being used as a WMD, there is still the potential for hijackings to occur. On average, since 9/11, there have been approximately 5–10 hijackings or hijacking attempts each year worldwide. Hijackings help meet many needs of the terrorist or criminal organization in terms of invoking fear and gaining publicity. Hijacking an aircraft and crashing it into a ground target invokes deep-seated

---

[2]If an aircraft is in flight outside U.S. jurisdiction when a threat is received, the aircraft operator must notify the appropriate authorities in the country where the aircraft is located and at the aircraft's destination if outside the United States. Notification to air traffic control is sufficient to meet this requirement as the International Civil Aviation Organization (ICAO) Annex 17 of the Chicago Convention requires that air traffic controllers pass along such information when it is received.

fears that many Americans have developed since 9/11, in addition to the massive loss of life, the destruction of property, and the huge economic impact that results from such an attack. The following narrative highlights an al-Qaeda plot to hijack aircraft and crash them into targets in the United States, Australia, the United Kingdom, and Italy. This passage was discovered in 2006, and it demonstrates that hijacking for the purpose of causing mass destruction is still a highly probable form of terrorism:

> The Department of Homeland Security report, dated 16 June 2006 and marked unclassified, was first reported by ABC News in the US and has since also been seen by the BBC. It makes clear that al-Qaeda remains interested in attacking aviation targets and "likely desires a successful repeat of a 2001 suicide hijacking against the United States." (Corera, 2006, p. 10)

The common strategy for dealing with hijackers before 9/11 was so ubiquitous it could have just as well been called the "commonly known strategy." The common strategy was essentially to cooperate with the hijackers and get the plane safely on the ground as soon as possible. With hundreds of hijackings bearing similar characteristics, the 9/11 hijackers relied on the flight crews and passengers executing the common strategy. The common strategy assumed that hijackers would not know the practices that pilots were instructed to take during a hijacking. Another assumption was that the hijackers would have little knowledge of the internal workings of an aircraft.[3]

The antihijacking strategy practiced today is essentially active resistance—also not a secret. Security practitioners can no longer assume that hijackers will *not* have knowledge of how an aircraft works. In fact, hijackers may understand the functions of various systems in the cockpit and know how to fly the plane—as did the 9/11 terrorists.

In 2005, the DHS rolled back some of the items on its prohibited items list because of the effectiveness of active resistance. There are numerous defenses against hijackings, including security screening; air marshals and other federal, local, and state law enforcement officers flying armed; passenger resistance; and airline personnel trained in self-defense. To continue to reduce the hijacking threat, consideration should be given to requiring airline personnel to receive hands-on self-defense training[4] and to allowing local and state law enforcement officers to carry their weapons onboard. Currently, local and state law enforcement officers must have a reason to carry their weapons onboard, such as prisoner transport or high-risk surveillance. Otherwise, if they are just traveling on vacation or even on business to attend a seminar or conference, they are required to store their firearms in checked baggage. Airlines operating without the support of a

---

[3]In the hijacking of TWA Flight 847, Captain John Testrake believed the situation would not have ended quite as well if the hijackers had a working knowledge of the aircraft's systems. Because of the hijackers' ignorance, Testrake and his crew were able to keep the aircraft on the ground (after a period of time) and prevent the hijackers from further risking the lives of the crew and those on the ground.

[4]The present requirements for self-defense training do not include hands-on training. It is optional on the part of the aircraft operator to provide such training.

government-sponsored air marshal program should consider employing their own armed and trained security personnel.

Managing a Hijacking Incident

Most strategies for managing hijacking incidents are confidential. Conventional wisdom and lessons learned are not a secret; if the incident occurs when the aircraft is still on the ground, the strategy is to do everything possible to keep it there. Once the aircraft is airborne, numerous variables come into play, including the potential that the aircraft will be used as a weapon or that it will be shot down by military aircraft to prevent just such an occurrence. Pilots should attempt to disable the aircraft either by escaping or through other means. Airport operators should consider parking snowplows, construction equipment, or hauling in Jersey barricades to prevent the aircraft from taking off. As called for in the ICAO standards, if possible, the aircraft should be relocated to an isolated parking position (IPP) on the airfield where the incident can be managed. This should occur only if the aircraft is otherwise prevented from accessing a runway or taxiway where it can take off as it transits to the IPP. The IPP is usually located away from the main terminal buildings and the runways and provides a location where negotiations can take place and tactical units can safely observe the aircraft.

Initially, state and local law enforcement, such as the airport police force or local sheriff deputy response, will contain the aircraft while waiting for the federal response. The TSA will be advised of the incident. However, the TSA does not typically have a law enforcement response team; therefore, the local law enforcement agency will manage the on-scene tactical issues until the FBI can arrive, assemble, and deploy. It should be noted that the federal response may be 30–45 minutes or more away, so local police and airport personnel must know how to handle such incidents in the meantime. Specific hijack management strategies should be discussed with federal, state, and local security personnel, including the FBI and the TSA, and training conducted for the first responders at the airport as the precise processes for managing a hijacking should not be made public.

If the aircraft hijacking occurs while the plane is in flight, the primary goal of the flight crew is to keep control of the aircraft, prevent unauthorized access to the cockpit, and get the aircraft on the ground as safely and quickly as possible. There are numerous other measures flight crew and flight attendants can take to communicate with air traffic control and prevent access to the cockpit, many of which are classified and taught to flight crews.

Airport Attack: Armed Assault

Although it has only occurred a few times in the course of aviation history, the armed assault of an airport terminal building ranks as the third most frequent kind of attack on aviation. Untrained individuals have committed these types of attacks, as well as trained assault teams using military tactics and weaponry.

In a coordinated attack by multiple, trained operatives, the results can be deadly, resulting in a massive loss of life and a major disruption to operations. In the case of an airport, the closure of the terminal building for an extended period could also result, with a consequentially large economic impact.

At many U.S. airports, the law enforcement agencies charged with patrolling the terminal areas are lightly armed, as compared to the level of armament used by assailants; they are thus limited in their response capabilities. Some exceptions include Los Angeles International Airport, which has special response teams on duty, and Boston Logan International Airport, which carries out patrols by law enforcement personnel carrying submachine guns. London Heathrow International Airport also employs Metro police armed with submachine guns, in addition to the standard police sidearm.

A coordinated attack on an airport terminal is likely to involve terrorist operatives who have been trained in military tactics and who are using military weapons and explosives. These operatives typically use automatic guns and have explosives for use during the initial strike and for arming access points such as doors to kill rescuers. It should be assumed that these types of terrorists do not intend to survive the assault and will target hostages if a rescue attempt is made. It should also be assumed that an assault team will comprise a large contingent of attackers, possibly 30 or more.

A highly visible and heavily armed police presence can be an effective deterrent to such attacks. An airport is particularly vulnerable when there are long lines at the screening checkpoints or when there are weather closures, resulting in thousands of people crowded into the terminal building and in security and ticketing lines. Airport operators should deploy additional law enforcement during these times. A rapid response by law enforcement is required for any such assault, particularly an assault where the assailants are intent on taking hostages. Rapid intervention may disrupt a planned seizure. Airport terminals should be equipped with panic alarms and emergency assistance buttons, along with closed-caption television (CCTV) with two-way audio capability, so that security personnel can monitor the public areas of the airport. Airport operators and local law enforcement entities can be proactive in preventing this type of attack through the tactics and strategies outlined in the "Airport Policing" section in chapter 12. These types of assaults usually involve prior surveillance and local coordination. Therefore, the activities of the terrorists may be detected before the attack.

Emerging Conventional Threats

Most emerging conventional threats have already occurred against nonaviation targets, such as attacks on military facilities, embassies, hotels, and businesses. It should be assumed that these forms of attack will eventually be used against aviation. Some emerging conventional attacks have already occurred against aviation in various forms including man-portable air-defense (MANPAD) attacks and airport perimeter assaults.

Airport Attack: Improvised Explosive Device

Bombs continue to be among the most common weapons used by terrorists and extreme criminals. Even the threat of a bomb, without an actual device, is often enough to disrupt the daily operations of a business or public entity. Commercial explosives are readily available, along with materials such as fertilizer (nitrates) and diesel fuel or fuel oil, to create an explosive, such as was used in the bombing of the Murrah Federal Building in Oklahoma City, OK. Military-grade explosives are available both for legitimate sale by certain companies and countries and through the black market. Beginning in 1996, U.S. airport operators had to start announcing over their public address systems that if bags are left unattended, they will be removed and destroyed. This warning stemmed from measures taken after the crash of TWA Flight 800 to deter the placement of a remote-controlled or timing device–actuated bomb. The warning has also reduced baggage theft, although this was not the primary reason the announcement was required.

There have been numerous attacks against society using explosives placed in a briefcase or other container that can be triggered remotely or via a timing device. Although fewer of these forms of attack have occurred in aviation, the potential remains high that they will be used. An unattended package or bag can cause a significant disruption to normal operations, delaying flights, inconveniencing passengers, and shutting down businesses in the airport. A detonation can cause massive loss of life, damage to property, and economic loss.

One of the most significant challenges with respect to improvised explosive devices (IEDs) is properly identifying an object as an IED. This is particularly difficult at an airport where, despite public announcements, baggage and briefcases are routinely left unattended (even shortly) every day. An unattended bag is an item that has been observed or reported not to be in the custody of an individual. When a bag is discovered to be unattended, airport security, law enforcement, or operations personnel should conduct a series of evaluations to determine whether additional action is necessary. One of the first steps is to note the location. Unattended bags tend to be on curbside passenger pickup and drop-off areas and near ticket counters and baggage claim areas. Bags are also routinely left unattended near airport restaurants and in gift shops.

An unattended bag should be carefully examined for the travel date, identification information, and general condition. Stains, oil or watermarks, and tears could be signs of tampering. The examination, however, should take place without touching or moving the bag. Next, security should make a public announcement asking for the owner of the bag to identify him- or herself or asking if anyone can identify who may have left the bag or item. Nearby passengers and employees should be questioned quickly. If an owner has not been found, officials should consider elevating the unattended bag to the category of a suspicious device and notify law enforcement.

The law enforcement notification should take place away from the unattended item (at least 50 feet away). Radios and cell phones should not be used around the item as they may trigger a detonation, particularly if the IED has a remote detonation mechanism triggered by a cell phone or radio frequency. Blasting caps can be triggered by radio frequency emissions. The area around the suspected IED should be immediately evacuated and cordoned off until law enforcement arrives.

The quickest way to check for an IED in an airport is to use a K-9 explosives detection team. Other methods include a portable trace detection device or a portable X-ray device.

If there is detection, by K-9 or through other means, of a possible IED inside the container, the evacuation area should increase to at least 1,000 feet. Floors above and below the suspected device should also be evacuated, and an explosive ordinance disposal (EOD) team should be notified immediately. Some airports have onsite EOD teams. Many large commercial-service airports have purchased and maintain a mobile explosive container device used to relocate and safely detonate the item away from the public areas. Unless properly trained and equipped, the number one rule for handling a suspicious item is not to handle it in anyway. The EOD should also be notified immediately if an aviation employee spots an IED such as a pipe bomb or similar device, hidden in a planter, trash can, or other areas that provide natural concealment.

Personnel handling incoming mail should be trained on what to look for to detect whether a parcel or letter contains an IED or a chemical, biological, or radiological agent. Signs such as excessive string or tape on a parcel, lopsided or uneven parcels, rigid or bulky parcels with the package clearly too small for the contents, oily stains or discoloration, wrong name and address or wrong title, and strange odors are all signs of danger. Letters with restrictive markings such as "personal" or "only to be opened by" written or printed on them, badly typed or written addresses, excessive postage, packages that have been mailed from a foreign country, misspelled words, and the absence of a return address are signs of a possibly dangerous letter. Some airports may consider installing small X-ray devices in their mailrooms to scan all incoming mail.

Vehicle-Borne Improvised Explosive Device

The use of a vehicle-borne improvised explosive device (VBIED), commonly called a car or truck bomb, was proven effective in the 1983 bombing of the U.S. Marine Corps barracks in Beirut, Lebanon. The basic concept is to fill a car or truck with a large quantity of an explosive, drive it to the target area, then detonate it, either from inside the vehicle (suicide attack) or by a remote command or timing device, such as occurred at the Murrah Federal Building in Oklahoma City, OK. VBIEDs have been used many times since the 1970s throughout the world.

In 1999, Ahmed Ressam attempted to use a VBIED to attack Los Angeles International Airport. He was discovered while traveling to California by a Port Angeles, WA, customs inspector. If he had been successful, the resulting explosion could have resulted in the deaths of hundreds, in both the initial explosion and in the falling debris in the collapse of the Bradley Terminal Building. In 1995, the FAA implemented the 300-foot rule, which at the time prevented unattended vehicles being parked within 300 feet of the terminal building, and police and security personnel were instructed to immediately tow any unattended vehicle left in the passenger pickup and drop-off lanes. These actions were eventually relaxed, but they were reestablished after 9/11. Again, the restrictions were slowly lifted at many airports, particularly those that conducted vulnerability assessments and integrated reinforced materials into their terminal remodeling. Many airports also conduct continuous police patrols of the landside areas looking for suspicious vehicles and individuals. Realistically, these measures also do little to prevent a suicide bomber from driving to the terminal building and immediately detonating a VBIED. Profiling can help identify reconnaissance personnel and may allow officials to spot a bomber shortly before carrying out a detonation, but it is likely there will not be enough time to intercede in the attack.

One solution is to create vehicle checkpoints on the roads accessing the airport. A standard checkpoint with personnel to stop and search all inbound traffic will quickly bring traffic into and out of an airport to a halt and cause numerous delays. Many airports are located within a city and do not have adequate standoff distances in which to create large vehicle checkpoints. Therefore, any checkpoint should first be able to distinguish among vehicles that do not pose a significant threat and those that require further scrutiny. Second, any technology or process should be able to scan vehicles rapidly for signs of explosive devices. Technological solutions such as millimeter wave devices that can emit radio waves into inbound airport traffic lanes, causing a device to either be detected or detonated before it reaches the terminal building, are one solution. A checkpoint using a scale matches the vehicle to a profile of what the vehicle is "expected" to weigh. Other technologies are also being developed that will detect trace chemical, nuclear, and other explosive elements. Technologies to scan maritime and aviation cargo, such as the use of infrared, thermal, and backscatter X-ray imaging, should also be utilized at vehicle checkpoints.

Airport design can also help mitigate a VBIED attack. Airports with one terminal building are particularly vulnerable as one VBIED may be able to destroy or disable the entire facility, whereas multiple VBIEDs would be required to disable airports with multiple terminals. Airports should also create natural standoff distances for vehicles in the landside access lanes. Airports should also install bollards next to the terminal building to prevent or deter vehicles from being driven or crashed into infrastructure. Safety glass and reinforced materials should be used within the structure of the terminal building. A critical component for protecting the airport is the ability to return to operation as quickly as possible, thus mitigating the impact of an attack on operations and to the economy. Rapid response by Urban Search and Rescue (USAR) teams, along with readily available backup facilities, should be included in any VBIED response plan.

Airport Attack: Suicide or Homicide Bomber

In 2006, a suicide bomber ran into the lobby of the Islamabad, Pakistan, airport, where he was confronted by police. He shot three officers then dropped a hand grenade he was carrying, which killed him. Although there have been several cases of suicide bombers boarding aircraft, this was one of the few attacks in an airport by a suicide bomber. What makes the suicide bomber one of the most deadly forms of attack used by terrorists in recent years is that it is essentially a "bomb with a brain." Unlike command-detonated IEDs, suicide bombers adjust to the environment, identify law enforcement and security measures, flee and return another day, or relocate into a crowd or key infrastructure to increase destruction. A suicide bomber can also quickly move into position and detonate. An IED has to be positioned by an operative. Then the operative must move from the IED, leaving the device unattended for at least several moments before detonation. During this period, there may be enough time to cause suspicion and prompt a law enforcement response. However, remember that a suicide bomber also does *not* have to worry about an escape route!

Besides the advantage a suicide bomber has as related to tactics, another advantage is a common belief that a suicide bomber cannot be stopped. A frequent comment is, "You cannot stop people who are so dedicated that they will kill themselves to commit a terrorist act." This is not true, and to adopt such a belief provides suicide bombers with their biggest advantage of all—the belief that they cannot be stopped. Suicide bomber operations often involve dozens of individuals; these include those involved in recruiting and preparing the operative, planners, safe house operatives, reconnaissance personnel, bomb builders, and others. Throughout this preparation time, others can become aware of the pending attack and potentially inform law enforcement agencies or accidentally leak information to friends and family. Suicide bombers are committed to dying in the attack. However, they do not want the attack to be a wasted effort and thus will spend considerable time planning for the attack.

Once a suicide bomber is loaded up and walking to the target area, stopping the attack is very difficult—but not impossible. In Israel, police and citizens are trained in tactics to defeat suicide bombers. These have been shown to be about 80% effective when employed by someone trained in the proper techniques.

Suicide bombers typically carry 2–30 pounds of plastic explosives attached to a firing trigger kept in their hand, pocket, or chest area. Pushing a button or toggling a switch completes the circuit and detonates the bomb. Sometimes, the bomber pushes the button to arm the device. When the button is released, the bomb explodes. This technology makes it difficult for law enforcement to stop a detonation. Bombs can be hidden in backpacks, vests, undergarments, briefcases, musical instrument cases, luggage, inside vehicles and boats, and on bicycles and motorcycles. For women, sometimes bombs are placed around the stomach area where it is disguised to make the woman appear pregnant. Suicide bombers have also used stolen trucks, ambulances, and police vehicles, and at times, large hazardous material or gasoline trucks have been stolen and used in suicide attacks. Occasionally, nails and bits of metal are wrapped with the IED to increase the damage caused by the blast. There have even been rumors of suicide bombers injecting themselves with the AIDS virus in an attempt to spread the disease during the blast. Chemical and biological elements may also be mixed into an explosive. However, one challenge to this concept is that the temperatures created by the explosion may vaporize these elements. In the 1993 bombing of the World Trade Center, it was reported that Ramzi Yousef attempted to put a chemical agent with the bomb but the element vaporized in the blast.

Soft targets, such as publicly accessible areas, are favorite locations for suicide bombers. Security checkpoints are also prime targets as are caravanning vehicles. At an airport, the primary target for a suicide bomber may be an airplane, as seen in Russia in 2004, or more likely the publicly accessible terminal building. Security checkpoints may also be targets at an airport, along with critical infrastructure such as the air traffic control tower. When security lines and ticket counters are jammed with passengers, the higher density of passengers creates a probable target for the suicide bomber. When the liquid bomb plot was discovered in London in 2006, thousands of travelers were packed into the public terminal areas at Heathrow International Airport, creating an extraordinary target. Military and law enforcement assets should be heavily deployed in such circumstances.

Airport Attack: Perimeter Breach and Standoff Weapons

In addition to the fence surrounding an airfield, the perimeter includes the gates and access control systems, which prevent unauthorized personnel from accessing the airfield. The Government Accountability Office (GAO) characterized the airport perimeter in a 2004 assessment as airfield fencing and access gates, access controls restricting unauthorized access to secured areas, and security measures pertaining to individuals who work at airports (Berrick, 2004).

Low priority in security concerning airport perimeters has been caused by a relatively low frequency of attack on airports or aircraft through or over a perimeter fence. Although there have been attacks at some foreign airports, a U.S. airport has yet to experience a known attack of this nature. With the changing nature of the terrorist threat, along with security gaps being addressed in passenger and baggage screening, the possibility of an attack through or over a perimeter boundary increases.

An airport perimeter can be exploited in many ways. For example, an armed assault by aggressors can drive through a perimeter gate, or individuals posing as airport security guards could access the airport perimeter without creating notice. When considering threats coming through the airport perimeter, offsite attacks must also be considered—for instance, a rocket-propelled grenade attack, automatic weapons fired through the fence at aircraft, or an IED detonated in the flight path of a landing plane.

Although many airports have limited fencing, 7 feet of chain-link fence with 3 feet of strands of barbed wire at the top is the minimum recommendation. Some airports are bordered by water but cannot have perimeter fences, as it would create safety issues for aircraft landing or taking off.[5] Some airports are bordered by densely wooded areas or densely populated areas, which makes it difficult for officials to observe large areas of the perimeter and detect intruders while there is still enough time left to respond. Another

---

[5]Certain regulations under Title 14 CFR Part 139 prohibit objects from being constructed near runways. Additionally, fencing that is between a runway and a large body of water could inhibit fire/rescue personnel from accessing an aircraft that has slid off the runway, through the fence, and into the water. The fencing acts as a metal net wrapped around the fuselage, thereby requiring aircraft rescue firefighting crews to use bolt cutters, under water, to open escape paths for the aircraft occupants.

challenge in protecting the airport perimeter is the length of the perimeter surrounding an airport, making constant physical surveillance by security guard patrols not economically feasible. New technology is addressing these problems.

Some foreign airports are switching to stronger fencing and including detection monitoring capabilities such as seismic sensors triggered when an individual touches the fence. Considering that replacement of airport fencing, particularly with larger and stronger materials, is cost prohibitive, many airport operators are looking at combining CCTV with smart software that can keep track of the entire perimeter. The software will discriminate against many nuisance alarms such as the wind, branches, and other objects blowing into the fence or the detection area, and small animals whose movement may cause a detection system to alarm. When these systems detect a valid alarm, the CCTV camera captures a recording of the intrusion attempt and alarms an operator. The operator can then implement a proper response, such as dispatching a security guard to check for a possible transient or drunk who has decided to jump the fence, or to issue a full law enforcement response if it looks like armed intruders are attempting to gain access to the airport. There are a variety of sensors currently in use and being developed, including infrared and thermal sensors, underground and fence-mounted seismic sensors, and video motion detection systems.

Penetrating the airport perimeter is one form of attack to consider, but equally important are threats that occur just outside of the airport's perimeter fencing, such as a rocket-propelled grenade (RPG) attack, an aerial IED, or an automatic weapon attack on an aircraft that is taking off or landing. In 2004, new phenomena highlighted these off-airport threats when several individuals thought it would be fun to shine laser lights into the cockpits of aircraft landing at airports. The lasers temporarily blinded the pilots. There was speculation that terrorists could exploit this use of lasers. However, the laser lights created more of a safety issue than a viable terrorist threat.

On landing, a civilian aircraft is operating "low and slow," just above its stall speed and in a vulnerable state. It cannot maneuver as well as it can at higher speeds as the aircraft may stall completely, with little distance between the aircraft and the ground. Pilots must also deal with winds, lightning, and other elements while landing. Being fired upon by an RPG or being shot at by attackers on the ground with high-powered weapons creates a situation that could easily cause the pilots to lose control of the aircraft or cause the aircraft to crash immediately.

The effectiveness of using a RPG against an aerial target was demonstrated in 1993 over Somalia as militiamen downed two U.S. helicopters in what is known as the Black Hawk Down incident. RPGs are widely available and in use around the world, most notably in Afghanistan and Iraq where they are used effectively against aerial and ground targets. An attacker off-airport could fire an RPG from the back of a truck, a boat, or from a position of concealment and quickly disappear. Although an RPG does not have a seeker capability, as a MANPAD does, a civilian airliner makes for a stable target because of its well-established flight path. Thus, the nature of the target increases the possibility of a successful hit from an RPG. An RPG can be fired at aircraft on the tarmac, either parked or taxiing. The high explosives carried by an RPG would certainly be capable of destroying an aircraft on the ground if it hit the right location.

One other potential standoff attack has been brought up in the media, but its actual use remains in question. In Iraq, some observers have discussed the use of so-called aerial IEDs. An aerial IED is a shaped-charge explosive that is placed along the known flight paths of a helicopter. When the helicopter flies over the IED, the IED is detonated, sending a cloud of shrapnel into the flight path. The helicopter quickly ingests the shrapnel, causing the engine to flame out at a high speed and low altitude. The helicopter crashes before the crew has time to react. With the known flight paths around an airport, this type of device could be very effective.

In response to standoff attacks, video detection technology combined with security and law enforcement patrols may be effective deterrents. Aviation security practitioners should also work with adjacent law enforcement communities and Neighborhood Watch programs to enlist the eyes and ears of the surrounding communities in watching for suspicious activity.

Aircraft Attack: MANPAD

Since 9/11, a great deal of attention has been focused on the surface-to-air (SAM) missile threat. Known as MANPAD systems, which distinguishes them from vehicle, vessel, or land-based air defense systems, they can be launched by one person to strike an aerial target at altitudes up to 18,000 feet. They do not need to be fired near an airport to hit an aircraft, as long as the aircraft is within the threat envelope of the missile.

In 2003, Secretary of State Colin Powell stated that there is no threat more serious to aviation than MANPADs (Beveridge, 2003). The Taliban definitely believed in the effectiveness of MANPADs, particularly U.S.-built Stinger missiles, so much that the Stinger is considered largely responsible for turning the tide of the war in Afghanistan against the former Soviet Union.

Present-day MANPADs are heat-seeking missiles, although some radar homing and laser beam "riding" SAMs are being developed, and there are radar homing SAMs that can be launched from a vehicle or vessel (Figure 11.2). MANPADs are individually launched from a person's shoulder; some missiles explode only on impact, whereas others can explode when in the proximity of a target.

Depending on the version, MANPADs can reach speeds of Mach 2 and altitudes up to 18,000 feet. MANPADs offer a method of attack that works well with the goals and operating strategies of al-Qaeda. Terrorist organizations try to kill and destroy in impressive yet simple ways. A MANPAD offers the ability to do both—it creates a spectacular attack that is relatively inexpensive and highly lethal. It is also difficult to detect and defend against a MANPAD attack.

Another challenge to the MANPAD threat is that it does not have to be fired on an airport or even near an airport to ensure a hit. A RAND study concluded that the envelope for firing a MANPAD was 870 square miles around Los Angeles International Airport

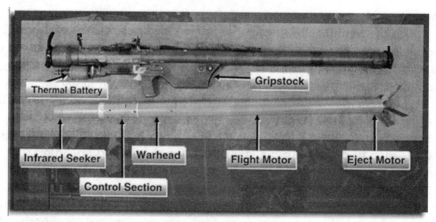

FIGURE 11.2 Components of a MANPAD. *(Source: Department of State.)*

(Chow et al., 2005, p. 14). Furthermore, protecting civil aviation from a MANPAD attack is an issue not just for U.S. airport operators but for aircraft operators conducting operations throughout the world. Today, it is more likely that a missile attack would be conducted outside of the United States rather than within.

Since 9/11, there have been two notable MANPAD attacks. The first was on an El Al charter flight departing Mombasa, Kenya, and the second was on a DHL cargo flight departing Baghdad, Iraq. In the first incident, the missiles missed the charter flight. In the second, two missiles were fired. The first one hit and caused significant damage to the aircraft, but the pilots were able to land that plane and survive the incident. Historically, in 35 attacks on commercial airliners, 24 have been brought down. However, most of those were smaller turbo-prop, piston engine, or helicopter aircraft. In the 5 known attacks on jets, only 2 were brought down (Chow et al., 2005, p. 5).

Other factors to consider in relation to MANPADs are whether an attacker can actually hit the target and whether the missile with a small warhead (less than 5 pounds) can cause enough damage to take down a large commercial airliner. Although a MANPAD may be fired from virtually anywhere, there are environmental factors that can contribute to the success or failure of the shot. Ideally, terrorists want favorable terrain and predictable flight paths, both of which exist around a commercial-service airport.

A missile can be deterred by providing its tracking sensors (seeker head) with something else to chase or by confusing its sensors. With the seeker head not knowing which direction to go, the missile "goes stupid" and either self-destructs or falls to the ground. Antimissile technologies used to protect aircraft include flares, laser jammers, and high-energy lasers. Flare systems were tested in the United States, but there are safety issues with putting flares on commercial aircraft. Commercial laser jammers are being developed that are both aircraft based and ground based. The aircraft-based systems are being adapted from military antimissile systems, but they cost about $1 million per airplane.

Airport security officials can assist with MANPAD prevention efforts. In 2003 and 2004, security officials and other private organizations conducted research at the nation's airports to determine the most advantageous locations from which to fire a MANPAD. These results were released to airport operators to assist security and law enforcement agencies in the surveillance of these locations.

Emerging Asymmetrical Threats

Asymmetrical threats include threats that either have not occurred against aviation or have occurred against aviation and are highly irregular, such as attacks using chemical weapons, cyber-attacks, and attacks using general aviation aircraft. These types generally fall outside the normal threat paths familiar to aviation security practitioners.

CBRN

CBRN is an acronym describing chemical/biological/radiological/nuclear forms of attacks, and all four are generally considered weapons of mass destruction. Investigations have found that both domestic and transnational terrorist organizations have formed plots to use chemical and biological weapons. A few examples of these plots are:

- In 1985, federal law enforcement authorities discovered that a small survivalist group in the Ozark Mountains of Arkansas known as the Covenant, the Sword, and the Arm of the Lord (CSA) had acquired a drum containing 30 gallons of potassium cyanide, with the apparent intent to poison water supplies in New York, Chicago, and Washington, DC.
- In 1992, German police thwarted the release of cyanide in a synagogue by neo-Nazis.
- In 1995, the FBI thwarted a possible sarin attack at Disneyland.
- From April through June 1995, there were several copycat attacks in Japan of the Aum Shinrikyo attack, using cyanide, phosgene, and pepper spray.
- In May 1995, a member of the Aryan Nation was arrested for ordering plague bacteria.
- In 1995, an Arkansas man possessing 130 grams of ricin was arrested.
- In 1997, two chlorine bombs were activated in a crowded shopping mall in Australia, injuring 14 and causing the evacuation of 500 shoppers.

Chemical or biological terrorism has become more popular because of its heightened fear factor. It is difficult to detect, most often easily and anonymously spread by air, and requires extensive decontamination, rendering facilities unusable for long periods. Terrorists do not need large quantities of chemicals or biological agents, particularly if their target is in an enclosed space such as a shopping mall, convention hall, or airport. In fact, airports can be ideal targets for chemical and biological agents, as they are high-profile enclosed spaces where there are many people.

Chemical

A chemical weapon is any weapon that uses a manufactured chemical to kill people. Chemical weapons are broken down into five categories: nerve agents, blistering agents, blood agents, choking agents, and irritants. Aviation security practitioners should understand the symptoms associated with each and train first-response personnel on these signs.

Chemical agents can be dispersed in a number of ways, but the most effective method is through the air. First responders arriving at the scene of a potential chemical attack may note patients in respiratory distress or patients with severe eye irritations, skin redness, or other symptoms. Unfortunately, eye irritation and distress can also be signs of smoke inhalation, shock, and numerous other situations that do not involve chemical warfare elements. Blistering, skin legions, tightening of the chest, and accumulations of fluid in the chest or larynx may indicate a chemical attack.

Biological

Aviation security practitioners are at a severe disadvantage in defending against biological agents. The aviation system makes possible the rapid and widespread distribution of a biological agent once an infection has begun. Further, it may take several days before symptoms of an infection manifest, during which time thousands may have been exposed. The only way to detect an outbreak is through a clinical presentation, which is generally after the fact. The good news is that many biological weapons are difficult to manufacture, more so than chemical weapons. From a practical point-of-view, reducing preventable losses has a greater impact on reducing causalities than does reducing the possibility of an attack (Christen and Maniscalco, 2002).

There are several types of biotoxins including botulinum toxins, ricin, saxitoxin, anthrax, cholera, Ebola virus, and the plague. Some toxins are related to common food biological problems such as Staphylococcal Enterotoxin (food poisoning), Cholera, and saxitoxin, a biotoxin transmitted by shellfish. Ricin is produced by the castor bean plant and is available throughout the world; it has been used in assassinations.

Biotoxins can have another impact on the aviation community, even if they are not used in a terrorist attack. The impact of SARS and the bird flu on the aviation community have been seen with periodic quarantines to certain countries. If such a disease were to break out in the United States, airport and airline managers could lose a large percentage of their workforce for several weeks until the disease runs its course, during which time the ability to continue to provide for the security and law enforcement needs of the aviation community will be considerably diminished.

Radiological

A radiological dispersal device (RDD), commonly called a *dirty bomb*, combines a conventional explosive, such as dynamite, with radioactive material. Some have called this type of device a weapon of mass *disruption*, not destruction. At an airport, such a weapon could shut down large areas of a terminal building for months, or years, and possibly even require the destruction of the contaminated portion of the facility.

A variety of radioactive materials, including Cesium-137, Strontium-90, and Cobalt-60, are commonly available and could be used in an RDD attack. Hospitals, universities, factories, construction companies, and laboratories are possible sources. According to the Nuclear Regulatory Commission, the levels or radiation possible in a dirty bomb from these sources would not be sufficient to kill anyone or even cause severe illness.

Certain other radioactive materials dispersed in the air could contaminate up to several city blocks, creating fear and possibly panic and costly cleanup. Prompt, accurate, nonemotional public information might prevent the panic sought by terrorists. A second type of RDD might not involve an explosive but could be a powerful radioactive source hidden in a public place, such as a trash receptacle in a busy train or subway station where people passing close to the source might get a significant dose of radiation. Of particular concern to the DHS is the possibility that someone could attempt to bring a radiological device into the United States via aircraft, specifically a general aviation aircraft. To that end, the agency has deployed Customs and Border Protection (CBP) agents to scan incoming flights arriving at GA airports and in GA areas of commercial-service airports. If they identify an aircraft that is emitting radiation, the airport operator must then handle the incident.

First responders must be aware that radiation is invisible; may exist in a liquid, solid, or gaseous state; and that someone does not have to come into direct contact with a radioactive substance to be exposed to the effects of radiation (Christen and Maniscalco, 2002). How much exposure to radiation is too much is a subject of considerable debate, but generally, high levels of exposure will quickly lead to incapacitation of the nervous system, the breakdown of bone marrow, and, eventually, death. For first responders, the first priority when responding to an incident where an RDD has been deployed is to establish a large perimeter and administer to the medical needs of those affected. Individuals who have been exposed to radiation do not pose a threat to responders, but individuals who have been contaminated will represent a health threat to responders (Christen and Maniscalco, 2002). First responders and EMS providers should not bring those who are contaminated directly to a hospital emergency room unless the hospital has been notified in advance and has had time to establish a proper receiving area. Both victims and rescuers must be decontaminated before being released.

Nuclear

The 2004 presidential candidates, President George W. Bush and Senator John Kerry, both agreed that the biggest threat facing the United States is the availability of nuclear materials and the possibility that they will fall into the hands of terrorists and be employed as WMD (Ervin, 2006). In *The Forgotten Homeland,* Clarke et al. (2006) warned of the ramifications of a nuclear attack on U.S. soil:

> *A nuclear detonation in a U.S. city would have profound and severe consequences for national security. The ripple effects would quickly extend beyond the destroyed city and could topple the government as American's could lose confidence in their government's ability to protect them against a horrific weapon of mass destruction. (p. 177)*

Although there is a higher risk of a nuclear device entering the United States by sea (Ervin, 2006), the threat of nuclear weapons being employed in the United States against a major city is also of considerable importance to aviation security practitioners. A nuclear device could be delivered by a variety of methods (e.g., truck, rail, maritime), but the risk of a nuclear device being placed on an aircraft is significant. An aircraft can access areas that ground-based vehicles cannot. An aircraft can carry a nuclear device to an optimum detonation position that could ensure the widest dispersion of radiation, electromagnetic pulse, and blast wave pressure.

There are many threats from nuclear devices to the aviation system. One is the use of an aircraft, most likely a GA airplane, to deliver a nuclear bomb to a U.S. city or target. A nuclear device could also be loaded onto a commercial airliner via cargo and timed to detonate as the aircraft flies over a major city. Terrorists could use an aircraft as a WMD and attempt to crash into a nuclear facility (as hijackers attempted in Oak Ridge, TN, in 1972).

Another consideration for airport security managers concerning a nuclear strike is the possibility that the device may be detonated away from the immediate airport area. In this situation, airport security coordinators and the airport management will most likely assume responsibility for disaster management at the airport. During a nuclear attack on a city, the airport will represent one of the primary locations for evacuation and for acceptance of outside supplies. The airport may also become a location for city officials to regroup and coordinate disaster relief efforts. Airport security managers will be faced with trying to maintain a secure operating environment. Although it will certainly be awhile before normal commercial flights resume, security personnel need to be aware of further attempts by terrorists to destroy or disrupt evacuation flights or even additional conventional attacks on the airport itself. Airport security managers will also be faced with fulfilling these duties with reduced staff, as some may have perished in the nuclear attack. Remaining staff will likely suffer posttraumatic stress as they may have lost loved ones in the attack and from the sheer magnitude of the destruction. Fortunately, nuclear attack remains a nightmare scenario. However, if the airport is still operational, as we have seen in numerous natural disasters, people will migrate to the airport for shelter, food, medical care, and a sense of order that someone is still in charge.

The Future Is Here: Insider Threat and Directed-Energy Weapons

A disturbing trend has occurred in the past 10 years that affects both aviation safety and security issues—the use of lasers pointed at aircraft cockpits. The result is a light show within the cockpit, which distracts the pilots at best, and when the pilots are hit directly with the laser, temporary blindness and permanent eye damage can occur. Finding perpetrators is extraordinarily difficult; in 2011, the FAA said that it pursued legal penalties against 28 people for pointing lasers at airplanes, but in that same year, over 2,800 laser pointing incidents were reported by pilots (Laing, 2012).

An FBI memo stated that al-Qaeda had explored lasers as a weapon, including a possibility of lasing aircraft cockpits to interfere with pilots' ability to operate their aircraft safely. The most vulnerable portion of a flight to such an attack is in landing (Bertolli and Pannone, 2012). Aircraft have been lased many times both before and after 9/11. In one incident a Russian vessel was suspected of lasing a U.S. military aircraft, resulting in pilot eye injuries; two U.S. helicopter pilots suffered eye injuries in Bosnia after lasers were aimed at them from the ground; and North Korea is suspected of using the Chinese-made ZM-87 military lasers on U.S. helicopters in the Korean demilitarized zone (Bertolli and Pannone, 2012). The ZM-87 lasers have the capability of flash-blindness and even tissue damage at 10 kilometers (over 6 miles). Nonmilitary incidents include uses of lasers at landing aircraft cities throughout the world, as well as lasings of police helicopters (Bertolli and Pannone, 2012).

While most people are familiar with the lasers used in presentations, there is a large market for commercial lasers, which are much more powerful. These are used in home improvements as distance measuring tools, and in medical, scientific, and industrial applications. As early as the mid-1990s the Japanese Aum Shinrikyo cult failed in its attempt to use a laser weapon against police, and there have been attempts by Mexican drug cartels and foreign militaries to use lasers, also referred to as directed-energy weapons. While lasers that can actually do physical damage to an individual presently remain the domain of the military forces and are still in development, stand-alone directed-energy weapons, such as the Chinese ZM-87 laser blinder, are in use to disrupt and blind cameras and other electro-optical equipment, or even start fires at long distances (Sloan and Bunker, 2011). Directed-energy weapons are often used in conjunction with traditional firearms to blind responding forces, or their weapons and equipment.

The threat to aviation from directed-energy weapons is clearly the danger of blinding the pilots, which itself is hazardous enough, but when combined with a follow-up rocket-propelled grenade or SAM attack, puts the pilots in a highly disadvantageous position to take evasive action. Also, should a terrorist group acquire a military-grade blinding laser, it could be used in an active shooter scenario to blind passengers, first responders, and CCTV cameras, making an effective response difficult.

Another new form of attack is the use of remotely operated weapons, which are used by the U.S. military in static, ground robot, or unmanned aerial vehicles. The cellular communication networks and even WiFi networks allow communication with the bomb or weapon, which can be detonated or fired remotely, giving terrorists a standoff capability. The success of the IED attacks throughout Iraq and Afghanistan point to the effectiveness of such weapons, however, many civilian facilities and vehicles are not equipped to handle such attacks.

As an example of these new types of threats, in 2012 a Massachusetts man admitted to plotting to attack the U.S. Capitol with remote-controlled model airplanes filled with explosives (NBC News Wire Services, 2012). He was arrested after taking delivery of plastic explosives, hand grenades, and six assault rifles from undercover FBI agents that he believed were members of al-Qaeda (NBC News Wire Services, 2012). Part of his plan was to blow up the Pentagon and U.S. Capitol using remote-controlled planes and to kill American soldiers in Iraq and Afghanistan using IEDs detonated by modified cellphones

(NBC News Wire Services, 2012). Experts said that it would be nearly impossible to inflict large-scale damage using model planes, but the man did manage to construct an IED device that was given over to undercover FBI agents who lied and told the man that the device was successful in killing U.S. soldiers in Iraq.

The additional threat here points to the fact that a U.S. citizen was "radicalized" into taking actions against the United States. This is indicative of a new type of threat—U.S. citizens. One of the main defenses against attacks in the United States is the protection provided by our border and transportation security agencies: CBP, Immigrations and Customs Enforcement (ICE), the TSA, and the U.S. Coast Guard. However, when the attacker is already inside the country, combined with knowing how our nation operates, our culture, our values, and our protections under the Constitution, this provides a context within which to operate. One of the most difficult types of attack to stop and respond to is a contextual attack, where attacks and attackers mimic their surroundings (Sloan and Bunker, 2011). These types of attacks have the lowest probability of being viewed as an anomaly or threat as the individual is within the context of his or her surroundings in terms of dress, ethnic profile, mannerisms and behaviors, cover stories, and actions (Sloan and Bunker, 2011). A disturbing demonstration of the effectiveness of a contextual attack is the Colorado theater shootings in July 2012. The individual entered the theater wearing a mask, helmet, and protective gear, which in any other film would be cause for concern, but the film playing was *Batman: The Dark Knight Rises*, and it was not unusual for some theatergoers to show up in costume. In fact, several early press reports quoted individuals who were inside the theater as saying they thought the individual was part of a stunt, or that the man was simply in costume.

Anwar al-Awlaki was a radical American-born Muslim cleric, who was raised in New Mexico and attended Colorado State University, and who was, for a period of time, al-Qaeda's chief propaganda agent. With his understanding of Western culture and the media, he was able to carry the message of violent jihad against the United States over the Internet, through online lectures. In an article in *The New York Times*, al-Awlaki was called possibly the most prominent English-speaking advocate of violent jihad against the United States (*The New York Times*, 2012). al-Awlaki is linked to the U.S. Army psychiatrist accused of shooting 13 people at Ft. Hood, TX, Umar Farouk Abdulmutallab (the underwear bomber), and Faisal Shahzad who attempted to set off a car bomb in Times Square in 2010 (*The New York Times*, 2012).

Al-Awlaki was killed by a Hellfire missile, fired by a CIA-operated Predator drone in Yemen in September 2011. The strike set off a debate as it represented the first time since 9/11 when a U.S. citizen had been deliberately targeted and killed by American forces. Also killed in the attack was Smir Khan, another U.S. citizen born in Pakistan who was traveling with al-Awlaki. Kahn was the editor for the online jihadist magazine *Inspire* (*The New York Times*, 2012). A month later, al-Awlaki's son, who was born in Colorado, was killed in another drone strike in Yemen.

Interestingly enough, in an old article published back in 2001, shortly after the 9/11 attacks, the FAA noted that it is "likely" the organization that staged the attacks, "has sought or will seek to place members in position at airports to facilitate future attacks, or that it will attempt to co-opt individuals already in such positions" (CNN, 2001). The article referenced the new requirements for aviation employees to undergo the fingerprint-based criminal history record checks. Combined with the recent attacks or attempted attacks by U.S. citizens and gaps identified by the DHS Office of the Inspector General in the credentialing process, this represents a weakness that aviation security practitioners should continually seek to resolve.

Attacks from within the aviation community are the most difficult to deter as insiders—including pilots, flight attendants, airport workers, and airline staff—have access throughout much of the aviation system. Many of the security layers in place to prevent outsiders from successfully attacking aviation can be bypassed through the normal course of business for an aviation employee. In 2007, an airport worker in Miami, FL, was arrested for smuggling large quantities of firearms onboard commercial airliners. Insiders may attempt to sabotage aircraft or circumvent airport access control systems to facilitate an attack. Fortunately, there have been few cases where an airline mechanic has sabotaged an aircraft resulting in it crashing. With the forensic abilities of the National Transportation Safety Board (NTSB) and the FBI, any such attempt would certainly be traced back to the perpetrator. However, numerous other workers do have daily access to aircraft, thus security practitioners must remain on the alert for this threat.

Another potential threat from within the aviation system lies in the GA community. As previously discussed, GA aircraft are more accessible than commercial-service aircraft. However, the majority of GA aircraft serve as inferior tools for destruction. With the exception of large business jets, most are too light and do not carry enough fuel onboard to make them an effective weapon.[6]

The most damaging threat from the use of GA aircraft would probably result from the use of a large corporate aircraft or light GA aircraft filled with explosive material or a CBRN agent and flown into a ground target. In a study conducted by the U.S. Office of Technology Assessment, the agency estimated that a small private plane, on a windless night, loaded with 220 pounds of anthrax spores and flying over Washington, DC, could kill between 1 million and 3 million people and render the city uninhabitable for several years (Christen and Maniscalco, 2002). Other studies have looked at the impact of a small aircraft filled with several hundred pounds of conventional explosives crashing into an open-air sports stadium. The immediate impact would result in many lives being lost, but the resulting panic would continue to take the lives of others as spectators scramble away from the flames. Local hospitals and burn centers would quickly be overwhelmed, resulting in more casualties as people waited for treatment. Some studies have estimated the loss from such an event to be in the thousands, and most certainly in the hundreds.

If a GA plane were used to create mass destruction, the resulting flight restrictions and congressional lawmaking would probably cause the economic demise of GA. The cost of

---

[6]Cases of this scenario include the intentional crash of a Cessna 172 into a Tampa, FL, bank building in 2002 and the accidental crash of a Cirrus aircraft into an apartment complex in New York City in 2006.

flying GA aircraft would increase significantly. There would be a resultant impact as fewer young people pursued aviation as a career, causing the pipeline for airline and military pilots to diminish. The cost of operating GA businesses and airports would escalate, forcing many GA airports to close. The business flights that would normally use GA airports would be forced to switch to commercial-service airports, increasing congestion and flight delays.

Infrastructure

Attacks on the critical infrastructure of the country relate to the aviation community in a variety of ways. Aviation and airports are considered critical infrastructure, along with the communication channels needed to interact with aircraft in flight. Ervin is the former inspector general for the DHS and states:

> Certain sites and sectors are defined as "critical infrastructure" because their functionality is necessary to the security, prosperity, and psyche of a nation. The Administration has designated the following sectors as critical infrastructure: agriculture and food, water, public health, emergency services, the defense industrial base, telecommunications, energy, transportation, banking and finance, chemicals and hazardous materials, postal and shipping services. In addition, certain national monuments and icons such as the Washington Monument and the Statue of Liberty, certain government facilities and certain private buildings are defined as "key assets" because they are likewise critical to the psychic well-being and stability of the nation. (Ervin, 2006, p. 145)

An attack on a power plant could leave an airport without the power necessary to distribute to air traffic control, runway lighting access control, and CCTV systems. Most commercial-service airports and some GA airports have backup generators for essential areas, such as air traffic control and runway lights. However, these generators are normally not capable of running the remaining areas needed for an airport to function, such as heating and air conditioning, lighting for the terminal buildings, and power to the baggage movement and security screening systems.

Much of the U.S. infrastructure, including air traffic control, emergency first responders, the financial sector, and airport security systems, rely on computers. A cyber-attack could produce devastating results to these and many other systems. Much of the world's air traffic control system is based on computer information and communication systems.

Within the digital age, the protection of computer systems is essential. Police, fire, and EMS agencies are notified and dispatched through the telecommunications systems and through computer-aided dispatch computers with satellite information systems. According to the 2000 Computer Crime and Security Survey, 90% of respondents, mostly large corporations and government agencies, detected computer security breaches within a

12-month period (Christen and Maniscalco, 2002). There has been a long history of attacks on computer systems, within both the government and the private sector. The U.S. DOD now includes information warfare as the fourth battlefield, joining land, sea, and air.

In *The Edge of Disaster*, Stephen Flynn (2007) provided four suggestions for protecting critical infrastructure. These suggestions are also helpful for protecting against cyber-attacks.

1. Constructing physical defenses, such as cement barricades, target hardening, access control systems, standoff distances, and so on.
2. Providing redundancy, such as backup computers and data files (located offsite), or supplemental systems that automatically start when primary systems fail. Redundancy in facilities and other infrastructure that are easily accessible to security personnel should be considered.
3. Rapidly repairing damage. A quickly reparable system or other form of infrastructure is less valuable to terrorists and criminals.
4. Reengineering to improve security by relocating infrastructure or systems to safer areas. Reengineering also includes designing resiliency into structures and systems.

Ultimately, protecting the aviation system from asymmetrical attacks will take foresight and action before an attack occurs. Aviation security practitioners should assess the areas that would be impacted by the various attacks noted here and develop contingencies for such attacks.

Counterterrorism: Terrorism Defense Planning

This section discusses the methods and options for aviation security practitioners to use before, during, and after a security emergency. It is largely based on published emergency management models and the National Incident Management System (NIMS). Airport security personnel along with law enforcement and other emergency workers are expected to be trained in the NIMS functions. To benefit fully from this section, it is recommended that you complete the NIMS training modules available at *www.fema.gov*.

Emergency Management Cycle

There are three elements to the emergency management cycle: mitigation/preparedness, response, and recovery.

Mitigation/Preparedness
The mitigation phase encompasses the actions taken to either prevent an attack or mitigate the effects of an attack. Mitigation strategies as they relate to the aviation security function can include the following:

- Conducting threat and vulnerability analyses of facilities and infrastructure
- Reinforcing structures such as the terminal building, installing safety glass, and locating parking away from the terminal building or airline administrative offices
- Monitoring new hazards and intelligence information
- Keeping abreast of security activity through industry publications, security newsletters, and communication with the federal security director
- Following codes and ordinances (building and zoning, fire, and hazmat)
- Imposing financial penalties or offering incentives to airport tenants, aircraft operators, vendors, contractors, and employees who do not adhere to security rules
- Consistently enforcing rules, inspections, patrols, and violation notices
- Providing security awareness training for airport and aircraft operator personnel
- Building partnerships and relationships with other airport tenants and emergency responders
- Ensuring sprinkler and alarm systems are fully functional and have backups

A critical mitigation function is the threat and vulnerability assessment. Aviation security practitioners should conduct a hazard identification to determine the full range of natural disasters and manufactured incidents that can affect their facility. Natural disasters can have a severe effect on security management. Outside resources should be identified as part of this assessment, along with their capabilities.

Preparedness consists of leadership, training, readiness, exercise support, and technical and financial assistance to strengthen emergency workers and the community as they prepare for disasters. Training, exercises, and acquiring the resources to support emergency operations are core functions of preparedness in an emergency operations plan (EOP).

Once the EOP is drafted, it must be exercised. It is typical for airport operators to exercise only the hijack section of an EOP. Aviation security practitioners should also focus on exercising other elements of the plan, such as those related to a bomb threat in the terminal or onboard an aircraft, an armed assault in the terminal building, or an attempted penetration via the airport perimeter. Federal regulations require that EOP exercises are conducted at least once a year and that a full-scale exercise is conducted every third year.

Response

The response component of emergency management encompasses those actions taken during an emergency. Response begins when an agency is notified of an emergency and is usually characterized by the activation of the incident command system. Response covers four stages:

- Alerting and notifying other emergency response agencies
- Warning the public
- Protecting citizens and property
- Providing for the public welfare, followed by beginning the restoration of normal services

Recovery

Recovery involves those activities that are necessary to restore normal operations. Recovery is divided into two phases: short term and long term. Short-term recovery often overlaps with the response phase as agencies begin to restore interrupted public services and reestablish transportation routes, such as resuming flights. Long-term recovery may continue for months or years while facilities are reconstructed and the impact of the event is analyzed. During long-term recovery, processes to mitigate future occurrences need to be developed.

Another component of the recovery phase is financial assistance and recovering money and resources lost during the event. It is imperative that a careful accounting is made of monies spent, resources used, and personnel from outside agencies that provided assistance, so that insurance companies and the Federal Emergency Management Agency (FEMA) can provide adequate funds to cover losses.

National Response Plan

The National Response Plan (NRP) applies to natural disasters and terrorist attacks as defined in the Robert T. Stafford Disaster Relief Assistance Act.[7] It can also be invoked when the U.S. president determines that federal assistance is needed to respond to a local event. The plan applies to numerous federal agencies and the American Red Cross. FEMA has the responsibility for managing the response plan and for conducting federal preparedness, planning and management, and disaster assistance.

National Incident Management System

In 2003, Homeland Security Presidential Directive-5 (HSPD-5) was issued, which directed the secretary of the DHS to develop and administer a National Incident Management System. NIMS provides a consistent nationwide template to enable all governmental, private-sector, and nongovernmental organizations to work together during domestic incidents. NIMS lays out the basic concepts of incident management and includes comprehensive sections on preparedness, resource management, communications, supporting technologies, and the Incident Command System (ICS) structure: operations, planning, logistics, administration, and command.

Conclusion

As threats from terrorist and criminals continue to evolve, so must our responses to those threats. Before 9/11, our defense strategies against terrorists or criminals remained essentially the same. Since 9/11, the progressive and unconventional methods of criminals, especially terrorists, have caused the aviation security industry to evolve in planning

---

[7]NRP is also known as the Federal Response Plan. It is more akin to a resource guide than a specific plan.

and implementing new security measures that can mitigate future attacks. As a result, aviation security is now a highly dynamic and complex system of layers containing policies, strategies, tools, and processes. We must now look to the future and use these resources to anticipate emerging threats. Our prime goal in aviation security is to employ security professionals who strive for maximum effectiveness in their areas of responsibilities. Above all, it is essential that aviation security professionals work toward the continuous development of sustainable methods and technologies that can detect, deter, and respond to existing and new threats.

References

Associated Press (AP), Nov. 10, 2006. U.K. Tracking 30 Terror Plots, 1,600 Suspects. MSNBC.com, retrieved Aug. 16, 2006, from www.msnbc.msn.com/id/15646571.

Berrick, C., 2004. Aviation Security: Further Steps Needed to Strengthen the Security of Commercial Airport Perimeters and Access Controls, Comments from the Department of Homeland Security, Appendix III (Publication No. GAO-04-728), retrieved Aug. 16, 2008, from www.gao.gov/new.items/d04728.pdf.

Bertolli, E.R., Pannone, R.D., July 24, 2012. Lasers: Unconventional Weapons of Criminals and Terrorists. The Police Chief, retrieved July 24, 2012, from http://www.policechiefmagazine.org/magazine/index.cfm?fuseaction=display_arch&article_id=1731&issue_id=22009.

Beveridge, D., 2003. APEC Nations Agree to Limit Missile Sales. Associated Press, Man-Portable Air Defense Systems (MANPADS) Proliferation Federation of American Scientists (FAS), retrieved Aug. 16, 2008, from www.fas.org/programs/ssp/asmp/MANPADS.html.

Chow, J., Chiesa, J., Dreyer, P., Eisman, M., Karasik, T., Kvitky, J., et al., Infrastructure, Safety and Environment, 2005. Protecting commercial aviation against the shoulder-fired missile threat. Retrieved from RAND Company website: http://www.rand.org/pubs/occasional_papers/2005/RAND_OP106.pdf.

Christen, H.T., Maniscalco, P., 2002. Understanding Terrorism and Public Health: A Balanced Approach to Strengthening Systems and Protecting People. Oxford, New York.

Clarke, R.A., Allen, C.A., Bullock, J.A., Eddy, R.P., Ferguson, C.D., Flynn, S.E., et al., 2006. The Forgotten Homeland: A Century Foundation Task Force Report. Century Foundation Press, New York.

CNN, Dec. 5, 2001. Airport Workers Face Stiffer FAA Rules. CNN.com, retrieved July 25, 2012, from http://edition.cnn.com/2001/TRAVEL/NEWS/12/04/rec.airport.security/index.html.

Corera, G., 2006. US Confirms Heathrow Hijack Plot. BBC News, retrieved Aug. 16, 2008, from http://news.bbc.co.uk/2/hi/uk_news/5104672.stm.

Ervin, C.K., 2006. Open Target: Where America Is Vulnerable to Attack. Palgrave MacMillan, New York.

Flynn, S., 2007. The Edge of Disaster. Random House, New York.

Gonzales, A., 2006. Transcript of Press Conference Announcing Florida Terrorism Indictments. U.S. Department of Justice, retrieved Aug. 16, 2006, from www.usdoj.gov/archive/ag/speeches/2006/ag_speech_0606231.html.

Graff, G., 2011. The threat matrix: the FBI at war in the age of global terror. Hachette Book Group, New York.

Laing, K., May 16, 2012. FAA to Pursue Stiffer Penalties for Laser Pointing Incidents, retrieved July 24, 2012, from http://thehill.com/blogs/transportation-report/aviation/227821-faa-to-pursue-stiffer-penalties-for-airplane-laser-pointers.

McRaven, W.H., 1996. Spec Ops: Case Studies in Special Operations Warfare: Theory and Practice. Presidio Press.

NBC News Wire Services, July 21, 2012. Would-be Model Plane Bomber Ferdaus Admits Plan to Attack Pentagon, Capitol. NBC News.com, retrieved July 25, 2012, from http://usnews.msnbc.msn.com/_news/2012/07/21/12870497-would-be-model-plane-bomber-ferdaus-admits-plan-to-attack-pentagon-capitol?lite.

New York Times, The, July 18, 2012. Anwar al-Awlaki, retrieved July 25, 2012, from http://topics.nytimes.com/topics/reference/timestopics/people/a/anwar_al_awlaki/index.html.

Ronczkowski, M.R., 2012. Terrorism and Organized Crime: Intelligence Gathering, Analysis and Investigations. CRC Press, Boca Raton, FL.

Sloan, S., Bunker, R., 2011. Red Teams and Counterterrorism Training. University of Oklahoma Press, Norman.

Winter, J., July 22, 2012. Exclusive: Massacre Suspect James Holmes' Gun-range Application Drew Red Flag. FOX News, retrieved July 24, 2012, from http://www.foxnews.com/us/2012/07/22/massacre-suspect-james-holmes-gun-range-application-drew-red-flag-for-owner/.

12 ⸬

Security Operations

Objectives

In this chapter, we describe methods and theories related to security operations, including airport policing, private security officers in the aviation environment, and security functions of airport operations and other personnel. Aviation security is a multilayered system, but there are key individuals who are responsible for the day-to-day monitoring and response to threats. Within the aviation environment, security and law enforcement personnel must also recognize that "airports mean business," which infers that there is an economic and public relations component, as well as a protection and response component.

Introduction

Being able to counter existing and emerging threats requires developing processes to detect and deter a threat. Implementing systems to respond to an exercised threat is also a required concern of aviation security. Some security practitioners consider that there is little difference between countering and responding to a terrorist attack as compared to more traditional criminal activities. The strategies for deterring, detecting, and responding to each are largely similar; the difference is primarily one of scale. Typically, traditional criminal activity produces limited damage or destruction to life and property. In contrast, a large coordinated terrorist attack conducted by highly trained individuals can cause far more damage and require larger response measures and longer recovery periods. As globalization and technologies enable knowledge and information to disseminate, terrorists are able to adapt to new security measures. Security practitioners must continuously be prepared to anticipate the capabilities of each new generation of terrorist.

Aviation Policing Strategies

"Airports Mean Business" is a motto once printed on the side of patrol cars at the Los Angeles International Airport in California. It is a reminder to the hundreds of dedicated security and law enforcement personnel at the airport that while threats require an immediate response, how they enforce the laws and rules at the airport can impact the bottom line of the facility. It is a difficult job when they have to respond to a suspicious bag, as the cop inside of them says to evacuate, take no chances, and call in explosive ordinance disposal (EOD), while the airport manager inside of them has to mitigate that response with the likelihood that the unattended bag is probably not a bomb, and that any evacuation will shut down businesses, cause flight delays and cancellations, and have a lasting and exponential impact on the bottom line of not only the airport but tenants and air carriers.

A significant part of the day-to-day concerns of airport security and airline security personnel focuses on routine crimes, such as theft and assault. Airports are in many ways similar to small cities. A major commercial-service airport has its own police force, fire department, and emergency medical and first aid services. It has businesses, tenants, and thousands of stakeholders who are not necessarily passengers (e.g., airport, airline, and government personnel; baggage handlers; fast-food workers; janitorial staff). Airports often host tens of thousands of transient passengers daily. Like any town or city, an airport must respond to its own criminal elements, both internal and visiting. Criminals are adept at exploiting vulnerabilities and whenever there are valuable goods for the taking, there will inevitably be criminal activities (Table 12.1).

Property theft is by far the most prevalent crime at an airport and includes baggage theft, vehicle theft (from the parking lot or landside areas), and burglaries. Criminals rely on passengers being nervous. Although in unfamiliar surroundings, passengers will frequently leave their luggage unattended, which is the criminal's desired outcome.

Duane McGray, executive director of the Airport Law Enforcement Agencies Network (ALEAN), has written extensively on the differences between airport law enforcement and municipal police departments. McGray notes that municipal police departments focus on three primary areas—calls for service, violent crime, and public service— whereas airport policing involves more police service, physical security, and customer service (McGray):

Urban police officers are frequently placed on the front lines in the war against drugs, violent crime and cultural diversity. The requirements of the job require personalities that tend to be adrenalin driven. Former military personnel normally are well suited

Table 12.1 Types of Crimes Experienced at Airports

• Robberies of cash facilities such as banks and ATMs	• Possession of a prohibited item (either prohibited by federal law, as in trying to gain access through a screening area checkpoint, or prohibited by local statue)
• Armed robbery of individuals and airport tenants outside of the sterile area, including in the parking lots	
• Vandalism	• Solicitation, often in the form of unauthorized ground transportation providers such as taxicabs and limousine operators
• Pick pocketing	
• Disturbing the peace	• Prostitution
• Drunk and disorderly conduct	• Drug smuggling
• Forgery of documents (primarily related to the application of an airport access/ID credential)	• Human trafficking
	• Shoplifting
• Theft from catering and supply trucks (often by employees)	• Identity theft—usually in association with another crime such as stealing a laptop, personal digital assistant, cell phone, wallet, or purse
• Tool and parts theft from aircraft maintenance hangars (again, often by employees)	• Rare: homicides, rapes, and kidnappings

for the urban police culture. The paramilitary demands of the job tend to reinforce personality types who see themselves as warriors fighting on a domestic front. They believe they are the "Thin Blue Line." Police departments have developed Special Weapons and Tactics Teams (SWAT), Street Narcotics Units and Tactical Patrol Units as tools to cope with their challenges. Police are expected to be service providers, protectors, social workers, security guards, psychologists, baby sitters, sanitary workers and anything else needed at a given time. Airport law enforcement tends to deal more with white-collar crime such as credit card fraud, ticket fraud, pickpockets, and distraction thieves, employees of airport businesses who victimize their employers. The issues addressed by airport police tend to be commercial in nature rather than the violent crimes confronted by their municipal counterparts. A very large component of airport policing is by nature devoted to public assistance, which is directed toward customers and therefore is more specific than generalized public service delivered at the municipal level. Expectations tend to be along the lines of traditional security roles. (McGray, n.d., p. 15)

With the increased threat of terrorism in aviation, airport law enforcement personnel must be focused on terrorist surveillance and attack methods, and trained in counterterrorism methods such as suspicious awareness and active shooter. Whereas a report of a suspicious person for a municipal officer is a common occurrence that may not always receive the highest priority, a report of a suspicious individual at an airport may have more serious consequences.

Airport policing also requires a type of officer who is more aligned with protecting critical infrastructure, responding to security violations at the checkpoint and on inbound aircraft, and irate passengers (McGray, n.d., p.16). At some airports, like Los Angeles International, airport police officers must be both airport oriented and municipal oriented as their jurisdiction extends into the immediate surrounding community. At the Port Authority of New York and New Jersey, Port Authority law enforcement officers are also cross-trained in aircraft rescue and firefighting and rotate assignments as first responders in both capacities (the organization is informally referred to as "Guns and Hoses").

Airport police are often thrust into a customer-service role as uniformed police officers are often the most visible symbols of authority who passengers can approach for assistance. While attempting to provide customer service, airport police must also attempt to distinguish between a regular passenger and someone who may be a terrorist, mentally ill, or a wanted criminal.

With the unique skill set required by airport police officers, the ideal officer may not be the one who recently graduated from the police academy and is looking for a lot of action (what police officers often refer to as "big badge" or "blue light fever"). Officers with good customer-service skills, good observation skills, and who can distinguish between threatening behavior and upset passengers can be effective candidates for airport policing. Airport police chiefs must also work to develop an organizational climate where police officers believe they are part of a larger team. While the airport police department is an important part of the entire airport organization, it is not more important than the other parts (McGray, n.d., p. 18)—airport fire responds to emergency situations to save lives as do airport operations personnel. Airport maintenance personnel ensure the airfield is safe for landing and departing aircraft, while planners, engineers, IT professionals, and others all serve vital roles in the smooth, safe, secure, and efficient operation of an airport.

McGray suggests a profile for recruiting officers who would do better in an airport, including:

1. Selecting officers who have a demonstrated work history of customer service.
2. Selecting officers with maturity and life experience.
3. Match potential airport police officers with finding individuals who will fit into the organizational culture and become like family members, similar to the Southwest Airlines hiring model, where individuals are hired who can not only do the job, but fit into the corporate culture.
4. Use personality tests to identify steady personality with excellent people skills, rather than thrill seekers.
5. Using an interview process that ferrets out negative people.

Another issue related to airport policing is the vulnerability assessment process and the prioritization of certain elements of the airport that should be more heavily protected, to ensure the safety of passengers and the continuity of business. McGray implemented a model at Nashville International Airport in Tennessee, conducting extensive risk analysis and dividing the entire airport into three critical zones: the Red Zone, the Amber Zone, and the Green Zone (McGray, n.d., pp. 36–37).

- The Red Zone includes all assets critical so there is uninterrupted operation of passenger operations.
- The Amber Zone includes areas involving airport assets and infrastructure, but that are not critical to the airport's core business, such as parking lots, general aviation, and cargo areas.
- The Green Zone includes all airport-owned assets that have no direct impact on commercial passenger flights.

A formal program was then designed that included sworn and armed police officers and civilian security officer personnel to inspect the zones based on a formula aimed at setting priorities to protect critical assets.

Community-based policing is another strategy that McGray suggests as an important tool in both crime prevention and the fight against terrorist activity. An airport should have an effective Airport Watch program and working partnerships between airport police and tenants, which encourages training of personnel in basic security and crime prevention techniques such as landside vehicle operators, tenants, retail clerks, airline employees, screeners, etc., who can all become the eyes and ears of the airport.

Security operations airport policing has often been compared to community policing. Community policing is a philosophy that supports the use of partnerships and problem-solving techniques to address the conditions that give rise to criminal activity (Fennelly, 2012). A good community partnership is a collaborative agreement between law enforcement, individuals, and organizations in the community to develop solutions to problems and increase trust in the police. Translated to the airport environment, the airlines, airport management, tenants, vendors, contractors, and even passengers are all members of an integrated community and all have a vested interest in the protection against both criminal and terrorist threats (Fennelly, 2012).

Community policing stresses proactive problem solving instead of reacting to problems, which makes it a perfect environment for the airport (Fennelly, 2012). One method of this, in terms of crime prevention, is to study calls for service, or analyze patterns of criminal activity, such as theft in certain parking lots or work areas. This strategy can also be used by airport security personnel in analyzing security violations, such as doors left open repeatedly or excessive piggybacking, allowing the security department to focus on those trouble spots. Another example can be with particular tenants, such as contractors, who seem to have high levels of violations, which can point to a training, management, or contract performance issue.

The SARA problem solving model is a common formula for solving problems in law enforcement agencies. SARA stands for scanning (identifying and prioritizing problems), analysis (researching what is known about the problem), response (developing solutions), and assessment (evaluating the success of the program) (Fennelly, 2012). Underlying causes are also examined, leading to find positive impacts on a problem such as the total elimination of the problem, fewer incidents, less serious incidents, improved response to the problem, and shift of the problem to someone more able to address it, like a gang intervention unit or training or social programs.

In addition to a community policing paradigm, airport security can also incorporate concepts from the crime prevention through environmental design (CPTED) body of work. The rationale behind CPTED is that there are five methods that affect the decision-making process of an offender or potential offender (Fennelly, 2012):

1. Increasing the effort needed to commit a crime (target hardening, access controls, screening, and reduce access ability for offenders).
2. Increasing the risks of the crime (additional patrols and surveillance, closed-caption television [CCTV], reduce anonymity, and alarms).
3. Reducing the rewards of the crime (conceal or remove targets of opportunity, identify property, disrupt trade of the illegal good, and deny benefits through the disabling of technology once illegally obtained—, that is, Lojack).
4. Reducing provocations (reduce frustration and stress, avoid disputes, reduce temptation and arousal, neutralize peer pressure, and discourage imitation).
5. Removing excuses enabling crime opportunity (set rules, post instructions, alert conscience [roadside speed boards is one good example], assist in compliance, and control drugs and alcohol).

Security Management Systems

Risk is a subjective concept that needs to be qualified on an individual basis (Sennewald, 2011). The risk management program quantifies, qualifies, and mitigates concerns for an organization, such as an airport or air carrier (Sennewald, 2011). In the aviation industry risk is a shared responsibility among airports, airlines, and the Transportation Security Administration (TSA). While it is an ethical responsibility of every airport security manager to conduct a vulnerability assessment, the TSA has also conducted a variety of other assessments locally, nationally, and even on specific types of attacks, such as surface-to-air missiles, and has developed response plans and contingencies. Airport and air carrier security personnel should be familiar with these contingencies and incident response plans to ensure they do not conflict with their own plans.

The industry has been adopting safety management systems (SMSs) as a way to increase safety throughout the aviation industry. Although it is a relatively new term, another international concept that is slowly gaining widespread acceptance is SeMSs, security management systems.

Using the concept of SMSs, there are four elements to an SeMS system:

1. Security policy
2. Security risk management
3. Security assurance
4. Security promotion

Each of these elements relates to the fundamentals of risk management, which are risk analysis, risk assessment and tracking, risk mitigation, and risk reporting, but each have been recouched in the SeMS vocabulary.

Security policy is management's written commitment to provide a secure environment. More than just a mission statement, this written commitment represents, ideally, the provision of adequate funding and focus on security. One actual example where management has made a commitment to security and consistently follows up is the daily 8:30 a.m. meeting at Boston Logan International Airport in Massachusetts, which is focused on security.

Security risk management is the process of identifying potential risks, qualifying each risk as to whether it actually exists, determining the probability of it occurring, and determining its impact and the total effect. Once risks have been identified, this process seeks to either eliminate or reduce the potential of the occurrence or reduce the damage through the implementation of security measures. Vulnerability assessments are common tools to assess risk at the airport.

Security assurance consists of conducting audits, tests of the system, and frequent reviews to determine systems and procedures are functioning correctly. The security audit process also includes enforcement and an anonymous reporting system. Enforcement tools, such as violation notices to employees for violating an airport security rule or regulation, and airport challenge programs where employees challenge others they see not wearing an airport access/ID badge, are good examples. Even better is the incentive-based challenge program where authorized members of the airport security department conduct random tests to determine if employees are indeed challenging.

Security promotion includes both physical promotions, such as newsletters, posters, and other visual reminders, but also promotes the culture of security within the organization. The See Something, Say Something and Airport Watch programs are good examples.

Qualifying the risk involves determining whether the facility has been prone to a particular type of incident in the past or whether there have been other occurrences of the same type at similar facilities. In the aviation sector, the significant risks are well documented (bombings, hijackings, etc.), but lesser crimes can also be assessed in this manner, and determining whether an airport or air carrier is likely to experience a significant attack can also be linked to whether other airports or aircraft are experiencing those types of attacks. For example, when a suicide bomber detonated a device within the terminal building in Moscow's Domodedevo airport, that could increase the risk to other airports of experiencing the same type of attack.

According to the American Society of Industrial Security's (ASIS) *General Security Risk Assessment Guidelines* (ASIS, 2003), security assessments follow a nine-step process. Throughout these steps, risks are determined, assessed for their possible consequences and potential frequency, and mitigation and preparedness steps are taken. The assessment loop closes with reevaluation of the entire process. The process in the aviation domain must be viewed from a system's perspective, with the additional knowledge that threats to aviation are often less associated with loss prevention, rather than with the loss of life or critical infrastructure to the United States. The steps are:

1. Identify assets
2. Specify loss events (types of hazards, human-made, natural)
3. Determine the potential frequency of such events
4. Determine the impact of the events
5. Identify options to mitigate (i.e., prevent, minimize damage, or accelerate recovery cycle)
6. Determine the feasibility of options
7. Perform cost-benefit analysis
8. Decide which options to pursue
9. Reassessment

ASIS recommends that the first step in any risk assessment is to develop an understanding of the people, the assets at risk, and the system in place (ASIS, 2003, p. 6). Security personnel must understand that protecting an airport and aircraft is markedly different than protecting an office building, nuclear facility, or Department of Defense (DOD) site. An airport is effectively a building with a huge back door. Security personnel have limited control over who can access the airfield (i.e., what aircraft land) and how those individuals have been screened or what materials or items they may be bringing into the facility.

Additionally, an airport is designed as a throughput process, not a closed-loop. In many nonaviation-related site assessments it is typically not required that the assessor have a complete understanding of how the entire business functions. In the aviation domain it is critical to understand three key components:

1. The purpose of the transportation system and how people, baggage, materials, and personnel enter and exit the facility.
2. The layered security system and the ways that airports, airlines, and the TSA function symbiotically to protect a complex supply-chain system.
3. The purpose of commercial aviation is to facilitate commerce, so policies and procedures that eliminate the ability of the system to perform its core function may achieve the result the terrorists are looking for with their attacks.

Understanding the organization includes the assets: people, property, core business, networks, and information (ASIS, 2003, p. 6). The security practitioner must understand the hours of operation of the airport, including the early-morning and late-night cargo flights, and the various times of day that international traffic departs and arrives along with the domestic flight banks. They must also understand the nature of the tenants, vendors, caterers, the general aviation facility, the fuel farm, and airline maintenance operations, along with any private facilities such as corporate or charter aircraft operations.

The vulnerability assessment should include both traditional forms of aviation terrorism, bombings, hijackings, active shooter, and criminal activity. The vulnerability assessment should also take into account the history of previous incidents and occurrences, along with police reports, crime rates, and intelligence and information on threats and criminal activities.

Upon identification, each risk must be assessed a potential consequence and a probability. The highest probable events with the greatest consequence should naturally be addressed first. There should be options to mitigate risk, such as more patrols, physical security measures, CCTV monitoring, and additional insurance. Cost-benefit analyses are then conducted on each measure, and measures are selected, implemented, and evaluated for effectiveness.

ASIS suggests the use of formulas to determine the probability of an event, its consequences, its frequency, and so on. In the aviation domain, particularly with respect to terrorist activities, these models may need to be adjusted based more on the potential threat than on actual prior incidents. Had a risk analysis been conducted on the probability of individuals hijacking four aircraft and crashing them into structures, the probability and frequency would likely have been so low that mitigation steps would not even be considered. However, the fact that these types of attacks are now part of the national threat matrix reminds security personnel that even though the probability and frequency of a certain type of attack may be nominal, the consequences are significant and appropriate measures should still be taken, regardless of the outcomes of the formulas.

Another component to a SeMS is the implementation of a daily airport self-inspection security procedure. Under Title 14 CFR Part 139, commercial-service airports perform a variety of self-inspections to ensure the airport meets the regulatory safety requirements. Considering that airport operations personnel, individuals who are generally most familiar with the day-to-day operations of the facility, often conduct this self-inspection, it may be beneficial to provide them with a security self-inspection checklist. Airport police and security personnel can also use the list to supplement their patrols.

The Part 139 self-inspection process includes four types of inspections: regular (performed at least once a day), continuous (performed any time they are in the airfield), periodic (detailed inspections of certain facilities, such as the fuel farm), and special (performed anytime there is an incident or condition other than normal operations, such as snow removal, wildlife on the airfield, or foreign object debris on the runway).

The regular security inspection should include all areas of the airport perimeter, gates, and key access points, both in the airfield and within the terminal building. Screening checkpoints should be at landside loading and unloading areas, special-use areas such as general aviation and air cargo, U.S. post office access points, employee access gates, loading dock areas, and any other area identified by a vulnerability assessment. The regular security inspection checklist can start as a baseline but should be developed for each facility, keeping in mind that this type of inspection should only take a small part of time, not the entire shift.

Continuous inspection items would include checking personnel for wearing the proper identification in the secure identification area (SIDA), monitoring construction activities to ensure individuals are staying within predefined areas, ensuring personnel without proper identification are being properly escorted, monitoring ramp-side passenger loading and unloading operations (typically in the regional/commuter area where passengers are accessing aircraft by walking across a ramp), and the observation of individuals moving through the facility.

Periodic inspections, more likely conducted by security personnel rather than airport operations personnel, should include close inspection of the airport perimeter fence, gate operation, and other areas as defined in a vulnerability assessment that are critical to the operation of the security system. Special inspections should be conducted any time there has been a security breach of a checkpoint, door, or gate that accesses a security area; for reports of suspicious individuals or activities; and before the end of any construction activity for the day.

International Civil Aviation Organization's (ICAO) Standards and Recommended Practices provides a lengthy security audit checklist that should be consulted in the development of a security self-inspection program.

Airport Crimes

Airports also experience theft and a variety of other crimes. Airport theft primarily occurs in the terminal and concourse areas, at screening checkpoints, and in the parking lots. Many airport thefts are preventable opportunity thefts, which happen when victims have either left vehicles unlocked or left baggage unattended. Purses, laptops and other electronics, jewelry, credit cards, and cash are items commonly stolen from baggage. In addition to the vehicle itself, installed electronic equipment is commonly stolen in parking areas. Distraction crimes and unattended baggage crimes are common in terminal buildings. When passengers leave bags unattended or their attention is focused elsewhere, their personal belongings are at risk of theft.

Baggage Theft

Opportunities for baggage theft, particularly by aviation employees, are frequent. Airport and airline employees with sufficient time to search a bag's contents can handle a checked bag multiple times. Because of the frequent theft of contents placed in checked baggage, passengers believe that if you do not want to lose it and it cannot be replaced, do not put it in checked baggage. Security regulations are requiring more items to be placed in checked baggage. Therefore, it is becoming increasingly difficult for travelers to protect their belongings.

Shortly after the airline liquid bomb plot was discovered in 2006 in London, the restrictions placed on items that could be carried onboard were so onerous that the amount of checked baggage increased by an average of 20% per airline worldwide. In Great Britain, where the carry-on restrictions forced travelers to check laptop computers, cell phones, and even car keys, the fallout by passengers was tremendous. Proprietary and confidential company information is often compromised when a laptop computer is stolen. The theft of cell phones and car keys can lead criminals to the homes and property of victims. At the very least, the theft or loss of car keys can leave a traveler stranded at his or her home airport. The increased value of the items left in lost luggage shortly after the restrictions went into effect caused a significant increase in looting by airline and airport workers (Rose, 2006).

Employees within the airport infrastructure, including airline baggage handlers and security screening personnel, commit most baggage and cargo thefts. What makes detecting this kind of crime difficult is that these people have a legitimate need to access baggage, particularly security screening personnel who have the authority to open a bag if they believe that a physical search of the contents is warranted to ensure the security of a flight. Here are examples of theft by aviation workers:

- In February 2004, Toronto, Canada, police cracked a multimillion-dollar theft ring with more than a dozen airline employees being arrested for theft. The stolen goods included jewelry and electronic equipment. Before the theft ring was shut down, the criminals had sold approximately $500,000 in stolen goods through Web sales.
- In 2007, two baggage handlers in Seattle, WA, were arrested for stealing laptop computers, DVD players, and video cameras from passengers' baggage.

Theft by federal screeners, although rare, also occurs. More than 170,000 travelers had filed baggage theft claims with the TSA by 2004. By 2005, at least 60 of the TSA's 43,000 screeners had been terminated for theft. Although the percentage losing jobs for stealing is miniscule, the perception by the public is that federal security personnel charged with protecting passengers from terrorism are also stealing from their baggage.

Prevention of baggage theft is a high priority of airline security personnel. Airline employees have the most unfettered access to passenger baggage and cargo. Baggage missing an assigned flight is stored in an airline baggage holding office, which means that it is particularly susceptible to theft as it is out of public view. Theft from air cargo shipments is also a problem, because airline employees often have unsupervised time in which to browse through cargo shipments before the items are transferred to the aircraft.

Strategies for preventing baggage and cargo theft include better background checks of personnel, the use of CCTV in baggage-handling and storage areas, and employee awareness programs.

Air Cargo Theft

Protecting air cargo from theft and tampering starts with educating the shipper on security practices. A standard preventive measure is to use packing procedures that immediately show tampering. Sealing an item in plastic wrap and tightly packing are low-cost anti-tampering strategies. Radio-frequency electronic seals with GPS tracking capability to enable continual monitoring and anti-tamper detection are being developed.

The theft of cargo occurs most commonly in the trucking phase, either by truck drivers or loaders working together or with others. Cargo is at greatest risk when being loaded or stored in a cargo facility. Access to cargo areas should be strictly controlled using computerized control systems that prevent unauthorized access to the facility and log authorized personnel when entering and exiting. CCTV systems should be used in cargo storage areas, together with frequent law enforcement and security patrols through the storage and load/offload areas.

Drug Smuggling

Drug smuggling is a major form of crime in aviation. Smugglers frequently use airports and commercial aircraft for conducting their trade. Drug couriers often carry narcotics on their person or in carry-on luggage. Drugs are also smuggled in cargo holds through checked baggage or loaded as consigned air cargo. People working in the aviation industry have opportunity to exploit their access to smuggle drugs. In 2006, two federal air marshals were arrested for using their positions and access to smuggle cocaine from Bush Intercontinental Airport in Houston, TX, to Las Vegas, NV. In 1999, agents from the U.S. Customs and Border Protection (CBP), Drug Enforcement Administration (DEA), and Bureau of Alcohol, Tobacco, and Firearms (ATF) agencies arrested more than 55 airline workers at the Miami International Airport in Florida for being involved in a smuggling network that was bringing narcotics to the United States on commercial flights.

Preventing drug smuggling via airlines requires a solid background check prehiring program that includes a check for misdemeanor offenses, CCTV surveillance of employee work areas, and confidential reporting programs such as anonymous tip lines. Employee education on what to look for can help employees spot potential drug trafficking activity.[1]

The DEA has been providing training to airport police and security officers under the auspices of a program called Operation Jetway, which predates the Screening of Passengers by Observation Techniques (SPOT) and Behavior Detection Officer (BDO) programs in effectively identifying drug traffickers, money launderers, and couriers of various types moving through the airport environment. The DHS and the TSA have requested DEA expertise in Operation Jetway training, and they regularly provide it to airport and local police responsible for policing airports.

Human Trafficking

Human trafficking is a very serious crime that makes use of aviation transportation. According to the Polaris Project, a website and organization focused on the prevention of human trafficking, it is the third largest and fastest growing criminal industry in the world, victimizing millions of people and reaping billions in profit for the traffickers. Human trafficking can include selling victims into the sex trade industry, forced labor in factories, domestic servitude, forcing a victim to become someone's spouse, or for the harvesting of human organs (see *www.polarisproject.org* for additional information regarding human trafficking). The aviation industry is an efficient form of transportation for the trafficking of humans worldwide, but awareness of the indicators that human trafficking is taking place can be an effective deterrent.

In the United Kingdom, Operation Pentameter (see *www.pentameter.police.uk/news.php?id=4*) is a high-profile campaign aimed at countering human trafficking through educational programs and awareness posters in and around airports. In February 2006, the program launched a joint effort between the 55 police forces of the United Kingdom, Scotland, Wales, Ireland, and Channels Island, along with the United Kingdom Immigration Service, Serious and Organized Crime Agency (SOCA), and Crown Prosecution Service. During its three-month operational phase, more than 188 victims of human trafficking were rescued, including 12 between the ages of 14 and 17. In this instance, 232 people were also arrested.

Indicators of human trafficking include missing or altered passports and other identification credentials. Verification of a traveler's identity is an important way to deter and apprehend human traffickers. U.S. citizens are now required to have a passport whenever traveling outside the United States, regardless of age or destination (including Canada and Mexico). Placing posters around airports encouraging women and children who are being

---

[1]The CBP provides antidrug smuggling training to airlines through the Carrier Initiative Program (CIP). The CBP offers two training programs: a two-day antidrug smuggling seminar for senior-level managers, and a one-day antidrug smuggling seminar for midlevel managers, supervisors, and front-line personnel. The CIP teaches airline employees the basics of narcotics search techniques, risk assessments, concealment techniques, document review, physical and procedural security, personnel hiring, drug source countries, drug characteristics, and internal smuggling conspiracies.

held against their will to report their concerns and seek help is also an effective security measure. Other indicators of trafficking can be revealed during security questioning by asking travelers if they know where their journey is leading, if they have a passport, if they know whom they are meeting, and if they arranged their own travel. A well-trained security operative or law enforcement officer should be able to detect signs of deception or duress during this interview process.

Ticket Fraud

Ticket fraud is another common crime at airports. Fraud costs the airlines about $1.5 billion annually. Airline employees can also facilitate ticket fraud. In 2003, eight Southwest Airlines employees in El Paso, TX, were arrested for allegedly stealing more than $1 million by pocketing money from tickets bought with cash and selling voided tickets. The FBI suspected that the fraud operation had been going on since 1996.

Other Crime

The airport operator is concerned with thefts primarily from vehicles in parking lots and parking garages. Other forms of crime such as assault, rape, and robbery can also occur in parking garages and at outlying lots. Airport parking lots are located varying distances from the terminal building. Being able to manage the detection and response to crime under these conditions is a prime concern to the security practitioner.

Airport Policing Strategies

Counterterrorism

There are approximately 5,000 FBI counterterrorism agents in the United States compared to more than 870,000 state and local law enforcement officers. After 9/11, the FBI deployed just 7,000 agents to canvass the nation. In comparison, the New York City Police Department (NYPD) has nearly 38,000 officers to cover five boroughs (Clarke et al., 2006). In terms of intelligence gathering, few FBI agents are trained in Islamic radicalism, the Arabic language, the history of terrorism, and related subjects. Again in contrast, the NYPD has put together a counterterrorism unit that includes members of nearly every ethnic group in the world (Flynn, 2007). The United States does not have a federal force solely devoted to counterterrorism, and the agency most closely charged with the counterterrorism mission, the FBI, routinely transfers its agents throughout the country and the world. State and local law enforcement officers are fully integrated into the communities in which they work and live (Clarke et al., 2006).

Terrorist attacks in the United States often require preparation, establishing safe houses, purchasing equipment and supplies, creating false documents, recruiting accomplices, surveillance, and practice sessions. Most of the requirements and related activities are potentially detectable by law enforcement agencies. In this regard, it is more likely that a terrorist operative will have a chance encounter with a local or state law enforcement officer than with a federal agent.

The underlying security strategy in New York City is building local and state law enforcement personnel into "first preventers"—as described by R. P. Eddy, a counterterrorism expert—not just first responders (Flynn, 2007). To build a first preventer mentality, police officers need to be trained to spot the support structures and activities of terrorists such as bomb making, preoperational surveillance, and religious radicalism (Clarke et al., 2006).

Although cities like New York and Los Angeles have resources to fund counterterrorism efforts, such as hiring personnel and creating their own intelligence agencies,[2] many smaller cities, and particularly airports, do not. Training local and state law enforcement officers, especially airport police, to spot signs of terrorist preplanning activities can help turn first responders into first preventers, with little capital outlay.

Terrorist operatives frequently plan or facilitate an attack while traveling by air. A police officer may spot an operative as the individual conducts surveillance on future targets. Routine training of police should focus on subjects such as recognizing chemicals used to construct explosive devices, materials needed for constructing weapons, and emerging threats. Officers should attend seminars by terrorism experts and relay that knowledge to their colleagues.

The role of the small airport operator should not be ignored in counterterrorism training. Terrorists may use small commercial-service airports to slip through security and gain access to a larger commercial-service airport (as elements of the Japanese Red Army did in an attack on Israel's Ben Gurion International Airport in 1972). Small commercial-service and general aviation (GA) airports have aircraft, fuel, and fire trucks that could be used in a terrorist attack and need to be kept secure. Training is a cost-effective measure to prevent terrorist attacks.

Anticrime

Policing an airport requires skills and knowledge that are usually not part of a police officer's training. Although police generally respond to people having problems or breaking the law, the large majority of people an airport police officer encounters are law-abiding citizens, many of whom are nervous since they are out of their element and who may be more anxious than usual because of the stresses associated with air travel. When dealing with most of the travelers at airports, police need to interact with the traveling public with decorum and tact—which is a primary difference between "life on the streets" for a police officer and interacting with individuals at an airport.

Robert Raffel (2001), writing for the FBI's *Law Enforcement Bulletin*, noted that many officers newly assigned to the airport rarely understand the complexities of the job. Airport

---

[2]In a study, the FBI criticized New York City for running an independent counterterrorism operation (Clarke et al., 2006).

police must handle a wide variety of crime, are often called on to resolve disputes or confiscate prohibited items at screening checkpoints, and must verify the identity of other law enforcement officers traveling on official business and carrying firearms. They must also be on the lookout for and respond to potential violent criminal or terrorist incidents and still provide a regulatory enforcement function under the airport security program (ASP).

Looking out for potential threats within the airport environment is a challenging responsibility for a police officer. In addition to employees and passengers, airports are filled with friends and family of passengers. Airports also attract some of the lost and lonely who seek airports as safe environments. Outside the airport, there are ground transportation companies, taxi and limo drivers, bus drivers, hotel shuttle van operators, and others conducting business. Individuals who enjoy watching aircraft may also be scattered just outside the airport's outer perimeter. Intermixed in this highly complex environment are individuals who may look and act like any of the aforementioned individuals at an airport. However, these individuals may be planning crime, conducting surveillance for a terrorist attack, or in the process of executing an attack.

Passengers generally fall into two categories: business travelers and vacationers. There are some smaller subgroups, such as those traveling because of a family emergency, a death in the family, or other sort of tragedy or urgency. Business travelers tend to know the system, ask few questions, are dressed in business or business-casual attire, and carry little luggage. What luggage they do have is usually carry-on and not checked baggage. Vacationers often dress casually. They have checked luggage but little carry-on luggage (except for families with small children). Vacationers usually travel in groups or as a family, and they seem to smile more than the business travelers do. Vacationers usually have a relaxed, easy-going manner—unless they experience circumstances such as a delayed schedule or lost luggage. The third group—those traveling because of an emergency—tend to have less baggage and be focused inwardly. These travelers may exhibit what is known as a thousand-yard stare. Emergency travelers may pray or speak quietly to themselves and seem more anxious than business or vacationer travelers. This is problematic because these behaviors are potential indicators of a suicide bomber.

Police officers should temper the response to an upset passenger by first attempting to understand the problem and then defuse the situation before an arrest becomes necessary. For example, a mother who is being told her infant daughter is on the no-fly list will be understandably upset when she is forced to go through additional time and effort to prove that her daughter is not the person whose name is on the no-fly list. This situation requires a great deal of understanding on the part of responding law enforcement personnel. Bad weather, maintenance problems with aircraft, and other routine aviation issues that are beyond the control of the passengers but are causing some to miss vacations or others to miss expensive business meetings can severely increase the passengers' stress levels. Therefore, law enforcement personnel must exercise understanding and diplomacy to diffuse these situations.

Through Title 49 CFR Part 1542, the airport police have a responsibility within the ASP to enforce the rules and regulations at their assigned airport. This provides the airport security coordinator (ASC) with a respectable level of input as to how police are used at the airport. Deployment of police will depend on recent threat assessments. Generally, more police should be deployed to watch the landside areas, for traffic congestion, and for signs of an attack via a vehicle-borne improvised explosive device (VBIED) or a suicide bomber entering the terminal building. Police should be deployed throughout the public areas of the airport, as these areas are open to anyone without being subjected to security screening. Police in the concourse and sterile areas can help prevent other types of crime, such as theft and shoplifting. Officers should be aware that an airport employee could have brought a weapon into the sterile area, and therefore they should always use caution.

Many airport police departments have taken an active role in reaching out to the airlines and tenants and in providing their personnel with additional training in airport law enforcement practices. The Austin Bergstrom Airport in Texas has 55 sworn law enforcement officers who are required to undergo a three-week orientation to the airport environment after being assigned to the unit. This is followed up with 100 hours of additional training every year in job-related subjects, including emergency medical training. At the London Gatwick Airport, the police commander spends a lot of his time interacting with the airport tenants, learning about their problems, and building relationships. Michael Burris (2005) stated that good security-related customer-service skills are not easily learned or developed. These skills must be retaught and honed as daily pressures can exasperate even the most seasoned professionals.

Because airports have large areas for police to cover, much of which is not amenable to vehicle use, some airports have equipped their patrol officers with bicycles. Bicycles allow officers to patrol more area, go where a patrol car cannot, catch a fleeing suspect, and approach a situation quickly and silently when necessary. Bicycles are much cheaper than a patrol vehicle.

Some airports have adopted the Segway for their patrol officer or security guard use. A Segway has the advantage of placing an officer up to 8 inches higher than the crowd and allows the officer to maneuver through a crowd more easily than on a bicycle and to not have to dismount to go through a door. A Segway provides high visibility and faster response times than the officer would have on foot.

Patrol vehicles are still needed to cover the airfield perimeter, to transfer people, and to carry equipment. Additional police vehicles should include response platforms able to move into position on an aircraft rapidly and with minimal waiting time to deploy the boarding ladder. These so-called raider vehicles can be custom built and stored at the airport for use by law enforcement or federal response agencies such as the FBI Hostage Rescue Team. Other support vehicles should be acquired as necessary, such as on-scene command platforms and armored vehicles.

Officers should receive extensive training on airfield driving practices and regulations so that they are able to enforce those practices while on the airfield and can avoid driving onto the runways and taxiways or in front of moving aircraft. Aircraft always have the right of way because of the limited visibility afforded the pilots and the inability of an aircraft to move as adeptly as a land vehicle.

Some organizations, such as the American Association of Airport Executives (AAAE) and the International Air Transport Association (IATA), offer courses on airport law enforcement techniques and practices. ALEAN provides an excellent support network for airport police and sheriff departments and has airport staffing duties to share information and best practices.

Crisis management can be another area new to a law enforcement officer. When there is a security incident at an airport, numerous parties are involved, and many have some level of authority. The airport director is responsible under federal regulation Title 14 CFR Part 139.325 Airport Emergency Plan. The ASC is responsible under Title 49 CFR Part 1543.307 Incident Management. The TSA's federal security director also shares responsibility for the management of the incident, and, depending on the presence of other agencies, the FBI and the local law enforcement agency may also share responsibility. Who is actually in charge of moving assets and making decisions largely depends on the nature of the incident and the responsibilities as assigned in the airport emergency plan (AEP) and the ASP. Law enforcement officers should be well versed in their roles under both of these documents.

Other tactics can help policing at an airport, including the use of process matrices. At the Boston Logan International Airport in Massachusetts, security staff have developed and employ a tracking system to help determine which areas of the airport need more attention in terms of reducing theft or other problems. The tracking system also helps determine how officers are deployed, including special SMG patrols,[3] which are used as deterrence to a terminal shooting or bombing attack and as a rapid-response force to any attack. Theft activities in the baggage claim areas can be tracked using matrices, along with theft activities in the parking structures and deployment of security assets.

Policing General Aviation

Policing a commercial-service airport differs greatly from policing a GA airport. A prime challenge to securing GA airports is that their environments vary from commercial airport environments and among the populace of GA airports. For example, the major tenants at a commercial-service airport are usually airlines. At a GA airport, the major tenants are typically fixed-base operators (FBOs). Except at some of the larger and busiest GA airports, there may not be a fence with access gates around the airport. This makes it difficult to determine which vehicles and personnel are authorized to be on the airfield. At many GA airports, there may not be an onsite airport manager. Frequently, the FBO staff also provides airport operational and management functions and, in some cases, air traffic advisory services.

Larger GA airports usually have access control systems, air traffic control towers, management, operational, and maintenance staff. Some GA airports cater specifically to

---

[3]SMG, in this case, is an acronym for submachine gun.

agricultural traffic, corporate business jet traffic, aircraft and flight training, or many other types of GA-related flight activity (including various combinations).

Individuals at commercial-service airports are primarily passengers, airport and airport businesses employees, and those employed by the government. At GA airports, stakeholders are commonly those who keep their privately owned aircraft at the airport or work in flight schools, aircraft maintenances centers, restaurants, FBOs, or corporate aircraft operations. Regardless of the size, most GA airports do not have their own police force. Law enforcement for GA airports is usually provided by the law enforcement agency responsible for the geographic area in which the airport is located. After 9/11, some cities and counties assigned more personnel or frequent patrols to GA airports in their jurisdictions. Rarely does a GA airport have its own police officer assigned to it.

The first action a police officer with a GA airport within his or her patrol area should do is introduce him- or herself to the airport manager and the tenants. A police officer may have to contact the local city or county to determine the identity and contact information of the GA airport manager. The officer should have a map of the airfield noting key entrance and exit points. The officer should request an airfield familiarization with airport staff to understand airfield markings and how to avoid driving on runways or taxiways when responding or patrolling.

Police officers should obtain an airport directory noting the locations of the various structures, hangars, and businesses to make responding to an incident easier. Many hangars and airport facilities look alike, so without a proper numbering system on the outside, it can be difficult for police officers to know where to respond to based on a street address alone. If airport buildings do not have visible numbers, police should request that they be numbered. Police officers should have point-of-contact information for the airport management or operations staff—whoever should be notified in the event of a problem at the airport.

Officers with GA airports as part of their routine patrol area should get to know some of the tenants and not just the major businesses. Often there will be a contingent of locals who enjoy visiting the airport as much as they enjoy flying their aircraft. These people are adept at identifying strangers or odd occurrences at the airport. Officers should try to ensure that these individuals have appropriate emergency contact information.

When patrolling a GA airport, officers should look for things out of the ordinary. Although it is relatively easy to access the airfield of a GA airport, it is difficult for people to look like they have a purpose for being there when they do not. On the FBO ramp, an officer would likely see fuel trucks operating in and around aircraft and frequently personnel assisting passengers and pilots on and off their flights. Pilots of corporate aircraft are usually dressed in a uniform or some sort of business attire. Pilots around small private aircraft could be dressed casually. Passengers on an FBO ramp generally tend to stay near their assigned aircraft. However, pilots will wander freely around other aircraft. They will not normally walk out onto the taxiway or runway.

Near the runways and taxiways, there should only be airfield maintenance personnel, airport operations or management personnel conducting airfield inspections, or Federal

Aviation Administration (FAA) maintenance personnel. When there is an aircraft emergency, firefighting personnel and equipment may also be in these areas.

Flight schools are frequently part of or associated with FBOs. Law officers should be aware that there are likely to be students and instructors and others pilots on the flight school ramp. Flight students conduct inspections of their assigned aircraft and often do so without the instructor present. Passengers may also be on the ramp but will generally be close to the aircraft; if they wander off, they tend not to go too far. If they are wandering too far from the area, police might consider stopping them to ask questions.

Corporations often own larger hangars, and officers will likely see larger business jet aircraft in these areas. It is not unusual for a corporation to have its own aircraft fuel facility that should be secured when not in use. Small hangars are usually leased or owned by private aircraft owners and pilots and may be a place that pilots and visitors use for socialization.

Aircraft charter operations are common at many GA airports. They can operate similar to a commercial airline operation. Many charter operators screen their passengers for prohibited items and closely control the boarding. Officers should watch these ramp areas to ensure that people do not wander from the charter aircraft loading/unloading area and that others do not enter the loading/unloading area without being properly screened.

Other critical security points on a GA airfield include the control tower, the airport or FBO fueling facilities, aircraft fuel trucks, onsite aircraft rescue and firefighting trucks and equipment and, in some cases, high-profile hangar and office areas. Some corporations are targets for crime and terrorism, and there have been incidents of corporate officers and assets, such as the aircraft or hangar facilities, being in danger of damage or destruction.

Professional Security Officers

In addition to law enforcement officers, many commercial-service airports have a contingent of unarmed security officers. Security personnel today run the gamut from companies that provide a virtual private army to protect U.S. contractors in war zones, to the classic night watchman walking through a warehouse armed only with a radio and watch clock. Today, the industry includes trained and professional security officers, loss prevention officers, and facility security managers, often meeting various higher levels of training and, in some cases, certification. Unfortunately, prior to 9/11 these trained and educated professionals had gone largely unnoticed by the aviation community, but their use is becoming more widespread, particularly as airports discover their cost effectiveness and their ability to fulfill many roles. Before 9/11, there was a general feeling in the aviation community that to be an effective ASC, one had to come up through the ranks. After 9/11, the aviation security industry was flooded with former federal, state, and local law enforcement officers and retired military officers. This influx was driven by the misperception that the skills of a federal agent or police or a military officer were directly transferable and immediately applicable to the civilian aviation industry. While taking nothing away from the perspective and background of these people, the professional security manager should be considered when deciding the personnel needed to protect an airport or airline operation from crime and terrorism.

The American Society for Industrial Security International (ASIS) is the world's largest organization for security professionals, with more than 36,000 members. ASIS has numerous educational programs to enable its members to stay ahead of threats and understand countermeasures. It provides a comprehensive certification program for security professionals to earn the certified protection professional (CPP) designation. This certification should be considered an asset for anyone being hired as an airport or aircraft security manager and something those already working in the field should consider obtaining.

Security professionals who already have this skill set should consider obtaining certification in an aviation area, such as the certified member status of the AAAE, which will help the security professional understand the nature of airport operations.

An unarmed and unsworn private security officer does not help an airport or aircraft operator meet regulatory requirements for law enforcement coverage, but this individual can free up law enforcement personnel for other patrol and regulatory enforcement duties. Security officers do help airport operators fulfill TSA-mandated minimum security staffing requirements for airfield personnel. Security officers can be used in a variety of areas including those listed in Table 12.2.

Table 12.2   Typical Duties Performed by Security Officers

- Staffing airport perimeter gates and conducting vehicle and personnel searches of vendors and others accessing the airfield.
- Monitoring CCTV camera operations in a security operations center.
- Staffing airfield access doors, particularly high-activity doors, such as those used by flight crews and airport workers. These have a high rate of violations as employees tend to hold a door open for one another (as both have access/ID badges) or airline flight crews carry luggage and have a tendency to lag the luggage behind them, setting off people count* alarms.
- Responding to incidents and contingencies, such as increases to the color-coded Department of Homeland Security (DHS) Alert System. Security guards often provide additional services during these increases, such as staffing incoming personal vehicle checkpoints in landside operations areas and providing additional security patrols throughout the concourse and terminal areas.
- Doing airfield vehicle patrols to watch for intrusions of the airfield perimeter; monitoring aviation employee activities including compliance with the ASP.
- Searching, evacuating, and providing resterilization assistance when a breach of security has occurred.
- Patrolling the internal terminal and concourse to monitor security (and safety) events and activities, including compliance with the ASP.
- Issuing notices of violations of the ASP to aviation personnel working on the airfield.
- Protecting the exit lanes of a screening checkpoint, particularly when the screening checkpoints are not in operation.**

*People count is a term used in reference to an access control system. A people count occurs when more than one individual accesses a door but only one has presented an access/ID badge to the reading device. Many people count violations occur when the electronic eye detects a piece of luggage or a tool that the individual is carrying, rather than an actual person.
*This was primarily a TSA duty.

Security officers provide a diverse array of services for their employers and clients, including staffing security operations centers, customer service, airfield escorts, and additional security for airport construction projects. Security officers are part of a holistic ASP, not a stand-alone resource, and where the regulations allow, there are in some cases security officers being used to replace police officers, freeing up armed law enforcement personnel for other duties.

Security officers also play an important public relations role and are often the first contact visitors, customers, vendors, or employees have with an organization (ASIS, 2011).

The number of officers required for an airport is determined by the physical complexity and size of the airport, the number of employees, the work requirements (security operations center, here Filtrol, staffing gates for responding to alarms), the amount of escorts[4] and special assignments required, and the number of access points (ASIS, 2011).

The basic functions of security officers are to control access throughout the facility; patrol buildings and the airfield; escort personnel into security areas; look for security problems, fire, hazardous material issues, or other safety issues; provide emergency response and in many cases respond to door alarms and breaches (ASIS, 2011).

Security officers can frequently control access to a facility and provide credentialing and visitor badging services. They can identify and question individuals who remove property from the facility, and monitor materials, individuals, and vehicles entering and leaving the airport (ASIS, 2011).

Security officers should be trained in the reporting of hazards, proper patrol techniques, the physical layout of the facility including all shortcuts and dead ends, and how to handle employees and passengers who have either broken the law or violated the airport's rules and regulations (ASIS, 2011). Further, they should be advised of unique activities such as construction work, emergency exercises or training, or airport emergencies such as an aircraft incident or accident.

Security officers generally operate per documentation known as post orders. Post orders outline the specific procedures security guards follow at a given position. For example, the post orders at an airfield perimeter gate outline the process for allowing or denying access of personnel and vehicles into and out of the security areas. Vehicle search procedures and visitor badge requirements may be included with various protocols on the notification of the airport operations department of a potential security issue. Post orders are usually very specific and should be reviewed frequently to ensure consistency with airfield security practices and relevance to new security requirements.

A large part of a security officer's success depends on the ability to observe. Kane noted that "most mission statements for security organizations define their principle duties as deterrence, detection and reporting" (2000, p. 85). Observation skills play a critical role in deterrence and detection. A security officer must be able to spot a suspicious

---

[4]The term escort is used to describe a situation when an individual does not have an authorized access/ID media for an airport security area and must be escorted by authorized personnel, such as a security officer, airport operations or airport police personnel.

person or item and recognize that individual or item as a specific threat. Many security positions are characterized by repetitive, often monotonous tasks that can lead to a lack of vigilance. To become a prudent observer, security personnel must develop "reactive" observation skills (Kane, 2000). These skills are necessary to not just detect and deter potential criminal or terrorist activity, but also to be an expert witness to an event and able to provide an accurate description of an event and the parties involved to law enforcement agencies.

Proactive observation includes such skills as recognizing something out of the ordinary. Unusual observations might include vehicles without license plates or a vehicle heavily loaded beyond normal capacity, an access/ID badge that is not valid, a door left open, or an unattended bag. A common method of improving observation skills is a training exercise such as requiring security personnel to memorize a specific pattern of items in a room, or a set of dissimilar objects, then involve them in a "distracter" activity for a period, after which they are asked to re-create visually or on paper the information previously memorized. Another method is having security personnel watch recorded coverage of an actual incident and then write a narrative report on what they have seen. Kane (2000) recommended that this activity be administered unexpectedly in the middle of a lecture on another subject, to disrupt the routine and better simulate a real-world situation. To reinforce the learning, the instructor should go back over the recorded event and discuss with trainees the most important elements (Kane, 2000).

Red Teaming

Red teaming is used to reveal weaknesses in security programs. Red teams usually operate outside standard testing protocols. They are used to identify vulnerabilities or provide feedback before criminal or terrorist activity. According to the Sandia National Laboratory:

> *Red Teaming is a tool for analyzing systems from a malevolent adversarial perspective. It is often used as an objective tool to optimize systems by eliminating their weaknesses in a design/assess cycle. It can also be applied to guide system development, train system operators, develop adversarial profiles and footprints, and analyze intelligence data from a particular perspective. (Sandia, 2007, p. 1)*

Before 9/11, the FAA frequently used red teams to attempt to penetrate airport and airline security systems, and they were frequently successful. Since 9/11, the inspector general for the DHS has continued to conduct red team testing. Airports and aircraft operators can also conduct their own red team testing. However, several parameters must be observed to ensure that the operation of the test is safe and that it is conducted in a manner that provides the best feedback for the agency being tested. Unlike formal security exercises, red team tests often take place without notification to the entity being tested and personnel often use unorthodox tactics.

Red team testing can be used to test for penetration of an access control system, such as a door, gate, or security screening checkpoint. It can also be used to determine if a screener is spotting prohibited items during a security inspection, to determine if a perimeter breach is detected, and to determine if an adversary's intelligence-gathering attempts are successful. The challenge program called for under Title 49 CFR Part 1542, where the ASC is required to establish a program that specifically tests whether airport workers are challenging anyone they see in the SIDA without a proper access/ID badge, is similar to a red team test in its most basic form.

Typical local red team testing includes approved airport personnel wearing an access/ID badge other than their own or wearing an expired access/ID badge and attempting access through a checkpoint. It can include approved testing personnel pushing open an alarmed door and testing the response time of security personnel, and testing personnel entering the security areas and attempting to elude security personnel. Many airport and airline operators do not have the extra personnel needed to plan and conduct red team testing, so consideration should be given to contracting out some testing or working with local military or law enforcement organizations to provide a volunteer testing force.

Red team members should find ways to defeat any access control system, personnel identification system, or screening system, but the results of their testing should be carefully considered in the overall security of the airport. Many testers have an unorthodox knowledge of the security workings of an airport or an airline, beyond that of most criminals or terrorists (Cheston, 2006). This should not prevent airports and airlines from paying close attention to the results, particularly if the testers employed tactics that required little knowledge of the aviation system. One unfortunate result of red team testing is that it can embarrass administration and management personnel with the results of the test being "pushed under the carpet" and future testing ignored. Another reason red teaming may be avoided is that once a threat is identified, an airport or airline operator is often compelled to act on it. If funds are not available to mitigate the discovered threat, then some administrators feel they should not conduct a red team exercise. This approach is not intelligent on the part of administrators. Failing to identify the risks and hazards is also a huge liability. Red teams can be used to develop a comprehensive threat matrix for an airport or airline (Cheston, 2006), from which intelligent spending decisions can be made. A solid matrix may also be used to argue for additional spending or grant money to fix the largest holes in the security system.

Conclusion

The nature of aviation security operations must continue to evolve to both meet new threats and to adapt to the unique nature of a transportation system. Airport police must focus not just on crime prevention but also understand basic facility security principles. Aviation security professionals must remember that the purpose of an airport and aircraft is to move people and materials, and that the vast majority of the traveling public, and their associated baggage and cargo, present little to no risk to the system. Airport police officers and security professionals also fulfill an important role as ambassadors to the airport, the city, and the community that relies on the economic benefits of air travel.

However, aviation presents different opportunities for crime, such as baggage and cargo theft, and as a transportation system allows for the facilitation of human trafficking and drugs and weapons smuggling. Security professionals must be trained in the indicators and how to interdict these types of activities.

A new focus in many areas of the aviation industry is the development of comprehensive risk management and mitigation systems. The identification of risks and the implementation of mitigation and preparation strategies is an important part of aviation security. Security management systems take advantage of the research into this developing field, and when tied into vulnerability assessments, may prevent or mitigate the damage from future attacks.

References

ASIS, 2003. General Security Risk Assessment Guidelines. ASIS, Alexandria, VA.

ASIS, 2011. Security Officer Operations. Williams, T.L. (Ed.). ASIS, Alexandria, VA.

Burris, M., 2005. Airport Policing: Management and Response to Airport Crime. Aviation Security International 11, 30–33.

Cheston, M., June/July 2006. Planning a Red Team Strategy. Airport Magazine 28–31, retrieved Aug. 16, 2008, from www.faithgroupllc.com/news-pub/am0604junjul.pdf.

Clarke, R.A., Allen, C.A., Bullock, J.A., Eddy, R.P., Ferguson, C.D., Flynn, S.E., et al., 2006. The Forgotten Homeland: A Century Foundation Task Force Report. Century Foundation Press, New York.

Fennelly, L.J., 2012. Handbook of Loss Prevention and Crime Prevention, fifth ed. Elsevier, Boston.

Flynn, S., 2007. The Edge of Disaster. Random House, New York.

Kane, P., 2000. Practical Security Training. Butterworth-Heinemann, Boston.

McGray, D., n.d. Airport Law Enforcement Leadership. TN: Nashville.

Raffel, R., Sept. 2001. Airport Policing: Training Issues and Options. Law Enforcement Bulletin 70 (9), 25–28, retrieved Aug. 16, 2008, from www.fbi.gov/publications/leb/2001/september2001/sept01p26.htm.

Rose, D., Aug. 16, 2006. Rich Pickings for Looters as Lost Luggage Gets Stuck in the System. Timesonline retrieved Aug. 16, 2008, from www.timesonline.co.uk/tol/news/uk/article1084617.ece.

Sandia, 2007. Radiological and Nuclear Countermeasures: Red Teaming and Risk Assessment. Sandia National Laboratories, retrieved Aug. 16, 2008, from www.sandia.gov/mission/homeland/programs/radnuc/redteamRisk.html.

Sennewald, C.A., 2011. Effective Security Management, fifth ed. Butterworth-Heinemann, Boston.

Index

Note: Page numbers followed by *b* indicate boxes, *f* indicate figures, *t* indicate tables and *np* indicate footnotes.